Understanding Urban Ecosystems

Springer
New York
Berlin
Heidelberg
Hong Kong
London
Milan
Paris
Tokyo

Alan R. Berkowitz Charles H. Nilon
Karen S. Hollweg

Editors

Understanding Urban Ecosystems

A New Frontier for Science and Education

With 49 Illustrations

Springer

Alan R. Berkowitz
Institute of Ecosystem Studies
Millbrook, NY 12545
USA
berkowitza@ecostudies.org

Charles H. Nilon
School of Natural Resources
University of Missouri—Columbia
Columbia, MO 65211
USA
nilonc@missouri.edu

Karen S. Hollweg
The National Academies' National
Research Council
Washington, DC 20001
USA
khollweg@stanfordalumni.org

Cover illustration: *Long Beach, California*, photograph © 1999 by William W. Fuller of Payson, Arizona.

Library of Congress Cataloging-in-Publication Data

Cary Conference (8th: 1999: Institute of Ecosystem Studies)
Understanding urban ecosystems: a new frontier for science and education/editors, Alan R. Berkowitz, Charles H. Nilon, Karen S. Hollweg.
p. cm.
Includes bibliographical references and index.
ISBN 0-387-95496-1 (alk. paper)—ISBN 0-387-95237-3 (pbk.: alk. paper)
1. Urban ecology—Congresses. 2. Ecosystem management—Congresses. 3. Biotic communities—Study and teaching—Congresses. I. Berkowitz, Alan R. II. Nilon, Charles H., 1956– III. Hollweg, Karen S. IV. Title.
HT241 .C37 1999
307.76—dc21 2002070474

ISBN 0-387-95496-1 (hardcover) ISBN 0-387-95237-3 (softcover) Printed on acid-free paper.

Printed in the United States of America.

9 8 7 6 5 4 3 2 1 SPIN 10877695 (hardcover) SPIN 10793061 (softcover)

www.springer-ny.com

Springer-Verlag New York Berlin Heidelberg
A member of BertelsmannSpringer Science+Business Media GmbH

To the people who were instrumental in turning us on to cities—to our mothers and fathers and our grandparents who encouraged us to explore Philadelphia, Vancouver, Chicago, Denver, Boulder, and other wonderful cities in our formative years.

Cary Conference VIII, 1999

1. Helen C. Thompson; 2. Noa Avriel-Avni; 3. Moshe Shachak; 4. Juan J. Armesto; 5. Lucille Barrera; 6. Jacqueline M. Carrera; 7. Kate Macneale; 8. Erika Latty; 9. Susan Kavy; 10. Kristen H. Desmarais; 11. Vicki O. Fabiyi; 12. Susan Mockenhaupt; 13. Celestine H. Pea; 14. David B. Campbell; 15. Shoshana Keiny; 16. Justin Wright; 17. Mark R. Walbridge; 18. Jo Ellen Roseman; 19. Lalit Pande; 20. Bora Simmons; 21. Susan Musante; 22. Bunyan Bryant; 23. Theresa Heyer; 24. Rusong Wang; 25. Anne Whiston Spirn; 26. Mary Lane; 27. Kathleen Hogan; 28. Steward T.A. Pickett; 29. Kathleen C. Weathers; 30. Garry Hamilton; 31. Pamela H. Templer; 32. William R. Burch, Jr; 33. Gary M. Lovett; 34. Charles Hopkins; 35. Louise Chawla; 36. Nancy B. Grimm; 37. Tammy Bird; 38. Randall E. Raymond; 39. Jack K. Shu; 40. Charles H. Nilon; 41. Richard V. Pouyat; 42. Lawrence E. Band; 43. Carolyn Mattoon; 44. Lisa LaRocque; 45. Karen E. Hinson; 46. Mary J. Leou; 47. Carolyn Harrison; 48. Gary C. Smith; 49. R. Mark Davis; 50. Alan R. Berkowitz; 51. William E. Rees; 52. Frank B. Golley; 53. Daniel Strauss; 54. David L. Strayer; 55. Jonah Smith; 56. Bruce P. Hayden; 57. Louis V. Verchot; 58. James Kohlmoos; 59. Debra C. Roberts; 60. Gene E. Likens; 61. Daniel Baron; 62. Carol Fialkowski; 63. Bruce W. Grant; 64. Peter Cullen; 65. Maciej Luniak; 66. John B. Wolford; 67. Karen S. Hollweg; 68. Julian Agyeman; 69. John Callewaert; 70. Joseph Poracsky; 71. Rosalyn McKeown; 72. William Robertson IV; 73. Paul H. Gobster; 74. J. Morgan Grove; 75. Joseph S. Warner; 76. Marc A. Breslav; 77. Seth W. Bigelow; 78. Anthony D. Bradshaw; 79. Richard S. Ostfeld; 80. Henry Campa III; 81. William S. Carlsen; 82. Clive G. Jones. Absent from photo Rodger W. Bybee, Peter M. Groffman, Nahid Khazenie, Francis P. Pandolfi, Ken A. Schmidt.

Preface

In what ways is a conference like an urban ecosystem? People come together for many of the same reasons they migrate to cities—for jobs; because they know someone there, perhaps someone they feel can help them achieve their goals; for the promise of a better life. Cities, and conferences, by placing people in close proximity, give us an opportunity to work together on a combination of individual and common goals. Both build on efficiencies of transportation and communication; both produce something of substance and something more of spirit. A conference is indeed much more short-lived than a city, but let us examine the considerable number of similarities—especially with regard to the focus of the conference from which this book sprang.

A conference and an urban ecosystem function ecologically in much the same way. Food is brought in from afar, but wastes are disposed of locally; people's movements are facilitated and constrained by the built environment; the nonhuman organisms in the environment can become invisible to the residents or participants (even if they are ecologists!); and in the case of the Cary Conferences at the Institute of Ecosystem Studies (IES), people are housed in "urban" (near the conference center) and "suburban" (slightly farther away) locations and depend on different sorts of transportation accordingly. The eighth Cary Conference, held at IES April 27–29, 1999, with the topic of "Understanding Urban Ecosystems," mimicked in a small way the cities it was designed to discuss in the social dimension, too. A diverse assemblage of thinkers and doers shared the same rich benefits and challenges of human diversity—creating social capital but also social strife, grappling with communication challenges and worldview conflicts, and evolving in a short time to become something more than just the sum of its parts.

At yet another level the conference mimicked the real world arena we were discussing—that of the challenge of advancing the field and practice of urban ecosystem education. In this case, because of the need to come up with a single, linear schedule for the conference we necessarily ran amuck of the real complexities and cyclical nature of the knowledge-creation →

knowledge-dissemination → new-question-generation process that underlies urban ecosystem research and education. The logic we tried to breathe into the conference, flowing through a series of seemingly straightforward questions—Why is understanding urban ecosystems important? What do we mean by understanding urban ecosystems? How do people develop such understandings? What would a system that fosters such understandings look like?—was stretched at virtually every step by the very real and very vexing challenges of answering one of these without also addressing the others and by the incompleteness of our current knowledge and experience.

Ultimately, history will judge cities and conferences by many of the same metrics. Do the benefits outweigh the costs? Were unique and enduring social goods created? In answering these questions for either city or conference, we must remember to look for both products *and* relationships. Books such as this are just a small part of the desired outcomes of a social gathering like a working conference. Like many other books to spring forth from conferences, this book speaks in many voices and, hopefully, to as diverse an audience and more. Frontiers abound in the complex knot of urban ecosystem education—do we know yet what we mean by "the city is an ecosystem"? How do people acquire such knowledge at the individual, cognitive level? How does information flow through the collective social system? How serious can we be in asking our informal and formal education systems to truly serve the common good that demands such deeper understandings as we hurtle into a new century, still exponentially increasing our resource use and utilizing other destructive practices? Hopefully, ecologists interested in cities will take away some of the rare perspectives gained when they and their colleagues are forced to distill a complex subject into its most important elements. Also gleaned might be new perspectives from other scientists who think about cities in different ways. For educators, we hope to have identified exciting new intellectual frontiers and practical challenges. But perhaps most importantly, we hope to stimulate a cross-fertilization of thinking and cooperation between practitioners in the two fields.

The biennial Cary Conferences were inaugurated by IES in 1985, with each conference examining a fundamental issue in ecology to advance the field and foster synthesis. The conferences are designed to promote critical discussion with minimal distraction, and the agenda structured to allow time for discussion and debate. This, the eighth Cary Conference, was the first to focus on ecology education, and a first-of-its-kind effort to bring together leaders in the biological, physical, and social dimensions of urban ecosystem research with leading education researchers, administrators and practitioners. Eighty-six people participated; nearly half were educators (including 11 education researchers and 29 practitioners, with 6 K–12 teachers, 3 K–12 administrators and the rest being higher education, informal, or community educators) and the balance were scientists (including 36 natural scientists and 10 social scientists).

The conference focused on urban ecosystem education as an important and exciting frontier for researchers and educators interested in understanding urban areas from an ecological perspective. The central premise for the conference was that all people—decision makers, managers and citizens, and not just scientists and educators—need a much better understanding of how cities work as ecological systems. With more and more of our population living in urban areas, and the lion's share of resource use taking place there, people need this understanding in order to make cities healthier places for people and the other organisms that live in and near them, and to rein in and minimize the enormous impacts that cities have on surrounding and distant ecosystems. Such knowledge, broadly embraced and exercised, will be vital in building an ecologically and economically sustainable future. Given the dynamic and multifaceted nature of cities and their immediate environs, the conference was further predicated on the idea that a human ecosystem approach—integrating biological, physical, and social factors and embracing historical and geographical dimensions—may provide our best hope for coping with the complexity of cities.

Ultimately, urban ecosystem education seeks to foster a broad, ecosystem-based understanding of cities among all people. Taking an ecosystem approach gives us the tools we need to integrate the many relevant disciplines and make more understandable: (1) the *complexity of cities*, especially as we integrate sociology, anthropology, economics, and history with the full suite of biological and physical concepts; (2) the *dynamic nature of cities* as places where change is the norm and is driven by a multitude of interacting forces and conditions; and (3) the vital roles played by *spatial relationships*, *human and other disturbances*, and *historical influences* in shaping the urban environment.

In developing a broad understanding of cities as ecosystems we face numerous challenges, both intellectual and practical. Until recently, many ecologists ignored cities as places for serious ecological study. The strong tradition of urban research and education focused on the conservation of green spaces and natural areas in cities. There is increasing attention being paid to urban ecology, however, including new initiatives aimed at understanding cities as ecosystems. Our education systems do only a spotty job of teaching systems and interdisciplinary thinking, and they have a hard time developing truly integrative themes that run across the subjects and through the years in the curriculum. Fortunately, national and many state standards for learning—in natural science, math, geography, social science—are calling for student-centered inquiry, and for teachers and schools to provide a rich range of opportunities for students to engage in genuine investigations of the real world around them. The Cary Conference and this book aim, in part, to crystallize the new frameworks that are emerging from research about urban ecosystems and from research on how people teach and learn about complex systems like cities.

As in any human ecosystem, each participant in the conference filled various roles. Everyone participated in discussion groups that provided authors rich grist for thought. Authors of the chapters in this book held telephone discussions and received feedback from two reviewers before the conference, and presented their papers or posters to a lively audience replete with comments and question during the conference. They received extensive feedback from the discussion groups they participated in and from summaries of all discussions of each paper sent to them after the conference. Thus, this book is more than a compilation of the papers presented at the conference, but rather a combination of the individual authors' insights and the input they received before, during and after the conference. We believe that colleagues of those involved in the conference's deliberation will gain insights and share some of the enthusiasm generated through our discussion by reading this volume. In this way, we hope that our ephemeral conference ecosystem might have as one key and lasting output some useful input to the ideas and practice of the nascent field of urban ecosystem education.

Alan R. Berkowitz
Charles H. Nilon
Karen S. Hollweg

Acknowledgments

This book originated from the eighth Cary Conference held in April 1999 at the Institute of Ecosystem Studies in Millbrook, New York. The conference would not have been possible without the hard work of many people, playing many different roles, and we are delighted to acknowledge and thank them for their contributions.

Our steering committee of William Burch, Rodger Bybee, Diane Ebert-May, Carol Fialkowski, Gary Heath, Shoshana Keiny, Dan Kincaid, Gene Likens, Richard Ostfeld, Steward Pickett, Jack Shu, and Bora Simmons made valuable suggestions on topics, speakers, and participants, especially in the formative stages of planning for the conference.

We thank our authors who worked so hard before, during, and well after the conference, bringing their critical thinking and creativity to bear on a challenging and vital topic. We thank also the participants who reviewed abstracts and summaries of each paper in the months leading up to the conference, providing invaluable feedback to the authors and editors. During the conference, many people served as facilitators of the synthesis discussion groups, as moderators of the plenary sessions, and as leaders of the impromptu "next steps" discussion groups that generated the key recommendations from the conference (see Table 30.1). In this way, virtually everyone at the conference had several roles, and all were engaged and made significant contributions for which we are extremely grateful. Their names and affiliations at the time of the conference are listed in the Participants section.

We thank Frank Golley, research professor at the University of Georgia, for delivering the conference keynote address, and Peter Cullen for giving the wrap-up synthesis talk at the end of the conference. Their wisdom and insights inspired and challenged us all.

The conference was supported by grants from the National Science Foundation, National Aeronautics and Space Administration, Environmental Protection Agency (Office of Environmental Education), Surdna Foundation, USDA Forest Service (Urban and Community Forestry Program), Nathan Cummings Foundation and the Institute of

Ecosystem Studies. We thank all for their interest in urban ecosystem education and their support of the Cary Conference process.

The hard work of many made the conference possible. The staff of IES was crucial in attending to the needs of the conference and the conferees both prior to the event and during the crucial days of the meeting. Although we cannot mention all, we would like to especially thank IES graduate students and staff who served as conference assistants—Kristen Desmarais, Erika Latty, Kate Macneale, Jonah Smith, Pamela Templer, Helen Thompson, and Justin Wright—for their tireless efforts transporting participants and seeing that they were comfortable and happy in our ephemeral conference ecosystem. Susan Eberth, Heather Dahl, Pamela Freeman, Janet Traweek, Jean Martell, and Deborah Fargione skillfully prepared many documents for the conference and this book. Finally, Susan Kavy coordinated the conference—her ability to manage the complexity of the undertaking, her imagination, attention to detail, and home-baked-cookies-in-every-room flair, made our jobs easy, and the conference a pleasure from start to finish. Thank you all!

Alan R. Berkowitz
Karen S. Hollweg
Charles H. Nilon

Contents

Participants

The 86 participants in Cary Conference VIII, held at the Insitute of Ecosystem Studies in Millbrook, NY, April 27–29, 1999, are listed below, along with their affiliations at the time of the conference.

Julian Agyeman
Slippery Rock University of Pennsylvania

Juan J. Armesto
Universidad de Chile, Laboratorio de Sistematica & Ecologia Vegetal, Facultad de Ciencias

Noa Avriel-Avni
Ben-Gurion University of the Negev, Israel

Lawrence E. Band
University of North Carolina

Daniel Baron
Harmony School Education Center, Bloomington, IN

Lucille Barrera
Houston Independent School District

Alan R. Berkowitz
Institute of Ecosystem Studies

Seth W. Bigelow
Institute of Ecosystem Studies

Tammy Bird
Crenshaw High School, Los Angeles, CA

Anthony D. Bradshaw
University of Liverpool

Marc A. Breslav
Breslav Public Relations, Cold Spring, NY

Bunyan Bryant
University of Michigan

William R. Burch Jr.
Yale University

Rodger W. Bybee
National Research Council

John Callewaert
University of Michigan

Henry (Rique) Campa III
Michigan State University

David B. Campbell
National Science Foundation

William S. Carlsen
Cornell University

Jacqueline M. Carrera
Parks and People Foundation, Baltimore, MD

Louise Chawla
Kentucky State University

Peter Cullen
University of Canberra

Mark R. Davis
Earth Conservation Corps, Washington, DC

Kristen H. Desmarais
Institute of Ecosystem Studies

Vicki O. Fabiyi
Institute of Ecosystem Studies

Carol Fialkowski
The Field Museum of Natural History, Chicago, IL

Paul H. Gobster
USDA Forest Service

Frank B. Golley
University of Georgia

Bruce W. Grant
Widener University, Chester, PA

Nancy B. Grimm
Arizona State University

Peter M. Groffman
Institute of Ecosystem Studies

J. Morgan Grove
USDA Forest Service, South Burlington, VT

Garry Hamilton
New Scientist, Seattle, WA

Carolyn Harrison
University College London

Bruce P. Hayden
National Science Foundation

Theresa Heyer
USDA Forest Service

Karen E. Hinson
Western School of Technology and Environmental Science, Baltimore, MD

Kathleen Hogan
Institute of Ecosystem Studies

Karen S. Hollweg
North American Association for Environmental Education

Charles Hopkins
United Nations Educational, Scientific and Cultural Organization (UNESCO)

Clive G. Jones
Institute of Ecosystem Studies

Shoshana Keiny
Ben-Gurion University of the Negev, Israel

Nahid Khazenie
NASA

James Kohlmoos
The Implementation Group, Inc., Washington, DC

Mary Lane
Komachin Middle School, Lacey, WA

Lisa LaRocque
Project del Rio, Las Cruces, NM

Erika Latty
Cornell University/Institute of Ecosystem Studies

Mary J. Leou
City Parks Foundation, New York

Gene E. Likens
Institute of Ecosystem Studies

Gary M. Lovett
Institute of Ecosystem Studies

Maciej Luniak
Museum/Institute of Zoology of the Polish Academy of Sciences

Kate Macneale
Cornell University/Institute of Ecosystem Studies

Carolyn (Lyn) Mattoon
The Hotchkiss School, Lakeville, CT

Rosalyn McKeown
University of Tennessee

Susan Mockenhaupt
USDA Forest Service, Washington, DC

Susan Musante
Ecological Society of America

Charles H. Nilon
University of Missouri—Columbia

Richard S. Ostfeld
Institute of Ecosystem Studies

Lalit Pande
Uttarakhand Environment Education Centre, Uttarakhand Seva Nidhi, India

Francis P. Pandolfi
National Environmental Education and Training Foundation, Briarcliff Manor, NY

Celestine H. Pea
National Science Foundation

Steward T.A. Pickett
Institute of Ecosystem Studies

Joseph Poracsky
Portland State University

Richard V. Pouyat
USDA Forest Service

Randall E. Raymond
Detroit Public Schools

William E. Rees
University of British Columbia

Debra C. Roberts
Durban Metropolitan Council, South Africa

William Robertson IV
The Andrew W. Mellon Foundation, New York

Jo Ellen Roseman
American Association for the Advancement of Science

Ken A. Schmidt
Institute of Ecosystem Studies

Moshe Shachak
Ben-Gurion University of the Negev, Israel

Jack K. Shu
California State Parks

Bora Simmons
Northern Illinois University

Gary C. Smith
California Department of Education

Jonah Smith
Rutgers University

Anne Whiston Spirn
University of Pennsylvania

Daniel Strauss
The High School for Environmental Studies, New York

David L. Strayer
Institute of Ecosystem Studies

Pamela H. Templer
Cornell University

Helen C. Thompson
Rutgers University

Louis V. Verchot
Institute of Ecosystem Studies

Mark R. Walbridge
George Mason University

Rusong Wang
Research Center for Eco-Environmental Sciences, Chinese Academy of Sciences

Joseph S. Warner
Institute of Ecosystem Studies

Kathleen C. Weathers
Institute of Ecosystem Studies

John B. Wolford
Missouri Historical Society

Justin Wright
Cornell University

Contributors

Julian Agyeman
Slippery Rock University of Pennsylvania. *Current address*: Department of Urban and Environmental Policy and Planning, Tufts University, Medford, MA 02155, USA. julianagyeman@tufts.edu

Noa Avriel-Avni
Education and Ecology Departments Ben-Gurion University of the Negev, Mitzpe-Ramon 80600, Israel. avrielav@bgumail.bgu.ac.il

Lawrence J. Baker
Baker Consulting, Tempe, AZ 85287, USA. bakerenv@earthlink.net

Alan R. Berkowitz
Institute of Ecosystem Studies, Millbrook, NY 12545, USA. berkowitza@ecostudies.org

Anthony D. Bradshaw
School of Biological Sciences, University of Liverpool, Liverpool, England L69 3BX, UK. tonybradshaw@cableinet.co.uk

Bunyan Bryant
The School of Natural Resources and Environment, University of Michigan, Ann Arbor, MI 48109, USA. bbryant@umich.edu

William R. Burch, Jr.
School of Forestry and Environmental Studies, Yale University, New Haven, CT 06511, USA. williamburch@yale.edu

Jacquie Burgess
Department of Geography, University College London, London WC1H, 0AP, UK. j.burgess@geog.ucl.ac.uk

Rodger W. Bybee
National Research Council. *Current address*: Biological Sciences Curriculum Study (BSCS), Colorado Springs, CO 80918-3842, USA.
rbybee@bscs.org

John Callewaert
University of Michigan. *Current address*: Institute for Community and Environment, Colby-Sawyer College, New London, NH 03257, USA.
jcallewaert@colby-sawyer.edu

Henry (Rique) Campa III
Department of Fisheries and Wildlife, Michigan State University, East Lansing, MI 48824, USA. campa@pilot.msu.edu.

Jacqueline M. Carrera
Parks and People Foundation, Baltimore, MD 21211, USA.
jackie.carrera@parksandpeople.org

Louise Chawla
Kentucky State University, Frankfort, KY 40601, USA. chawla393@aol.com

Peter Cullen
Cooperative Research Centre for Freshwater Ecology, University of Canberra, ACT 2601, Australia. cullen@lake.canberra.edu.au

Carol Fialkowski
The Field Museum of Natural History, Chicago, IL 60605, USA.
cfialkowski@fmnh.org

Frank B. Golley
Institute of Ecology, University of Georgia, Athens, GA 30602, USA.
fgolley@arches.uga.edu

Bruce W. Grant
Department of Biology, Widener University, Chester, PA 19013, USA.
grant@popl.science.widener.edu

Nancy B. Grimm
Department of Biology, Arizona State University, Tempe, AZ 85287, USA.
nbgrimm@asu.edu

J. Morgan Grove
USDA Forest Service, South Burlington, VT 05402, USA. mgrove@fs.fed.us

Christine Hanaburgh
Michigan State University, East Lansing, MI 48824, USA.
hanaburc@michigan.gov

Carolyn Harrison
Department of Geography, University College London, London, England WC1H 0AP, UK. c.harrison@geog.ucl.ac.uk

Karen E. Hinson
Western School of Technology and Environmental Science, Baltimore, MD 21228, USA. *Current address*: Carver Center for Arts and Technology, Towson, MD 21204, USA. keh11@lycos.com

Kathleen Hogan
Institute of Ecosystem Studies, Millbrook, NY 12545, USA.
hogank@ecostudies.org

Karen S. Hollweg
North American Association for Environmental Education. *Current address*: The National Academies' National Research Council, Washington, DC 20001, USA. khollweg@stanfordalumni.org

Diane Hope
Center for Environmental Studies, Arizona State University, Tempe, AZ 85287, USA. di.hope@asu.edu

Shoshana Keiny
Department of Education, Ben-Gurion University of the Negev, 84105 Beer-Sheva, Israel. shoshk@bgumail.bgu.ac.il

Martin V. Melosi
University of Houston, Houston, TX 77204, USA. mmelosi@uh.edu

Charles H. Nilon
The School of Natural Resources, University of Missouri—Columbia, Columbia, MO 65211, USA. nilonc@missouri.edu

Robert J. Northrop
Maryland Department of Natural Resources, Forest Service, North East, MD 21901, USA. rnorthrop@dol.net

Zhiyun Ouyang
Research Center for Eco-Environmental Sciences, Chinese Academy of Sciences, Beijing 100080, China. ouyang@panda.ioz.ac.cn

Celestine H. Pea
Education Reform Division, National Science Foundation, Arlington, VA 22230, USA. cpea@nsf.gov

Steward T.A. Pickett
Institute of Ecosystem Studies, Millbrook, NY 12545, USA.
picketts@ecostudies.org

Delia F. Raymer
Department of Fisheries and Wildlife, Michigan State University, East Lansing MI, 48824-1222, USA. draymer@raymer.cs.wright.edu

William E. Rees
School of Community and Regional Planning, University of British Columbia, Vancouver, BC V6T 1Z4, Canada. wrees@interchange.ubc.ca

Debra C. Roberts
Development Planning Department, Durban Metropolitan Council, Durban 4000, South Africa. robertsd@cesu.durban.gov.za

Jo Ellen Roseman
American Association for the Advancement of Science, Washington, DC 20005, USA. jroseman@aaas.org

Ilaria Salvadori
College of Environmental Design, University of California—Berkeley, Berkeley, CA 94705, USA. *Current address*: Project for Public Spaces, Inc., New York, NY 10014, USA. ilaria@pps.org

Moshe Shachak
Marco and Louise Department of Desert Ecology, The Jacob Blaustein Institute for Desert Research, Ben-Gurion University of the Negev, Sede Boqer Campus, 84990 Israel. shachak@bgumail.bgu.ac.il

Jack K. Shu
California State Parks, San Diego, CA 92108, USA. jkshu@parks.ca.gov

Bora Simmons
Department of Curriculum and Instruction, Northern Illinois University, DeKalb, IL 60115, USA. boras@niu.edu

Gary C. Smith
California Department of Education, Anaheim, CA 92806, USA. *Current address*: Katella High School, Anaheim, CA 92806, USA.
gcsmith@earthlink.net

Anne Whiston Spirn
University of Pennsylvania, Philadelphia, PA 19104, USA. *Current address*: School of Architecture and Planning, Massachusetts Institute of Technology, Cambridge, MA 02139, USA. spirn@mit.edu

Luli Stern
American Association for the Advancement of Science, Washington, DC 20005, USA. Lstern@techunix.technion.ac.il

Rusong Wang
Research Center for Eco-Environmental Sciences, Chinese Academy of Sciences, Beijing 100080, China. wangrs@sun.ihep.ac.cn

Kathleen C. Weathers
Institute of Ecosystem Studies, Millbrook, NY 12545, USA.
weathersk@ecostudies.org

John B. Wolford
Missouri Historical Society, St. Louis, MO 63112, USA.
john_wolford@umsl.edu

1 Introduction: Ecosystem Understanding Is a Key to Understanding Cities

Charles H. Nilon, Alan R. Berkowitz, and Karen S. Hollweg

In June 2000 the *Baltimore Afro-American* newspaper ran an article on the threat of mosquito-borne diseases to urban residents (Thompson 2000). The article highlighted the life history of mosquitoes, explained how people and their activities contribute to the distribution and abundance of mosquitoes, and suggested some simple, clear steps that residents could use to reduce the number of mosquitoes around their homes. Rather than describing vector-borne diseases as an impending threat to public health or high mosquito populations as an environmental catastrophe, the article clearly explained the issue within the context of the day-to-day lives of Baltimore residents, the way they manage the area immediately around their homes, and the organisms that share this environment with them. It presented mosquitoes as part of a system that links people, other organisms, and the built and natural environments. The article reached more than just the relatively small group of people who make decisions about mosquito control in Baltimore. It was not targeted at the somewhat larger audience that is concerned with broader environmental or health issues and belongs to environmental, conservation, or health organizations. It was specifically aimed at the *Baltimore Afro-American*'s readers and provided these residents with practical information relevant to their daily lives, yet is important in understanding how Baltimore works as an ecosystem.

This book is about why it is important to develop an understanding of cities as ecosystems among people who live in and care about the world's cities.

Some Definitions

What Is Urban?

Definitions of "urban" vary among countries and often are specific to the political, social, and economic context in which they are utilized. The United States Census Bureau defines urban areas as populated regions with a density of 1,600 people/km^2 or greater and a minimum population of 2,500. Most U.S. cities fit within a second category of "metropolitan areas": these consist of a central city with a minimum population of 50,000, the county in which least 50% of the population of the central city lives, and outlying counties with well-defined links to the central county or counties based on commuting patterns (Office of Management and Budget 2000). The census bureau in the Republic of South Africa defines urban areas as built-up areas, including vacant space, within a proclaimed municipal or local authority boundary, with various structures—houses, flats, hotels, boarding houses, old age homes, caravan parks, school and university hostels—built according to municipal bylaws. Informal urban areas, often squatter areas, are found within a proclaimed urban area but consist mainly of informal dwellings. Census South Africa also identifies a third "other urban area" category of mines, factories, municipal hostels, hospitals, prisons, and other institutions within a local authority boundary (Statistics South Africa 1998). Key to both the U.S. and South African definitions is how we will define urban areas in this book: identifiable places with defined or fixed boundaries and a high human population density.

Urban ecosystems are shaped by the process of urbanization, which involves the conversion of rural and other areas due to increases in the urban population *or* to the spatial spreading of cities or both. Nowadays, in developed countries urbanization is influenced by economic changes associated with the transition from an industrial to a service economy, the decentralization of employment, the stratification of the labor market into high- and low-paying jobs, technological changes related to information management, and resulting changes in family structure, culture, and politics (Knox 1991). Sprawl, which can actually result in an overall decrease in a metropolitan area's density while it spreads in space, is an important urbanizing process in many developed countries. In developing countries, urbanization is driven in part by rural people moving to cities and moving into and building formal and informal settlements (Celecia 2000). These changes mean that the area covered by urban and urbanizing landscapes is increasing (Knox 1991).

What Is an Urban Ecosystem?

Urban ecosystem models are based on the interaction of the social, biological, and physical components of a city (Figure 1.1). This interaction can best be understood by recognizing that urban ecosystems are dynamic and influenced by different types of driving forces. The idea that people and their activities influence the ecology of urban areas has a relatively long history. Urban ecology has been associated with a focus on solving problems of cities. Andrew Hurley's 1997 environmental history of the St. Louis metropolitan region notes that since the early 1900s there have been efforts to control floods, vector-borne diseases, and toxic waste within the context

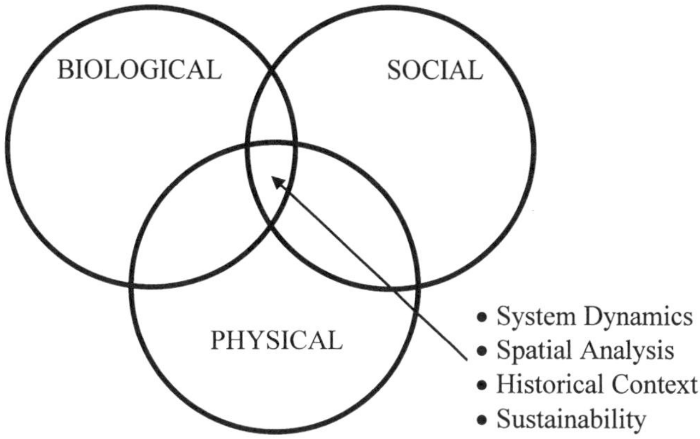

FIGURE 1.1. A simplified model for understanding urban ecosystems as the interaction of the social, biological, and physical components of a city. Several of the many possible integrating frameworks at the overlap of the three disciplines or perspectives are shown.

of viewing the city as a system. Current research initiatives in urban ecosystems build on this problem-solving tradition and a strong foundation of urban ecology research dating from the 1960s, when ecologists recognized that cities and the agricultural and forested areas that surround them are unique places ecologically and worthy of study (Wagoner and Ovington 1962; Numata 1973). In the 1960s and 1970s the International Biological Program initiated a study of Brussels, and the UNESCO Man and the Biosphere (MAB) Program began urban ecosystem projects in Hong Kong, Tokyo, Sydney, and Rome (Sukopp 1990). These studies developed models of energy flow and mass balance in cities (Douglas 1983), and sought to develop a concept of human ecology that was applicable to urban ecosystems, including research on cognitive psychology, environmental perception, and learning (Bonnes 1987). Significantly, these projects viewed urban ecosystems from a problem-solving perspective that differed from approaches to ecosystem ecology favored by mainstream ecologists. The study by Boyden et al. (1981) of the Hong Kong ecosystem was driven by questions about human health and well-being. Other MAB projects used an ecosystem approach to identify strategies for managing air quality, rapid urbanization, and other planning concerns (Celecia 2000).

Urban areas concentrate enormous amounts of energy and resources, creating unique ecosystems and causing innumerable impacts to adjacent and distant lands and waters (Sukopp 1990; Nowak 1994; Costanza and Greer 1995). Cities also provide social and environmental benefits. Knox (1994) describes the "catalytic quality" of urban areas that lead to innovation and expression for individuals and groups. Douglas (1983) noted that the health of residents of cities is often better than that of residents of adjacent rural areas. Spirn (1984) describes how the efforts of the urban architects of the 1890s led to land management and urban design projects in Boston, Denver, and other U.S. cities that focused on controlling the pollution of rivers and streams and that resulted in declines in urban mortality from typhoid fever and other infectious diseases.

The interaction between the social, biological, and physical components of an urban ecosystem can be studied using any of a number of frameworks (Figure 1.1). One framework, a geographic or spatial approach, considers the particular spatial setting and arrangement of key components of the city. Ecosystems also can be studied using an historical framework, recognizing that system dynamics and spatial context are influenced by past events. Either framework can be applied to better understand urban ecosystems, whether they are entire metropolitan areas or smaller areas within cities (e.g., small watersheds and neighborhoods).

What Do We Mean by Understanding?

We view understanding as knowledge of factual information and the ability to apply that information in the context of an individual's day-to-day life. Since Kellert's (1976) study of the attitudes, knowledge and behavior of the American public toward animals and nature, ecologists, educators, and environmentalists have been concerned with the public's apparent lack of ecological knowledge. Kellert (1976) found that some urban residents were less knowledgeable than others about animals and environmental issues, and that these differences in knowledge were influenced by gender, race, income, location of childhood residence, and level of education. Two decades later, the National Environmental Education and Training Foundation (NEETF), which conducts an annual survey of the environmental knowledge of the U.S. public, found that on average only 22% of survey respondents were able to answer a set of factual questions about environmental issues (National Environmental Education and Training Foundation 1998, Table 1.1). In an encouraging note, the NEETF surveys have found a positive correlation between respondents' knowledge of environmental concepts and their self-reported level of activity on behalf of the environment (National Environmental Education and Training Foundation 1998; 1999).

Understanding urban ecosystems means more than acquiring knowledge of specific facts about environmental issues. To us, understanding means that people living in and around cities are aware that cities are ecological systems and can apply that concept in their thinking and actions. Developing such an understanding among the area's residents requires involvement of formal and informal education, local government, the media, and the various institutions that urban people rely on for educational experiences

TABLE 1.1. Environmental knowledge (% answering question correctly) of adults (18 or older, n = 2000) surveyed in the United States in May 1998 for the Seventh Annual National Report Card on Environmental Knowledge, Attitudes, and Behaviors.

Question and *correct answer*	Total	Urban	Rural	Variation
What is the definition of a watershed? *Land area that drains into a specified body of water.*	41	36	43	−7
The government tests tap water. *False*	35	36	36	0
What are the main sources of ozone-depleting CFCs? *Auto air conditioners and refrigerators*	33	31	32	−1
The government tests household chemicals. *False*	27	31	24	+7
How is most electricity in the US produced? *Burning coal*	27	26	25	+1
What is the largest single source of waste in landfills? *Paper products*	23	21	23	−2
What is the leading cause of water pollution in US? *Surface water runoff*	22	17	22	−5
How do we dispose of spent nuclear fuel? *Store on-site at power plants*	17	17	15	+2
What is main source of oil pollution in water? *Improper disposal of motor oil*	16	16	16	0
What is leading cause of animal entanglement? *Improper disposal of fishing line*	10	8	9	−1
What is the leading cause of child death worldwide? *Microorganisms in water (pollution)*	9	8	9	−1
Average % answering questions correctly	**22**	**22**	**23**	**−1**

Source: National Environmental Education & Training Foundation 1998, and Pandolfi and Coyle 1999 *personal communication.*
Total column gives results for whole sample; Urban includes respondents indicating they live in a large, medium-size or small city; Rural includes respondents indicating they live in a small town or rural/farm area. Other categories were suburban town or small town. The difference (Urban − Rural) is shown in the final column. Questions are sorted by the % correct responses from most to least.

and information. Developing an understanding of urban ecosystems also requires the recognition that "understanding" is a participatory and deliberative process, not just a one-way exchange of facts and information about ecological, physical, and social systems between experts and the public (Harrison and Burgess 1994). Understanding must take into account the range of experiences, perceptions of place, and local knowledge that urban residents possess (Handley, et al. 1998; MacFarlane, et al. 2000).

Guiding Questions

This book looks at several questions that we find basic to understanding urban ecosystems: (1) Why is understanding urban ecosystems important? (2) What do we mean by understanding urban ecosystems? (3) How do people acquire such understanding? (4) What practical strategies can we employ to achieve broad understanding of cities as ecosystems in the future? Here we touch briefly on our rationale for framing the subject in this way.

Why Is It Important for People to Understand That Cities Are Ecosystems?

Our premise is that residents of cities, people who work for institutions and organizations in cities, and ecologists all can benefit from an understanding that cities are ecosystems, and that part of understanding is the ability to use information to answer questions about cities and solve problems in cities.

Examples like the situation described in the *Baltimore Afro-American* illustrate that residents of cities can benefit from understanding the relationship between people, their activities, and the environment. Residents of Baltimore and other cities concerned about diseases carried by mosquitoes are often warned to avoid places with standing water. A more thorough knowledge of cities as ecosystems might reveal why some neighborhoods have more of these wet areas than others. Residents might learn how the hydrologic cycle of the city has been altered over years of human settlement. Douglas (1983) presented an easy-to-understand model of an urban watershed showing how the hydrologic cycle is impacted by a range of human activities. These activities influence water quality, water quantity, and stream characteristics. By studying the inputs, outputs, and flows of water throughout the city, and using their knowledge of their neighborhood, residents would learn that the management of land in cities influences where open, stagnant water will develop, and how the distribution of these areas is tied to social and economic patterns in their city.

As you read the chapters in Section I, consider the perspectives of managers and applied ecologists who work in agencies, and of ecologists working in academia. How can they benefit from an ecosystem-focused understanding of their city? Public and private institutions and organizations that provide services and support to urban residents will benefit from being able to cultivate an ecological understanding of a number of crucial issues faced by urban residents. Many of these issues are not viewed as environmental or ecological in nature but rather as public health or policy concerns. For example, local housing authorities hoping to restore abandoned houses and rebuild vacant lots often deal with issues of lead contamination. Understanding how lead cycles through the urban ecosystem would certainly improve the ability of local authorities to assess the risk of lead exposure to different groups of people. A systems approach could help applied ecologists better understand the mechanisms that influence how cities work, especially as these forces operate over longer time frames and larger spatial scales than their problem-solving focus normally allows.

Ecologists and the field of ecology will benefit from an understanding of cities as ecosystems. Cities have not traditionally been considered objects of study by mainstream ecologists nor objects of teaching by ecological or environmental educators (McDonnell and Pickett 1990). However, there is increasing attention being paid to urban ecology, including new initiatives aimed at understanding cities as ecosystems (McDonnell and Pickett 1993). There is also a recognition that cities are complex, and understanding them requires not just the full suite of biological and physical concepts, but also ways of integrating these with understandings from sociology, anthropology, economics, and history (Cronon 1991; Pickett, et al. 1997). Could ecologists' understanding of cities not benefit from an approach that studies the role of people and their activities as driving forces influencing the structure and function of ecosystems? If they do, ecologists may gain an understanding that ecological research in cities requires a new approach that is participatory and involves urban residents in asking research questions, developing hypotheses, collecting data, and interpreting and utilizing research results.

What Are the Important Concepts About Urban Ecosystems That People Should Understand?

Understanding urban ecosystems starts with the recognition that cities are ecosystems. This requires knowledge of the ecosystem concept and how it applies to urban areas. The resulting conceptual model includes an understanding of nutrient cycling and energy flow in cities, but also of the social dynamics of cities and how these interact with the biological and physical dimensions (Douglas 1983; Spirn 1984; Pickett, et al. 1997). The critical test is whether knowledge of the key concepts from ecosystem ecology contributes to an understanding of how cities work. This type of knowledge might lead residents to ask, "How does the movement of water in the city influence mosquito abundance?" or "How have changes in our neighborhood shaped where we find vacant lots with open, stagnant water?"

Understanding urban ecosystems also means being aware of the ecology of cities as complex habitats for the plant and animal species found in cities (Sukopp 1990; Luniak and Pisarski 1994; Nilon and Pais 1997). Knowledge of this kind might lead city officials to ask, "What kinds of mosquitoes are found in cities? In what part of the city are they? Are they more abundant or less abundant than fifty years ago or five years ago? What are the characteristics of areas that attract mosquitoes?"

Finally, understanding urban ecosystems means understanding the key concepts from the social sciences that constrain and control ecosystem processes. Knowledgeable members of a neighborhood association might ask, "How is the pattern of vacant lots in my neighborhood linked to patterns of home ownership and income?" and "As our neighborhood association starts cleaning up these vacant lots, how will the kinds of animals and plants we see around our homes change?" In the chapters in Section II, researchers who study cities describe the most important concepts they believe people should understand to have an appreciation of urban ecosystems and to be able to make decisions and solve problems within those systems.

How Do People Learn These Key Concepts About Urban Ecosystems?

We feel that people best learn about cities by learning key concepts about ecology and the social sciences in the context of exploring urban issues. In this way, they can come to understand how cities are ecosystems, learn about the ecology of cities, and become knowledgeable about how people and their activities play a dominant role in shaping all of the earth's ecosystems, especially cities. Relevant examples of concepts and state-of-the-art applications to cities are important.

We know that people are educated through a variety of means: family interactions, the media, outreach and communications efforts by agencies and decision makers, and schools and informal educational institutions. In the short run, the media and the traditional communication channels between decision makers and experts serve this function, but in the long run, it is also our schools and informal educational institutions that will shape how we think and what we know about the city as an ecosystem.

Schools are responsible for educating children and young adults. Developing an understanding of urban ecosystems among all people requires attention to the ways that education in science and other disciplines takes place. Can application of research about teaching and learning guide our efforts in changing how and at what ages students are taught key concepts? Can an examination of the many different skills and broad and diverse priorities in our educational systems lead us to uncovering a place for learning about urban ecosystems, even in an era with tremendous emphasis on students' performances on standardized tests?

What role can be played by informal institutions and the many forms of communication that reach children and adults beyond the schools? Children may visit nature centers and museums, participate in 4H groups, scouts, and other activities, or simply watch television. Their parents may participate in master gardener programs, or belong to churches and community organizations that deal with local issues. In Section III, authors present ways in which both formal and informal educational institutions incorporate ecological understanding into their programs, and play a positive role in educating people about their cities.

Can We Have an Urban Ecosystem Education Practice That Builds Partnerships Among Scientists, Educators, and Communities Where Everyone Benefits?

Urban ecosystem education also means involving people in ways that go beyond simply understanding concepts that describe the city as a system. Understanding also means being able to use and apply this information. How do urban residents, planners, politicians, and scientists get the information they need to understand cities and make hard decisions about their future? Figure 1.2 highlights the parallels between the pathways of inquiry and knowledge generation and use among scientists, managers, policy makers, and citizens. Inquiring or investigating springs from questions that arise as we identify gaps in our current knowledge and as we grapple with practical problems and strive to achieve our aspirations. The participatory approach we envision for understanding urban ecosystems will involve an overlap of questions and a blending of the investigating and learning processes as people work together to improve the quality of life in and around cities.

How might this actually work? Baltimore, St. Louis, and other cities across the U.S. and around the world have pioneered the development of participatory approaches to research and education. They inspire urban residents, agencies, and researchers to collaborate and seek common ground and mutual benefits. Such approaches provoked stimulating discussions at the Cary Conference that led us to envision the numerous opportunities for the future of urban ecosystem research and education touched upon in Section IV. The University of Illinois's East St. Louis Action Research Project (Reardon 1998) is just one of many models for linking university researchers, teachers, students, and local communities. In that project, local residents and community-based organizations provide the research questions and goals for the studies done by university faculty and students that are geared toward changing the environment of East St. Louis.

In other models of partnerships, community-based organizations or the media lead efforts to use information about urban areas and involve people in changing their day-to-day environment and thereby make changes in the environment of their cities. Hurley (1997) describes how local activists linked environmental concerns to civil rights issues in St. Louis, Missouri in

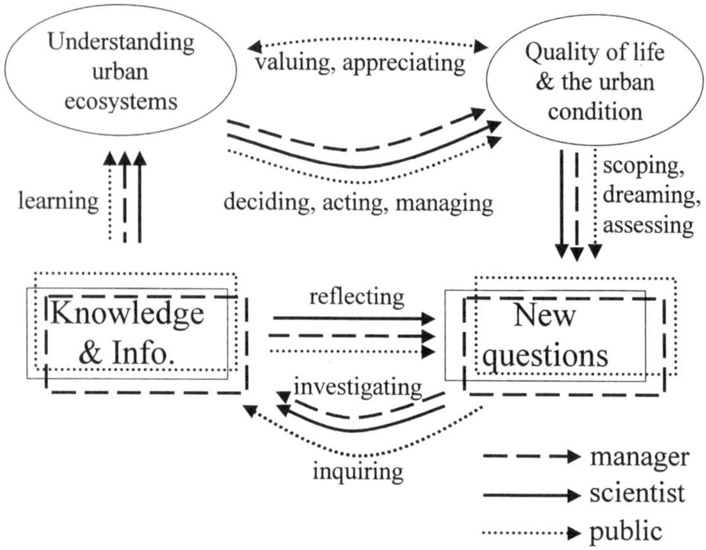

FIGURE 1.2. A schematic view of the interplay between the pathways by which knowledge is developed and used by scientists, managers (or policy makers), and citizens. Questions arise from identified gaps in knowledge and/or by identifying practical needs and problems. There may, or may not be, overlap and interaction between the kinds of questions asked by different people in this scheme, and the processes of inquiry, learning and application of knowledge they employ. Benefits to all accrue from fostering genuine interaction and from learning from each other how better to pursue each part of the process.

the 1940s. Frances Murphy, editor of the *Baltimore Afro-American*, founded a Clean Block Campaign in 1934 to involve children in cleaning up their neighborhood. Currently the paper bills this as the "oldest environmental program in the United States" (AFRO-Charities Inc. 2000).

The Book's Organization and Intended Audiences

This book is based on our vision of how the process of building and using an understanding of urban ecosystems might occur (Figure 1.2), addressing the different sorts of actors involved—scientists, educators (in the broadest sense of the term) and the students and citizens they work with, and those charged with making decisions and managing cities. We recognize that the knowledge-building enterprise is carried out in part by institutions, both formal and informal, that provide information on ecology and the environment, but that also go beyond information transfer and teaching to emphasize learning in its deepest sense. Urban ecosystem education is also carried out by people, groups, and communities that use this knowledge as they wrestle with problems and possibilities in cities. In this book we explore many facets of this complex and vital process, giving particular voice to the notion that it is much more than a one-way process of "transmitting information or knowledge" to people.

The book is organized in four sections, each addressing one of the key questions introduced above. We hope that our book will be useful to three groups of readers.

Ecologists and social scientists will benefit from an overview of the key concepts from ecology and the social sciences that underlie the study of urban ecosystems. They will benefit from case studies of important work being done by those who work in and study urban ecosystems. They will also gain an understanding of how people learn about these concepts in both formal and informal ways, better understand how these concepts are being applied by those who work in and live in cities, and be stimulated by the examples of partnerships we provide.

Educators will get up-to-date insights into ecology and other sciences that form our understanding of cities and urbanizing areas; gain practical ideas about what works, what doesn't work, and why; help forge a new vision for their work in urban ecology education; and formulate ideas for building new partnerships. Education researchers will gain insights into the challenges and opportunities faced in teaching about urban ecosystems, leading to new avenues for research in teaching and learning, and new partnerships with scientists and education practitioners.

We also hope that this book will reach the people who live in and are concerned about cities. This book is about the process of how people identify key concepts from the ecological and social sciences; how these concepts are learned, understood, and interpreted; and finally, how these concepts may be used by people in cities. We have asked the authors of each chapter to write for an audience that is concerned about cities and the physical, biological, and social factors that define them. Perhaps the readers of the *Baltimore Afro-American*, who care about their city and who have been supporters of the "oldest environmental program in the country," will be among our readers.

References

AFRO-Charities Inc. 2000. *AFRO-Clean Block*. AFRO-American Newspapers, Baltimore. <http://www.afro.com/culture/cleanblock/cleanblock.html>.

Bonnes, M. 1987. *Urban ecology applied to the city of Rome*. Italia MAB Project 11 Progress. Report N. 3. University of Rome, Italy.

Boyden, S., S. Millar, K. Newcombe, and B. O'Neill. 1981. *The ecology of a city and its people: the case of Hong Kong*. Australian National University Press, Canberra.

Celecia, J. 2000. *UNESCO'S Man and the Biosphere Programme and urban ecosystem research: a brief overview of the evolution and challenges of a three-decade international experience.* MAB ad hoc working group to explore the application of the biosphere reserve concept to urban areas and their hinterlands. UNESCO, Paris. <http://www.unesco.org/mab/urban/jc1doc.pdf>

Costanza, R., L. Wainger, C. Folke, and K.G. Mäler. 1993. Modeling complex economic systems: toward an evolutionary, dynamic understanding of people and nature. *BioScience* 43:545–555.

Costanza, R., and J. Greer. 1995. The Chesapeake Bay and its watershed: a model for sustainable ecosystem management? Pages 169–213 in H.L. Gunderson, C.S. Holling, and S.S. Light, eds. *Barriers and bridges to the renewal of ecosystems and institutions*. Columbia University Press, New York.

Cronon, W. 1991. *Nature's metropolis: Chicago and the great west*. Norton, New York.

Douglas, I. 1983. *The urban environment*. Edward Arnold, London.

Greenwood, E.F., ed. 1999. *Ecology and landscape development: a history of the Mersey Basin*. Liverpool University Press, Liverpool.

Grove, J.M., and W.R. Burch, Jr. 1997. A social ecology approach and applications of urban ecosystem and landscape analyses: a case study of Baltimore, Maryland. *Urban Ecosystems* 1:259–273.

Handley, J.F., E.J. Griffiths, S.L. Hill, and J.M. Howe. 1998. Land restoration using an ecologically informed and participative approach. Pages 171–185 in H.R. Fox, H.M. Moore, and A.D. McIntosh, eds. *Land reclamation: achieving sustainable benefits*. A.A. Balkema, Rotterdam.

Harrison, C.M., and J. Burgess. 1994. Social constructions of nature: a case study of conflicts over the development of Rainham Marshes. *Transactions Institute of British Geographers* 19:291–310.

Hurley, A. 1997. Floods, rats, and toxic waste: allocating environmental hazards since World War II. Pages 242–261 in A. Hurley, ed. *Common fields: an environmental history of St. Louis*. Missouri Historical Society Press, St. Louis.

Kellert, S.R. 1976. Perceptions of animals in American society. *Transactions of the North American Wildlife and Natural Resources Conference* 41:533–545.

Knox, P. L. 1991. The restless urban landscape: economic and socio-cultural change and the transformation of Washington, DC. *Annals Association of American Geographers* 81:181–209.

Knox, P.L. 1994. *Urbanization: an introduction to urban geography*. Prentice Hall, New York.

Luniak, M., and B. Pisarski. 1994. State of research into the fauna of Warsaw (up to 1990). *Memorabilia Zoologica* 49:155–165.

McDonnell, M.J., S.T.A. Pickett, P. Groffman, P. Bohlen, R.V. Pouyat, W.C. Zipperer, R.W. Parmelee, M.M. Carreiro, and K. Medley. 1997. Ecosystem processes along an urban-to-rural gradient. *Urban Ecosystems* 1:21–36.

McDonnell, M.J., and S.T.A. Pickett. 1990. Ecosystem structure and function along urban-rural gradients: an unexploited opportunity for ecology. *Ecology* 71:1232–1237.

McDonnell, M.J., and S.T.A. Pickett, eds. 1993. *Humans as components of ecosystems: the ecology of subtle human effects and populated areas*. Springer-Verlag, New York.

MacFarlane, R., D. Fuller, and M. Jeffries. 2000. Outsiders in the urban landscape? An analysis of ethnic minority landscapes. In J. Benson and M. Roe, eds. *Urban lifestyles: spaces, places, people*. A.A. Balkema, Rotterdam.

National Environmental Education and Training Foundation. 1998. *The seventh annual national report card on environmental attitudes, knowledge, and behavior*. National Environmental Education and Training Foundation. Roper Starch Worldwide, Washington, DC.

National Environmental Education and Training Foundation. 1999. *Environmental readiness for the 21st century. The eighth annual national report card on environmental attitudes, knowledge, and behavior*. National Environmental Education and Training Foundation. Roper Starch Worldwide, Washington, DC.

Nilon, C.H., and R.C. Pais. 1997. Terrestrial vertebrates in urban ecosystems: Developing hypotheses for the Gwynns Falls watershed in Baltimore, Maryland. *Urban Ecosystems* 1:247–257.

Nowak, D.J. 1994. Urban forest structure: the state of Chicago's urban forest. Pages 140–164 in E.G. McPherson, D.J. Nowak, and R.A. Rowntree, eds. *Chicago's urban forest ecosystem: Results of the Chicago urban forest climate project*. Volume Gen. Tech. Rep. NE–186. USDA Forest Service, Radnor, PA.

Numata, M., ed. 1973. *Fundamental studies in the characteristics of urban ecosystems*. Chiba University, Chiba, Japan.

Office of Management and Budget. 2000. Standards for defining metropolitan and micropolitan statistical areas. *Federal Register* 65(249):82228–82238.

Pickett, S.T.A., W.R. Burch, Jr., S.E. Dalton, T.W. Foresman, J.M. Grove, and R.A. Rowntree. 1997. A conceptual framework for the study of human ecosystems in urban areas. *Urban Ecosystems* 1:185–199.

Reardon, K.M. 1998. Enhancing the capacity of community-based organizations in East St. Louis. *Journal of Planning Education and Research* 17(4):323–333.

Spirn, A.W. 1984. *The granite garden*. Basic Books, New York.

Spirn, A.W. 1998. *The language of landscape*. Yale, New Haven.

Statistics South Africa. 1998. *The people of South Africa: population census 1996, the count and how it was done*. Statistics South Africa, Pretoria.

Sukopp, H. 1990. Urban ecology and its application in Europe. Pages 1–22 in H. Sukopp, S. Hejny, and I. Kowarik, eds. *Urban ecology: plants and plant communities in urban environments*. SPB Academic Publishers, The Hague.

Thompson, A. 2000. The mosquito as urban problem. *Baltimore Afro-American* 108(47):A10.

Wagoner, P.E., and J.D. Ovington. 1962. Proceedings of the Lockwood conference on the suburban forest and ecology. *Connecticut Agricultural Experiment Station Bulletin* 652.

Section I
The Importance of Understanding Urban Ecosystems: Themes

ALAN R. BERKOWITZ, CHARLES H. NILON, and KAREN S. HOLLWEG

Section I sets out a rationale for why understanding urban ecosystems is important, recognizing and celebrating the diversity of perspectives needed to adequately address this question. Each author places urban ecosystem education into the context of other important activities—efforts to reform education in general and in cities in particular; linking environmental education to community development, especially in underserved urban areas; the growing movements for environmental justice and sustainability; and current scientific efforts in urban ecosystem research. Where does urban ecosystem education fit into these larger enterprises, how can it contribute to achievement of these goals, and what insights and constraints are imposed on efforts to foster an understanding of cities as ecosystems?

The fundamental premise, clearly, is that understanding *is* useful and valuable—for guiding action, identifying problems, designing solutions, making policies and decisions, and for enriching people's lives. What, then, are the most important problems that understanding urban ecosystems will help us resolve, and how do these questions vary among the different kinds of people living in and concerned about cities, whether educators, scientists, policy and decision makers, or citizens? How can people use an understanding of urban ecosystems? What kinds of improvements can we expect from the increased understanding we hope to foster? The answers to these questions should motivate and guide the work of both academics and practitioners in urban ecosystem education. Knowing *why* understanding is important for different people also helps inform our thinking about *what* people need to know and *how* they can acquire this knowledge, themes that are developed in the rest of the book.

Consider, by way of analogy, the growth in understanding of *forests* as ecosystems, something many people have an easier time with than with the idea that cities are ecosystems. As a result of this increased understanding of forests, policy makers approach them differently now than they did several decades ago, bringing more experts to the table and appreciating forests' complexity more. Managers have developed whole new approaches springing, in part, from increased ecosystem understanding—ecosystem management, stakeholder participatory approaches, and the like. And the average citizen has benefited, from increased appreciation of the forests' wonders and from the identification of ways they can act on their values concerning the forest, whether for protection, wise use, or some other objective. The parallel advantages that would obtain with increased understanding of urban ecosystems are many.

Four Voices—Science Educators, Environmental Educators, Social Scientists, and Ecological Scientists—with Distinct Messages

Education reform is an enormously important concern today. Urban school systems face particular challenges, many springing from the very diversity that makes cities so rich, and from inequitable distribution of wealth and resources within and between urban and other systems. Urban areas also have proven challenging for environmental education in the past, with many perceived and in some cases real challenges regarding the unavailability of suitable sites for teaching and learning. Hollweg and coauthors (Chapter 2) place urban ecosystem education into the context of education reform, and emphasize important ways that reform can inform, guide, and constrain its practice. Teaching about urban ecosystems can be a compelling, integrating theme for schools everywhere, and can help city educators teach ecology, social science, geography, history, and science and technology using the local environment in its rich biological, physical, and social entirety.

Environmental education aimed at community development is another important enterprise of increasing importance worldwide, with urban community development of particular interest. Shu (Chapter 3) discusses how traditional environmental education is shifting to place more attention on humans as part of ecosystems, and how human ecosystems are becoming central foci of environmental education. Such efforts can empower communities to care for their residents and the other parts of the urban ecosystems surrounding them, leading to tangible improvements in community vitality and people's quality of life.

Social scientists and many others are paying increased attention to the causes and consequences of inequitable distribution of environmental benefits and ills as these phenomena play out in cities. Their work contributes to growing efforts for environmental justice (emphasizing the need for equity "in space," or among different peoples) and for sustainability (emphasizing the need for equity "in time," or for future generations). Bryant and Callewaert (Chapter 4) argue for the importance of understanding urban ecosystems as a key contributor to these efforts toward environmental justice and sustainability, helping to reverse declines in environmental quality that especially impact minority and poor communities in cities.

Pickett (Chapter 5) places urban ecosystem education into the context of the scientific enterprise itself, focusing on current efforts to bring a multidisciplinary research effort to bear on cities as ecological systems. He argues that public understanding of science, especially of urban ecosystem science, has direct and profound benefits to science. Public support of science is critical, both in general (e.g., by, but not limited to, providing funding) and in unique ways in cities and suburbs where science requires the formal and informal permission and assistance of a large and diverse collection of citizens. Furthermore, urban areas are commonly encountered by the majority of people in many countries, and thus a science-based education using these ecosystems as classrooms is an essential opportunity for developing public understanding of ecology and the environment.

Four Voices Touching on Common Themes

Despite their diverse backgrounds, all of the authors in Section I argue that it is essential to focus on the needs and interests of people who live in and near cities, especially those who have traditionally borne the brunt of environmental degradation and social inequities, in establishing the importance of urban ecosystem education. They also assert that the importance of understanding urban ecosystems goes beyond the need for city folks to understand their own homes to the need for people everywhere to understand cities and their role in the global commons. Understanding is portrayed by authors here and elsewhere in the book as a very dynamic process, rather than as a static body of facts or ideas. This places people inside the intellectual and social enterprise of research and learning, rather than outside as passive recipients of the fruits of science.

Finally, the authors here, and elsewhere in the book, wrestle with the complex relationship between understanding and action. None, for instance, argue that there is a simple, causal pathway from understanding to sound environmental action or behavior. However, each author within her or his own context implies, and in many cases makes explicit, very tangible benefits that increased understanding of urban ecosystems can yield, while recognizing that other forces are often at work in shaping our choices, decisions, and behaviors.

2
Why Is Understanding Urban Ecosystems an Important Frontier for Education and Educators?

Karen S. Hollweg, Celestine H. Pea, and Alan R. Berkowitz

The world is changing rapidly, making new demands on our schools and expecting them to prepare citizens with new understandings, skills, and abilities. For many reasons, schools (at least here in the United States) have been accused of failing to meet these demands. As a result, there is great attention to reforming schools and schooling. In the United States, the education system is currently in the midst of the longest sustained era of reform in the country's history. In this chapter we show how engaging students in real-world learning experiences in the human-dominated ecosystems in which they live will enable us to achieve many of the already established goals of education reform. Urban ecosystem education can help address goals directly pertaining to ecology, geography, and other disciplines, and we assert that it also may be effective in achieving other goals such as literacy and numeracy. These goals are pertinent for all students, regardless of where they live. All citizens need to understand urban ecosystems, and schools must play an integral role in fostering this understanding.

An understanding of the metropolitan area as a system is also helpful for all those who strive to implement systemic reform in urban schools. In addition, for students in and around cities, the urban environment is both their outdoor classroom and an important vehicle for improving education. For the 70% of the people in the U.S. who live in urban areas, their immediate environment is where they can easily go to learn firsthand from real problems in biology, chemistry, physics, sociology, economics, geography, and other fields. Education reform cannot proceed in these places unless it embraces local problems, resources, and opportunities for teaching and learning. An education agenda that helps urban residents understand their local ecosystem is one that gives real meaning to our commitment to purposeful education aimed at helping people improve their lives.

In this chapter, we start by putting the terms *important* and *frontier* into an educator's perspective. We discuss the education reform movement in the United States as an example of a large-scale, defining context that all education innovations—like urban ecosystem education—must work within. We then describe particular aspects of urban school systems and the nature of education reform there, concluding this section with lessons learned for urban ecosystem education. In the next section, we address the central question posed in our title, thereby shedding light on why urban ecosystems are an important and useful focus for educators and education reform. This is followed by some thoughts on how urban ecosystems are a frontier challenging and beckoning educators onward, and we end with a portrait of what urban ecosystem education might look like in practice and some final remarks.

The Importance and Frontier Aspect of Urban Ecosystems from an Educator's Perspective

What Do We Mean by "Education Reform?"

In September 1989, for the first time in U.S. history, a president and the nation's governors met to focus on how to improve the quality of American education. This National Education Summit arose because of economic concerns. The scores of U.S. high school students on most standardized achievement tests had been declining (National Education Goals Panel 1999) and their scores on international mathematics and science assessments were low in comparison to those of students in other countries (National Research Council 1999). Business leaders, the media, educators and others asked: If our students cannot do better than this, will our country be able to compete in our new global economy?

By 1994, the governors, Congress, and the president had agreed on eight National Education Goals (see Table 2.1) and a National Education Goals Panel had been formed to report national and state progress toward the goals, identify promising practices for improving education, and build a nationwide consensus to achieve the goals.

The hallmarks of this "education reform movement" in the United States have been and continue to be *clearly defined goal*s and *accountability for reaching those goals*. The initial goal-setting by our political leaders has been paralleled by educators. Between 1989 and 1997, professional associations or nonprofit organizations led the consensus-building process that resulted in the publication of voluntary national "standards" for many different disciplines that describe what students should know and be able to do in each discipline (National Council of Teachers of Mathematics 1989; American Association for the Advancement of Science 1989, 1993; American Geological Institute 1991; Geographic Education Standards Project 1994; Center for Civic Education 1994; Consortium of National Arts Education Associations 1994; National Council for the Social Studies 1994; National Center for History in the Schools 1994a, 1994b, 1996; Joint Committee on National Health Education Standards 1995; National Research Council 1996). These consensus documents provide a vision and

TABLE 2.1. The National Education Goals.

1. Ready to Learn—By the year 2000, all children in America will start school ready to learn.
2. School Completion—By the year 2000, the high school graduation rate will increase to at least 90 percent.
3. Student Achievement and Citizenship—By the year 2000, all students will leave grades 4, 8, and 12 having demonstrated competency over challenging subject matter including English, mathematics, science, foreign languages, civics and government, economics, arts, history, and geography, and every school in America will ensure that all students learn to use their minds well, so they may be prepared for responsible citizenship, further learning, and productive employment in our nation's modern economy.
4. Teacher Education and Professional Development—By the year 2000, the nation's teaching force will have access to programs for the continued improvement of their professional skills and the opportunity to acquire the knowledge and skills needed to instruct and prepare all American students for the next century.
5. Mathematics and Science—By the year 2000, United States students will be first in the world in mathematics and science achievement.
6. Adult Literacy and Lifelong Learning—By the year 2000, every adult American will be literate and will possess the knowledge and skills necessary to compete in a global economy and exercise the rights and responsibilities of citizenship.
7. Safe, Disciplined, and Alcohol- and Drug-free Schools—By the year 2000, every school in the United States will be free of drugs, violence, and the unauthorized presence of firearms and alcohol and will offer a disciplined environment conducive to learning.
8. Parental Participation—By the year 2000, every school will promote partnerships that will increase parental involvement and participation in promoting the social, emotional, and academic growth of children.

Source: From National Education Goals Panel 1999, page vi.

have served as models or resources for the development of state and local standards. As of 1999, 40 states had established standards in the four core subjects of English, mathematics, science, and social studies, and 39 states reported that they have aligned their assessments in one or more subject areas to measure progress against their standards (*Education Week*, January 11, 1999). It is important to note that there are among educators, policy makers, politicians, and others differing views on the quality of the state standards and the extent to which tests used by states and districts measure students' achievement of those standards.

Reforming the nation's education system and achieving our goals and standards has been compared with changing the direction of a huge ocean liner. To appreciate the complexity of the challenge and the investment needed, one must understand something about the size and nature of the system. In the United States, for example, the "school system" includes more than 89,000 public and private schools with 46.8 million students and over 3 million teachers governed by 50 states and 17,000 school districts (National Center for Education Statistics 2000, and see description of the *No Child Left Behind Act of 2001, http://www.whitehouse.gov/infocus/*

TABLE 2.2. Makeup of teacher and student populations in the United States.

	Teachers, all U.S.	Students, all U.S.	Students, NSF Urban Systemic Initiative Schools	Students, Great City Schools
White (%)	87.0	63.5	15.0	21.0
African American (%)	6.7	17.0	40.0	40.0
Hispanic (%)	4.1	14.4	38.0	30.0
Other (%)	1.8	5.1	7.0	7.0
Limited English Proficiency (%)		5.0		21.0
Eligible for Free or Reduced Lunch (%)		33.2		60.5
Total number	3,030,000	52,000,000	3,878,000	6,500,000

Percentages and the total number are shown for all teachers in the U.S. (National Center for Education Statistics 1997a) and for public school students in the U.S. (National Center for Education Statistics 1997b and 2000), in the National Science Foundation (NSF)-supported Urban Systemic Initiative schools (from the NSF's Core Data Elements, National Science Foundation 2001a, *http://www.ehr.nsf.gov/gpra/anrpt/2000/ESR2000anrpt.html*), and in the Great City Schools (The Urban Teacher Collaborative 2000).

education/). Demographic data for teachers and students is included in Table 2.2.

The U.S. federal government as well as private foundations have invested substantial amounts of money in this education reform effort. For example, the Title I program, one of a comprehensive set of programs that provides over $10.4 billion annually in federal aid to disadvantaged children to address the problems of poor urban and rural areas, has been increased (U.S. Department of Education 2002, *http://www.ed.gov/offices/OUS/budget02/summary/chapter1.html*). Between 1991 and 2002, the National Science Foundation's (NSF) Systemic Initiative Programs in the Directorate for Education and Human Resources (EHR) Division of Educational System Reform (ESR) invested approximately $800 million in efforts specifically designed to improve selected state, urban and rural education systems (National Science Foundation 1999; National Science Foundation 2001a and *http://www.ehr.nsf.gov/EHR/ESR/*). Each of these programs heightened the focus on standards and on improving the achievement levels of all students.

Addressing these numerous challenges will require not only a massive effort, but also a sustained one. Although past education reform efforts have not had much staying power, it seems like the commitments to our current educational improvement efforts are both broad-based and long-lasting. Despite changes in the presidency, governorships, and congressional leadership, there is still bipartisan support for continued reform. This support was evident in 2001 when Congress and a new administration voted for and signed into law increased federal funding for education (*http://www.whitehouse.gov/infocus/education/*). Americans consistently list education improvements among the most important national issues, and poll results show that the public continues to believe that the improvements called for in the National Education Goals are important and that achieving them would benefit the nation and their communities (Johnson and Aulicino 1998).

Cities and Urban Education Challenges and Strategies

Magnitude and Nature of the Challenge

The difficulty in achieving the National Education Goals and improving education for all students, particularly in urban school districts, can be linked to many factors. Research points to an uneven allocation of resources, resulting in a lack of high quality curriculum materials, equipment, facilities, and role models in urban districts. Rapid turnover in administrators and teachers, conflicts with teachers' unions, disengaged or angry parents, and apathy from state lawmakers add to the mix of conditions (*Education Week*, January 8, 1998).

Perhaps most important are factors linked to the teaching profession that concern certification, qualifications, and out-of-field assignments, the ethnic disparity that exists between the student and teaching populations, the need for ongoing professional development, and the operation of supply and demand. The 1996 National Assessment of Educational Progress (NAEP) survey showed that students with higher science and mathematics scores were more likely to have teachers who are certified with more than 5 years of experience (National Science Foundation 2000b). Similar positive relationships between student performance in science assessments and their teachers' training in science were found for some grade levels in the 2000 National Assessment of Educational Progress (National Center for Education Statistics 2002, *http://nces.ed.gov/nationsreportcard/*). Though the Third International Mathematics and Science Study (TIMSS) and National Centers for Education Statistics (NCES) data indicate that U.S. teachers completed more years of college as a whole than their counterparts in other countries, many science and mathematics teachers do not have degrees and/or lack certification in their fields and this problem is worst in urban districts (National Science Foundation 2000b; Darling-Hammond and Ball 1998). For instance, in high-poverty schools, 22 percent of teachers are not certified, compared with only 11 percent in low-poverty schools (National Center for Education Statistics 1996). This out-of-field teaching clearly hampers science teaching in cities.

Research over the last 30 years consistently shows that student results are better in schools where students are well known to their teachers (Miles and Darling-Hammond 1998). As illustrated in Table 2.2, however, there is a significant disparity between the ethnic composition of the U.S. student population as a whole and the teacher population, especially in urban areas. The supply of teachers required to meet the instructional needs of the rapidly changing student population is woefully inadequate. African American and Hispanic teachers make up only a small fraction of the workforce. One survey of the U.S. teaching profession showed that even though African American and Hispanic students made up 27.8 percent and 21 percent of the student population in "central city" schools, respectively, teachers from the same minorities comprised only 16.7 percent and 7.3 percent of the workforce (National Center for Education Statistics 1997a). As a result, few minority students have the opportunity to learn with minority teachers, especially in science and mathematics; only 14 percent of biology and mathematics students, 10 percent of chemistry students, and 7 percent of physics students have minority teachers.

The need for minority teachers is also a language issue. Although diversity adds richness to the learning environment, it also presents special challenges. Poor and minority students with limited English proficiency are more likely to experience difficulty in early grades, to repeat a grade, to need special education services, or to leave school without a diploma (National Science Foundation 2000b). Estimates of the number of students in the United States in need of bilingual instruction range between 3.5 and 6.4 million and they need teachers that speak their first language to ensure they are successful (Cummins 1989). Moreover, cultural and linguistic identification between student and teacher is desirable since teachers can serve as role models. Furthermore, the ability to provide "supportive environments for children" in which the validity and integrity of the home culture of the student can be confirmed is considered educationally enhancing (Delgado-Gaitan 1987, p. 131). Thus, obtaining the appropriate number of teachers who speak the language of the students they teach is becoming more of an imperative.

The National Science Board (National Research Council 1999) pointed out that responding to challenges such as those related to race, ethnicity, gender, language, or economics may be the most difficult task faced by schools and teachers in the twenty-first century. Prospects for improvement, at least for increasing the numbers of African American teachers, are uncertain; a 1993–1994 survey found an even lower representation of African Americans in entry-level teaching positions (National Center for Education Statistics 1997a). The proportion of the teaching profession that is white has remained essentially constant at 87–88 percent since 1976 (National Center for Education Statistics 1997a).

To improve student learning, attention must be focused on the support and professional development of teachers—enabling them to improve both their content knowledge and teaching skills. In the 1990s, states, districts, universities, and funders increased their investments in professional development for our teachers, but the demand for even more teachers remains tremendous. In the next decade government officials and educators project that up to two million teachers will be needed due to the increase in the student population, class size reductions, and the aging of the current teaching population (*New York Times*, July 11, 1998). Congress, national and state policy makers, higher education institutions, and voters are beginning to respond in a way that will meet this need, because the call for action is becoming louder and clearer. In the 2002 education budget, improving teacher education became a top priority with the allocation of $2.6 billion to assist states in developing a high-quality teaching workforce. Although overall the focus on improving the teaching workforce increased significantly, funds specifically targeted for science and mathematics did not fare as well. The federal Eisenhower Professional Development program which spent approximately $400 million in 2001 was eliminated and replaced by the Math and Science partnerships (MSP). Only $12.5 million was appropriated for the MSPs in the 2002 budget (U.S. Department of Education 2002, *http://www.ed.gov/offices/OUS/budget02/summary/chapter1.html*).

In January 2000, results of a survey of forty large city school districts in the U.S. were released (The Urban Teacher Collaborative 2000). Despite the fact that these districts are using a full range of targeted teacher recruitment strategies and 68 percent are offering incentives, 98 percent reported immediate demand for science and special education teachers; 95 percent, immediate demand for mathematics teachers; and 73 percent, immediate demand for bilingual teachers and teachers of color. In the coming decade our nation's large urban districts must find and hire some 700,000 new teachers (The Urban Teacher Collaborative 2000). By 2008 the number of students in kindergarten through grade 12 is projected to reach approximately 53.4 million, up from roughly 52 million today (National Center for Education Statistics 2000).

The U.S. National Science Foundation's Urban Systemic Initiative Program

In 1994, the National Science Foundation's (NSF) Division of Educational System Reform developed a K–12 Urban Systemic Initiative (USI) to promote systemic reform (O'Day and Smith 1993) of science and mathematics education for all students, especially those in the urban school districts with the highest number of high-poverty schools and the largest number of poor children. The program was designed initially to assist the twenty-eight cities with the largest number of school-age children between the ages of five and seventeen living in economic poverty, as determined by the 1990 census. The program was redesigned in 1999 to include approximately two hundred urban districts that serve "central cities" as defined by metropolitan statistical areas established by the National Center for Education Statistics (National Science Foundation 2000c, *http://www.nsf.gov/cgi-bin/getpub?nsf0034*). Nearly 69 percent of the students in these districts are eligible for free or reduced lunch (National Science Foundation 2000a, *http://www.ehr.nsf.gov/gpra/anrpt/2000/ESR2000anrpt.html*; Kim, et al. 2001) and schools serving nearly 60 percent of the students are considered "high-poverty" schools (*Education Week*, January 8, 1998). Through FY 2000, NSF had spent more than $400 million on its urban program, with an additional $500 million contributed by other sources. To implement broad-scale reform has required coordination and collaboration with other NSF-funded programs, U.S. Department of Education programs such as Title I and Title II, the Department of Energy National Laboratories, and the National Institutes of Health. The urban districts also sought and received financial support from national, state, and local entities including Texas Instruments, IBM, DOW Chemical, Annenberg, Danforth, and others. Moreover, hundreds of thousands of volunteer hours (more than 500,000 per year) also supported reform activities (Kim, et al. 2001; *www.siurbanstudy.org/newspublication*).

According to the Core Data Elements reported in the annual report for the Educational System Reform program (National Science Foundation 2001a, *http://www.ehr.nsf.gov/gpra/anrpt/2000/ESR2000anrpt.html*), funds were used to make significant strides in reforming the teaching and learning of science and mathematics via activities that included: (1) improving teacher education by providing professional development for in-service teachers and for connecting to and influencing pre-service preparation programs; (2) securing instructional materials to assist these teachers in implementing what they learned; (3) supporting cadres of lead teachers for follow-up and support at the classroom level; (4) establishing student support programs in meeting the higher requirements in the new standards; (5) securing services from colleges and universities; and (6) improving curricula and assessment measures. Some of the most far-reaching assistance came from the schools of education and the departments of arts and sciences at local colleges and universities. They assisted the districts in developing new science and mathematics courses, and with the alignment of science and mathematics curricula with standards. They also provided professional development, developed graduate-level programs to increase the number of content specialists, and helped address issues such as the science and mathematics teacher shortage, and certification and re-certification requirements. Colleges and universities also provided research, mentoring, tutoring, and specific problem-based projects through the use of graduate students.

Preliminary findings from an evaluative study entitled *Academic Excellence for all Urban Students: Their Accomplishments in Science an Mathematics*, conducted by Jason Kim and colleagues at Systemic Research, Inc., show that student achievement increased steadily in the urban school districts that participated in the USI for the longest period of time (Kim, et al. 2001). The report presented evidence that the USI districts substantially increased student enrollment and completion rates in higher level and gatekeeping courses. Enrollment gains of minority students were greater than those of nonminorities, resulting in reduced disparities with their peers during a given time period. Results from assessment tests showed that minority students made gains in science and mathematics achievement, which led to a reduction in the achievement gap among racial/ethnic groups. Kim and co-authors (Kim, et al. 2001) point to these gains in enrollment and achievement as evidence that urban districts are developing a solid learning infrastructure for bringing about and sustaining reform in science and mathematics.

Increasingly, these districts began to use research-based strategies proven to be effective in what constitutes good teaching, and, more importantly, what comprises good teaching for urban students. More than 80 percent of the schools participating in the reform programs are implementing standards-based curricula that reflect their state and local standards. Almost all of the districts now require the teaching of science and mathematics at the K–12 levels. New standards for schools, teachers, administrators, and counselors have brought about changes in participation in professional development, have increased certification and recertification requirements for teachers and graduation requirements for students, and heightened accountability at all levels. Teams of teachers, school and district administrators, principals, representatives from local colleges and universities, national consultants, parents, and members from the broader community are increasingly collaborating to improve the quality of education for all students.

Lessons Learned for Urban Ecosystem Education

The experience in the United States of the education reform movement in general, and the urban systemic reform efforts in particular, provides some valuable lessons for the proponents of urban ecosystem education. These include:

1. Reform efforts must be long-term, sustained, and systemic in nature. This necessarily means that they require broad-scale public understanding and support. Outcomes from specific programs suggest that high-quality mathematics and science programs, including strong curricula, instruction, and assessment, *can* bear results. Moreover, the building of solid leadership and expertise at all levels, including the school principal as a necessary leader, can promote and sustain reform over time. However, the first ten years of the current education reform era have demonstrated that a decade is not long enough to change our educational systems and reach the ambitious Year 2000 goals we set for ourselves (National Education Goals Panel 1999).

2. Education reform hinges on successfully recruiting, training, and supporting teachers and on their continued professional development. In addi-

tion, professional development linked with improved curriculum is most likely to succeed. For instance, in California, professional development that was centered on a new student curriculum led to both classroom practices that were oriented to the new state framework and to significantly higher student achievement. New curricula not accompanied by adequate professional development or professional development not grounded in academic content was less likely to have constructive effects (Cohen and Hill 1998).

3. We must continually assess our efforts and honestly report progress (or lack thereof) toward our goals. The improvement of teaching and learning must be informed by research. Those responsible for achieving the goals must get the feedback they need to determine what is working and ways to improve their efforts, and those providing the funds and "political will" must receive updated information to use in judging whether or not the work deserves their continued support. Such an approach must be used if education reform and curriculum innovations, like urban ecosystem education, are to be integrated into our education system.

Why Is Urban Ecosystem Education an Important Frontier for Educators?

In this section, we describe four ways in which we believe urban ecosystem education can help achieve the goals of education reform. We argue that urban ecosystem education is an important frontier for educators because: (1) it is a subject around which new, improved curricula can be built; (2) it fosters the development of school–community partnerships that enhance learning; (3) it provides avenues for integrating computer and other information technologies into schools; and (4) it offers real situations and issues that can be used to teach citizenship and content identified as important in the standards of many different disciplines.

Urban Ecosystems and New Curricula

Urban ecosystems can be the subject of new curricula that are both focused and coherent, that address key student learning goals and that build on what research tells us. A review of the standards in just one discipline, science (National Research Council 1996), shows key content that can readily be taught by focusing on urban ecosystems. For example, students could develop the following abilities, identified as key outcomes in the National Science Education Standards:

- conduct and understand scientific inquiry;
- develop understandings of concepts such as the flow of matter and energy, populations and ecosystems;

- understand, make decisions and act on social and personal issues, such as environmental quality and natural and human-induced hazards; and
- understand the nature of science by applying it to real problems.

In addition, skills and concepts integral to urban ecosystem education are well established in many of the other national and state standards. Urban ecosystem understanding can be an integrating context for addressing these standards and can provide opportunities for problem-based learning approaches. An interdisciplinary approach will require both the development of new curricula and the implementation of those curricula in schools that have for decades used traditional disciplines to organize their days, curricula, and faculties; so this approach brings with it considerable challenges. At the same time, it provides an opportunity for bringing more focus and coherence to our schools and for pursuing approaches that may lead to the achievement of our national goals.

New urban ecosystem curricula could be designed to address such content while drawing on research findings to offer teachers and students improved approaches for teaching and learning. In the Third International Mathematics and Science Study (TIMSS), analyses of teachers' instructional time and subsequent analyses of science and mathematics textbooks, indicate that U.S. students are exposed to a larger number of topics during the course of a school year and to more repetition of topics from one year to the next than students in other countries (NRC 1999; National Science Foundation 2000b). These data and the relatively low scores of U.S. students suggest that youngsters may benefit from more focused curricula that cover fewer topics in greater depth and build on ideas and skills learned in previous years. According to a study conducted by the Organization for Economic Cooperation and Development, several other countries are pursuing innovations that emphasize cross-disciplinary approaches (Atkin and Black 1997). Urban ecosystem education could provide the rich subject matter for teaching and learning the age-appropriate ideas and skills set forth in the standards in an interdisciplinary way, and could foster deeper understanding and higher-level abilities as increasingly complex material is explored in subsequent years.

In the United States, innovative curricula such as the QUASAR (Quantitative Understanding: Amplifying Student Achievement and Reasoning) project, have shown that challenging tasks that are especially relevant to students' life experiences, interests, and cultural heritage produce high-level cognitive outcomes, as set forth in the standards (Silver 1995). Garcia (1991), in a review of educational practices used successfully with linguistically and culturally diverse students, reported that collaboration and communication are key elements of effective instructional practice, especially when the curriculum blends challenging and basic content. These and other similar findings suggest that standards-based curricula and classroom practices that enable students to engage in collaborative problem solving in the complex urban ecosystems in which they live will enable them to achieve reform-minded goals.

Partnerships

Urban ecosystem education requires and nurtures the development of partnerships between schools and parents, the community, higher education, government, and the private sector. Such partnerships can bring more resources to bear on youngsters' learning, involve adults in lifelong learning, and bridge the formal and informal education sectors.

If students are to investigate the urban ecosystems in which they live, their parents, neighbors, and people in their community's businesses and agencies will, by definition, become involved in their educational activities. A large body of research shows that students of all ages benefit from their parents' involvement in their education (gaining higher grades, greater academic achievement, and more positive attitudes), and that children from low-income and minority families have the most to gain (Henderson 1989). More recently, the NSF's urban program has seen an increase in the number of partnerships and their level of involvement in local education reform initiatives. Partners that were satisfied adding a logo to the school bulletin or newspaper 10 years ago now want their contributions to support the content standards directly and they want to see evidence that their efforts are affecting student outcomes (National Science Foundation 1999; 2000a).

In Fresno, California, partnerships between the school district and several community organizations in both formal and informal settings have assisted in building children's knowledge of ecological principles in an urban setting. An Environmental Education Leadership Institute has been held annually for Fresno teachers, with follow-up activities provided to assist teachers in integrating lessons learned into their classroom practice. The institutes have been cosponsored by the San Joaquin Air Resources Board, the Central California Science Leadership Association, the City of Fresno, the Central California Environmental Education Collaborative, and California State University, Fresno. Other partnerships, such as ones between the school district and the Discovery Center and the city's zoo, have provided teacher workshops, student field studies, and family education activities.

Urban Environmental Education in Detroit (UEEID) is another example of partnerships that have arisen in urban ecosystem education (Raymond 1999). The program has been a collaborative initiative of the Detroit Public Schools and Wayne State University. High school students learn to apply advanced features of Geographic Information System (GIS) software to local problems. An extension of this project was the Work/Site Alliance—Community Based GIS Education program, a partnership with Eastern Michigan University and Henry Ford Community College. At each of four high schools in the Metropolitan Detroit region, interdisciplinary teams of four teachers and four students per teacher have worked on real, community-based projects, addressing urban issues that affect the communities such as public safety, crime reduction, brownfield redevelopment, and environmental justice. Wayne County actually has contracted with the teachers and students for work taking place as part of the curriculum during the regular school day. Two project examples are: (1) mapping out all of the neighborhood gardens in the Detroit Agriculture Network, with students relating data on garden soils to the living conditions in the neighborhoods; and (2) mapping the distribution of the 5,396 school age children who tested positive for blood lead poisoning in 1998, examining spatial relationships, and asking questions about how these conditions might be corrected.

Technology

Understanding urban ecosystems will require both the understanding of technology's role in shaping ecosystems, and the use of such technologies as GIS, Internet-facilitated information exchange, and scientific research equipment. In using these technologies to learn about cities, students will gain important understandings, learn life skills and experience vocational options.

Standards in mathematics, geography, social studies, and science specifically call for the use of technology to collect, store, organize, and display information and to create models and simulations. They also expect learners to evaluate the social and environmental impacts of various technologies and technological systems. A few examples show ways that innovative uses of technologies are enabling schools to address the standards and give students experiences that promote their understanding of their cities as ecosystems and develop workplace skills. In some places, partnerships are facilitating the use of computer and other information technologies.

A notable partnership that arose through NSF's urban program has been a technology-based effort that involves the Detroit Public Schools and the University of Michigan. The program has been centered on project-based science units (for 10–12 weeks) designed for middle school students. Teachers, students, graduate students, and university faculty members engage in inquiry-based investigations through web-based interactive activities and classroom project research. The collaboration is facilitated through the Center for Highly Interactive Computing in Education and the Center for Learning Technology in Urban Schools (*http://hi-ce.org*).

This Detroit program has provided avenues for students and teachers to find answers to questions about the world around them. One of the units has involved the River Rouge that meanders through the city. Students have collected environmental data and have used a computer program to analyze it (Manzo 1998). Such hands-on activities offer a powerful entree into lessons that prompt students to go beyond the obvious conclusions and to probe into the deeper causes and effects of natural phenomena. Technology becomes the tool used by students to investigate, collaborate, and access information. The final result is a series of artifacts or products that address questions or problems the students investigate. The program also assisted teachers in their day-to-day practice (Kracjik 1993). The UEEID program described above and this program are two examples where students use technology to learn about their urban world while acquiring job skills.

Citizenship

Citizenship is a recognized goal of many education systems and standards. Insights into what citizenship entails in the context of our schools can be gleaned from nationally developed standards. The history, social studies, geography, civics, and government standards, for example, speak to students developing abilities, such as:

- recognizing citizens' rights and responsibilities and the importance of exercising them
- identifying and evaluating alternative solutions and courses of action for particular situations and issues
- evaluating whether action is needed and whether they should become involved
- accepting personal responsibility for their actions and evaluating the results of their actions

Certainly, the sorts of situations or issues that K–12 students will be aware of, motivated to address, and capable of influencing will be local in scope.

To be able to understand, analyze, and propose realistic ways to address these issues, people need to understand their locale and its ecology. Most of our students live in metropolitan areas, and they need to understand their home communities. In addition, to be effective citizens, they need to understand their communities and the surrounding areas as systems, or nested sets of systems, and must grasp how their actions link them to the ecology of distant ecosystems in vital and pervasive ways. Urban ecosystem education programs that foster such citizenship often can involve service learning and other strategies for involving youth in genuine work in their communities, that are being developed throughout the world (Hart 1992; 1997). We are convinced that the citizens of the twenty-first century need to understand urban ecosystems, perhaps more than any other type of ecosystem on the globe.

Why Should Educators Embrace the Urban Ecosystem Education Frontier?

There are several reasons why urban ecosystems pose a vexing but alluring challenge for educators. Perhaps most obviously, we are just beginning to understand these incredibly complex systems from just one perspective or from two perspectives at once. The kind of multi-dimensional understanding that scientists are striving for, integrating across the biological, social, and physical dimensions of the urban world, lies more in the realm of the future than in the concrete present. This makes it both an exciting frontier where one can learn the concepts and processes of science, the use of appropriate technology, and the roles and responsibilities of citizens, and, as an evolving field, makes it difficult to elaborate clear learning outcomes, curriculum guidelines, teaching strategies, and assessment measures. However, placing urban educators directly at the lab bench of discovery with scientists, and enabling their students to be on the cutting edge of a nascent field that is directly relevant to them and to the majority of the world's citizens is to us an opportunity we must seize. Educators and scientists together can help shape the understanding itself, rather than just being passive "consumers" of the knowledge derived by others.

In an education world wrestling with accountability, we acknowledge that it will not be easy to measure some of the outcomes of urban ecosystem education. Consider, for example, the difficulty of assessing a systems-level understanding that crosses disciplinary lines. Or, as another example, imagine the difficulty in assessing a student's understanding of the way specific history and neighborhood effects determine the nature of a local ecosystem. This same difficulty in measuring outcomes plagues our efforts to conduct research into teaching and learning about urban ecosystems. In our very attempts to simplify enough to arrive at an answerable question, we risk losing the most important, emergent properties of the complex urban system. In addition, high-stakes tests currently used to measure student achievement are discipline focused. To enable testing companies and analysts to compare test results from one year to the next, the tests tend to remain quite stable from year to year. Thus, including questions that reflect a new multi-disciplinary subject in standardized tests presents additional challenges.

Urban ecosystem education requires a kind of cross-age, cross-disciplinary, community-based and collaborative teaching and learning model that is very challenging for schools to implement. Kids move, teachers move, principals move, and school boards change, all thwarting efforts to develop progressive curricula across years and among parts of a complex school system and its surrounding communities.

At the same time, urban ecosystem education carries with it the promise of genuine participation by students, teachers, and schools in real world problems and their solutions. Is education ready to embrace, fully, this kind of immediate, relevant learning experience? Do others see, as we do, the potential for engaging students who might otherwise be disenfranchised or uninterested in education, not only in opportunities to learn the kinds of ecological concepts and problem-solving skills we've alluded to, but also to see real value in knowing basic skills and participating in schoolwork?

An Example of Urban Ecosystem Education in Action

There are many examples of how the various facets of urban ecosystem education can be put together to advance student learning and work towards achievement of several of our country's education goals. We select one from Philadelphia that also is touched upon by Anne Winston Spirn in Chapter 13 of this volume. Since 1995, students and teachers at Sulzberger Middle School in West Philadelphia and students and faculty from the University of Pennsylvania have been collaboratively working and learning in the youngsters' neighborhood. Since Mill Creek runs through this long-established, racially diverse, middle- and lower class neighborhood, it is an ideal place for studying hydrologic processes at work, engaging in neighborhood development, and tackling water resource management challenges. The partnership has enabled university professors and classroom teachers to jointly plan, implement, and assess a new interdisciplinary curriculum.

Working together, the university and middle school students began by observing and mapping the landscape; collecting old maps, photographs, tax records, census tables, and city plans to trace its past; interviewing and sharing what they were learning with their families, local residents, and the larger community; and envisioning the future. Federal and municipal agencies, the university, private foundations, and corporations have funded this effort. Officials from the city planning commission, water department, and non-profit organizations have also been involved. Increasing numbers of students and teachers have joined in.

They have organized collections of primary documents and made them available at the school. And they have created a digital database of the area with text, statistics, maps, and graphics, available on a regional, block, or individual property scale. It can be accessed from a personal computer by individuals and by large institutions and government agencies.

Sulzberger and Penn students have investigated, reflected upon, developed understandings and solved real-world problems in their own neighborhood/university community. For example, they have investigated possible correlations between respiratory illness, damp basements, and house locations by surveying residents. The city has pursued construction of a stormwater detention facility/wetland/outdoor classroom next door to the school based on the work and designs of Penn and Sulzberger students. City monitoring will show the project's effectiveness in reducing combined sewer overflows into the nearby Schuylkill River.

The ecosystem concept provides participating learners with a powerful tool for understanding the urban environment; permits every individual to perceive his or her cumulative impact on the city; and furnishes a framework for examining all levels of living things, perceiving the effect of human activities and their interrelationships, and weighing the costs and benefits of alternative actions (Spirn 1999). Community-based, interdisciplinary learning is enabling students to construct new knowledge about their neighborhood, urban planning and design, and computer technology. And most importantly, according to one of the middle school teachers, it allows students to teach each other, use cultural exploration to shatter cultural and academic myths, create powerful partnerships, and develop a genuine sense of mutual respect for one another. (See West Philadelphia Landscape Project site at: *http://www.upenn.edu/wplp/home.htm*)

Conclusions

There is tremendous potential to capitalize on the efficiency that a multidisciplinary, multi-faceted theme like urban ecosystems can provide to education. Several topics can be covered at once and in depth, if we look across the traditional disciplinary boundaries. Students get to work with a varied group of professionals, peers, and role models. By bridging to the community and a diverse pool of local resources, the schools expand and diversify their sources of financial and other forms of support. And the convenience of local ecosystems as a site for frequent, repeated and in depth study, without expensive field trips, cannot be overestimated. In urban ecosystem studies there is the potential for some truly remarkable, innovative, and exciting education to take place.

Cities are incredibly important features of the modern world for many reasons, and it is our obligation as educators to help students understand these important places as best we can. Modern ecology, social science, geography, and other fields all are forging new approaches to understanding metropolitan areas and the dynamic processes that shape them and their interactions with the rest of the globe, and it is incumbent upon us as educators to stay at this cutting edge of inquiry with our students.

Given the limited resources available for solving the complex problems in our educational systems and our communities, this kind of multifaceted approach makes sense for both improving educational opportunities for our students and making our communities more ecologically sustainable and livable. It brings together a diverse array of perspectives, skills, and expertise, and combines funding from multiple sources. It shows that students, indeed learners of all ages, need not venture to far-away places to become immersed in fascinating learning environments. In addition, it engages people in shaping their own future as they learn large, integrating concepts and life-long skills for working and communicating with others and for effecting change in the places they live. Since we believe this kind of learning will enable students to become productive citizens of the twenty-first century, we encourage urban ecosystem education advocates to join forces with education reformers and promote expanded use of this type of approach.

References

American Association for the Advancement of Science (AAAS). 1989. *Science for all Americans*. Oxford University Press, New York.

American Association for the Advancement of Science (AAAS). 1993. Project 2061. *Benchmarks for science literacy*. American Association for the Advancement of Science, Washington, DC.

American Geological Institute (AGI). 1991. *Earth science content guidelines for grades K–12*. American Geological Institute, Alexandria, VA.

Atkin, J.M., and P. Black. 1997. Policy perils of international comparisons: The TIMSS case. *Phi Delta Kappan* 79(1):22–28.

Center for Civic Education. 1994. *National standards for civics and government.* Calabasas, CA: Author.

Cohen, D.K., and H.C. Hill. 1998. *State policy and classroom performance: mathematics reform in California*. CPRE Policy Briefs. Consortium for Policy Research in Education, University of Philadelphia, Philadelphia, PA.

Consortium of National Arts Education Associations. 1994. *National standards for arts education: What every young American should know and be able to do in the arts.* Reston, VA: Music Educators National Conference.

Cummins, J. 1989. *Empowering minority students*. California Association for Bilingual Education, Sacramento, CA.

Darling-Hammond, L., and D.L. Ball. 1998. *Teaching for high standards: what policymakers need to know and be able to do*. CPRE Joint report Series, co-published with the National Commission on Teaching and America's Future, JRE-04.

Delgado-Gaitan, C. 1987. Parent perceptions of school: supportive environment for children. In H. Trueba, ed. *Success or failure? Learning and the language minority student*. Newbury House Publishers, Cambridge, MA.

Education Week. 1998. *Quality counts: the urban challenge, public education in the 50 states.* In Collaboration with the Pew Charitable Trusts. 17(17), January 8, 1998.

Education Week. 1999. *Quality counts '99: rewarding results, punishing failure*. 18(17), January 11, 1999.

Garcia, E. 1991. *Education of linguistically and culturally diverse students: effective instructional practices (Educational Practice Report 1)*. University of California, Santa Cruz, National Center for Research on Cultural Diversity and Second Language Learning, Santa Cruz, CA.

Geography Education Standards Project. 1994. *Geography for life*. National Geographic Research and Exploration, Washington, DC.

Hart, R.A. 1992. *Children's participation: from tokenism to citizenship*. Innocenti Essays No. 4. UNICEF International Child Development Centre, Florence, Italy.

Hart, R. 1997. *Children's participation: the theory and practice of involving young citizens in community development and environmental care*. UNICEF, New York.

Henderson, A.T., ed. 1989. *The evidence continues to grow: parent involvement improves student achievement*. National Committee for Citizens in Education, Columbia, MD.

Johnson, J., and C. Aulicino. 1998. *Summing it up: a review of survey data on education and the National Education Goals*. A report from Public Agenda. Paper prepared for the National Education Goals Panel, Washington, DC.

Joint Committee on National Health Education Standards. (1995). *National health education standards: Achieving health literacy.* Association for the Advancement of Health Education. Reston, VA.

Kim, J., L.M. Crasco, R.B. Smith, G. Johnson, A. Karantonis, and D.J. Leavitt. 2001. *Academic excellence for all urban students. Their accomplishments in science and mathematics*. How reform works: an evaluative study of NSF's Urban Systemic Initiatives. Systemic Research, Inc. Norwood, MA. *http://www.siurbanstudy.org/newspublication* or *http://systemic.xohost.com/usi/Booklet.pdf*

Krajcik, J.S. 1993. Learning science by doing science. In R. Yager, ed. *What research says to the science teacher: science, society, and technology*. National Science Teacher Association. Washington, DC. *http://hi-ce.org.*

Manzo, K.K. 1998. Making learning authentic: lessons from a dirty river. *Education Week*, October 1, 1998, 38–39.

Mathews, J. 1999. Teacher training debate extends to colleges. *Washington Post*, December 6, 1999.

Miles, K.H., and L. Darling-Hammond. 1998. Rethinking the reallocation of teaching resources: some lessons from high performing schools. *Educational Evaluation and Policy Analysis*, Spring 1998, 20(1):9–29.

National Center for Education Statistics (NCES). 2000. *Quality of elementary and secondary school environments: teachers[1] perspectives and quality of public school teachers*. Available at *http://www.nces.ed.gov/programs/coe/2000/section4/indicator47.html*

National Center for Education Statistics (NCES). 1997a. *America's teachers: profile of a profession, 1993–1994.* U.S. Department of Education, Office of Educational Research and Improvement. NCES 97–460.

National Center for Education Statistics (NCES). 1997b. *The condition of education 1997.* U.S. Department of Education. Washington, DC.

National Center for Education Statistics (NCES). 2000. *Digest of education statistics 1999.* U.S. Department of Education. Washington, DC.

National Center for Education Statistics (NCES). 2002. *The nation's report card. 2000 science assessment results*. U.S. Department of Education, Washington, DC. *http://nces.ed.gov/nationsreportcard*

National Center for History in the Schools. (1994a). *National standards for United States history: exploring the American Experience.* (Expanded ed.). Los Angeles: Author.

National Center for History in the Schools. (1994b). *National standards for world history: Exploring paths to the present.* (Expanded ed.). Los Angeles: Author.

National Center for History in the Schools. (1996). *National standards for history.* (Basic ed.). Los Angeles: Author.

National Council for the Social Studies. (1994). *Expectations of excellence: Curriculum standards for social studies.* Washington, DC: Author.

National Council of Teachers of Mathematics (NCTM). 1989. *Curriculum and evaluation standards for school mathematics*. National Council of Teachers of Mathematics, Reston, VA.

National Education Goals Panel. 1999. *The national education goals report: building a nation of learners*. U.S. Government Printing Office, Washington, DC.

National Research Council (NRC). 1999. *Global perspective for local action: using TIMSS to improve U.S. mathematics and science education*. National Academy Press, Washington, DC.

National Research Council (NRC). 1996. *National science education standards*. National Academy Press, Washington, DC.

National Science Foundation (NSF). 2000a. *Urban systemic program in science, mathematics, and technology education: a foundation for K–12 science and mathematics educational system reform. http://www.nsf.gov/cgi-bin/getpub?nsf0034*

National Science Foundation (NSF). 1999. *Annual Report for FY 1999*. Education and Human Resources (EHR), Division of Educational System Reform (ESR).

National Science Foundation (NSF). 2001a. *Annual Report for FY 2000*. Education and Human Resources (EHR), Division of Educational System Reform (ESR). *http://www.ehr.nsf.gov/gpra/anrpt/2000/ESR2000anrpt.html*

National Science Foundation (NSF). 2000b. *Science and Engineering Indicators 2000 Vol. I*. NSB-00-1, National Science Foundation.

New York Times. 1998. *Low on teachers, New York recruits in Austria*. Jacques Steinberg, July 11, 1998.

O'Day, J.A., and M.S. Smith. 1993. *Systemic reform and educational opportunity*. Pages 250–312 in S.H. Fuhrman, ed. Designing Coherent Policy: Improving the System. Jossey-Bass Publishers, San Francisco.

Raymond, R.E. 1999. *Connecting the schools and community through the education of students in geographic information systems*. Cary Conference VIII, April 27–29, 2000, Millbrook, New York.
http://www.ecostudies.org/cary8/raymond/raymond.html. 11 August 2000.

Silver, E.A. 1995. *Shuffling the deck to ensure fairness in dealing: a commentary on some issues of equity and mathematics education from the perspective of the QUASAR project*. A paper presented at the Seventeenth Annual Meeting for the Psychology of Mathematics Education (North American Chapter), October 21–24, 1995. Available from the Educational Resources Information Center (document number ED 389 538).

The Urban Teacher Collaborative. 2000. *The urban teacher challenge: teacher demand and supply in the great city schools*. Recruiting New Teachers, Inc., Belmont, MA.

U.S. Department of Education. 2002. *Budget Summary—Elementary and Secondary, FY 2002*. U.S. Department of Education, Washington, DC. *http://www.ed.gov/offices/OUS/budget02/summary/chapter1.html*

3
The Role of Understanding Urban Ecosystems in Community Development

Jack K. Shu

Our football team had just won when the bricks and bottles started to drop out of the sky. The girls' drill team, in their brightly colored uniforms, were lined up getting ready to leave as bottles shattered around them and rocks bounced off their yellow bus. At this inner city high school, it was a common practice to attack members of a visiting high school at the end of a game, especially if the visiting team had won. I was from the visiting high school and watched this frightening event unfold from the other side of a field. It was a form of madness, a group of youth trying to hurt anyone that looked like they were from the other school. Dozens of police officers soon arrived and the incident ended without any serious injuries to anyone; however, the event did leave a memorable impression in my mind about the reality of human communities.

What does this incident have to do with the importance of understanding urban ecosystems? Nothing, if our perspective of urban ecosystems simply involves tasks like measuring energy flows and goals like reducing energy consumption. But if our perspective deals with people and the places where they live and the goal is to improve the quality of life in the ecosystem, then the realities of the various dimensions of human communities are very significant.

Another major incident may clarify this point further. Many years after high school I was working in South Central Los Angeles and employed a few youths during the summer to conduct various conservation projects. One Monday I was told that one of the youths had been shot and killed over the weekend. He was shot in front of his home, possibly related to an old gang conflict he was involved with. This was another instance of senseless violence. In this case, it was in the context of a program to improve the urban environment. He was watering trees in a new urban park a day or two before he was killed. We used the most current outdoor education curriculum; we worked to "green" the city, promote positive action, and improve the environment. However, did we really understand the whole urban ecosystem? Was it an idealistic crusade based on a simple notion that nature is good and that putting some of it in an urban setting will save people?

I recently learned of a project by the Earth Conservation Corps, employing youth from Washington, D.C., to restore bald eagles to the Anacostia River basin. The river had suffered from industrial pollution and was restored to the point that it was ready to become a home for the national bird again. Over the course of the project, corps members worked to improve the habitat for the eagles, helped to release young eagles, and monitored their progress. As the eagles were coming back along the river, corps members were being killed in their home neighborhoods. Five of the eagles that were successfully reintroduced to the area were named after corps members who died during the term of the project. It seems that we understand how to fix polluted rivers and restore large raptors next to cities, but do not know how to find peace in "the Hood."

We need to understand urban ecosystems far beyond the point of believing that making them "greener" will be enough. The model of separating natural and developed environments, where one is compared with the other, falls short of providing solutions; we must bring them together as one system. A greater understanding of urban ecosystems could provide new and more effective approaches to improve urban communities. Consideration of all of the elements of a system, even the violent conflicts between people, must be part of this understanding. It would be an holistic and inclusive view.

A Different Perspective Is Needed

Perhaps a paradigm change is needed in the field of studying ecosystems, a shift from focusing on natural systems to one with greater emphasis on human interactions and effects. This will require a different perspective in the science of ecosystems, a point of view that places people as part of the system.

To illustrate how we can view urban or human ecosystems differently, it may be useful to use a common science project as a metaphor. A bowl of pond water, full of typical plants and animals, is often used by science teachers to study ecosystems. Students are asked to track the flow of energy in the system or observe change in the population of organisms over time. If the bowl is placed in the dark, the system will react. If pollutants are introduced, some organisms will die. The bowl is a convenient study tool, for it places the student as an outside observer with the ability to change the system. This tool has introduced many young ecologists to an understanding of ecosystems. This view of ecosystems, from *outside* the system, seems to be the perspective many scientists and advocates for the environment retain as they deal with real-world systems. How well does such a traditional perspective work involving the people in the ecosystem and implementing solutions to problems?

Frances Moore Lappe, author of *Diet for a Small Planet*, was interviewed for an article, *The Art of Democracy* (Hildegarde 1999), in which she reinforces her belief that solutions must come from the bottom up. She commented, "Democracy is not only the structure of government; it's what we do, how we relate to one another in schools, human-service agencies, or workplaces." If we were to use this philosophy in developing our understanding of urban ecosystems, the issues that need attention would be defined by those in the community. Solutions for problems would be derived through an informed democratic process.

Empowering communities to care for their ecosystem is consistent with a bioregional model for environmental education. In the bioregional approach, those who are affected learn about their region and then take appropriate action. This is an alternative to working within geopolitical jurisdictions that may not have any relationship to the resources or people that are affected by the issue. There are hundreds of examples where groups are active within their bioregion (Kahn 1995). Bioregionalism emphasizes a study of place, clearly placing the student or participant within the ecosystem (Traina 1995).

With a community-centric approach, many disciplines not traditionally included in ecology will need to be brought into the discussion to understand urban ecosystems. Teams that study urban ecosystems would involve social scientists that can consider cultural differences and conflicts when defining issues and developing solutions. Public health experts may be needed for their knowledge of how information is spread or determining the best way to change human behavior in a community. Economists can conduct economic feasibility studies, politicians can factor in political realities of dealing with laws and public funding; the list could go on and on. The point is that there is a lot more to understanding ecosystems when people are a major part of those systems. Interdisciplinary teams are essential when dealing with such difficult urban environmental issues as toxic metals in soils or urban sprawl. These issues need to be considered in the social context of racial conflicts, cultural differences, and economic needs as well as biological and physical sciences, if effective solutions are to be sought. For example, through such a viewpoint urban sprawl may not be a problem; rather, it could be a solution to finding safe and affordable communities in which to live. The focus can then shift to determining the best solutions to a basic human need rather than responding to the loss of open space or freeway congestion.

Recognizing that the word *urban* is often used as a metaphor for multicultural or minority communities, it may be important to address the issue of racial and cultural diversity. Lewis and James' (1995) article, *Whose Voice Sets the Agenda for Environmental Education? Misconceptions Inhibiting Racial and Cultural Diversity* points out a number of misconceptions about minority groups. African American and Hispanic communities have a history of addressing and being involved with environmental issues. Similar to the way much of American history is told from a Eurocentric perspective and omits the stories and contributions of minorities, environmental education has left out people of color. It follows that the selection of topics or issues for environmental education is often done by the dominant society and therefore is not inclusive of all people. For example, how important is wilderness protection to African Americans when such a disproportionately small number of them visit wilderness areas? Unfortunately, the way environmentalists often answer that question is with an assimilationist approach: "Let's get more Blacks to use the wilderness so that they will learn to appreciate and protect it." This usually entails arranging a handful of visits to wilderness areas by groups of inner city kids, or placing a number of token representatives in the wilderness. This may be well intentioned, but is the primary purpose to serve the African-American community or to broaden support for wilderness protection? The alternative approach would be to look into the issues that the African-American community believes are important. Then see if some aspect of wild lands stewardship can address one of these issues and be a part of its solution. If wilderness is not relevant to what is of great consequence, then it really is not valuable. Thus, an element of understanding urban ecosystems is to keep our studies relevant to the lives of people of all cultures and ethnicities.

What Would This New Paradigm Be Like?

The move to view urban systems with multiple perspectives may not be so elusive. On September 5, 1999, the *Los Angeles Times* printed an article on the front page titled "New Tests Show Human Viruses in Beach Waters" (Cone 1999). In this one article by an investigative environmental writer, most of the elements of this human systems approach are included. Beaches are symbolic to Southern California. They are important to tourism and the local economy, and they represent one of the forms of "gold" in the state. For countless individuals, from surfers to poor subsistence fishermen, the water quality off the shore is important. Indeed, it affects many of the 13–15 million lives in the region. The article is not just about the detection of human viruses that can cause swimmers to get sick from ailments ranging from diarrhea to hepatitis. Nor is it limited to the shortcomings of human sewage treatment systems. It presents the problem within the context of an increased population, more paved-over lands, the lack of natural estuaries to cleanse water, aging sewage lines, human error, large homeless populations, and economic factors. The reporter interviewed microbiologists, stormwater managers, sanitation people, health department officials, sociologists, lifeguards, and business owners, just to name a few. High concentrations of bacteria harmful to people were traced to drainage from streets with concentrated homeless populations. A map of one of the local watersheds showed how runoff from one site could affect a beach ten to fifteen miles away five hours after a rainstorm. A diagram illustrated sources of pollution, which included everything from pet droppings to motor oil to zinc flaking off metal structures. The article made it easy for the reader to realize that a social worker helping a homeless family get back on track is also assisting the economy of a beach community ten miles away (Cone 1999).

The *Los Angeles Times* has a very large circulation, and after speaking to many of my friends and relatives in the region, it seems that a significant percentage of people noticed the article. We should not be so naïve, however, as to believe that many of its readers or even the writer of the article understood all of the connections between the various elements of the story. Do not expect the formation of new collaborations or coalitions to deal with these urban issues anytime soon. One reason is that we know and understand so little about these relationships and connections. Another reason may be that people are simply not trained to look for and form holistic understandings of systems. Progress will have to be measured in small steps. The reporter took such a step by looking at the big picture rather than investigating something smaller, like how making a wrong connection at a building can allow sewage to flow into a storm drain. We can all take such steps so that our understanding of urban ecosystems is inclusive of all factors.

Promoting the Understanding of Urban Ecosystems

In January 1991, 4-H in California published its action plan called *"Pride in a Past—Vision for a Future"* (University of California Division of Agriculture and Natural Resources 1991). The organization is well known for developing youth through projects; typically, raising farm animals. Through a strategic planning process, the organization outlined a program that would be more inclusive and relevant to the state without changing the organization's core purpose. The last of seven goals in the plan is particularly germane to promoting the understanding urban ecosystems. Here is the goal in its entirety:

Goal VII: The Image of 4-H Youth Development Program (4-HYDP)

The final goal of the action plan calls for a strong statewide plan to communicate the mission of 4-HYDP throughout California, with particular emphasis on reaching segments of the population who may be unaware of the benefits of participation. This would include cultural minorities now underrepresented in the program and those difficult to reach with conventional programs. Within the University community increased efforts will be made to inform those who are unaware that 4-HYDP is a program.

Another important purpose of this effort is to counter the misconception that 4-H is a program designed for rural youth. 4-HYDP will be promoted as a University-based program whose purpose is to serve the needs of a full spectrum of youth throughout the state (University of California Division of Agriculture and Natural Resources 1991).

Much of what the goal addresses could apply to environmental education and the field of ecology. Based on the demographics of the state as well as the issues the people were facing, the leadership of 4-H realized that they needed to change the organization's methods and focus. The project descriptions for 4-H now include resource sciences such as energy management and marine biology. Under the category of social science, topics such as citizenship, communications and community pride are included, with "exploring our community" and "heritage and culture" in bold print. Raising large animals like goats, cattle, and rabbits are still listed and may continue to be a major part of 4-H. But projects also could involve raising small animals like caged birds, rats, or guinea pigs. A young person could live in an apartment and still fully participate in a project.

Another example of how the understanding of urban ecosystems can be promoted comes from a very well known environmental education program. Project Learning Tree, one of the oldest providers of high quality curriculum for environmental education, developed a unit entitled *Focus on Risk* in 1998 (Project Learning Tree 1998). This unit for secondary education engages students to measure risks for many environmental conditions. The unit can be used throughout the country, with many of the lessons focused on issues which would affect mostly urban communities. Studying toxicity and the level of risks for hazards like radon and chlorine and placing these issues in the context of the students' communities makes environmental education more relevant. The lessons prepare students to deal with complex urban environmental problems. The program is no longer about the study of plants and animals in a distant forest. It is about the students' home environments and learning how to improve them. Project Learning Tree is now in the process of developing a high school unit centered on the place the students live in.

If other organizations follow the lead that 4-H and Project Learning Tree have taken, we would be moving ahead in promoting the understanding of urban ecosystems. For these two organizations, their future seems more promising because they are making the paradigm change (Project Learning Tree 1998).

Another significant way we can promote the understanding of urban ecosystems is through funding. Grant makers should seek to fund holistic programs with larger grants rather than funding many groups on a survival level (Shuman 1999). Grassroots organizations are important and should be supported with programs on asset building and community development. Both public and private funds can be directed in this manner.

A Vision for Improving Urban Ecosystems

Increasing the use of the urban ecosystem as a topic for student action projects would be an initial goal. As grassroots organizations and community leaders from urban communities gain more knowledge about urban ecosystems, they may be more effective at addressing issues. This should lead to more successful and effective efforts to reduce some of the most threatening problems in urban communities. More importantly, it will stimulate people to have pride in their own communities, reversing the tendency to idealize natural systems and see only the despair of older cities. The cumulative long-range goal would be the empowerment of whole communities to develop a positive vision for urban environments.

What is the importance of understanding urban ecosystems? I began this article describing the death of young urban youth. I have no doubt that urban violence is an indicator of ecological health, not just in inner cities but in all communities. If we are to achieve meaningful improvement in our urban ecosystems, we must take an holistic approach and address not only physical or biological resources but also human issues such as social, economic, and cultural factors. These elements may be finding employment, getting an education, keeping families together, staying free of lead contamination, building quiet neighborhoods, or having open space, but they all work together to form a healthy ecosystem. That is the kind of understanding we need.

References

Cone, M. 1999. New tests show human viruses in beach waters. *Los Angeles Times*, Sept 5, 1999, pp. A1,A14–A15.

Hildegarde, H. 1999. The art of democracy: an interview with Frances Moore Lappe. *A Field* 3:24–26.

Kahn, C. 1995. Bioregional education: knowing love and connectedness. Pages 49–54 in F. Traina, S. Darley-Hill, eds. *Perspectives in bioregional education.* NAAEE.

Lewis, S. and J. James. 1995. Whose voice sets the agenda for environmental education? misconceptions inhibiting racial and cultural diversity. *Journal of Environmental Education*, 26:5–12.

Project Learning Tree. 1998. *Focus on risk—exploring environmental issues.* American Forest Foundation.

Shuman, M. 1999. What's wrong with green funding in America? *A Field* 3:32–35.

Traina, F. 1995. Methods in bioregional education. Pages 93–101 in F. Traina, and S. Darley-Hill, eds. *Perspectives in bioregional education*. NAAEE.

University of California Division of Agriculture and Natural Resources. 1991. *Pride in a past—vision for a future—the 4-H Youth Development Program (Narrative Summary of the Action Plan).*

4
Why Is Understanding Urban Ecosystems Important to People Concerned About Environmental Justice?

BUNYAN BRYANT and JOHN CALLEWAERT

Ecosystems, and more specifically urban ecosystems, represent important models for understanding particular places, environments, or regions. Even though ecologists generally view ecosystems as functional and geographic units, we suggest that ecosystems should also be viewed as cultural constructs. By this we mean that understandings of ecosystems exist within a cultural context, and meanings assigned to ecosystems cannot help but reflect this cultural context. Thus, understandings of nature are themselves cultural constructions, even though their referents have independent standing as biological realities (Kirsch 1999).

Environmental justice is both a field of study and a social movement that seeks to address the unequal distribution of environmental benefits and harms and asks whether procedures and impacts of environmental decision making are fair to the people they affect. A primary issue for people concerned about environmental justice is that some groups, most often communities of color and low-income communities, face a disproportionate exposure to environmental health risks such as air and water pollution, and environmental hazards such as landfills, incinerators, sewage treatment plants, and polluting industries. As with ecosystems, environmental justice can also be understood as a cultural construct—one that focuses on the class and racial aspects of environmental concerns.

This chapter begins by examining in more detail the perspective of ecosystems and environmental justice as cultural constructs. Understanding the connections between urban ecosystems and environmental justice concerns is an important first step and will prove helpful in identifying common areas of knowledge in supported sustainability. Following these conceptual perspectives, specific reasons are presented as to why an understanding of urban ecosystems is important to people with environmental justice concerns. Finally, three strategies are offered to strengthen the connection between an understanding of urban ecosystems and environmental justice.

Understanding Ecosystems and Environmental Justice as Cultural Constructs

Urban Ecosystems

While it is true that biological realities such as species present, the amount of water available, climatic conditions, flows and patterns of resource exchange, and so on ultimately set the limit for a region's political, economic, and social institutions, we hypothesize that if ecosystems, be they urban or rural, are not understood within a cultural context, then we fail to fully understand them. An ecosystem as a culturally defined construct says more about ourselves than perhaps about ecosystems, and we must therefore understand the values and belief systems that shape and motivate behavior toward ecosystems, particularly if we hope to explain why people protect or exploit the Earth (Cronon 1996).

Cultural constructs may be defined as mental representations of external reality that are unique to the human species (White 1949). Humans have an extraordinary ability to construct, symbolize, and name the world. Language or combinations of symbolic constructions are used to organize thoughts for understanding and meaning, for organizing behavior and management, and for envisioning and planning the future. Humans name elements of the world for specific purposes. Terms such as wildlife, park, virgin forest, externality, carbon sink, and brownfield are examples of how we construct conceptions of the world. Such conceptions are often for the self-interests of certain groups and their use or application can influence the building and maintaining of urban ecosystems.

Our speech, our work, our play, and our social life, our ideas about ourselves and nature all exist within a cultural context that is historically, geographically, and culturally determined and cannot be understood apart from that context. Thus, the way we understand an ecosystem, the way we see and value an ecosystem is a construct of a particular culturally determined context (Cronon 1996). When we think of ecosystems or modify them, however,we think of nature—not culture. Cities are more visible cultural constructions; they are places where ecosystems have been transformed by humans to support urban habitats that bear little resemblance to nature.

We contend that conceptions of ecosystem education, management, policy, planning, and design are based in cultural values of efficiency, beauty, convenience, and utility. Decisions about ecosystems are therefore value-laden. Forests cannot be managed or planned unless decisions are made about whom they will serve. Will they serve industry, local human communities, or non-human species? More specifically will they serve the spotted owl or the English sparrow; hikers or hunters; naturalists or lumbermen or some combination of the above? Will forests be managed for native oaks or Norway maples, jack pines or walnuts (Cronon 1996)?

In an urban environment we also need to consider the parts of an ecosystem that are managed for affordable housing, the business community, industrial production, landfills, incinerators, sewage treatment plants, urban parks, and recreational facilities. What do the spatial relations of these entities say about other cultural constructs such as race, class, and gender? We must ask ourselves the question: Who benefits and who loses from these culturally defined constructions? The answers to these questions depend upon the cultural values and belief systems of a particular place and people. In essence, we need to deconstruct and examine our notions of ecosystems to discover their core meanings.

To understand the values and motivations that shape our actions toward an ecosystem and to explain our actions that abuse that system, we should be more concerned about the impact of culture. Many of our values and motivations are steeped in the marketplace and the immense power of the accumulation system. Culturally transformed and commodified ecosystems are another extension of the market, producing both "social goods" and "social bads" and alienation from the natural world in which we live. Externalities such as hazardous waste are traditionally ignored by the market system and often find their way into neighborhoods with high proportions of low-income residents or people of color; these communities, themselves struggle to be valued and fully respected by the market system.

Environmental Justice

Environmental justice as a cultural construct challenges the absolute authority of the market system and places emphasis on the interconnections between environmental quality, social justice, and civil rights. With a specific focus on distributional equity, environmental justice adds new layers of analysis to the field of environmental science. Just as environmental scientists examine how human actions can alter local, regional, and global ecological systems, environmental justice advocates call attention to the environmental repercussions of human actions that threaten and disrupt particular social systems.

Environmental injustice can cover a very broad range of environmental disparities and the unequal enforcement of environmental regulations (Goldman 1994; Lavelle and Coyle 1992). In an analysis of 64 empirical studies, Benjamin Goldman (1994) found an overwhelming body of empirical evidence that people of color and lower incomes face disproportionate environmental impacts in the United States. All but one of the 64 studies found environmental disparities either by race or income, regardless of the kind of environmental concern or the level of geographic specificity examined. One of the most influential investigations of environmental injustice was a national study on the distribution of hazardous waste sites that was conducted by the Commission for Racial Justice (CRJ) of the United

Church of Christ (1987). The CRJ study revealed that the proportion of minorities residing in communities with a commercial hazardous waste facility is about double the proportion of minorities in communities without such a facility. Where two or more facilities are located, the proportion of residents who are minorities is more than triple. Furthermore, the CRJ study and others have shown that race is often the single best predictor of where commercial hazardous waste facilities are located (Commission for Racial Justice of the United Church of Christ 1987; Bryant and Mohai 1992).

Today people of color and low-income communities across the country are rebelling against the siting of locally undesirable land uses in their communities (Taylor 2000; Tesh and Williams 1996). Through these struggles, people concerned about environmental justice are deconstructing the belief that such communities are valueless. They are seeking to make their communities safe, healthy, viable, and productive. Often these activists are focused on specific places within urban ecosystems that experience the brunt of toxic and hazardous waste and polluting industries; they decry environmental racism and distrust government and the scientific community because neither provides answers to their demands for certainty or immediate solutions. As a result, many community groups are doing their own research in order to find answers to their questions, and to reconstruct their communities to be more viable and livable places.

The struggle of two community groups—the Alum Crest Acres Association and the South Side Community Action Association—representing a predominantly middle-class African American neighborhood on the south side of Columbus, Ohio clearly demonstrates such concerns. Since the mid-1980s the community has voiced numerous environmental and health complaints about a Georgia-Pacific resins facility in the neighborhood. Community concern about the facility peaked in 1997 when chemicals were improperly mixed and exploded violently, leaving one worker dead, several others injured, parts of the facility in ruins, and many residents upset about property damage and a host of alleged health impacts (Edwards 1997). Frustrated with the lack of response from the Columbus Health Department, the community groups applied for and received funding from the United Way to conduct their own health study. The funding for the study, however, was temporarily suspended due to the influence of local government officials (Columbus Dispatch 1999). The community groups have also filed a complaint under Title VI of the 1964 Civil Rights Act with the Office of Civil Rights of the U.S. Environmental Protection Agency (USEPA) alleging a discriminatory impact from permit decisions by the Ohio Environmental Protection Agency concerning the Georgia-Pacific facility. The civil rights complaint was recently accepted for investigation by USEPA. Ohio EPA is also under investigation currently by USEPA for failing to adequately enforce environmental regulations (Edwards 2000). The above represents only one of many communities where people of color and low-income groups are disproportionately impacted by environmental hazards.

Connecting an Understanding of Urban Ecosystems with Concerns About Environmental Justice

A deeper and more comprehensive understanding of urban ecosystems will perhaps provide the incentive for a paradigm shift to knowledge that is more sustainable and that will change how we build and reconstruct healthy and livable urban ecosystems. When we speak of sustainable knowledge, we use "sustainable" as an adjective to describe knowledge just as others use the term in sustainable development. Sustainable knowledge is broader than sustainable development in that the former is knowledge that guides our behavior and our understanding of nature. When we speak of sustainable knowledge, it is not knowledge that will remain static, but it is knowledge that mimics nature. It is knowledge that is consistent with and not disruptive of the Earth's life cycles, and it is knowledge that will sustain plant and animal species (Hawken 1993). In nature, the waste of one life form becomes food for another life form. In the same way we need to create knowledge so that the waste from one industry will become the raw materials for another (Anderson 1998). Such a sustainable knowledge conception of urban ecosystems is needed to help eliminate the environmental injustices present in so many cities.

An urban ecosystem built upon injustice will not survive. When people are not allowed their fair share of market benefits but are saddled with more than their fair share of environmental burdens, an ecosystem view tells us that such disparities and imbalances will eventually create problems for the entire system. This emphasis on social dimensions such as race, class, and justice adds important new dimensions of analysis that have not yet been considered in current understandings of humans as components of ecosystems.

Environmental justice often involves the struggle of a particular neighborhood or community against a local polluting industry or facility. A better understanding of ecosystems can help environmental justice advocates connect their specific concerns to broader, regional issues that may reveal significant environmental and/or health concerns. For instance, besides having impact on people of color in a low-income neighborhood, emissions or waste from a facility also may be harming a preserved area or estuary. The work of Walsh, Warland, and Smith (1997) has shown that when environmental justice advocates establish coalitions and partnerships with other groups and institutions, they are much more successful than if they had only focused on the environmental justice aspects of the problem.

For people concerned about environmental justice, knowledge of an ecosystem's characteristics is very important. For example, after one community on the south side of Chicago learned how emissions from a proposed incinerator would combine with the prevailing wind patterns to disproportionately impact their neighborhood, a new environmental justice organization was formed (Schwab 1994). In the Columbus, Ohio, example cited earlier, there have been numerous concerns expressed about contamination of the underground aquifer. These concerns, however, have not been fully explored in terms of what an ecosystem perspective can reveal regarding water flows and other vital characteristics.

Another way to understand and develop the connections between urban ecosystems and environmental justice is through geographic information system (GIS) applications. Such techniques have become an important tool for those with environmental justice and ecosystem concerns. Combining economic, social and environmental data will support better-coordinated efforts by all involved parties. GIS can help environmental justice advocates better understand the characteristics and dimensions of ecosystems and it also can help ecosystem scientists become more fully aware of the important overlap between physical, ecological, and social dimensions of an ecosystem.

Strategies

In order to strengthen the connections between environmental justice and understanding ecosystems, we offer the following three strategies: (1) promoting community-based research initiatives; (2) incorporating environmental justice concerns within a sustainable knowledge construct of urban ecosystems; and (3) supporting the formation of a new type of professional that will be able to forge the connections between understanding urban ecosystems and concerns about environmental justice.

Promoting Community-Based Research

There must be a vigorous effort to increase community involvement in designing initiatives that promote the understanding of urban ecosystems and environmental justice. This emphasis on participatory research or community-based research is highlighted in the recent Institute of Medicine (1999) report, *Toward Environmental Justice: Research, Education, and Health Policy Needs* and has been supported by other leading research institutions. Our emphasis here on community-based research is not to exclude other research approaches, but to suggest that given particular settings and desired outcomes, some approaches are more appropriate than are others. Table 4.1 offers a modified version of Patton's (1990) typology of research purposes and explains some of the differences in research approaches based on a number of variables.

We emphasize a community-based research approach for the following three reasons: (1) it focuses the locus of control of knowledge within the community; (2) people feel they have more control over their lives by being

TABLE 4.1. A Typology of Research Purposes.

	Basic Research	Action Research	Community-Based Research
Focus of research	• Questions deemed important by one's discipline or personal intellectual interest	• Organization and community problems	• Solve problems and identify societal causes of problems
Goals	• Knowledge as an end in itself; discover truth	• Solve problems in a program, organization, or community	• Advance practical knowledge • Solve problems and create systemic change • Empower participants and strengthen capacities
Key assumptions	• The world is patterned; those patterns are knowable and explainable	• People in a setting can solve problems by studying themselves	• People in a setting can understand, confront, and change oppressive forces
Desired results	• Contribution to theory	• Solving problems as quickly as possible	• Changing societal structures that created problems
Investigator's relationship with providers of data	• Subjects/Objects • Detached and external	• Clients/subjects • Agency control • Internal or external	• Participant and researcher co-control • Responsive to community needs • Internal priority with external help
Utility of research for providers of data	• Low likelihood (at least not directly or soon)	• Low to medium depending on agency status and role	• High
Who benefits from research	• University • Scientific community or other researchers • "Trickle down" to policy makers	• Client agency • Clients of agency • Policymakers, community leaders	• Participants and community members • Total system (conflicting parts and interest groups) • Constituency

Source: Adapted from Patton (1990) and Chesler. *Personal communication.*

actively engaged in a democratic process of creating knowledge for sustainable and viable communities; and (3) by understanding the role of knowledge and culture. A fundamental difference between community-based research and both action research and basic research is that rather than seeking simply to resolve a problem or to expand knowledge, community-based research involves participants in challenging basic cultural constructs and knowledge that may support unsustainable practices or conditions. In analyzing data from a national study of community-based research in the United States, Sclove, Scammell, and Holland (1998) note that community-based research processes differ fundamentally from mainstream research in being coupled relatively tightly with community groups that are eager to know the research results and to use them in practical efforts to achieve constructive social change. Community-based research is not only usable, it is actually used and, more than that, used to good effect.

In many cases community groups concerned about environmental justice and involved in participatory research have been very successful in problem solving (Schafer, et al. 1993). This process does not mean, however, that they would do a better job than a researcher from a university community—this is hardly the point. The point is that they feel that have control over what happens in their community by being involved in a participatory process. Most importantly, community-based research provides the opportunity for people to learn about their communities (Israel, et al. 1998). This is particularly important in terms of understanding urban ecosystems as cultural constructs, with all strengths and weaknesses that such a concept presents.

Community-based research can also strengthen or build new social relationships and enhance social trust. This is essential in situations that are complex or involve controversial and value-laden issues. There is a long history of outside researchers producing work that has had devastating impacts on people of color such as the Tuskegee Study (Hatch, et al. 1993; Thomas 1991), Jensen's (1968) research on black children, Schockley's (1992) work on intelligence, and Moynihan's (1965) report on black families.

Community-based research, though, is not at present a prominent form of research in the United States. Figure 4.1 clearly shows that community-based research accounts for only a small fraction of research expenditures in the United States. It is not the type of research that usually gets funded and it may require many years of work in order to establish the necessary community trust and participation. Furthermore, many of the results of community-based research—such as community empowerment—are not standard research outcomes and are therefore difficult to quantify. Despite the lack of attention given to community-based research, we still believe it offers the most appropriate methodology that can enable people to deconstruct the cultural conceptions of urban ecosystems while empowering them to use an understanding of urban ecosystems to address environmental injustices. The Loka Institute in Amherst, Massachusetts has spent several years studying the idea of community-based research and suggests that the university-affiliated community research centers in Holland, popularly

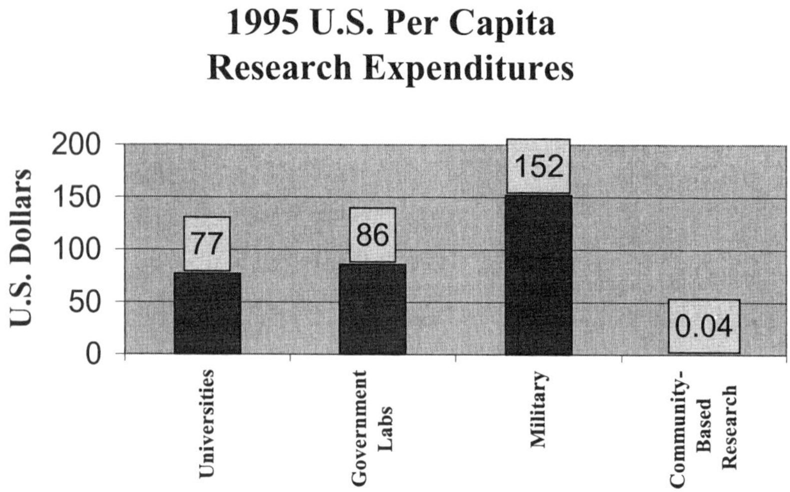

FIGURE 4.1. Comparison of Research Expenditures. Adapted from Sclove, Schammell, Holland (1998).

known as "Dutch Science Shops," offer one approach to more successfully promote community-based research in the United States. Through such centers, the Dutch are able to invest in community-based research at 37 times the U.S. rate (Sclove, et al. 1998).

Incorporating Environmental Justice in Urban Ecosystem Understandings

Our second strategy of incorporating environmental justice concerns within the context of understanding urban ecosystems builds directly on the opportunities for local learning emphasized with community-based research. Although people concerned about environmental justice often place health and survival issues as top community priorities, they must place these priorities in the context of the failure of urban ecosystems; they must make the connection between healthy ecosystems that mimic nature and just social systems. Those gathered at the First National People of Color Leadership Summit understood this when they established the 17 Principles of Environmental Justice and acknowledged that environmental justice affirms the ecological unity and interdependence of all species, and affirms the need for urban ecological policies to clean up and rebuild cities in balance with nature (Newton 1996). These principles challenge the unsustainable aspects of urban ecosystems and suggest ways in which such systems can be more sustainable.

When environmental justice struggles join with wider regional environmental coalitions, there is greater overall success than if each issue group works independently (Walsh, et al. 1997). It is also important for those working to advance the understanding of urban ecosystems to reach out to environmental justice advocates. Connecting an understanding of an urban ecosystem with a desire for environmental justice can help identify the wider social and environmental implications of a particular concern (landfill, incinerator, industrial facility, etc.). This connection can lead to stronger networks providing greater overall resource mobilization and support. An important tool to help incorporate environmental justice within an understanding of urban ecosystems is GIS (e.g., the "Environmental Mapper" website www.epa.gov/compliance/whereyoulive.html of the USEPA is one option for working with GIS that is accessible to anyone with access to the Internet). Having a visual representation of the overlap of social and environmental concerns is key to building these important partnerships.

A New Type of Professional

Our third strategy, calling for the formation of a new type of professional, is the most important one of all. This type of person is needed in communities, government agencies, and university research institutions. They will need to understand the culturally constructed dimensions of urban ecosystems and be able to forge connections with a variety of groups with environmental justice concerns. Only recently have humans been recognized as components of ecosystems (McDonnell and Pickett 1993). This recognition was seen as a fundamental shift in the understanding of ecosystems. A similar fundamental shift is now needed to promote sustainable knowledge and to fully appreciate the complexity of the human dimension of ecosystems.

Such professionals need to accept the challenges of working directly with communities and should be able to use participatory and community-based research methods to involve community members in the design, implementation, data collection, and analysis of research initiatives connecting environmental justice with a better understanding of urban ecosystems. Institutions also need to recognize the difficulty of such work as it reaches across disciplines and challenges many of the assumptions of scientific inquiry. In a recent analysis of adaptive strategies for ecosystem management, Aley, et al. (1999) provide helpful examples of how some natural resource professionals are successfully integrating social dimensions into natural resource initiatives.

Conclusions

We have attempted to explore the importance of understanding urban ecosystems from the perspective of people concerned about environmental justice. By understanding ecosystems as cultural constructs, we are pointed in the direction of intentional cultural change to help ameliorate environmentally unjust conditions. Understanding the complexities of race, class, and justice is key to understanding the complexity of urban ecosystems as culturally defined constructs. If we fail to fully understand urban ecosystems, the urban environment will continue to decline and be made more unhealthy by policy decisions that disproportionately affect people of color and low-income groups. An understanding of urban ecosystems also can provide opportunities for additional networking and information exchange that can be very helpful to environmental justice initiatives.

To achieve these ends we have stressed the need for participatory or community-based research initiatives, the importance of placing concerns about environmental justice within the context of urban ecosystems, and finally we have to called for a new type of professional that will be able to use sustainable knowledge to help us reconstruct urban ecosystems to be more livable. The results of such efforts would hopefully be better community-based initiatives that are informed by economic, social, and ecosystem realities. There would also be stronger, more successful coalitions working on environmental justice and expanding the understanding of urban ecosystems. The time for such action is now.

Acknowledgments. We are grateful for the insights on the issue of community-based research provided by Dr. Mark Chesler, Sociology Department, University of Michigan—Ann Arbor.

References

Aley, J., W.R. Burch, B. Conover, and D. Field. 1999. *Ecosystem management: adaptive strategies for natural resources organizations in the twenty-first century*. Taylor & Francis, Philadelphia, PA.

Anderson, R.C. 1998. *Mid-course correction. Toward a sustainable enterprise: The interface model*. Peregrinzilla Press. Atlanta, GA.

Bryant, B., and P. Mohai. 1992. *Race and the incidence of environmental hazards*. Westview Press, Boulder, CO.

Columbus Dispatch. 1999. Editorial and Comment May 15:13A.

Commission for Racial Justice of the United Church of Christ. 1987. *Toxic wastes and race in the United States: a national report on race and socio-economic characteristics of communities with hazardous waste sites*. Commission for Racial Justice of the United Church of Christ, New York.

Cronon, W. 1996. *Uncommon ground: rethinking the human place in nature*. W.W. Norton and Company, New York.

Edwards, R. 1997. Chemicals had ingredients for volatile reactions. *The Columbus Dispatch* September 11:4B.

Edwards, R. 2000. U.S. probe aimed at Ohio EPA: complaints say enforcement is lax. *The Columbus Dispatch* January 31:1A.

Hatch, J., N. Moss, A. Saran, L. Presley-Cantrell, and C. Mallory. 1993. Community research: partnership in black communities. *American Journal of Preventive Medicine* 6:27–31.

Hawken, P. 1993. *The ecology of commerce: a declaration of sustainability*. Harper-Collins, New York.
Israel, B., A.J. Schultz, E.A. Parker, and A.E. Becker. 1998. Key principles of community-based research. *Annual Review of Public Health* 19:173–202.
Institute of Medicine. 1999. *Toward environmental justice: research, education, and health policy needs.* National Academy Press, Washington, DC.
Jensen, A. 1968. Biogenic perspectives. Pages 7–10 in M. Deutsch, I. Katz, and A. Jensen, eds. *Social class, race and psychological development*. Holt, Rinehart and Winston, New York.
Kirsch, S. 1999. *Proposal for doctoral program in anthropology and natural resources and environment.* 3rd draft (unpublished proposal). University of Michigan, Ann Arbor, MI.
Lavelle, M., and M. Coyle. 1992. Unequal protection: the racial divide in environmental law. *National Law Journal* (Sept):S1.
McDonnell, M.J., and S.T.A. Pickett, eds. 1993. *Humans as components of ecosystems.* Springer-Verlag, New York.
Moynihan, D.P. 1965. *The Negro family: the case for national action*. Office of Policy Planning and Research, U.S. Department of Labor, Washington, DC.
Newton, D.E. 1996. *Environmental justice: A reference handbook*. ABC-CLIO, Santa Barbara, CA.
Patton, M.Q. 1990. *Qualitative evaluation and research methods*. Sage Publications, Newbury Park, CA.
Schafer, K., S. Blust, B. Lipsett, P. Newman, and R. Wiles. 1993. *What works: local solutions to toxic pollution*. The Environmental Exchange, Washington, DC.
Shockley, W.B. 1992. *Shockley on eugenics and race: the application of science to the solution of human problems*. Scott Townsend, Washington, DC.
Schwab, J. 1994. *Deeper shades of green: the rise of blue-collar and minority environmentalism in America*. Sierra Club, San Francisco, CA.
Sclove, R.E., M.L. Schammell, and B. Holland. 1998. *Community-based research in the United States: an introductory reconnaissance, including twelve organizational case studies and comparison with the Dutch science shops and the mainstream American research system*. The Loka Institute, Amherst, MA.
Tesh, S.N., and B.A. Williams. 1996. Identity politics, disinterested politics, and environmental justice. *Polity* 18:285–305.
Taylor, D.E. 2000. The rise of the environmental justice paradigm. *American Behavioral Scientist* 43:508–580.
Thomas, S.B. 1991. The Tuskegee study, 1932–1972: implications for HIV education and AIDS risk education programs in the black community. *American Journal of Public Health* 81:1498–1505.
Walsh, E.J., R. Warland, and D.C. Smith. 1997. *Don't burn it here: grassroots challenges to trash incinerators*. Pennsylvania State University Press, University Park, PA.
White, L.A. 1949. *The science of culture: a study of mankind and civilization*. Farrar, Straus, New York.

5 Why Is Developing a Broad Understanding of Urban Ecosystems Important to Science and Scientists?

STEWARD T.A. PICKETT

Urban systems are among the most recent to be seriously studied by ecologists (Barrett 1985). This chapter examines the benefits to science and scientists of developing an understanding of urban environments as ecological systems. I make four points: (1) public understanding of science benefits science; (2) studies of urban systems have value to science itself; (3) urban ecological systems have characteristics that make them particularly useful for enhancing public understanding of science; and (4) public understanding of urban systems has an unusually high potential to benefit science. These points emerge from the fact that the public supports science. In addition, research in cities and suburbs requires the formal and informal permission and assistance of a large and diverse collection of citizens. Furthermore, urban areas are an environment commonly encountered by the majority of people in many countries. Because the benefits to science rest on a dialog between science and the public, I lay out the benefits to both groups.

In this chapter, the public refers to the suite of individuals and institutions that reside in and influence a metropolitan area. The concept of institution refers to any aggregation of individuals, whether it be formal or informal, public or private (Perrow 1986). Examples of institutions include nuclear families, households, clubs, neighborhoods, schools, government agencies, religious congregations, firms, and political parties. Some institutions, such as corporations, can persist for long periods, while others, such as an action group focused on a zoning change, disband when a specific purpose is achieved.

Every institution makes decisions that affect resource use and therefore affects environmental processes locally or at a distance. Even the decision by an individual or a single household to make or defer a particular purchase has potentially far-reaching environmental consequences. There is thus a wide-ranging public discourse that affects environmental decisions. Throughout this chapter, I will use "the public" as shorthand for this diverse array of decision makers, ranging from individuals to large institutions. If science is to benefit from the public's understanding of the metropolis as a complex ecological system, the decision makers must appreciate the pı and products of science. This chapter explores the benefits of a ı awareness of ecology and the value of ecological studies in urban systems.

Beneficial Public Interactions with Science

The Public Pays

The most fundamental interaction between science and the public is that the public pays for science. The payment is made either through taxes, through the profits of businesses that invest in research, education, and training of scientists, or through the charitable contributions of corporations, trusts, and individuals. In each of these cases, the public supports science through personal or commercial choices or through legislatively and judicially supported policies. The continued support of science depends on the appreciation of a vast array of individuals and institutions, ranging over such contrasts as the president's budget advisors, judges, individual voters, and philanthropists.

The Public Uses Science

The public benefits from science by using the knowledge scientists generate to help make decisions. In using scientific knowledge, the public considers an array of additional influences, including taste, economics, cultural values, and political expediency. Science is one of the strands in the dialog that leads to a decision (Page 1992). Science introduces straightforward factual information into the dialog, and identifies values that may have uniquely arisen from scientific research and analysis. For example, the awareness of acid deposition, popularly called "acid rain," is a scientific discovery, and the public appreciation of the diversity of native lake organisms that are sensitive to acid rain is a value that emerges from scientific knowledge.

Public decisions express a mix of values held by the participants. Science too, expresses values. This fact has led some to conclude that science is merely a social construction, and that its voice is suspect. However, the charge that science is suspect in policy decisions because science does express values is too extreme (Hacking 2000). Science is *in part* a social construct, but it is answerable to the public amongst a diverse community of practitioners for the degree of fit between its conclusions and measurable features and processes in the observable world (Longino 1990). Social construction cannot hide from fit (Lloyd 1988). In other words, for any new proposition offered within science, some scientists will respond with criticism based on their experience in different systems, or even because of their political views. But ultimately if either the original proposition or the criticism fails to hold up when it is compared with measurements derived from the laboratory or the field, it will not survive. For example, some scientists have complained that concern with invasive exotic species is inappropriate because it is akin to stereotyping human groups. However, the argument is countered by recognizing the differences between statistical descriptions of organism behaviors that result from evolution and that are relevant to biological invasion, and the lack of evolutionary explanation for the differences in capability usually ascribed to human groups in stereotyping (Ehrenfeld 1999).

Values can be exposed in scientific research because they are often reflected in the fundamental assumptions that underlie scientific models and, hence, the conclusions that scientists reach. The test of the assumptions supported by one or a suite of values occurs when the model or the theoretical expectations derived from the assumptions are tested against the observable world. Although testing may be difficult and the results may meet with resistance from vested individuals, the persistent failure of a test to support assumptions leads to replacement or refinement of that model of the world. Three processes reduce the influence of values as a source of bias in science: (1) the application of broadly different classes of methods to a particular problem; (2) the analysis of problems by a diverse scientific community; and (3) the assessment of problems during application of the conclusions to real issues. Although contemporary philosophy, history, and sociology of science have shown how values can be dealt with in science (Mayr 1988; Pickett, et al. 1994), little of this is known among scientists. Therefore, scientists often erroneously defend scientific objectivity from the perspective of a supposedly unassailable normative and completely value-free procedure. The objectivity of science lies not in a superhuman disinterest, but rather in the critical and creative participation of a diverse community (Longino 1990). The diversity in fact relies in part on different practitioners holding different values that may cause them to propose or criticize different assumptions, methods, or applications. A better appreciation of how science works, how it contributes to public discourse, and how it relates to values internally and externally would all increase the ability of the public to use science.

The Public Enjoys Science

People like a good story. For all its seriousness and difficulty, science presents good stories. The material origin of the universe, the riddle of past events in speciation, or the unexpected indirect interactions in an ecological community all have great potential for public engagement. Effectively communicating to individuals, groups, and the media could enhance this public appreciation of science. There are several strategies for generating this benefit. First, the public needs to be educated about the core aspects of science that are uncontroversial and well established. It is often a struggle for the public to extract these nuggets from the debate that swirls at the controversial frontiers of a science (Ziman 1978). For instance, in the science of evolution descent with modification is a core generalization. The debate (Gould 1980; Eldridge 1985) about whether the process of species formation is gradual or abrupt over geological time scales occurs on a frontier of the study of evolution, and does not threaten the core. The difficulty of recognizing the secure core of any scientific discipline is exacerbated by the critical attitude that so many scientists possess. Many scientists seek fame by challenging relatively secure portions of their specialty.

The Public Allows Access

Ecology is a science of place, and access to property is fundamental to its success. The public, either through their public agencies and executives, corporate bodies, or individual landholders, regulates the access that ecologists have to the materials of their research. Lands in municipal parks, agency holdings, and private hands are key to the success of much ecological research. Continued access to land as a limiting resource for ecological research depends on at least three phenomena. First, the owners and tenants must appreciate research as either a practical or intellectual product. Those who control access to land may be fascinated by the actual or anticipated results of research, or they may need some of the information, or they may hope to see the solution of some problem. Note that those who control access may not be the landowners (Grove 1995). For example, having the permission of an owner off site may not ultimately ensure access when a researcher actually knocks on the door.

A second requirement for continued access is respectful behavior. Although this may seem straightforward, the vast range of cultures, conceptions of privacy and politeness, and formal regulations for approach and interaction can be daunting. Considerable effort may have to be expended to find out what the "rules" are in specific cases.

Finally, to ensure continued access to research sites, the benefits of research on the land in question, or places like it, must be communicated to the public. Such reporting requires knowing what the concerns of the tenants or managers are, and how to communicate the significant insights in a clear, simple, and expressive way.

All of these complexities of access are compounded in urban settings. For example, in the City of Baltimore, Maryland, there are some 250 recognized neighborhoods. Learning who the power brokers, community leaders, and gatekeepers are in a variety of neighborhoods, with their ethnic, demographic, and institutional contrasts is magnitudes more difficult than approaching the superintendent and rangers of a national forest for access to a single research site.

The Public Is the Source of New Scientists

The most precious resource for science is the pool of motivated, bright, thoughtful, and interactive people who would consider a scientific career. The value in having a robust community of practitioners is more than a simple amassing of hands or minds. Rather, the key is the diversity of approach, of style, and of creative or critical impulses that contribute to the productivity and objectivity of science (Grene 1985). Training new generations of scientists requires that an appreciation of science be generated among potential recruits. In addition, potential recruits must know that there are pathways that they can follow to become members of the scientific community. Not all potential recruits will be engaged by the same motivations nor encouraged by the same path of training, nor do mentors necessarily have to "look like" their protégés. Scientists, however, must communicate the diversity of stimuli and ways to pursue a scientific career.

The diversity of potential recruits to science in the United States is greatest in cities and older suburbs, where the diversity of human population and social situations is greatest. If science cannot learn successfully to engage this diversity of potential colleagues, the diversity of recruits that occurs by happenstance is not likely to be great. Ecologists currently constitute a stunningly homogeneous group. The opportunity for diversification is great, and urban systems are the vineyard where the harvest awaits.

Internal Benefits to Science

The benefits to science discussed so far derive from interactions built on the public understanding of science, whether in an urban or seemingly wild setting. In this section, I discuss benefits *within* science that accrue to working in metropolitan systems.

A New Frontier

The most obvious benefit of ecological research on metropolitan areas is the exploration of a new frontier for ecology. Most ecological research has focused on areas where people are absent, or where their effects are distant (McDonnell and Pickett 1993). For example, the photographs that most ecologists show at the beginning of their illustrated professional lectures are generally scenes that could be the subject of an American landscape painting of the nineteenth century frontier. Purple mountains, shaded brooks, vacant forests, waving prairies, and such, are the stuff of ecological illustration. Whatever the reasons for this bias, the absence of ecological information on metropolitan systems is profound. This limits the ability of ecology to understand such areas, to contribute to integrated studies in cities and their surroundings, and to inform the public. At a time when the extent of urban development and the human population of these areas are increasing both nationally and globally (Wetterer 1997), this neglect is one that ecology can ill afford. Planners and geographers have cited powerful cases in which ecological knowledge can improve city life (Spirn 1984). Where the streams ran before they were hidden in pipes and filled to support foundations, or how the winds that buffet pedestrians downtown might be buffered by vegetation are but two examples of useful ecological knowledge in the city. In other words, knowing how environmental processes interact with current and planned structures and uses in urban systems is crucial.

Integrating with Other Disciplines

If ecology is to contribute to an understanding of the metropolis, it must participate in integrated scientific studies (Pickett, et al. 1999). Because people and their institutions are key components of the comprehensive systems of the metropolis, a pressing need is to work with social scientists. Understanding economics, political decisions, organizational networks, and the spatial patterns of human tenure and action is required (Ehrlich 1997). For example, understanding how ecological processes such as vegetation sustainability affect the social capital in neighborhoods emerges from an interdisciplinary linkage. Even the seemingly straightforward concern with land use change is layered with human behaviors and institutional actions (Foresman, et al. 1997). For example, the maps of land use expose only the most general of categories. Mapping land as "low density residential" leaves out much ecological information. How much water or fertilizer is used on particular parcels or in specific neighborhoods? What social processes affect the specific ecological processes within a coarse land use class? Even the more obvious integrations with physical sciences, such as hydrology and atmospheric science also require further development in the new arena of the metropolis. Again, more refined characterizations of the physical environment that can be linked to ecological process are needed. Does the water draining from an area move more rapidly into storm sewers because of intact curbs, or is it more likely to infiltrate into the soil along unguarded road verges? How do these structures affect loading of the water with toxins or potentially disease-causing organisms?

Many of the models from the physical sciences on which ecologists currently depend have been developed in wild or agricultural lands, or refer to built infrastructure alone (Brun and Band 2000; Voinov, et al. 1999). How the potentially complex linkages among social, physical, and ecological processes function requires a fundamentally new integration. It is a gradient of blending wild, built, and managed systems that the field of urban ecological studies confronts. For example, the elements of hydrological models for the metropolis must deal well not only with the built infrastructure such as roads, buildings, and sewers, but also with lawns, tree canopies, and various types of soil. An example of the linkages still to be joined is the fact that most urban soils have not been classified and are often dismissively referred to as rubble or fill (Effland and Pouyat 1997).

Developing New Tools

Approaching a new realm of study often requires new tools and techniques. In the case of metropolitan areas, the integration required heightens the need for new tools. Although it is much too early for there to be a comprehensive technical theory of the metropolis, the development of frameworks that can support such a theory is underway. One example is the "Human Ecosystem Model" used in the Baltimore Ecosystem Study (Grove and Burch 1997). The model is based on the robust and widely applicable concept of ecosystem (Likens 1992). An ecosystem is an area, of any size, that contains organisms, the physical environment, and the interactions between them. A forest stand, with its interacting trees, shrubs, herbs, birds, mammals, insects, and arthropods in air and soil, fungi, mineral matter, water, and atmosphere above and below ground is an ecosystem. An ecologist chooses the boundaries for studying a given ecosystem. A forest stand may be demarcated by a watershed boundary, or by a management parcel.

Contemporary systems theory does not assume that all systems are closed or self-regulating, or that they have a single stable point, or that their dynamics are deterministic. Ecosystems may be studied from a variety of perspectives that focus on the organisms and the structures they make, or on materials and energy and how they flow through the system. The contemporary ecosystem concept is very flexible and remarkably free of narrow assumptions that would restrict it to only wild or pristine places (Pickett, et al. 1992). A vacant lot can be studied as an ecosystem as well as can a pristine prairie.

An important addition is needed when studying urban systems, however. Applying the basic ecosystem concept to human-dominated areas requires new components that natural scientists have not needed for their traditional studies (Machlis, et al. 1997). The human ecosystem model is a conceptual model of the components of a metropolis from a joint ecological, social, and physical perspective. It is amenable to exploring dynamics, and to informing more specific models, such as those that might explore human and social capital along with natural, economic, and built capital.

Broad inclusive frameworks such as the human ecosystem model must be complemented by more specific, "middle level theories." Such bodies of knowledge are neither the overarching theories of a discipline nor the very specific models from which narrow and focused forecasts can be made. The human ecosystem framework (Machlis, et al. 1997) is an example of a very general theory. To a non-specialist, this may seem to be a hopelessly complex catalogue of processes that might affect any human-dominated ecosystem. At the other extreme are specific models such as "RHESSys" (Brun and Band 1999), which mathematically describe the control of water flow along a pathway of different slope units. An intermediate level of theory is emerging that takes crucial kinds of social and human capital from the human ecosystem framework and combines it with the major physical and biological flows on slopes to generate a model that should be interpretable by both biophysical and social scientists. Furthermore, because the new, synthetic model uses quantitative parameters that can be interpreted in familiar terms of "capital," it should be readily interpretable in the social discourse on environmental quality. The middle level theories are thus general enough to allow the subject matter of different disciplines to be linked to motivate research questions, research designs, integrated databases, and integrated models.

The refinements to existing theories may benefit the disciplines from which they came as well as the emerging integration. Many theories in ecology are poorly articulated and not well understood outside the specialty that works with them (Pickett, et al. 1994). For example, in ecology there is a famously vague concern with diversity and stability. In order to link diversity with social controls and effects, exactly what diversity is and how it is generated and controlled at specific scales in ecology will have to be sorted out (Kinzig and Grove 2000). This will be a benefit to the field of ecology as well as permit the linkage with social processes.

To generate and run the new models, integrated data sets are required. They will likely include new features that no one discipline alone would require. In addition, such data sets will be accessible on different scales, so that the processes studied by different disciplines can be linked functionally. Each contributing discipline may have to work at scales beyond those it usually addresses.

Expression of a Contemporary Systems View

In order to advance the integration required to study urban ecological systems, there are new perspectives available. One of the most powerful of these is a hierarchical approach to systems. This is neither the single-scale holism of the past, nor blind reductionism that requires the lowest mechanistic common denominator (Auyang 1998). In the past, holism focused on the whole systems only, and rarely decomposed them into component parts wherein mechanism might lie. While valuable information on the behavior of large systems resulted from this approach, the understanding was incomplete. Equally incomplete is narrow reductionism, which insists on understanding systems by decomposing them to some very low level where an ultimate mechanism is sought. A contemporary, hierarchical approach to systems recognizes pattern to appear on a specific level or scale, and mechanism to be nested within that level, and constraint to derive from higher levels (Ahl and Allen 1996).

An example of a strictly holistic approach is to assume that ecological communities, such as forests, are so integrated as to move in lockstep during periods of climatic modification. At the reductionistic extreme is an approach that assumes that everything of ecological importance about a plant community can be discerned from looking at the behavior of the individual populations that it comprises. Empirical research, theory, and models that take a hierarchical perspective have produced the contemporary understanding of communities (Pickett, et al. 1987). The patterns that appear in the community as an aggregate, and the interactions that emerge from the behavior of individual populations both help explain how communities work. For example, forest edges affect ecological flows across them. Sometimes this is the result of their coarse scale structure as a mappable landscape feature, and thus understood from a holistic approach. In other ways, edges affect flows because of their fine-scaled architecture resulting from the size and shape of interacting plants at the boundary between forest and field, an understanding derived from a mechanistic perspective (Cadenasso and Pickett 2001).

Testing Contemporary Theory and Perspectives

A powerful test of theories, and the expectations derived from them, is to apply them in new settings and cities provide this opportunity for ecology. For example, the theory of patch dynamics can be tested in this way (Flores, et al. 1997). Patch dynamics takes ecological systems to be spatially structured, divisible into discrete patches or multiple gradients. An isolated forest, a stream corridor, or a soil type that dries out more readily than neighboring soils are examples of patches. Each component of the patchwork has different structures, functions, longevities, and compositions, and the whole array acts as a shifting mosaic. The dynamics of the mosaic are controlled by interactions within each patch as well as by flows of matter, energy, and information across the mosaic between patches. For example, the way a treefall gap in a city park changes over time will depend on what happens inside the patch (people trampling, breaking saplings, planting trees), and on adjacent patches (what seed-producing trees are nearby, etc.).

Because much of ecology has usually focused on local communities and ecosystems at small spatial scales, the extension to larger spatial context required by metropolitan systems can contribute to the growth of ecology. Because cities and their surrounding metropolitan areas have clear spatial structure (Shevky and Bell 1955; Hamm 1982; Bogue 1984), it is hard to imagine successful ecological approaches that do not account for structure. Simple averaging over the mosaics in metropolitan areas may obscure important controls on ecological and social functions. The metapopulation approach and the emerging functional concepts of landscape ecology are examples of ecological theories that apply to metropolitan areas within which ecological populations, communities, and ecosystems exist, change and interact.

Another contemporary perspective that can be tested in cities is the hypothesis that temperate ecosystems generally retain nutrients (Likens and Bormann 1995; Aber, et al. 1998). The general ability of ecosystems to retain materials is suggested by theories of thermodynamics and self-organization (Cousins and Rounsevell 1998). The apparent desire of people and institutions to speed the flow of wastes and unexploited resources through the infrastructure of metropolitan areas, however, suggests that metropolitan ecosystems may be less retentive than other managed ecosystems. The metropolis as a whole and the specific land cover–based ecosystems within it provide a gradient over which to test the predictions of ecosystem retention theory.

What Characteristics of Urban Systems Help Develop Public Understanding of Science?

The first two sections of this chapter have explored values of public appreciation of science, and of new scientific knowledge that is emerging from urban systems. Given those two complementary kinds of value to science of urban ecological studies, it is worth considering whether the metropolis offers some particular boost to those values.

Extent and Familiarity

If urban systems can contribute to a greater understanding of ecology, then the benefit can potentially be widespread. Urban areas are expanding both in population numbers and spatial extent (Berry 1990). If the public bases its understanding of ecological processes on its local environment, then extracting ecological knowledge from urban systems has the best chance of enhancing ecological understanding worldwide. Some 75 percent of the population in the United States already live in urban areas. These are the areas people encounter daily, although most people may not recognize the ecological component of the metropolis. Therefore, urban systems are also relevant to the environmental decision-making most people engage in. The accessibility of various ecological components of metropolitan systems should make them the most convenient platform for ecological education.

Dynamism

Ecological knowledge is readily conveyed in systems that are undergoing obvious change because such changes expose ecological processes. Because urban areas are expanding on their suburban fringes and leap-frogging into the farms and forests of the hinterlands via exurban development, there are a variety of situations in which the resulting changes in land cover and resource management can be understood ecologically. Second homes, telecommuting, and the replacement of older suburbs are examples of the fringe dynamics of cities. The changes extend into the wild and managed lands that the cities and their increasingly dispersed residents and recreationists exploit.

The dynamics within established cities are also fertile ground for ecological understanding (Foresman, et al. 1997). In many older cities, vacant lots proliferate and the intensity of investment in the green infrastructure of parks and the built infrastructure of water supply and processing, for example, are changing. Additional changes in modes and patterns of transportation, human population density, social resources, and management practices affect neighborhoods and business districts. The alteration to cities resulting from the growth of the national defense highway system—the interstates—is a well-known example. In some urban areas, the establishment of a new rail-based transportation system has made more recent social and ecological changes.

Finally, metropolitan systems embody a broad suite of ecological processes. The U.S. National Science Foundation's Long-Term Ecological Research (LTER) network in the United States was established in 1980 to study different ecosystems and their dynamics through time. It has mandated that all sites examine biological productivity, nutrient dynamics, soil processes, trophically important biological populations, and disturbances. That same mandate applies to the two urban LTER programs in Baltimore, Maryland and Phoenix, Arizona. Of course, the Baltimore Ecosystem Study and the Central Arizona-Phoenix LTERs are additionally mandated to integrate human components, land use, and civil infrastructure (Grimm, et al. 2000).

Multiplier Effects of Benefit to Science from the Metropolis

Visibility

Science in the city is visible to people. There are a large number of opportunities to expose the public to the scientific process and its insights. These include formal programs in the classroom, informal activities at schools and community centers, and interactions with government agencies and neighborhood associations. There is also the opportunity for serendipitous learning during encounters with the public in the course of doing research in parks and neighborhoods, for example. The question, "What are you doing?" presents a common "teachable moment" in metropolitan research.

New Research Approaches

Social sciences have developed a methodology called "participatory action research" (Whyte 1991). This approach recognizes that conducting research in inhabited systems requires the interest, permission, and support of the inhabitants and resident institutions. Without such an entree into the system, at worst the research effort is untenable, and at best it yields a biased or incomplete understanding of the roles and actions of people and institutions as dynamic elements within the system. The specific questions for research in inhabited areas can be guided and shaped by understanding the needs and concerns of the public.

New Constituencies

Ecology, from the evidence of the current ethnic composition of its practitioners (Holland, et al. 1992), has apparently engaged a relatively narrow spectrum of the U.S. population. There are two reasons to increase ethnic diversity and gender representation throughout the ranks of ecology. First, one major contributor to the objectivity of science is the participation of a diverse community of practitioners. Although we cannot assume that individuals who look different will necessarily think differently, the broad array of experiences, perspectives, motivations, and concerns that we can statistically expect people of diverse backgrounds will exhibit should enhance the diversity of science.

The second motivation for diversification of the ecological community reflects the changing diversity of the population of the United States. Over the coming decades, the ethnic mix of the North-American population will change dramatically. Even now, the ease with which ecological knowledge can be applied in and developed from areas that are inhabited largely by Native Americans, or by African-Americans, or Hispanics is limited. Of course, good and useful science can, demonstrably, be done by people of any identity. The message in general is that the diversification of science can help connect the process, practice, and products of science with a diversifying national population and power structure.

If ecological studies and ecological education can effectively increase their presence in urban settings, perhaps a new source of recruits to the science can be tapped and—citizens, understanding and appreciation of ecological research can be broadened. Both professional and public awareness of ecology are important. The more citizens understand ecology as a science, so much the better for the science itself and for the public discourse that affects environmental issues.

Acknowledgments. I thank M.L. Cadenasso and P.M. Groffman for comments and helpful input. I am grateful to the National Science Foundation

Long-Term Ecological Research (DEB 97114835), and the EPA-NSF Water and Watersheds (GAD R825792) for support. The USDA Forest Service Northeastern Research Station provided site management and in-kind services to the Baltimore Ecosystem Study, to which this paper is a contribution.

References

Aber, J., W. McDowell, K. Nadelhoffer, A. Magill, G. Berntson, M. Kamakea, S. McNulty, W. Currie, L. Rustad, and I. Fernandez. 1998. Nitrogen saturation in temperate forest ecosystems. *BioScience* 48:921–934.

Ahl, V., and T.H.F. Allen. 1996. *Hierarchy theory: a vision, vocabulary, and epistemology*. Columbia University Press, New York.

Auyang, S.Y. 1998. *Foundations of complex-systems theories in economics, evolutionary biology, and statistical physics*. Cambridge University Press, Cambridge.

Barrett, G.W. 1985. A problem-solving approach to resource management. *BioScience* 35:423–427.

Berry, B.J.L. 1990. Urbanization. Pages 103–120 in B.L. Turner II, W.C. Clark, R.W. Kates, J.F. Richards, J.T. Matthews, and W.B. Meyer, eds. *The earth as transformed by human action: global and regional changes in the biosphere over the past 300 years*. Cambridge University Press, New York.

Bogue, D.J. 1984. Procedure for delimiting ecological community areas. Pages 27–36 in D.J. Bogue and M.J. White, eds. *Essays in human ecology, 2nd Edition*. Volume 2. The Community and Family Study Center, University of Chicago, Chicago, IL.

Brun, S.E., and L.E. Band. 2000. Simulating runoff behavior in an urbanizing watershed. *Computers, Environment and Urban Systems* 24:5–22.

Cadenasso, M.L., and S.T.A. Pickett. 2001. Effects of edge structure on the flux of species into forest interiors. *Conservation Biology* 15:91–97.

Cousins, S., and M. Rounsevell. 1998. Case studies: soil as the interface of the ecosystem goal function and the Earth system goal function. Pages 255–268 in F. Müller and M. Leupelt, eds. *Eco targets, goal functions, and orientors*. Springer-Verlag, New York.

Effland, W.R., and R.V. Pouyat. 1997. The genesis, classification, and mapping of soils in urban areas. *Urban Ecosystems* 1:217–228.

Ehrenfeld, D. 1999. Andalusian bog hounds. *Orion* 18(4):9–11.

Ehrlich, P. 1997. *A world of wounds: ecologists and the human dilemma*. Ecology Institute, Oldendorf/Luhe, Germany.

Eldridge, N. 1985. *Unfinished synthesis: biological hierarchies and modern evolutionary thought*. Oxford University Press, New York.

Flores, A., S.T.A. Pickett, W.C. Zipperer, R.V. Pouyat, and R. Pirani. 1997. Adopting a modern ecological view of the metropolitan landscape: the case of a greenspace system for the New York City region. *Landscape and Urban Planning* 39: 295–308.

Foresman, T.W., S.T.A. Pickett, and W.C. Zipperer. 1997. Methods for spatial and temporal land use and land cover assessment for urban ecosystems and application in the greater Baltimore-Chesapeake region. *Urban Ecosystems* 1:201–216.

Gould, S.J. 1980. Is a new and general theory of evolution emerging? *Paleobiology* 6:119–130.
Grene, M. 1985. Perception, interpretation, and the sciences: toward a new philosophy of science. Pages 1–20 in D.J. Depew and S.H. Webber, eds. *Evolution at a crossroads: the new biology and the new philosophy of science*. MIT Press, Cambridge.
Grimm, N.B., J.M. Grove, S.T.A. Pickett, and C.L. Redman. 2000. Integrated approaches to long-term studies of urban ecological systems. *BioScience* 50: 571–584.
Grove, J.M. 1995. Excuse me, could I speak to the property owner please? *Common Property Resources Digest* 35:7–8.
Grove, J.M., and W.R. Burch Jr. 1997. A social ecology approach and application of urban ecosystem and landscape analyses: a case study of Baltimore, MD. *Urban Ecosystems* 1:259–275.
Hacking, I. 2000. *The social construction of what*? Harvard University Press, Cambridge.
Hamm, B. 1982. Social area analysis and factorial ecology: a review of substantive findings. Pages 316–337 in A. Theodorson, ed. *Urban patterns: studies in human ecology*. Pennsylvania State University Press, University Park, PA.
Holland, M.M., D.M. Lawrence, D.J. Morrin, C. Hunsaker, D. Inouye, A. Janetos, H.R. Pulliam, W. Robertson, and J. Wilson. 1992. *Profiles of ecologists: results of a survey of the membership of the Ecological Society of America*. Public Affairs Office, Ecological Society of America. Washington, DC.
Kinzig, A., and J.M. Grove. 2000. The Urban Environment. Pages 733–745 in S. Levin, ed. *Encyclopedia of biodiversity*. Academic Press, New York.
Likens, G.E. 1992. *Excellence in ecology, 3: the ecosystem approach: its use and abuse*. Ecology Institute, Oldendorf/Luhe, Germany.
Likens, G.E., and F.H. Bormann. 1995. *Biogeochemistry of a forested ecosystem, 2nd Edition*. Springer-Verlag, New York.
Lloyd, E.A. 1988. *The structure and confirmation of evolutionary theory*. Greenwood Press, New York.
Longino, H.E. 1990. *Science as social knowledge: values and objectivity in scientific inquiry*. Princeton University Press, Princeton, NJ.
Machlis, G.E., W.R. Burch, Jr., and J.E. Force. 1997. The human ecosystem part I: the human ecosystem as an organizing concept in ecosystem management. *Society and Natural Resources* 10:347–367.
Mayr, E. 1988. *Toward a new philosophy of biology: observations of an evolutionist*. The Belknap Press of Harvard University Press, Cambridge.
McDonnell, M.J., and S.T.A. Pickett, eds. 1993. *Humans as components of ecosystems: the ecology of subtle human effects and populated areas*. Springer-Verlag, New York.
Page, T. 1992. Environmental existentialism. Pages 97–123 in R. Costanza, B.G. Norton, and B.D. Haskell, eds. *Ecosystem health: new goals for environmental management*. Island Press, Washington, DC.
Perrow, C. 1986. *Complex organizations: a critical essay*. Random House, New York.
Pickett, S.T.A., S.L. Collins, and J.J. Armesto. 1987. Models, mechanisms and pathways of succession. *Botanical Review* 53:335–371.
Pickett, S.T.A., V.T. Parker, and P.L. Fiedler. 1992. The new paradigm in ecology: implications for conservation biology above the species level. Pages 65–88 in P.L.

Fiedler, ed. *Conservation Biology: The Theory and Practice of Nature Conservation, Preservation, and Management*. Chapman and Hall, New York.

Pickett, S.T.A., J. Kolasa, and C.G. Jones. 1994. *Ecological understanding: the nature of theory and the theory of nature*. Academic Press, San Diego, CA.

Pickett, S.T.A., W.R. Burch, Jr., and J.M. Grove. 1999. Interdisciplinary research: maintaining the constructive impulse in a culture of criticism. *Ecosystems* 2:302–307.

Shevky, E., and W. Bell. 1955. *Social area analysis: theory and illustrative application and computational procedure*. Stanford University Press, Stanford, CA.

Spirn, A.W. 1984. *The granite garden: urban nature and human design*. Basic Books, New York.

Voinov, A., R. Costanza, L. Wainger, R. Boumans, F. Villa, T. Maxwell, and H. Voinov. 1999. Patuxent landscape model: integrated ecological economic modelling of a watershed. *Journal of Environmental Modelling and Software* 14:473–491.

Wetterer, J.K. 1997. *Urban ecology*. Encyclopedia of Environmental Sciences, Chapman and Hall, New York.

Whyte, W.F. 1991. *Participatory action research*. Sage Publications, Newbury Park, CA.

Ziman, J. 1978. *Reliable knowledge: an exploration of the grounds for belief in science*. Cambridge University Press, Cambridge.

Section II
Foundations and Frontiers from the Natural and Social Sciences: Themes

Charles H. Nilon, Alan R. Berkowitz, and Karen S. Hollweg

The authors in Section One answered the question, "Why is it important that people understand urban ecosystems?" from the perspectives of people with an interest in education reform, environmental justice, community development, and science. But what do we mean by understanding urban ecosystems? What should an informed person actually know about urban ecosystems? Does "knowledge" mean a set of facts that people should know and recite or does it have a more complex meaning?

In 1989 the British Ecological Society conducted a survey of ecologists to determine the top 10 concepts in ecology (Cherrett 1989). The "ecosystem" was listed as the most important concept, yet as the authors in this section discuss, the term ecosystem has a variety of meanings to both ecologists and the general public. People living in cities obviously should have an understanding of the ecosystem as a framework for studying urban areas; however, the ecosystem concept and the systems approach to the study of cities have been applied in very different ways by ecologists. Social scientists' approaches to urban ecology and the human ecosystem are different still, providing different perspectives, and adding more complexity to "what people should know."

We maintain that understanding urban ecosystems starts with an understanding of the knowledge gained by previous studies of cities by ecologists and social scientists. Since the 1960s natural and social scientists have conducted research on urban ecosystems (Boyden 1981). The objectives of these studies were tied to a broader goal of understanding cities and using that information in the planning, design, and management of urban areas. In addition, there is a long history of research on the natural history of cities, and of studies by social scientists who have investigated the people and institutions of cities (Greenwood 1999). The body of literature from these studies has been important in developing the idea that cities are systems with physical, biological, and social components. And these studies provided a basic understanding of the role that people and their activities play in shaping the distribution and abundance of organisms, and in shaping a variety of ecological processes. Chapters in this section by Tony Bradshaw

(Chapter 6), Martin Melosi (Chapter 12), and Anne Spirn (Chapter 13) describe some of this earlier work done on urban ecosystems and identify some of the key concepts about cities that are important in understanding urban ecosystems.

This basic understanding of cities as systems is enhanced by a knowledge of the ecosystem approach, with its focus on pattern and process and ecosystem drivers, and the key concepts emerging from the application of this approach to cities. The last several years have seen the initiation of interdisciplinary ecological studies of urban areas through the Natural Environment Research Council's Urban Regeneration and the Environment (URGENT) program in the United Kingdom, the GSF-Research Center for Environment and Health's Ecological, Research in Urban Regions and Industrial Landscapes (Urban Ecology) in Germany, and the U.S. National Science Foundation's Long-Term Ecological Research Program. Contributions by Grimm, et al. (Chapter 7) and Grove, et al. (Chapter 11) describe how this contemporary approach to studying urban ecosystems is being developed and applied in Phoenix and Baltimore. The concepts in these chapters contribute to the type of understanding of urban ecosystems that we feel is important to all people.

A true understanding of cities as ecosystems requires a knowledge of the context in which the key concepts are applied. This means understanding the issues and factors that shape questions of concern to the managers and residents of cities—what are the best ways to maintain viable neighborhoods? How can we maintain sustainable cities? How can we better conserve and manage natural resources? Some of the most important work on urban ecosystems comes from studies and projects that address such pragmatic issues. Chapters by Rees (Chapter 8), Harrison and Burgess (Chapter 9), Wolford (Chapter 10), and Wang and Ouyang (Chapter 14) describe integrated studies of cities using concepts from the ecological *and* the social sciences. These chapters represent some of the state-of-the-art work done on urban ecosystems. They suggest ways to bridge the disciplines and offer an applied perspective on how an ecological understanding of cities is being applied to the range of issues discussed in Section I, such as environmental justice and community development.

Our focus in this section is on foundations and frontiers for understanding urban ecosystems from the sciences. We asked each author to identify key concepts from their own natural and social science disciplines that are essential in defining cities as systems. We also challenged them to explore the frontiers of work on urban ecosystems, showing how the concepts they identify as critical can be used to address issues that are important to people who live in and care about cities.

References

Boyden, S., S. Millar, K. Newcombe, and B. O'Neill. 1981. *The ecology of a city and its people: the case of Hong Kong*. Australian National University Press, Canberra.

Cherrett, J.M., ed. 1989. *Ecological concepts: the contribution of ecology to an understanding of the natural world*. Blackwell Scientific Publications, Oxford.

Greenwood, E.F., ed. 1999. *Ecology and landscape development: a history of the Mersey Basin*. Liverpool University Press, Liverpool.

6
Natural Ecosystems in Cities: A Model for Cities as Ecosystems

Anthony D. Bradshaw

There is an underlying harmony among living things, well understood by biologists. This implies that the principles that apply to one group of organisms can be applied to another. As a result the attributes of the second group can be better illuminated and understood. This can apply just as much to very different groups of living organisms as to very similar ones. If we are trying to understand more clearly how cities, which are a particular group of organisms living together in a system, work, there is every reason to look at equivalent systems in nature. We can do this conveniently, and appropriately, by examining the natural systems which actually occur within cities.

These natural systems can best be described as *ecosystems*. This word was originally coined by Tansley (1935) to refer to a group of organisms and their environment "with which they form one physical system." The emphasis was that, to understand living communities, all the things that occur in any place being studied—the organisms, the soil, the atmosphere, and all the individual materials—have to be considered together, and therefore should be covered by a single term. Tansley, a life-long ecologist and naturalist with outstanding field experience, realised that all these components interact together and that the behavior of one component could not be understood without reference to what was going on in others. The concept can be represented by a diagram of interactions (Figure 6.1).

As the use of the word ecosystem has grown, interest has developed not just in the interactions that could take place within an ecosystem, although these remain a very important area of study, but in the mechanisms of these interactions (references to the development of the concept are given in Calow 1998). If an ecosystem really is a system, then there will be specific functions and processes taking place between the components that are crucial for the development and maintenance of the ecosystem (Odum 1953). Materials will be flowing from one component to another. The fate of any important single element should be able to be analyzed, and both its effects and its transfers and points of accumulation understood. This will be

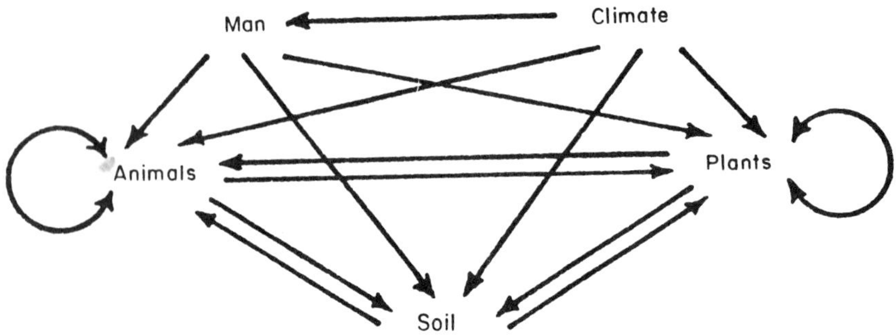

FIGURE 6.1. A diagrammatic representation of the interactions occurring in an ecosystem: a good beginning, but it does not tell us anything about the details of what is going on (Bradshaw 1983; reprinted by permission of Blackwell Science).

a way of understanding how the whole ecosystem works (Figure 6.2). An ecosystem should be able to be analyzed in terms of supply and demand, and even disposal, for particular resources, rather than in terms of some vague concept of "ecological interactions." As was wisely pointed out by Commoner (1971), "everything has to go somewhere."

The concept of the ecosystem has been immensely valuable. It has drawn us logically to study in detail the relationships and interactions occurring in communities, including the patterns of store and flow of essential materials. Because of this, the ecosystem is a concept that can appropriately be applied to cities. Cities are made up of living and interacting organisms, whose life and development depends on satisfactory supplies of many different materials and subsequent disposal of wastes (Bradshaw, et al. 1992). Surely we know this already? Thus, in what ways can an ecosystem approach reveal matters that would be valuable to an understanding of cities that we do not appreciate? Authors have suggested that cities are "super-organisms" (Girardet 1999; see also Melosi 2002, Chapter 12 in this volume). The concept of the city as an ecosystem is perhaps more appropriate.

The task of this chapter is to answer this question by looking at the natural ecosystems that exist alongside the human components of cities, and see what they have to tell us. We would certainly like to develop cities that are more sustainable than they are currently (Elkin, et al. 1991; Girardet 1999); that is, sustainable in the sense that they are self-sustainable—not requiring outside support. The natural ecosystems within cities manifest many attributes of sustainability, and are readily accessible and easy to study for investigators of all ages.

Natural Ecosystems in Cities

At first sight, nature and natural ecosystems may seem to have little place in the centers of cities. The original natural ecosystems have been destroyed, centuries or millennia before, and the obvious green elements (i.e., the parks, street trees, and gardens) seem to be almost completely artificial, with little place for anything very natural. Nature is adept, however, at exploiting any opportunity that is available to it and at developing new ecosystems in all sorts of natural or unnatural places. The raw moraines left by glaciers, or parcels of bare farmland deserted by their owners, which are so familiar in the Eastern United States, are very soon colonized by a wide range of plants. In cities it is remarkable how quickly nature can colonize vacant derelict sites. If these sites have retained their original soil, vegetation re-establishes in a few years; however, even where the soil has been lost and only raw stones, bricks, and broken concrete remain (e.g., in sites

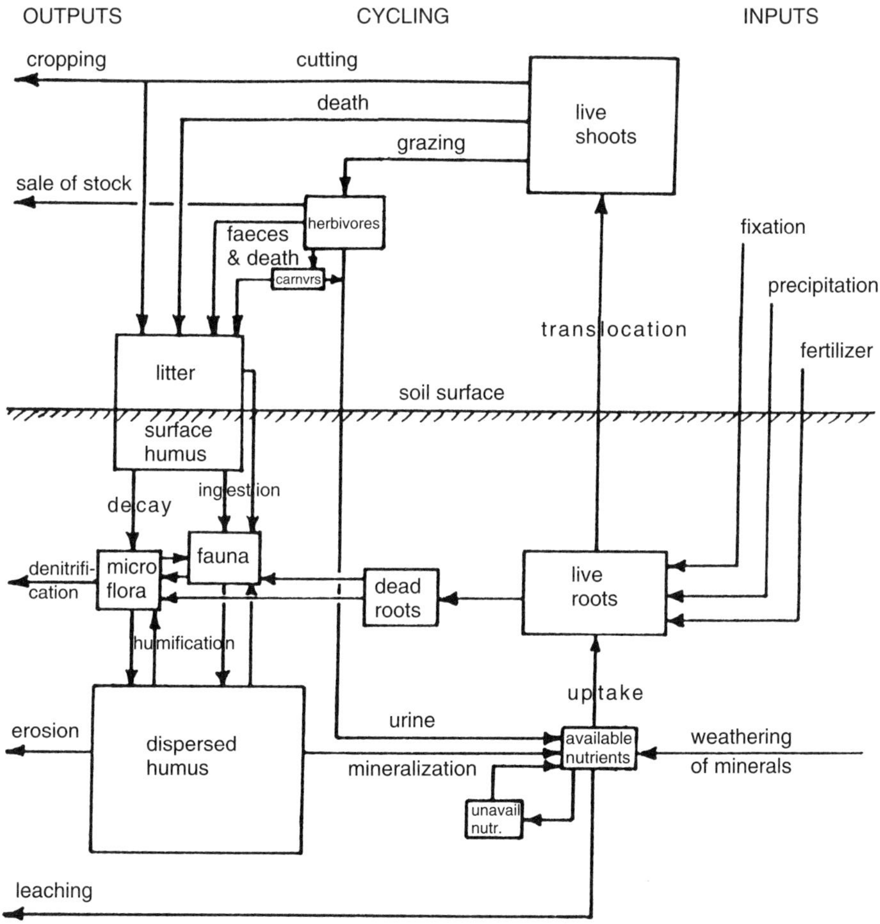

FIGURE 6.2. A diagrammatic representation of the flows and points of accumulation of a single important element, nitrogen, within an ecosystem: This provides a much more precise picture of what is going on. All the many other elements and supplies, however, need to be considered in the same way (Bradshaw 1983; reprinted by permission of Blackwell Science).

FIGURE 6.3. Natural succession on an urban site. Although the underlying material is brick rubble, a dense scrub has already developed within 15 years.

cleared of obsolete housing, vacant industrial sites, disused railway land and track, and disused paths and trackways), vegetation still establishes rapidly. Within 10 years or so, scrub and even trees become prominent, with a good complement of small mammals and birds (Bradshaw 1999) (Figure 6.3). This provides us with practical examples of the ecosystem processes that are fundamental in nature and relevant to what occurs in cities. These are processes that can be difficult to see and analyze. Having them on our doorstep makes study much easier.

Many of the ecosystems occurring in cities are natural in the sense that they have been made by nature, but they may well be semi-artificial in the sense that they may have been strongly affected by factors of human origin. This is not a problem, because it parallels what occurs in the wider world, where there are very few ecosystems not affected by human activity, whether burning, mowing or trampling, or just general disturbance. These effects give us an added area of study.

Ecosystem Development—Natural Succession

Natural ecosystems begin and develop through a process known as succession. This involves the arrival of species, their growth by the acquisition of resources, their interaction with other species, and the recycling of materials being produced by the growth and death of individual members of the species involved. This leads to a fully developed ecosystem. The final product can be termed a *climax community*, the end point of a complicated process. In areas with adequate moisture, this is usually some type of forest. It is widely considered that the climax is a stable state, in equilibrium with the climate. Ecologists, however, are now unsure whether this perceived stability is real. It certainly can be in the short term, but in the long term disturbances occur, the climate may change, and new species may be introduced from elsewhere, all upsetting the equilibrium. The climax is a useful idea, but is perhaps a theoretical rather than a practical concept. It is the processes that lead to ecosystem development that deserve our attention.

A number of distinctive, separately identifiable, processes are involved in succession (Table 6.1). The final, fully functional ecosystem depends on the contribution of each. If any one process fails, or contributes only weakly, then the final ecosystem will be limited in its characteristics. It may still be distinctive; indeed, it may be very distinctive because of its incomplete attributes. These can often tell what has gone "wrong."

The development of ecosystems and the processes involved are a study in their own right, but their utility for illuminating the development and the workings of cities is

TABLE 6.1. The essential steps in the process of natural succession in urban areas.

Ecosystem attribute	Processes involved
1. Colonization by species	Immigration of plant species Establishment of those plant species adapted to the local conditions
2. Growth and accumulation of resources	Surface stabilization and accumulation of fine mineral materials Accumulation of nutrients, particularly nitrogen
3. Development of the physical environment	Accumulation of organic matter Immigration of soil flora and fauna causing changes in soil structure and function
4. Development of recycling processes	Development of soil microflora and fauna Possible difficulties in urban areas
5. Occurrence of replacement processes	Negative interactions between species by competition Positive interactions by facilitation
6. Full development of the ecosystem	Further growth New immigration, including aliens
7. Arrested succession	Effect of external factors Reduction of development
8. Final diversification	The city as a mosaic of environments High biodiversity as a result

something we will explore. It will have to be in simple terms. Some of the analogies are fundamental and deep, however, and deserve more analysis than is possible in a short chapter.

Colonization by Species

If anything at all is to occur, then species have to get to the site. The way this occurs depends very much on the characteristics of the site and of the individual plant and animal species involved. Plants with light, windblown seeds get there first, (e.g., fireweed (*Chamaenerion angustifolium*) in London after the blitz of the Second World War); heavier-seeded species may take some time, unless there are external agents (e.g., birds) to carry them. Some examples are given in Table 6.2, but much depends on chance, particularly on whether the species are in the vicinity already. In the early stages, it is perfectly normal for some species that would be entirely suited to the situation to be missing. This in fact is the rule rather than the exception, so different sites may host very different plant communities.

TABLE 6.2. Plant species particularly characteristic of primary successions on wasteland in Liverpool, arranged in order of usual commonness, legumes indicated by (L); in many sites species indicated as belonging to different stages may occur together; due to chance, there may also be deviations from this list.

Calcareous wastes	Acid wastes
Early stages	
annual meadow grass, *Poa annua*	common bent, *Agrostis capillaris*
Oxford ragwort, *Senecio squalidus*	early hair grass, *Aira praecox*
yorkshire fog, *Holcus lanatus*	hawkbit, *Heiracium* species.
creeping bent, *Agrostis stolonifera*	birdsfoot trefoil, *Lotus corniculatus* (L)
white clover, *Trifolium repens* (L)	sheep's sorrel, *Rumex acetosella*
mayweed, *Matricaria recutita*	mosses, esp. *Polytrichum* sp.
suckling clover, *Trifolium dubium* (L)	lichens, esp. *Cladonia* sp.
Middle stages	
cocksfoot, *Dactylis glomerata*	birdsfoot trefoil, *Lotus corniculatus* (L)
buddleia, *Buddleia davidii*	wood rush, *Luzula campestris*
false-oat grass, *Arrhenatherum elatius*	wavy hairgrass, *Deschampsia flexuosa*
mugwort, *Artemisia vulgaris*	oval sedge, *Carex ovalis*
red clover, *Trifolium pratense* (L)	red fescue, *Festuca rubra*
bramble, *Rubus* species	rose-bay willow herb, *Chamaenerion angustifolium*
Late stages	
sallow, *Salix cinerea*	birch, *Betula pubescens/pendula*
birch, *Betula pubescens/pendula*	heather, *Calluna vulgaris*
common oak, *Quercus robur*	honeysuckle, *Lonicera periclymenum*
sycamore, *Acer pseudoplatanus*	common oak, *Quercus robur*
ash, *Fraxinus excelsior*	goat willow, *Salix caprea*
hawthorn, *Crataegus monogyna*	sallow, *Salix cinerea*

Source: From Bradshaw 1999.

In a similar way the cities and settlements of North America, and elsewhere, show that the development of human communities is the result of combinations of the characteristics of individuals with accidents of immigration.

This problem of colonization is readily observable through comparing what happens on different but similar sites (e.g., a set of urban clearance areas). It is sometimes easy to see that what has colonized is derived from a group of parental plants nearby. It is possible to see experimentally what happens when missing species are introduced artificially, easily done by adding seeds. Some spread rapidly in the new habitat, showing that the only limitation was their inability to get there (Ash, et al. 1994).

The species that arrive then have to establish; arrival is not enough. This requires the seedlings to be tolerant of the environmental conditions of the raw site. As a result, many species may arrive, but only a few establish. This can be tested again very easily by sowing a range of species. It is particularly interesting because a city offers a range of different environments. A site may be made up of very alkaline materials, full of lime and cement from demolished houses, or it may be nothing but acid materials, such as ashes originally used as ballast in a derelict railway yard. Some sites may be very fertile, others nearly sterile. Thus, although the same species may have arrived, very different species survive and develop, and contrasting sites can be very different, in ways that are not due to chance, but to ecological reasons (Table 6.2).

Similarly, human immigrants select particular areas in which to settle, related to their past experience or preferences. The historical ecology of settlement can throw considerable light on the origin and development of present day cities.

Growth and Accumulation of Resources

Plant immigrants cannot survive if they cannot accumulate resources and grow. The resources of carbon and oxygen are freely available from the air. Water is available from the soil, freely or intermittently. Nutrient elements (e.g., nitrogen, phosphorus, and potassium) are only available from the soil. Plants have extensive root systems by which they acquire the water and nutrients they need from the soil. They require different nutrients in different amounts. If any are in short supply they may not be able to flourish. In cities, surprisingly, important mineral nutrients such as calcium, magnesium, and potassium are not usually in short supply in city soils because they are common in building materials (Dutton and Bradshaw 1982).

In the same way, cities depend for their development on adequate resources being available, both for the development of their physical structure and for their trade and industry. The important difference from natural ecosystems is that human beings can forage at some distance from where they live and transport the resources they need back to their cities. It is this ability that allows the remarkable development of cities in comparison with natural ecosystems.

Nitrogen in natural ecosystems is a special case. It is not held in soil mineral matter, but only as a gas in air, and in soil organic matter, which itself is made up of the residues from plant growth. The plants cannot use either of these sources of nitrogen directly. The organic matter, however, decomposes slowly as a result of the activities of microorganisms, releasing small amounts of soluble mineral nitrogen in the form of nitrates and ammonium salts that the living plants can take up. Organic matter only accumulates where plants have been growing, and remains where it has been deposited on and in surface soils. Thus, neither organic matter nor the important nitrogen it contains are to be found in subsoils or in raw building materials. As a result, nitrogen is in short supply in most new urban sites, and is the resource most likely to limit growth.

This can readily be demonstrated where grass has established itself naturally, or been sown, on a cleared urban site and is growing poorly. The addition of nitrogen in a fertilizer produces immediate greening and increased growth. This growth is usually no better with complete fertilizer than with nitrogen alone, showing that nitrogen is the critical deficiency.

In many urban waste sites and lawns there are curious bright green patches. Sometimes these can be caused by an area of better soil. But usually they are compact patches—not more than 20 cm across—which would make this an unlikely explanation. The cause often is dogs (Figure 6.4). Fed on a high-protein diet, dogs have an excess intake of nitrogen, which is excreted in their urine and acts as a very effective fertilizer. Once dog patches have been seen and understood, their widespread occurrence shows how common a deficiency of nitrogen can be.

How, then, is the deficiency relieved naturally? It can be overcome artificially by adding nitrogen fertilizer or an organic manure. In nature the most important pathway is by nitrogen fixation—fixation of some of the abundant gaseous nitrogen in the air. If you wander over grassy areas in a city looking for dog patches, you will come across some green patches clearly associated with a plant. This is usually white clover (*Trifolium repens*) or another species belonging to the legume family. The plant itself is very green, but so is the associated grass. This is because the clover carries small nodules on its roots containing bacteria capable of fixing atmospheric nitrogen, transforming it into a soluble form that the plant can use. Such plants can fix as much as 100 kg N/ha/yr—as much as a good fertilizer dressing applied by a farmer. As the clover grows and then dies, this nitrogen accumulates in the soil organic matter until there is a substantial store which the developing ecosystem can draw on by the breakdown process already described. In temperate regions vegetation needs about 100 kg N/ha/yr to achieve reasonable growth. The organic matter breaks down at a rate of about 10 percent per year, so once the capital in the soil reaches about

FIGURE 6.4. Dog and dog patch. A conspicuous increase in growth due to dog urination, showing a serious deficiency in nitrogen in this urban grassland ecosystem.

1,000 kg N/ha, the ecosystem can be self-sustaining for nitrogen (Marrs, et al. 1983).

Nitrogen is a good model for the underlying importance of resources in the development of a viable city. If a city can find a resource—a material or skill—and develop its use, it can flourish in comparison with other cities that cannot find such a resource. In modern Europe, grants are made available from the European Union to its poorer areas to help them to regenerate—an equivalent of fertilizer. Nitrogen, however, is also a model for the more narrow characteristics of monetary wealth. As the ecosystem accumulates nitrogen by acquisitive processes it increases its capital, and it becomes more able to live on that capital and ultimately will be self-sustaining. Its nitrogen wealth accumulates until it becomes an ecosystem of independent means and no longer has to struggle.

All this suggests that the resource problem is always overcome in natural successions. This is far from true. Although hydrogen, oxygen, and nitrogen are available in nearly unlimited amounts in the air, their availability in biologically useful forms, and the availability of such mineral nutrients as phosphorus, potassium, and calcium derived from the soil can be inadequate in many situations, thus limiting ecosystem development in many places. Mineral nutrient deficiencies are difficult for a natural ecosystem to overcome, unless the nutrients are stored in the minerals making up the soil. If these minerals are not too resistant to chemical weathering, the nutrients will be released slowly by weathering and can be accumulated by plants (Likens, et al. 1977). Some plants, especially those growing in difficult situations, have remarkable systems to help this process. They have special relationships with certain fungi, known as mycorrhiza, in which the fungus lives either inside or just outside the root surface, gaining nourishment from the plant. In return the fungus, by ramifying through the soil, is able to collect such nutrients as phosphorus more effectively than the plant and transfer them back to the plant.

The analogy to a city way of life is obvious. All successful enterprises rely on agents to get out and find the commodities required, but, again, the mobility of human beings allows this to take place on a much more extensive scale than ever a plant or a whole ecosystem can. This unfortunately allows a well-endowed city to exploit a vast hinterland and make itself even richer, in a manner that a natural ecosystem cannot. The "ecological footprint" of London is 125 times London's surface area (Girardet 1999).

Development of the Physical Environment

The soil is a crucial vehicle for ecosystem development for obvious reasons. It has to be satisfactory not only in terms of its chemical makeup but also its physical properties. The raw soil of an urban clearance site on which natural succession is taking place may have serious physical problems—particularly compaction, although sometimes the opposite extreme, high porosity. The former prevents root penetration; the latter prevents the soil from retaining water.

In natural successions two things happen. Slowly, as plants begin to grow, organic matter accumulates at the surface from litter, and within the soil from the growth of roots. Quite apart from being a source of nitrogen this organic matter improves the soil structure, particularly by increasing water-holding capacity and by reducing compaction. This process is assisted considerably by the invasion of many different sorts of soil microorganisms—bacteria and fungi—and soil animals. These break up and decompose the organic matter. The larger soil animals disturb the soil particles and distribute the organic matter. The largest animals, the earthworms, are particularly effective and visible in urban successions. They make burrows and transport dead plant material down into the soil, and bring soil material to the surface at a rate of 4 mm/yr in a way well described by Darwin (1881). The products of the bacteria and fungi include mucilages that cement the soil particles into crumbs. As a result the soil becomes a loose, friable, crumbly material in which plants can root freely, which a few seconds with a spade will quickly reveal.

This is very much the model of a developing city because its infrastructure becomes built up by countless workers, providing services and supply routes, a large bulk of which are underground. Many people know only too well how difficult life can be in cities where these services have never been provided or have been destroyed by war.

Development of Recycling Processes

In a natural ecosystem it is inevitable that organisms die. Many parts of plants (e.g., leaves) are shed at the onset of winter. These all are the product of the growth process and therefore represent valuable resources, or sources of resources. If they were to remain unchanged the ecosystem would have to struggle to find more of the resources locked up in the dead material. But this does not happen. Piles of dead and discarded material do not accumulate in ecosystems. All new ecosystems progressively develop recycling systems, from the activities of the soil microorganisms we have already discussed (Swift, et al. 1979). Given time, all the organic matter decomposes, at an overall rate of about 10 percent per year in temperate regions. Although this varies with moisture and temperature, the rate is normally enough to ensure that the mineral elements locked up in the organic matter are released and reused on a continuous basis. The process can easily be studied by burying nylon bags containing litter within the soil and retrieving them at intervals. It is also not difficult to make comparative studies of carbon dioxide output produced by the breakdown process in different soils.

How different is this from what occurs in cities, where recycling even now is looked upon as something rather special, and recycling of not more than 20 percent of the total materials discarded is considered good (i.e., 80 percent never gets recycled but goes to landfill)? The major exception is iron and steel, where the total recycled is about 80 percent. Nature sets high standards, which now, at last, some individual countries are trying to follow.

The existence of a recycling system is visible in every developing ecosystem. Only a limited amount of organic matter is to be found accumulating on the ground in any ecosystem, and after the early development phase of the ecosystem is passed, no more organic matter accumulates. This is because the system reaches an equilibrium state when the rate of accumulation is equal to the product of the amount of accumulated material multiplied by its rate of decay.

This important equilibrium can, however, be upset if some factor disrupts the decay process. In cities this can happen where there is pollution, either due to acid rain from sulphur emissions or something more severe such as heavy metal contamination. Both of these types of pollution reduce microbial activity, and therefore decay rates, by 50 percent or more. As a result, organic matter accumulates at the soil surface. An interesting place to see this is in parks in old city centers, where one hundred years of acid rain have reduced soil pH from a normal 6 to a pH as low as 3 in some places. Under these circumstances a layer of peat can accumulate in the park's grassland due to lack of organic matter breakdown, and the grass itself may yellow and grow poorly because there is no recycling of the nutrients, especially nitrogen, that the grass needs. If the acidity is relieved by the addition of lime, for instance by the marking out of lines on athletic fields, the

FIGURE 6.5. Conspicuous green lines produced on poor acid grass turf in a park in Liverpool by the use of lime to mark out a football pitch. The lime has lessened accumulated acidity due to acid rain and allowed the release and recycling of nutrients in undecayed organic matter.

growth and colour of the grass is transformed as the accumulated organic matter begins to break down more rapidly, releasing its store of nutrients (Figure 6.5).

So although recycling is ubiquitous in nature, it can be upset, leading to serious consequences for the ecosystem. In modern cities recycling is still noticeable by its absence. The great man-made landfill sites are memorials to our inefficiency in running our city ecosystems. What has to be remembered, also, is that all that material has had to be replaced by material plundered from elsewhere. Lack of recycling has a double effect. Modern manufacturers are beginning to realize how all wastes represent an economic loss and are designing their factories to produce no waste at all (Frosch and Gallopoulos 1992).

Interaction Between Species—Replacement Processes

As time goes on, these urban ecosystems develop and change in species composition. What is most obvious is that the early colonists, mostly annual plants that can only persist by reseeding themselves every year, find no space to establish again. This space has been taken by perennial species, which each year hang on to the space they occupied in the previous year. Then larger growing perennial species squeeze out smaller varieties, very obvious when tree species arrive and develop. All this can readily be seen if an area of originally open ground is studied for a few years. Competition in natural succession is ruthless, and in the course of ecosystem development there is a progressive replacement of species.

Similar processes take place in human communities. Individuals and firms that do well early on in the life of a city rarely persist into a developed city. The traits that distinguish wild species, which are good colonists from those that are successful later, have been recognized and documented as *r*-traits, as opposed to *K*-traits (Begon, et al. 1986). Those who study cities could well examine the relevance of these traits to human beings.

Not all interactions between species, however, are competitive and negative. It is possible for some species to be helped by the presence of another—what is termed facilitation. In natural succession the effects of the nitrogen-fixing species such as clover on accompanying grasses, which we have already reviewed, are an excellent example. The facilitation provided by the clover may mean that the grass may grow so well that it suppresses the clover completely—facilitation leading to competition. Facilitation through N-accumulation can occur also among woody species. Alders (*Alnus* species), for instance, although not legumes, possess N-fixing microorganisms in conspicuous nodules on their roots and accumulate nitrogen rapidly. This can have a profound effect on the growth of other accompanying woody species (Kendle and Bradshaw 1992).

Another form of facilitation in the course of natural succession is the way woody species grow and create a hospitable environment for a variety of woodland species that find extreme open conditions difficult. The only problem is that although a favorable environment may have been created, the woodland species that could take advantage of it often are not present and/or they are poor dispersers. These problems have already been discussed.

In many ways the facilitation shown in natural ecosystems is less than that which can be found in human communities, whether cities or smaller communities. Cooperation and mutual help among human beings has always been a powerful force assisting development.

Full Development

The developing ecosystem progressively acquires more resources. The plants, especially the tree species, continue to grow and accumulate essential elements. An understory vegetation develops, comprised of shade tolerant shrubs and herbaceous species not found in the early stages. This can be observed on any developing site; however, the mixture of species is rarely as extensive as that found in a long established forest or woodlot. This is quite simply due to problems of immigration. Because of the urban conditions, there may be no sites containing these species nearby which could act as seed sources. At the same time many woodland species are well known for having only limited powers of dispersal.

Animals are different. Their mobility makes immigration easy, so if the habitat is suitable, they will come in. Small mammals such as shrews and voles arrive very early, leaving signs of their tracks as tunnels in the ground vegetation. These are followed by the animals that feed on them, such as foxes, badgers, and even owls. It is remarkable how quickly woodland birds come in—in the United Kingdom particularly robins, wrens, and blackbirds are favored by young scrub. It is remarkable the way some migrants, having flown three or four thousand miles from the south, are able to spot and select these developing woodlands even when they are firmly situated in industrial areas. In April developing stands of sallow (*Salix cinerea*), a species of willow very successful at colonizing urban wasteland in Britain, are alive with the sound of willow warblers staking out their territories after flying 3,000 miles from central Africa.

Immigrants are a conspicuous feature of human societies, adding greatly to their strength and diversity, and an important contributor to their development. Although human beings are better at building structures to protect themselves from summer or winter extremes, it is interesting to realize that in mountainous regions a whole system of summer immigration, transhumance, was widely practiced, which included most of the farm stock. In Wales it was from the "hendre"—the winter residence, to the "hafod"—the summer residence, usually 1,000–2,000 feet up in the hills.

Arrested Succession

All this suggests that the natural succession we can find in cities is a continuous process, always ending with woodland—the climax vegetation typical of most temperate regions. This is obviously not true, because large building developments are likely to be the endpoint of most city sites. But if we put that aside, and look at those sites where vegetation is allowed to remain, we can see many situations where the course of succession is interrupted. The two most common causes are fire and mowing. Fire can be accidental; mowing is the result of someone's decision to impose "tidiness." Both stop the development of woodland, and the succession usually remains as grassland. It is held at an "arrested" stage of succession. In natural conditions, or really semi-natural, this is a well known occurrence, not only where land is farmed and grazed intensively, but also in great areas like the prairies of the American upper Midwest, where succession was arrested by regular burning by the native population. When the burning ceased with the arrival of European settlers, the prairie, if not plowed, gradually disappeared into woodland or scrub.

Such arrested stages have an equilibrium and stability, although this is dependent on the occurrence of the arresting factor. They also have their own characteristic species, which can be different from those of a normal succession. Mowing, in particular, allows the colonization of species particularly adapted to it, such as daisies (*Bellis perennis*), plantain

(*Plantago lanceolata, P. major*), and creeping buttercup (*Ranunculus repens*) in temperate climates, familiar to many people as annoying weeds in their lawns.

In many ways, because human beings are so able to confront and overcome the difficulties they have had to face in their environment, arrested succession does not have a clear analogy in human societies, although it is obvious that societies that have developed in benign situations have done better than those in difficult situations. The success of freedom under a capitalist system has its model in the natural process of ecosystem development. But let us not forget that it is a somewhat ruthless process, with the species that are weaker for whatever cause being quietly eliminated or reduced to a subservient role.

We should also notice that in cities succession can be prevented by other, less natural, factors, particularly chemical pollution. Any site where little growth is occurring and there is no sign of a physical problem is suspicious. With higher environmental standards and controls, polluted sites are becoming rarer. But where they remain they are a source of considerable interest. Factories refining heavy metals such as lead, zinc, and copper have been built in many cities. The smelting involved released copious quantities of particles of the metals and their compounds into the air, which then were deposited on to the land surrounding the factory—usually with a radius of 1 km. These metals are poisonous to plants and can eliminate nearly all plants completely, leading to curious bare areas, which because the metals remain in the soil indefinitely can persist long after the factory has closed.

But a few plants are usually found in the bare areas. These are of species which have had the capacity to evolve tolerance to the metals. The actual populations in the polluted areas have a special tolerance not to be found in other populations of the same species growing in unpolluted areas. This can be demonstrated by trying to grow different samples of the same species in the polluted soil. The evolution of tolerance can occur quickly, in 5–20 years, and has become an excellent example of the way evolution by natural selection occurs (Macnair 1981). Similarly, when atmospheric conditions in cities were poor due to sulphur emissions, evolution of city populations of plants tolerant of sulphur dioxide has been shown. The ryegrass (*Lolium perenne*) from Central Park, New York, has actually been sold as a variety, Manhattan, useful for grass establishment in polluted parks.

Such evolution has no real analogy for human beings. Their evolution is too slow and protected, but that it can occur in city ecosystems is well known to pest managers ever since the brown rat, an all-too-successful denizen of cities, once easily controlled by the poison Warfarin, became resistant to it.

Diversity

All these processes lead to a great diversity of ecosystems in cities, often existing closely together, reviewed for Europe by Bornkamm, et al. (1980)

and Gilbert (1989). As they develop, each of these ecosystems tends to become more complex, even though their development may be restrained in certain directions. As a result there can be great diversity of ecological niches and resources in cities available for species to exploit. Many species (e.g., birds) are mobile enough to get in on their own. Others depend on chance introductions and may be aliens. Some quite alien species are brought in by people for special purposes, especially to beautify their gardens.

The result is that although wild and rather specialist species may be missing, cities are great havens for biodiversity, in terms of both ecology and species, even in industrial areas. London is an excellent example (Fitter 1945; Goode 1986). A survey of an area of Merseyside in England, which was the heart of the Industrial Revolution, has shown that, if all plant species are counted, biodiversity has actually increased over the last 50 years (Greenwood 1999a). At the same time this diversity is not static, but always changing as the result of the many changing factors influencing the landscape (Greenwood 1999b).

However, some of the alien species are not to be welcomed. Japanese knotweed (*Fallopia japonica*), for instance, is a dull, stemmy perennial plant about 2m tall that smothers and inhibits almost every species beneath it. It was introduced into Britain about 1880, and is now widespread in urban areas, carried from one place to another in soil and waste materials. By contrast, another alien, the butterfly bush (*Buddleja davidii*) from China, which is even more widespread as a colonist of wasteland, has long spikes of purple flowers very attractive to butterflies, and is much valued.

In a modern world widely influenced by transport of materials, alien species occur everywhere. North American cities are full of aliens from Europe. These species, often well adapted to city environments, add to diversity, even though they may have negative as well as positive effects. So it is with human societies, only more so. The success of the United States, as it was previously for Western Europe, is that it is a melting pot of people from different places, with different cultures and backgrounds and capabilities, woven together in a complex web in cities and neighborhoods. If plant and animal species are considered, however, we tend to be concerned about aliens because of the negative effects they can have on the indigenous species, sometimes to the extreme of extinction. This, of course, only mirrors the negative effects of the European immigrants on the indigenous people of North and South America, sometimes rather easily forgotten.

Conclusions

In the hurly-burly of modern existence, the wildlife in our cities is easily overlooked. It often is treated as something negative because it is untidy. Indeed urban wildlife sites can attract litter and rubbish, and be places for antisocial behavior. As a result it is constantly under threat, notably from those who seek for tidiness. Now, for seemingly good planning reasons, there is a concern to concentrate new development on old "brown field" areas. This is good for the regeneration of urban areas, but it eliminates important examples of what nature can do given the opportunity.

No policies are ever complete, however, and tidiness is not universal. There is also a continuous process of obsolescence and renewal in cities from which there is always a certain amount of unoccupied wasteland, in which all these steps in ecosystem development are available for study. Each small patch has within itself the characteristics of a human urban society. And the different patches with different histories and backgrounds can reveal the different ecosystem outcomes that can arise in comparative situations. From this a fascinating picture of the processes of action and interaction that are such an important part of natural communities, and that lead to biodiversity, can be built up. There has not been space to examine in detail how much these processes provide a provoking model for human urban societies, but perhaps the essentials have been illuminated. It would be interesting to examine elements that we ought more to incorporate in urban societies, and those we would prefer to reject.

Acknowledgments. I am grateful to the Institute of Ecosystem Studies for the support which enabled me to attend the eighth Cary Conference, and to many past students and colleagues without whose work and ideas this paper would not have been possible.

References

Ash, H.J., R.P. Gemmell, and A.D. Bradshaw. 1994. The introduction of native plant species on industrial waste heaps: a test of immigration and other factors affecting primary succession. *Journal of Applied Ecology* 31:74–84.

Begon, M., J.L. Harper, and C.R. Townsend. 1986. *Ecology: individuals, populations, and communities*. Blackwell, Oxford.

Bornkamm, R., J.A. Lee, and M.R.D. Seaward, eds. 1980. *Urban ecology*. Blackwell, Oxford.

Bradshaw, A.D. 1983. Ecological principles in landscape. Pages 15–36 in A.D. Bradshaw, D.A. Goode and E.H.P. Thorp, eds. *Ecology and design in landscape*. Blackwell, Oxford.

Bradshaw, A.D. 1999. Urban wastelands—new niches and primary succession. Pages 123–130 in E.F. Greenwood, ed. *Ecology and landscape development: a history of the Mersey Basin*. Liverpool University Press, Liverpool.

Bradshaw, A.D., R. Southwood, and F. Warner, eds. 1992. *The treatment and handling of wastes*. Chapman and Hall, London.

Calow, P., ed. 1998. *The encyclopedia of ecology and environmental management*. Blackwell Science, Oxford.

Commoner, B. 1971. *The closing circle*. Knopf, New York.

Darwin, C. 1881. *The formation of vegetable mould through the action of earthworms*. John Murray, London.

Dutton, R.A. and A.D. Bradshaw. 1982. *Land reclamation in cities*. HMSO, London.
Elkin, T., D. McLaren, and M. Hillman. 1991. *Reviving the city*. Friends of the Earth, London.
Fitter, R.S.R. 1945. *London's natural history*. Collins, London.
Frosch, R.A., and N.E. Gallopoulos, 1992. Pages 269–292 in A.D. Bradshaw, R. Southwood, and F. Warner, eds. *The treatment and handling of wastes*. Chapman and Hall, London.
Gilbert, O.L. 1989. *The ecology of urban habitats*. Chapman and Hall, London.
Girardet, H. 1999. *Creating sustainable cities*. Green Books, Dartington.
Goode, D. 1986. *Wild in London*. Michael Joseph, London.
Greenwood, E.F. 1999a. Vascular plants: a game of chance? Pages 195–211 in E.F. Greenwood, ed. *Ecology and landscape development: a history of the Mersey Basin*. Liverpool University Press, Liverpool.
Greenwood, E.F., ed. 1999b. *Ecology and landscape development: a history of the Mersey Basin*. Liverpool University Press, Liverpool.
Kendle, A.D., and A.D. Bradshaw. 1992. The role of soil nitrogen in the growth of trees on derelict land. *Arboricultural Journal* 16:103–122.
Likens, G.E., F.H. Bormann, R.S. Pierce, J.S. Eaton, and N.M. Johnson. 1977. *Biogeochemistry of a forested ecosystem*. Springer, New York.
Macnair, M.R. 1981. Tolerance of higher plants to toxic materials. Pages 177–208 in J.A. Bishop and L.M. Cook, eds. *Genetic consequences of man-made change*. Academic Press, London.
Marrs, R.H., R.D. Roberts, R.A. Skeffington, and A.D. Bradshaw. 1983. Nitrogen and the development of ecosystems. Pages 113–136 in J.A. Lee, S. McNeill, and I.H. Rorison, eds. *Nitrogen as an ecological factor*. Blackwell, Oxford.
Odum, E.P. 1953. *Fundamentals of ecology*. W.B. Saunders, Philadelphia, PA.
Swift, M.J., O.W. Heal, and J.M. Anderson. 1979. *Decomposition in terrestrial ecosystems*. Blackwell, Oxford.
Tansley, A.G. 1935. The use and abuse of vegetational concepts and terms. *Ecology* 16:284–307.

7
An Ecosystem Approach to Understanding Cities: Familiar Foundations and Uncharted Frontiers

NANCY B. GRIMM, LAWRENCE J. BAKER, and DIANE HOPE

The ecosystem concept has been one of the most useful in ecology, and also has been embraced by managers and the public in general (Likens 1992; Golley 1993). Even though there are disparities between ecologists and nonspecialists on exactly what constitutes an ecosystem, the potential utility of the concept when applied to urban systems where people live and work argues for redoubled efforts to bring the ecological concept of ecosystems, which is based on "systems thinking," into usage in education. We take the stance here that cities can be understood as ecosystems and that the ecosystem concept is highly appropriate to understanding both ecological and social dynamics (and their interactions) in cities. Our charge was to outline the conceptual foundations and explore the intellectual frontiers of urban ecosystem understanding, and to do this by describing what ecosystem ecologists mean by "city as ecosystem" and identifying the appropriate conceptual frameworks and their importance. Thus, our view emphasizes urban ecosystem research, although we will attempt where possible to point out the value of the approach to education. In particular, we will argue that certain key concepts—ecosystem; nutrients; input, output, and retention of materials; energy use; and heterogeneity—can be taught and learned based on material educators have close at hand: the urban ecosystem that surrounds them.

Our objective is to compare traditional ways of understanding ecosystems with the new perspectives that will be required to understand and study cities as ecosystems. We maintain that the ecosystem approach can be used to understand how cities work, how they interact with surrounding local and global ecosystems, and how expected changes in landscapes and regions resulting from increased urbanization will affect the future of Earth's systems. Moreover, we will argue that ecosystem study as we know it is necessary but not sufficient to understand urban ecosystems. Modifications of existing theory and practice will be required.

Ecologists often identify with one of two general approaches to their subject matter: a population-community approach or a process-functional (sometimes referred to as an ecosystem) approach (O'Neill, et al. 1986), although there has been great interest in merging these perspectives (Jones and Lawton 1994). In application to urban environments, one might distinguish between ecology *in* and ecology *of* cities in the same vein, with the appropriate caution that the two contrasts are not strictly analogous (Grimm, et al. 2000). Ecology *of* cities has to do with how aggregated parts sum, that is, how cities or parts thereof process energy or matter relative to their surroundings; whereas ecology *in* cities focuses on how ecological patterns and processes (especially populations and organismal interactions) vary within cities, or differ in cities compared with other environments. Whether one or the other approach applies can change as the scale of interest changes; for example, a study of the ecology *of* a schoolyard (an ecosystem in its own right) may become part of an investigation of ecology *in* a city when the larger scale is of primary interest. In contrast to the preceding chapter, here we adopt a conceptual framework of ecosystem science and use the ecology *of* cities approach. Specifically, we will ask two questions: How is energy use or consumption of a city or parts of a city dependent upon other ecosystems outside the boundaries under consideration? Is the city a source or a sink for nitrogen in the context of its surroundings, and what are the dominant inputs and outputs of this element?

Familiar Foundations: The Ecosystem Approach in Brief

What is an ecosystem? An *ecosystem* is a piece of earth of any size that contains biotic and abiotic elements, and has both intrasystem interactions and interactions with its surroundings. Necessary components of an ecosystem include boundaries, biota, and abiotic elements; ecosystem ecologists concern themselves with fluxes, interactions, and transformations of energy and materials, and controls of these processes. The concept of ecosystem is not free from controversy. The term was first coined in 1935 by English plant ecologist A.G. Tansley who, rejecting earlier notions of the "superorganism" promoted by Clements and Phillips, preferred to consider animals and plants as associations together with the physical factors of their surroundings as "systems" (Ricklefs 1990). Tansley (1935) outlined his concept of the ecosystem as follows:

> The more fundamental conception is, as it seems to me, the whole system (in the sense of physics), including not only the organism-complex, but also the whole complex of physical factors forming what we call the environment of the biome—the habitat factors in the widest sense. Though the organisms may claim our primary interest, when we are trying to think fundamentally we cannot separate them from their special environment, with which they form one physical system.

By the 1950s, the ecosystem concept had widely pervaded ecological thinking. Francis C. Evans (1956) provided this definition of ecosystem:

> In its fundamental aspects, an ecosystem involves the circulation, transformation, and accumulation of energy and matter through the medium of living things and their activities. . . . The ecologist . . . is primarily concerned with the quantities of matter and energy that pass through a given ecosystem and with the rates at which they do so.

This emphasis on the cycling of matter and the associated flux of energy is strongly associated today with the process-functional approach. Odum (1989) has further argued that an integral part of the ecosystem concept is a model of an open, thermodynamic nonequilibrium system, with the emphasis on the external environment.

Despite divergences and debates, Tansley's concept is still widely accepted, with the ecosystem having long been recognized as a fundamental organizational unit in ecology and a major structural unit of the biosphere (Krajina 1960). In modern ecology, we can distinguish between the *ecosystem concept* as defined in a widely used textbook (Begon, et al. 1990): "A holistic concept of the plants, the animals habitually associated with them, and all the physical and chemical components of the immediate environment or habitat which together form a recognizable self-contained entity," and an *ecosystem approach* (a particular branch of ecological research that emphasizes energy flow and material cycling and is characterized by systems thinking). Perhaps the fact that the ecosystem is an overarching and organizing concept that can encompass a variety of ideas within it, rather than being a single, coherent, tightly reasoned theory, makes it such a useful ecological paradigm (Kuhn 1962; Burns 1992).

Defining Ecosystem Boundaries, Structure, and Function

Ecosystem ecologists begin their studies of ecosystems by delimiting the *boundaries* of the system of interest. This may be relatively simple (e.g., the shoreline of a lake) or complicated by movements of organisms or materials (e.g., a stream). Alternatively, boundary definition may be accomplished with respect to the purpose of the study (e.g., a field or a forest patch of manageable size). One well-known example of boundary delimitation is that employed in the watershed approach (Likens and Bormann 1995). The watershed ecosystem is the area drained by a particular stream. Boundaries often are defined by identifying a discontinuity in physical, chemical, or biological processes (O'Neill, et al. 1986), and the watershed is a clear example of this method.

Despite the widespread adoption and use of the ecosystem concept, some have argued that it remains diffuse and ambiguous (O'Neill, et al. 1986), in particular because boundaries often are abstract (Sjors 1955; Fredericks 1958). There also has been debate about the question of spatial scale when defining ecosystems. Colinvaux (1973) argued that one could choose any size area, provided it has defined boundaries. Indeed, in landscape ecology the term has been applied across a range of spatial scales: "The ecosystem concept, which includes structure, function, and development, may be applied at any level of spatial scale, from the size of a rabbit dropping, to the planet" (Forman and Godron 1986). One development that may help to resolve this debate is *hierarchy* theory (e.g., Allen and Starr 1982; O'Neill, et al. 1986). In using the term *hierarchy*, ecologists most often are referring to multiple levels or multiple scales of ecological phenomena. Scientists studying the urban ecosystems of Phoenix and Baltimore in the context of two recently initiated, long-term ecological research (LTER) projects have espoused the importance of a hierarchical approach because it is capable of integrating across subject boundaries, as well as across spatial and temporal scales (Grimm, et al. 2000; Zipperer, et al. 2000; Grove, et al. 2002, Chapter 11 in this volume). By examining ecological phenomena in the context of a hierarchy, simultaneous attention to several scales or hierarchical levels is possible.

Once boundaries are established, the *structure* of the ecosystem is described, including the geophysical setting, plant and animal community structure, trophic relationships (i.e., who eats whom), soils and/or sediments, architecture (e.g., the layering of a forest or the shape, height, and arrangement of vegetation clumps in a desert), and storage pools of major elements. Measurement of biomass in different trophic levels, or of carbon storage in soil, plant, and animal matter, are examples of how structure may be quantified. Descriptions of ecosystem structure permit inferences about function or processes, although such inferences must be made with caution, accompanied by appropriate process measurements.

Ecosystem *function* refers to the processes that occur within ecosystems and the net result of those processes for the system as a whole. Questions that address function include: What are the key players in ecosystem processes? What factors control their rates? What diversity of processes is represented in the ecosystem? The two main elements of ecosystem function on which ecologists have focused their efforts are energy flow and material cycling, which in any ecosystem are governed by the laws of thermodynamics. In the realm of energy flow, for example, ecosystem ecologists measure rates of primary production (i.e., photosynthesis) or respiration, or secondary production of consumer organisms. Specific nutrient transformations within ecosystems, fluxes of materials across ecosystem boundaries, or retention of materials (i.e., the difference between inputs and outputs) may be the focus of material cycling studies.

In most early work on ecosystems, the system was viewed as spatially homogeneous, that is, as a "well-mixed reactor". Emergence of the field of landscape ecology, and integration of some of the ideas of landscape ecology into ecosystem studies, have changed this view. Landscape ecology focuses on patterns in heterogeneous tracts of land, and asks questions about both the causes and origins of those patterns and their consequences for processes (Turner 1989). Forman and Godron (1986) chose to distinguish between ecosystems and landscapes on the basis of a homogeneity criterion: "Although one may apply the ecosystem concept to a heterogeneous region, landscape, or landscape fragment, in this volume we basically limit its use to relatively homogeneous areas within a landscape." In this chapter, we adopt the position that the ecosystem approach is applicable both to a well-mixed reactor model and one that views the ecosystem as a more heterogeneous assemblage of parts or patches. The parts and patches themselves, for example, upland forest, riparian zone, stream, or wetland, might be viewed as ecosystems within larger ecosystems (watersheds). Thus, an ecosystem and its component parts could be treated as well-mixed reactors at some scales and heterogeneous systems at others. *Input–output budgets*, which ask whether an ecosystem retains (inputs > outputs) or releases (inputs < outputs) materials, are built on the well-mixed reactor model but also can be applied to different parts of a complex ecosystem and hence can yield information about *spatial heterogeneity* in material retention.

The Uncharted Frontiers of Urban Ecosystems

From this familiar ground, there are challenges at every step in applying the ecosystem approach to cities. For example, consider the structure of an ecosystem: A forest's architecture is a function of the growth forms of the mix of tree species that make up the forest and how they are constrained by topography, climate, soil fertility, and so forth. A city's structure is largely built and often designed. Even the "natural" components (e.g., trees in parks and in front and backyards) are subject to modification, rearrangement, and conscious or accidental design by humans. How can we apply a simple and elegant concept like the watershed to delineate urban ecosystems when flowpaths may be altered to such an extent as to be unrecognizable by conventional ecological techniques? Are urban streams so modified that they can no longer be reasonably compared with their "natural" counterparts using conventional ecological theory? If so, what changes in theory will be necessary?

The expansion of ecosystem research into ever more human-dominated environments, and in particular to cities as one extreme on a continuum from "pristine" to human-managed or human-defined ecosystems, represents an important test for the generality of the ecosystem concept itself. In some ways cities are like any other ecosystem: (1) the number of species, species diversity, and the number and types of species guilds is probably comparable to, or perhaps even higher than, surrounding ecosystems; (2) soils represent major storage pools of nitrogen and carbon relative to inputs; and (3) primary productivity (rate of photosynthesis), except in the most intensely urbanized parts of a city, is probably not appreciably different than it is in other ecosystems in the region. The following attributes, however, make cities unique: (1) they are heterotrophic (primary production <<< respiration) and extremely energy intensive; (2) they therefore require large inputs of energy and materials—the relative importance of external inputs to internal production and recycling is very high compared with all other types of ecosystems; (3) they produce copious amounts of waste compared with most ecosystems and often lack effective assimilation mechanisms to handle these wastes (or strain existing ones); (4) urban ecosystem function is controlled not just by biophysical factors but also by social and political forces (although this type of control now affects most ecosystems to some extent, it affects cities in a profound manner); and (5) one keystone species—humans—exerts overwhelming control on ecosystem processes. Because of these features, the study of urban ecosystems is likely to provide insights that will lead to refinement of many aspects of ecosystem theory.

As extreme examples of human-dominated ecosystems, can cities be defined adequately by physical attributes (e.g., landscape pattern), population densities, functional attributes (e.g., energy inputs per unit area), or some combination of these traditional variables? Do urban environments necessitate the development of an entirely different conception of the ecosystem, capable of integrating not only ecological but also socioeconomic, political, and cultural factors (Boyden 1977; Redman 1999)? Development of such integrated conceptual models in application to urban and other human-dominated ecosystems is proceeding on numerous fronts (Costanza 1996; Pickett, et al. 1997; Carpenter, et al. 1999; Grimm, et al. 2000), building upon a few notable urban investigations of the 1970s and 1980s that used an ecosystem approach (Boyden 1977; Boyden, et al. 1981), but a synthesis of this information is beyond the scope and intent of this chapter. We will identify here some of the obvious modifications that are needed to understand urban ecosystem boundaries, structure, and function, providing examples from our early experience in the Central Arizona–Phoenix (CAP) ecosystem.

Urban Ecosystem Boundaries

What are the boundaries of a city? Some features of urban boundaries are distinct and easily defined (e.g., the sharp edges that delineate new housing developments from desert in the Phoenix metropolitan area) (Figure 7.1). In other cases, where cities end and suburbs or rural lands begin is more difficult to ascertain. The Census Bureau (1995) defined "urban" for the 1990 census as comprising all territory, population, and housing units in urbanized areas and in places of 2,500 or more persons outside urbanized areas. Urbanized areas comprise one or more places ("central place") and the adjacent densely settled surrounding territory ("urban fringe") that together have a minimum of 50,000 persons (U.S. Census 1995). These population-based definitions may make little sense when considering, for

FIGURE 7.1. Phoenix's South Mountain, showing the distinct edge where urban development meets the desert. Photo by Ramón Arrowsmith, used with permission.

example, the mass balance of elements (e.g., the inputs minus outputs of carbon) for an ecosystem, but the most appropriate definition will of course depend on the question being asked in the study. Once the question is identified, however, an important first step is to define boundaries and to be consistent in their application to the question at hand. This may be a challenge when urban boundaries are rapidly changing due to urban growth or human migration patterns. In Phoenix's Maricopa County, the fastest-growing county in the United States, population has doubled twice since 1960, accompanied by an amoebalike spread of urban lands (Figure 7.2). Hence, boundaries of the CAP LTER area have been drawn far outside the current urban fringe, to account for anticipated future expansion. To construct a mass balance for nitrogen (N), for example (see later), the Salt River watershed was used, of which only 25 percent is urban or agricultural land (Baker, et al. 2001). This is akin to the treatment given Vancouver by Boyle and Lavkulich (1997) in their investigation of carbon storage in the lower Fraser River Basin in British Columbia, Canada: The urban ecosystem was subsumed within the larger watershed they considered.

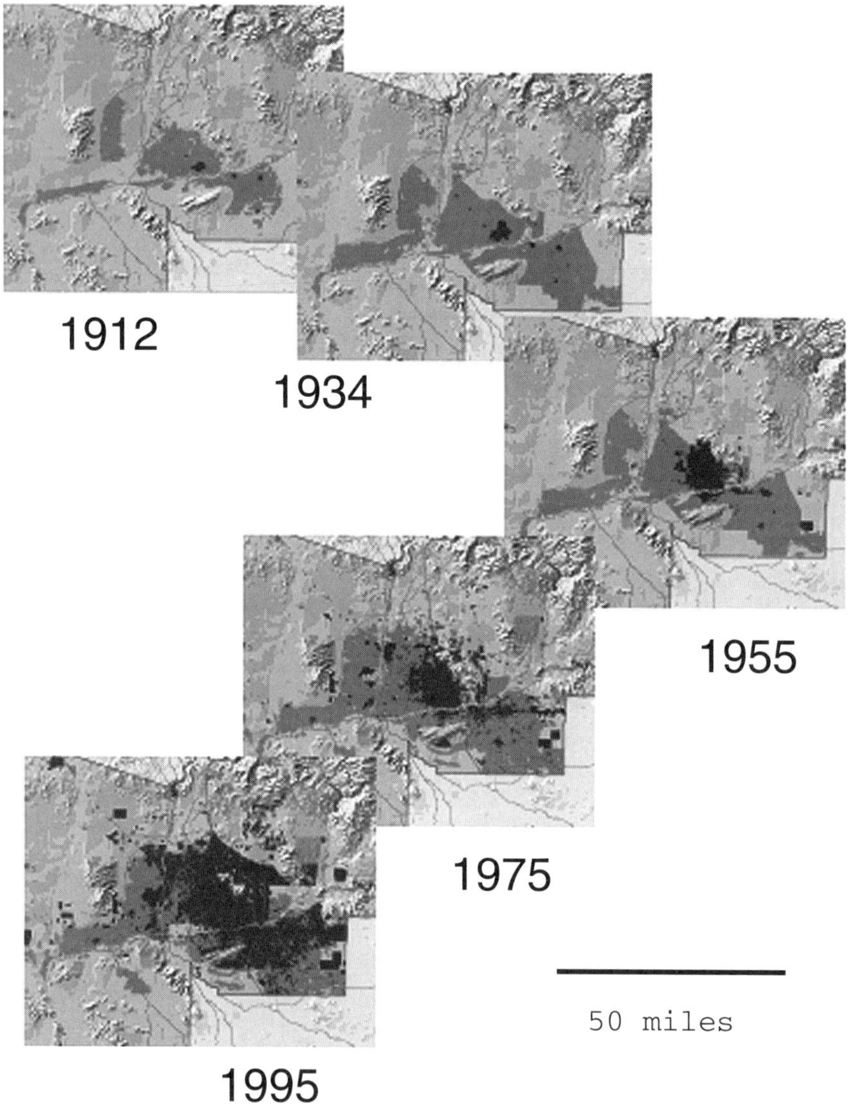

FIGURE 7.2. Changes in land use in the Central Arizona–Phoenix area from 1912 to 1995. Light gray—desert; medium gray—agricultural land; black—urban/suburban land. After Knowles-Yánez, et al. (1999).

Urban Ecosystem Structure

All of the measures of structure for nonurban ecosystems apply in cities, but there are additions; the most obvious of these is the built environment, including houses, buildings, roads, and service infrastructure (e.g., plumbing, wiring, and water delivery systems). These structural elements have characteristics that can influence heat budgets and material storage and transport in urban ecosystems (Newcombe, et al. 1978).

The so-called natural areas of cities, such as greenways, parks, preserves, and lakes and rivers, are often designed and intensively managed. Thus, they have characteristics that are unique in comparison to the rural hinterlands. Although the intention of landscape architects may be to create an environment that bears some resemblance to nonurban environments, often design features are selected because of their particular appeal to people, or because of some additional function that the design performs (such as flood control or retention of storm runoff). In Phoenix, many of the urban greenways and parks seem out of place in an arid environment, yet they share the label of urban "natural" areas with desert parks and preserves.

The demographic structure of the human population is another aspect of urban ecosystem structure that must be included. Such variables as age, sex, racial or ethnic group, and income are quantifiable from databases such as national censuses. Finally, there are unseen elements of structure with which ecologists have little experience: those associated with social institutions and culture. These might include, for instance, the economic system, the political system, cultural structures, and belief systems (Pickett, et al. 1997).

The urban landscape thus contains elements of natural ecosystem structure (species composition, trophic structure, vegetation architecture, soils, water), plus built structure, designed structure, and social structure. A promising way to deal with this complexity is through the landscape ecological approach of defining patches at a range of spatial scales (i.e., defining a patch hierarchy). *Hierarchical patch dynamics* models provide a relatively new way to look at complex systems that change through time (Wu and Loucks 1995). Processes are measured at a specific scale for fundamental units of the landscape at that specific scale. Those fundamental units are called patches, and their structure (sizes, arrangement, types) can be a major determinant of the processes. However, patch structure and arrangement also can change through time; hence, the models are dynamic. Furthermore, because patches can be identified at many scales, the models must also be hierarchical. Hierarchical patch dynamics models thus give us a way to consider patch dynamics simultaneously at multiple scales, rendering the complexity of urban systems more tractable. A fundamental question in both the Central Arizona–Phoenix and Baltimore urban LTER programs is: How does patch structure change with time and how does this in turn influence ecological patterns and the interaction between social and ecological spheres (Grimm, et al. 2000)? To answer this question, both projects are using hierarchical patch dynamics models as an important tool for integrative ecosystem research in urban settings (Zipperer, et al. 2000).

For element mass balances, characterization of flowpaths may serve as a useful measure of ecosystem structure. Much of the flow of water and materials is controlled within human-developed conduits that are often distinctly separated from natural flowpaths. In Phoenix, for example, all of the flow of the Salt River is diverted to canals for human utilization; the bed of the Salt River as it runs through Phoenix is now dry except during flood events (Graf 2000). Municipal water is distributed over the broad metropolitan area, and water that is not evaporated moves through sewer systems, where flows converge at a few large wastewater treatment plants and from there back to the natural river system. Agricultural water also is distributed widely, providing water to crops at a rate about 10 times higher than natural precipitation. Most of the agricultural water that is not evaporated (about 20% of input) enters the groundwater, providing most of the groundwater recharge. To prevent flooding, natural precipitation is collected and diverted to retention basins, where it too recharges aquifers. Thus, the paths of water flow in an urban ecosystem are very different from those of the natural system that preceded it.

Urban Ecosystem Function

Energy flow and nutrient cycling in urban ecosystems must conform to the same thermodynamic laws that apply to any other ecosystem (e.g., energy and matter cannot be created or destroyed, but they are transformed). A suite of social drivers, however, must also be considered and may prove to be as significant to ecosystem function as are biophysical variables. These include institutions and organizations, information flow, and cultural attitudes and perception (Grimm, et al. 2000). Perhaps the most obvious difference between urban and nonurban ecosystems is that urban ecosystems consume vastly more energy than they produce; that is, they are characterized by an extremely high energy expenditure (Newcombe, et al. 1978). Odum (1989) reported that energy consumption in urban–industrial ecosystems exceeds by one to two orders of magnitude that of even human-subsidized agricultural ecosystems. Why is energy expenditure so high in cities? It is high because in addition to plant, microbial, and animal (including human) respiration, cities have a hungry industrial metabolism. That metabolism is supported mainly by imported fossil fuels, as Stephen Pyne et al. (2001; p. 116) points out in a consideration of the ecology of urban fire:

> Modern cities remain fire-driven ecosystems. Fire's influence is everywhere, yet fire is almost everywhere invisible. . . . Cars, trucks, buses, motorcycles, tractors, backhoes, bulldozers, graders, generators, lawnmowers, the urban landscape overflows with a mechanical fauna that feeds on fossil fuels.

All of the human activities that lead to dependence on this imported energy, the economic, governmental, and social institutions that enable procurement of the needed energy, and finally, the connection between quality of life and the use of that energy, are rooted in social factors that fundamentally influence the function (metabolism) of urban ecosystems.

Accompanying this prodigious energy consumption in cities is production of wastes like CO_2, nitrogen oxides, sewage, solid wastes, and water and air pollutants (Boyden and Dovers 1992), in stoichiometric proportion to the materials imported and/or consumed. At the whole ecosystem scale, the magnitude of these flows of energy and matter is certainly a product of the activities of members of the dominant species (humans), but through collective behavior (i.e., the actions of social institutions), the species can modify material cycles beyond the summed total of those individual activities. In Phoenix, for example, the need to store and deliver water to meet human demand in this aridland city led to large-scale manipulation of the two large desert rivers that converge in the Phoenix basin. Manipulations included upstream impoundment for water storage, diversion into canals serving agricultural and municipal needs, and utilization of the river channel for gravel mining operations and as a recipient of treated wastewater. This tremendous alteration of hydrologic routing has no doubt altered the flows of materials in metropolitan Phoenix, and point additions of treated wastewater create a nutrient-enriched riverine system downstream from the city. A prevailing cultural attitude associated with these manipulations was that water development was essential for colonization of the American West (Reisner 1986; Gammage 1999), which led to implementation of policies at both local and national levels in strong support of large water projects such as those that made expansion of agriculture and later, the rapid population growth of Phoenix possible. An analysis of how social factors such as these might have influenced specific characteristics of material transport and cycling represents an important frontier for urban ecosystem understanding. Social factors also can ameliorate pollution effects. Changes in sewage treatment policy, for example, are clearly indicated in the long-term sediment record of Toolonlahti Bay, which receives inputs from the city of Helsinki, Finland (Tikkanen, et al. 1997).

The Ecosystem Approach and Potential Education Applications: Phoenix, Arizona

Energy expenditure in urban ecosystems is such that all cities can be considered to be extremely heterotrophic ecosystems (primary production < respiration), and therefore dependent upon import of energy. One well-known concept is that of the ecological footprint, an index that captures the essence of the dependence of a city on ecosystems outside it. The ecological footprint measures the productive land area required to continually produce all of the energy consumed in an ecosystem, without regard to where on Earth that production occurs (Wackernagel and Rees 1996). The ecological footprint has widespread appeal because of its apparent simplicity and comparability among cities, regions, or nations (see van den

TABLE 7.1. Approximate ecological footprint calculation for Phoenix.

Consumption of	Per cap. land consumption (ha)	Ecological footprint (km^2)
Food	1.55	34,000
Housing	1.06	23,000
Transportation	1.06	23,000
Consumer goods	1.06	23,000
Services	0.40	9,000
TOTAL	5.1	112,000

Source: Data on consumption are from Wackernagel and Rees (1996).
The area of the Central Arizona–Phoenix ecosystem (metropolitan area) is 2,000 km^2, thus, the ecological footprint for this ecosystem is 56 times its size.

Bergh, et al. 1999 for a critique). The ecological footprint of a city like Vancouver, BC, Canada, is 180 times the city area (Wackernagel and Rees 1996), whereas the ecological footprint of a well-known heterotrophic ecosystem, a forest stream (Bear Brook, NH), is just 31 percent of the stream's area (Collins, et al. 2000).

An estimate of the ecological footprint of Phoenix based on per capita data for the U.S. as a whole (Table 7.1) provides an interesting perspective on the dependence of Phoenix on production that occurs elsewhere. Because the current Phoenix metropolis covers approximately 2,000 km^2 of land area, a simple calculation based on published per capita energy use suggests that the ecological footprint may be some 56 times the size of the city itself. This is probably an underestimate, however, because in order to live in the arid southwest, Phoenicians expend vast amounts of energy to cool their homes and businesses and to bring water to the city. In addition, refinements of the ecological footprint concept that ask the question of how much land area is needed to absorb the wastes produced by a given population (e.g., Folke, et al. 1997) can extend the utility of the concept beyond energy consumption considerations (see also Chapter 8 in this volume; Luck, et al. 2001).

The ecological footprint is an heuristically useful tool, and for that reason it has seen widespread use by governments (Toronto, Canada; London, England; and all of the major cities of Australia are some examples, based on a cursory search of the World Wide Web) and has made its way into classroom curricula. One advantage of the approach from a teacher's point of view is its applicability at a range of scales, which can be defined based on social or ecological criteria (Table 7.2). Simple calculation procedures allow students to determine how changes in behavior at the individual, family, or neighborhood level can illustrate which are the critical variables in human energy use. For example, the effects of changes in eating habits might be compared with changes in driving patterns, revealing the much greater energy use (larger footprint) associated with the latter. Comparison of the ecological footprint among cities that differ in their climatic set-

TABLE 7.2. Scales of ecological footprints.

Geopolitical and cultural	Ecological
Individual	Individual
Household	Individual Property
Neighborhood	Land Use/Cover Patch[1]
School District	
City	Watershed
Metropolitan Area	
Region	Region or Large Watershed
Nation	Biome
Continent	Continent

[1] Patches defined on the basis of land use and cover might include residential areas (single family and multifamily), parks and preserves, industrial districts, urban core, agricultural areas, schools or other institutions.

tings (e.g., Phoenix and Baltimore), might show the greater summertime energy demand associated with air conditioning in Phoenix. More sophisticated exercises using the ecological footprint concept would examine the effects of different planning options (e.g., development of a public transportation system vs. freeway construction that encourages automobile use), or the effects of affluence (e.g., comparing industrialized to developing nations) on the dependence of regions or nations on external productivity. The point is that the ecological footprint, being transferable among scales and comparable among different situations, can illustrate the impact of individual human choices at a range of scales.

A Nitrogen Mass Balance for Phoenix

Mass balances are used by ecosystem ecologists to quantify inputs, outputs, and changes in storage pools of elements. In most terrestrial ecosystems, rates of input and output are small compared with rates of internal cycling, whereas in open ecosystems (e.g., streams), fluxes of nutrients across ecosystem boundaries are much larger than nutrient transformations within the ecosystem (e.g., Sprent 1987). A balance sheet of inputs and outputs for natural ecosystems includes atmospheric, hydrologic, and biologic vectors (Likens and Bormann 1995). The simplicity of the watershed approach is that hydrologic inputs are usually absent (i.e., there are no streams entering the watershed) and biologic inputs and outputs often are small; thus, retention can be measured as the difference between atmospheric deposition inputs and streamflow outputs.

Urban ecosystems introduce entirely new categories of inputs and outputs—those associated with human actions (Newcombe 1977). Baker, et al. (2001) constructed a nitrogen budget for the Central Arizona–Phoenix ecosystem that illustrates the dramatic quantitative and qualitative differ-

TABLE 7.3. Inputs and outputs of nitrogen for the CAP ecosystem.

Inputs
Natural Inputs
Atmospheric deposition
Surface water inflows
Biological N_2 fixation (desert vegetation)
Deliberate Human Inputs
Fertilizer
Human food
Animal feed
Fuels
Other imports
Biological N_2 fixation (alfalfa)
Human immigration
Inadvertent Human Inputs
NO_X production by fossil fuel combustion
Outputs
Deliberate Human Outputs
Crop exports
Meat and milk exports
Human emmigration
Inadvertent Outputs
Volatilization and denitrification
Surface water outflows
NO_X export in air
Change in Storage
Deliberate
Landfills
Vegetation (in part)
Built structure
Human population
Inadvertent
Groundwater and vadose zone
Vegetation and soils

ence in fluxes for an urban ecosystem compared with a nonurban ecosystem. One category of inputs is the *deliberate* import of materials, such as the import of nitrogenous fertilizer for agricultural production, which contrasts with natural inputs and outputs (Table 7.3). Furthermore, *inadvertent* inputs and outputs may make up a significant portion of the mass balance. For example, fixation of N_2 by fossil fuel combustion produces NO_X compounds. Even though the fate of this NO_X currently is unknown, it represents a huge input in comparison with the natural inputs of surface water inflow and atmospheric deposition for Phoenix (Figure 7.3), and probably for most cities.

One finding of the mass balance estimate for Phoenix (Figure 7.3; Baker, et al. 2001) is that the city must be accumulating nitrogen at a very high

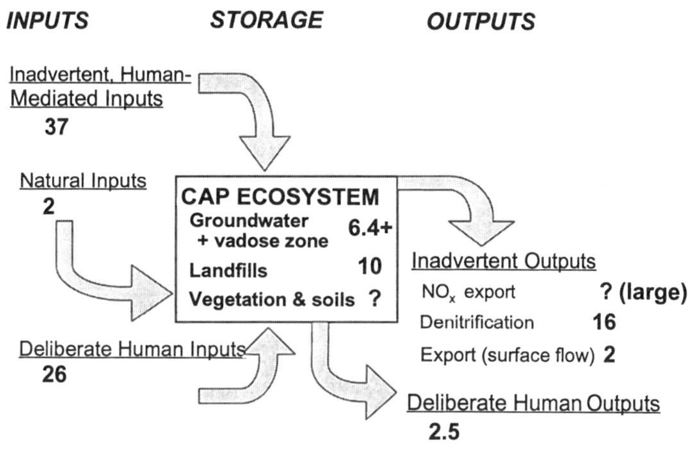

FIGURE 7.3. Nitrogen budget for the CAP ecosystem, showing major categories of inputs, outputs, and change in storage (see Table 7.3 for subcategories). All values in $kg\,N\,ha^{-1}\,y^{-1}$ (redrawn from data in Baker, et al. 2001).

rate (nearly $40\,kg\,ha^{-1}\,y^{-1}$, or $44\,lb\,acre^{-1}\,y^{-1}$). Some of this retention can be accounted for as increases in N stored in groundwater and landfills. The amount of N accumulation in other storage pools (especially vegetation and soils) is not yet known. Even if we do not consider the large, combustion-derived NO_X input, inputs still exceed outputs ($16.6\,kg\,N\,ha^{-1}\,y^{-1}$ or $18.6\,lb\,acre^{-1}\,y^{-1}$, compared with $39.3\,kg\,N\,ha^{-1}\,y^{-1}$ or $44.1\,lb\,acre^{-1}\,y^{-1}$ when NO_X inputs are included).

The history of development of the Phoenix metropolis and associated changes in land use in the region help to explain the causes and implications of the CAP ecosystem's present-day nitrogen accumulation. The Salt and Gila Rivers converge at the site of modern-day Phoenix, and, indeed, these rivers are the environmental feature that allowed establishment of a large ancient civilization (the Hohokam) in this hot, arid region (annual precipitation ~100 mm). The nitrogen budget was constructed for the lower portion of the watershed drained by the Salt River ($12{,}000\,km^2$). Since 1950 the human population has increased from 330,000 to more than 3 million inhabitants, largely due to immigration. Currently, 54 percent of water use in the CAP ecosystem is for irrigated agriculture, with 40 percent to municipal uses and the remainder to industrial uses (AZ Department of Environmental Quality 1994). Surface water from the Salt River supplies 48 percent of this water, whereas 27 percent is from groundwater and 22 percent from the Colorado River via the Central Arizona Project Canal.

Since the early 1900s, agriculture has played a key role in the development of Phoenix and surrounding municipalities (Gammage 1999). Urban expansion occurred largely at the expense of desert until the most recent 20 years, when agricultural land use has begun to decline as residential and commercial uses have expanded (Figure 7.2). As in most areas of the developed world, fertilizer use has increased dramatically since World War II, and, not surprisingly, increases in nitrate concentration have been observed in many groundwater wells in the region: concentration exceeds 10 mg/L nitrate-N throughout most of the area, and is as high as 50 mg/L in some wells. This increase in groundwater N contamination is cause for concern because levels exceeding 10 mg/L are considered a threat to human health. One potential use of the nitrogen mass balance, therefore, is to identify the major sources and accumulation zones for this element, so that changes in policy or behavior can be guided by scientifically based understanding. For example, an understanding of groundwater N accumulation could aid cotton production while at the same time reducing deleterious effects of fertilizer use. Irrigating crops with groundwater (approximately 27 percent of the water use in Phoenix is from groundwater sources, and most of that is used by agriculture) at a concentration of 20 mg nitrate-N/L, when supplied at a rate of 1.5 m/year (typical irrigation rate for cotton), would provide 300 kg N ha^{-1} y^{-1}. This is about 150% of the fertilization requirement (~200 kg N ha^{-1} y^{-1}) for cotton. Nitrate supplied in excess of crop requirements will leach back into aquifers, adding to net accumulation. Overfertilization also costs the cotton farmer, not just because of the expense of fertilizer and fuel, but also because high N levels inhibit boll formation and reduce the cotton crop.

The N mass balance also can be a powerful educational tool. We consider here the impact of a simple change in individual behavior on the nitrogen budget. Everyone who has children can appreciate the difficulty of convincing them not to waste their milk: It seems that the behavior of pouring one's milk down the kitchen sink when Mom or Dad is not looking is indeed one with a long tradition. It is a national problem—about 32 percent of the milk produced in this country is not consumed by humans and is therefore considered to be wasted somewhere between the dairy and a consumer's mouth (Kantor, et al. 1997). What impact would reducing this wastage have on one's personal nitrogen budget?

If milk wastage were reduced to zero, we could produce 32 percent less milk to satisfy our needs. This means that we would have 32 percent fewer cows and 32 percent less cow manure leached to aquifers. It would also mean that we could produce less high-protein grain concentrates to feed the cows, which in turn would mean that we would reduce fertilizer consumption, and therefore fertilizer leaching, by 32 percent. Overall, the effect would be to reduce the amount of N pollution created by a milk-drinking individual by about 2 kg/year (Table 7.4).

For comparison, per capita output of N to sewers in the CAP ecosystem, which includes human waste, detergents, and garbage grinder wastes, is about 7 kg N/year (Lauver and Baker 2000); however, this waste is treated to remove nitrogen before it is discharged to the Salt River channel. In the CAP ecosystem, the overall treatment efficiency for N removal in waste-

TABLE 7.4. Calculating the effect of avoiding milk wastage on an individual's N balance.

Budget term and reference	Current[1]	No waste[1]
Milk consumed (USDA 1999)	3.3	3.3
Wastage, as fraction (Kantor et al. 1997)	0.32	0.00
Milk produced	4.9	3.3
Feed required	13.3	9.0
Manure produced (N in feed that does not become milk)	8.4	5.7
Manure recycled as fertilizer[2]	4.2	2.8
Manure N leached to groundwater	2.1	1.4
N from alfalfa (33% of feed; Ensminger 1993)	5.0	3.4
N from concentrates (rest of feed)	8.2	5.6
Fertilizer (manure + chemical)	16.4	11.2
Chemical fertilizer (total fertilizer minus manure N)	12.3	8.3
Fertilizer N leached to groundwater	4.1	2.8
Total N leached to groundwater	6.2	4.2

[1] Units in kg N cap^{-1} y^{-1} unless otherwise indicated.
[2] Assuming 50% loss by volatilization and leaching.

water is about 75 percent, which means that only 1.75 kg N cap^{-1} y^{-1} actually reaches the river (Lauver and Baker 2000). Thus, if one individual stopped flushing the toilet, taking showers, and washing dishes, the reduction in N output to the environment would be 1.75 kg N/year.

This analysis shows that the apparently trivial action of eliminating retailer and household milk wastage would be more effective at reducing an individual's output of N to the environment (by 2 kg N/year) than would entirely eliminating his or her production of wastewater (1.75 kg N/year). Further analysis of the N budget is likely to reveal other simple and inexpensive methods of reducing N contamination of the environment that would not have been recognized by intuition or targeted by government pollution reduction programs. Moreover, the power of this approach is that it can be developed in the classroom and related directly to pupils' and their families' everyday lives.

Conclusions

We have presented examples from the Central Arizona–Phoenix urban ecosystem, promoting the view that we can apply familiar techniques of ecosystem ecology to cities, just as we would to any ecosystem. Given our charge to explore the intellectual frontiers of urban ecosystem understanding, it may be useful to consider whether our initial approach to mass balance should be modified. In particular, do we need to incorporate models of human behavior or economic drivers? The answer here is probably *yes* because the largest inputs are a consequence of human behaviors (e.g., driving patterns, fixation via combustion) and economics (e.g., imports of food, animal feed, and fuels). Would a different view of ecosystem structure improve our ability to put the mass balance to use in informing policy that will promote environmental protection? More fundamentally, would it improve our ability to predict the major points of production, accumulation, and transformation of nitrogen in the ecosystem? Again, we answer in the affirmative: Understanding how humans have manipulated paths of water flow within cities may be key to unlocking the dynamics of transport and transformation of materials.

We suggest that the next step in understanding urban ecosystems is to begin to incorporate social scientific explanations, controls, and mechanisms into our existing ecosystem models. Just as ecologists learned to speak the language of physical scientists when an understanding of climatic controls and changes was required for ecological explanations, we must now engage in a dialogue and sharing of conceptual models with the social sciences. With our new emphasis on the urban extreme along a spectrum of human-dominated ecosystems, the time is right to develop a more comprehensive ecosystem theory.

Acknowledgments. We thank the organizers for the invitation to participate in this stimulating Cary Conference. Suggestions and critique by reviewer participants from the conference were helpful in framing this contribution. The manuscript benefited from review by K. Hollweg and C. Nilon. Support from the National Science Foundation under grant number DEB 9714833 (CAP LTER) is gratefully acknowledged.

References

Allen, T.F.H., and T.B. Starr. 1982. Hierarchy. *Perspectives for ecological complexity*. University of Chicago Press, Chicago, IL.

Arizona Department of Water Resources. 1994. *Arizona water resources assessment*, Phoenix Active Management Area. Arizona Department of Water Resources, Phoenix, AZ.

Baker, L.A., D. Hope, Y. Xu, J. Edmonds, and L. Lauver. 2001. Nitrogen balance for the central Arizona–Phoenix (CAP) ecosystem. *Ecosystems* 4:582–602.

Begon, M., J.L. Harper, and C.R. Townsend. 1990. *Ecology—individuals, populations, and communities. 2nd Edition.* Blackwell Scientific Publications, Cambridge, MA.

Boyden, S., and S. Dovers. 1992. Natural-resource consumption and its environmental impacts in the western world. Impacts of increasing per capita consumption. *Ambio* 21:63–69.

Boyden, S., S. Millar, K. Newcombe, and B. O'Neill. 1981. *The ecology of a city and its people: the case of Hong Kong.* Australian National University Press, Canberra, Australia.

Boyden, S.V. 1977. Integrated ecological studies of human settlements. *Impact of Science on Society* 27:159–169.

Boyle, C.A., and L. Lavkulich. 1997. Carbon pool dynamics in the Lower Fraser Basin from 1827 to 1990. *Environmental Management* 21:443–455.

Burns, T.P. 1992. Ecosystem: a powerful concept and paradigm for ecology. *Bulletin of the Ecological Society of America* 73:39–43.

Carpenter, S., W. Brock, and P. Hanson. 1999. Ecological and social dynamics in simple models of ecosystem management. *Conservation Ecology* 3:4.

Colinvaux, P.A. 1973. *Introduction to ecology*. John Wiley & Sons, New York.

Collins, J.P., A.P. Kinzig, N.B. Grimm, W.F. Fagan, D. Hope, J. Wu, and E.T. Borer. 2000. A new urban ecology. *American Scientist* 88:416–425.

Costanza, R. 1996. Ecological economics: reintegrating the study of humans and nature. *Ecological Applications* 6:978–990.

Ensminger, M.E. 1993. *Dairy cattle science*. Interstate Publishers, Danville, IL.

Evans, F.C. 1956. Ecosystems as the basic unit in ecology. *Science* 123:1127–1128.

Folke, C.A., J. Jansson, and R. Costanza. 1997. Ecosystem appropriation by cities. *Ambio* 26:167–172.

Forman, R.T.T., and M. Godron. 1986. *Landscape ecology*. John Wiley & Sons, New York.

Fredericks, K. 1958. A definition of ecology and some thoughts about basic concepts. *Ecology* 39:154–159.

Gammage, G., Jr. 1999. *Phoenix in perspective*. Herberger Center for Design Excellence, Arizona State University College of Architecture and Environmental Design, Tempe, AZ.

Golley, F.B. 1993. *A history of the ecosystem concept in ecology*. Yale University Press, New Haven.

Graf, W.L. 2000. Locational probability for a dammed, urbanizing stream, Salt River, AZ. *Environmental Management* 25:321–335.

Grimm, N.B., J.M. Grove, C.L. Redman, and S.T.A. Pickett. 2000. Integrated approaches to long-term studies of urban ecological systems. *BioScience* 70: 571–584.

Jones, C.G., and J.H. Lawton, eds. 1994. *Linking species and ecosystems*. Chapman and Hall, Inc., New York.

Kantor L.S., K. Lipton, A. Manchester, and V. Oliveira. 1997. Estimating and addressing America's food losses. *Food Review* 20(1).

Knowles-Yánez, K., C. Moritz, J. Fry, C.L. Redman, M. Bucchin, and P.H. McCartney. 1999. *Historic land use: Phase I report on generalized land use*. Central Arizona–Phoenix Long-Term Ecological Research Contribution No. 1. Center for Environmental Studies, Arizona State University, Tempe, AZ.

Krajina, V. 1960. Ecosystem classification of forests. *Silva Fennica* 105:107–110.

Kuhn, T.S. 1962. *The structure of scientific revolutions*. University of Chicago Press, Chicago, IL.

Lauver L., and L.A. Baker. 2000. Wastewater reuse: Implications for nitrogen export and retention in the Phoenix–Central Arizona Project ecosystem. *Water Research* 34:2734–2760.

Likens, G.E., and F.H. Bormann. 1995. *Biogeochemistry of a forested ecosystem, 2nd edition*. Springer-Verlag, New York.

Likens, G.E. 1992. *The ecosystem approach: its use and abuse*. Ecology Institute, Oldendorf/Luhe, Germany.

Luck, M., G.D. Jenerette, and J. Wu. 2001. The urban funnel model and the spatially heterogeneous ecological footprint. *Ecosystems* 4:782–796.

Newcombe, K. 1977. Nutrient flow in a major urban settlement: Hong Kong. *Human Ecology* 5:179–208.
Newcombe, K., J.D. Kalma, and A.R. Aston. 1978. The metabolism of a city: the case of Hong Kong. *Ambio* 7:3–15.
O'Neil, R.V., D.L. DeAngelis, J.B. Waide, and T.F.H. Allen. 1986. *A hierarchical concept of ecosystems*. Princeton University Press, NJ.
Odum, E.P. 1989. *Ecology and our endangered life-support systems*. Sinauer Associates Inc., Sunderland, MA.
Pickett, S.T.A., W.R. Burch, Jr., S.E. Dalton, T.W. Foresman, J.M. Grove, and R. Rowntree. 1997. A conceptual framework for the study of human ecosystems in urban areas. *Urban Ecosystems* 1:185–199.
Pyne, S.J. 2001. Fire: A brief history. University of Washington Press, Seattle, WA.
Redman, C.R. 1999. Human dimensions of ecosystem studies. *Ecosystems* 2:296–298.
Reisner, M. 1986. *Cadillac desert: the American west and its disappearing water*. Viking Press, New York.
Ricklefs, R.E. 1990. *Ecology. 3rd edition*. W. H. Freeman and Company, New York.
Sjors, H. 1955. Remarks on ecosystems. *Svensk Botanisk Tidskrift* 49:155–169.
Sprent, J.I. 1987. *The ecology of the nitrogen cycle*. Cambridge University Press, Cambridge.
Tansley, A.G. 1935. The use and abuse of vegetational concepts and terms. *Ecology* 16:204–307.
Tikkanen, M., A. Korhola, H. Seppä, and J. Virkanen. 1997. A long-term record of human impacts on an urban ecosystem in the sediments of Töölönlahti Bay in Helsinki, Finland. *Environmental Conservation* 24:326–337.
Turner, M.G. 1989. Landscape ecology: the effect of pattern on process. *Annual Review Ecology and Systematics* 20:171–197.
U.S. Census Bureau. 1995. *Urban and rural definitions*. United States Census Bureau, Population Division.
USDA. 1999. *Continuing survey of food intakes by individuals*. U.S. Department of Agriculture, Beltsville, MD.
van den Bergh, J.C.J.M., and H. Verbruggen. 1999. Spatial sustainability, trade and indicators: an evaluation of the 'ecological footprint'. *Ecological Economics* 29: 61–72.
Wackernagel, M., and W. Rees. 1996. *Our ecological footprint: reducing human impact on the earth*. New Society Publishers, Philadelphia, PA.
Zipperer, W.C., J. Wu, R.V. Pouyat, and S.T.A. Pickett. 2000. The application of ecological principles to urban and urbanizing landscapes. *Ecological Applications* 10:685–688.

8
Understanding Urban Ecosystems: An Ecological Economics Perspective

William E. Rees

Framing the Analysis

Is an anthill an ecosystem? Is a cattle feedlot? These might seem like strange questions to begin a discussion of so-called urban ecosystems, but if we think about it for a moment there is more than mere metaphor in comparing cities with anthills and feedlots.

Anthills, feedlots, and cities share important systemic characteristics. Each of these entities is associated mainly with a single species—anthills with ants, feedlots with cattle, and cities with humans. All three are characterized by large populations and extraordinary densities of their keystone inhabitants. Most importantly, from the perspective of their defining species, none is ecologically self-contained nor self-sufficient—anthills, feedlots, and cities all are sustained by biophysical processes that occur mainly outside the boundaries of the named entities themselves. What, then, are the relevant ecosystems for ants, cattle, and people? This chapter addresses this question for people, particularly urban dwellers, from the perspective of human ecology and "ecological economics."

What Do We Mean by "Urban Ecosystems?"

We increasingly hear the term *urban ecosystem*, and there is now even a scientific journal called *Urban Ecosystems*. Nevertheless, the very idea remains ambiguous as illustrated by virtually any copy of the journal itself. Despite the fact that human beings are clearly the dominant species in the urban environment, a majority of the articles published focus on nonhuman plants and animals or on remnant "natural" ecosystems within the city. These studies analyze such things as changes in earthworm species composition with soil contamination, the pattern of vegetation across the urban density gradient, the habitat characteristics of urban scorpions or coyotes, or the fate of urban wetlands.

Such studies clearly cast the city as a somewhat unnatural habitat for nonhuman organisms. Indeed, it is the presumptive unnaturalness that makes the urban habitat interesting. To many ecologists, the "urban ecosystem" consists of the assemblage of nonhuman species *in the city*, and the purpose of inquiry is to determine how these species have adapted to the structural and chemical vagaries characteristic of the "built environment." People are implicitly involved, but merely as the external causative agents, the doers of the damage or the creators of the habitat to which the other species are adapting. In this sense, many of the papers are variations on impact ecology—the scientific issue is: How do the qualities of the built environment affect the structure and function of the (nonhuman) ecosystems it contains?

Other articles reverse the orientation and study the impacts of nonhuman ecosystems (e.g., urban forests and green open space) on the quality of the urban environment, or on urban micro-climates as they affect people. Still others examine the human economic or aesthetic values ascribable to nature in the city. These all are valid research perspectives and the questions they raise are clearly important, but they by no means tell the full story. Most significantly, they do not examine humans as integral components of urban ecosystems, nor do they study the role of "the city" in human ecology.

In this chapter, therefore, I adopt an explicitly human-centered ecological perspective in an exploration of so-called urban ecosystems. After all, cities are among the most spectacular of human creations, and many people see cities as *the* principal habitat of sophisticated modern "man". Almost half of humanity was living in cities by the year 2000 and the world's (human) urban population is expected to swell by an additional 2.1 billion people by 2025 (UN 1995; UNDP 1998). The urban expansion anticipated in the first quarter of the new century is the equivalent of the entire human population attained by the early 1930s. The scientific questions here are: What is the role of the city in human ecology? How should this be reflected in urban land use planning? What are the implications of urbanization—local and global—for ecological security and socioeconomic sustainability?

Defining Ecosystems

Understanding *urban* ecosystems requires some exploration of the ecosystems concept itself, and this leads immediately into difficult territory. Ecosystems are not discrete entities like apples or oranges. To some extent, they are a product of the human mind, constructed for purposes of analytic convenience. For example, boundaries between ecosystems are often indistinct and seemingly arbitrary. Although anyone can distinguish the deep forest from the open prairie, the line of demarcation between them (the "ecotone") may be diffuse, especially if there are no marked changes in topography. Ecosystems are also constantly in flux. Their species composition and structure changes over time in a natural "succession" of plant and animal communities, particularly after being perturbed by fire or human action such as clear-cutting (see Chapter 6). Where we draw the line between ecosystem types in space and time is therefore more often determined by human purpose than by any abrupt discontinuities in nature.

All this serves to emphasize that ecosystems should be understood more in terms of characteristic structural properties and functional relationships than by gross morphology. All ecosystems comprise living organisms existing in obligatory relationships with each other and with certain nonliving components of their physical environment. Any biotic community interacting with its environment such that the flow and dissipation of energy results in a defined trophic (feeding) structure, the emergence of biodiversity, and characteristic material cycles between the living and nonliving components may be considered an ecosystem (Odum 1971).

The living component of an ecosystem includes both producer and consumer organisms. The "producer" or autotrophic (self-feeding) organisms, mainly green plants, are able to fix solar energy and use it to manufacture complex chemicals (food) from simple inorganic chemicals (e.g., carbon dioxide and water) and a few nutrients (e.g., nitrate, phosphate). By contrast, "consumer" or heterotrophic (other-feeding) organisms obtain their energy and material requirements by feeding on other organisms or on organic detritus (i.e., by rearranging and degrading the complex chemicals originally photosynthesized by the autotrophs). The heterotrophs can be further divided into macroconsumers, mainly animals that eat other plants or animals; and microconsumers, mainly bacteria and fungi. The latter break down the complex compounds of dead (sometimes living) organisms, extracting some of the energy and material for their own growth and reproduction, and releasing organic nutrients back into the systems for reuse by the autotrophs.

The major nonliving components of most ecosystems include simple organic and inorganic substances, solar energy, a physical substrate, and the climatic regime (temperature and other physical factors).

The various components of ecosystems typically interact in a manner that results in the emergence of consistent and predictable functional relationships. The key relationships can be analyzed in terms of energy flows, food chains or food webs, and material (nutrient) cycles. Ecosystem relationships also produce discernable developmental and evolutionary patterns in time and space.

Two points from the preceding are particularly critical to interpreting ecosystem structure and function. First, a universal feature of ecosystems is the continuous recycling of nutrients between the autotrophic and heterotrophic organisms. Second is the generally high degree of interdependence, particularly causal linkages and obligatory relationships, among ecosystem components. Thus, even though there is often minor spatial or structural separation between system components in natural ecosystems (e.g., the vertical stratification between producers and consumers in forests and lakes), in functional terms, critical components are operationally inseparable from each other and the whole (Odum 1971).

What makes these latter qualities critical is that they characterize complete ecosystems as "autopoietic" or self-producing systems. "Living systems . . . [are] organized in a closed causal circular process that allows for evolutionary change in the way the circularity is maintained, but *not for the loss of the circularity itself*" (Maturana 1970, emphasis added). Thus, autopoiesis results from a network of production processes in which each component participates in the production of other components of the network through the relationships that specify the system (Maturana and Varella 1980; 1987).

Thinking Ecologically: Beyond Cartesian Dualism

With these concepts in mind, contemplate the following simple experiment designed to determine how typical urban residents conceive of the city. Ask a random sample of adults on the street to define "city." Hypothesis: Virtually no one will characterize the city as an ecosystem or even as part of the total ecosystem complex of which humans—and their cities—are a part.

Almost invariably, people describe cities as places characterized by great (human) population densities or as spaces dominated by buildings, streets, and other human-made artifacts (this is the architects' "built environment"). Some may think first of the city as a political unit with a defined boundary over which the municipal government has jurisdiction; the artistically inclined might see the city as a concentration of cultural, social, and educational facilities that would simply not be possible in a town or village; and the economically minded will describe the city as a place of intense exchange among individuals and firms, as the engine of national economic production and growth. (Indeed, Jane Jacobs [1984] famously described cities the basis for the "wealth of nations.") Only rarely, however, does anyone characterize the city in terms of its ecological structure or function, as an ecosystem, and certainly not as part of the human ecosystem. Although these qualities are as real and arguably as important to the human condition as the demographic, social, or economic dimensions of cities they go unremarked by the average citizen today. Modern humans are unaccustomed to thinking of themselves as ecological or even biological entities.

This observation illustrates the inordinate power of "Cartesian dualism" in maintaining the psychoseparation of humans from their natural roots. The legacy of the Enlightenment in western culture is a reductionist mindset that sees the human enterprise as somehow separate from and above the natural world (Hayward 1994). Such humanity/nature apartheid is even evident in most of our academic disciplines. For example, as previously noted, ecologists studying urban ecosystems tend to focus on nonhuman species as if people, the major inhabitants of the city, were not even part of the system. Similarly, neoliberal or neoclassical economists tend to treat the economy and "the environment" as separate systems, with the former functioning more or less independently of the latter (Daly 1991a) (Figure 8.1a).

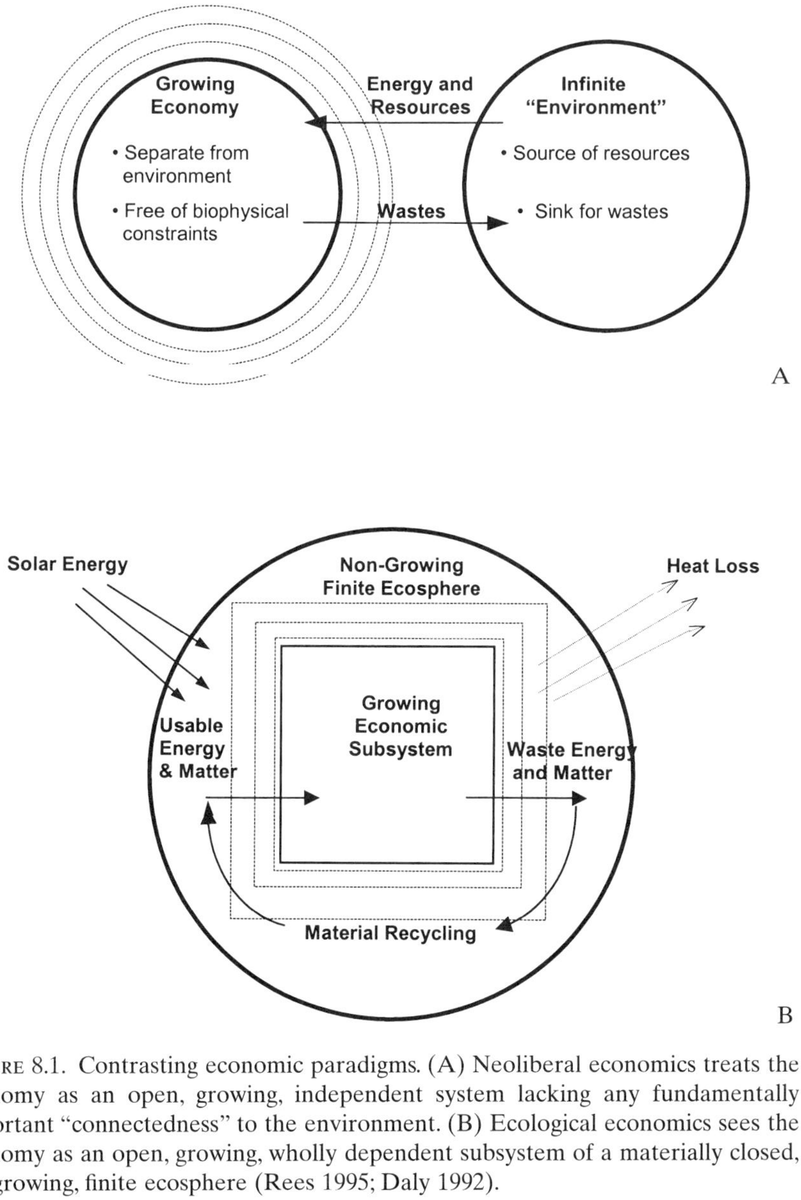

FIGURE 8.1. Contrasting economic paradigms. (A) Neoliberal economics treats the economy as an open, growing, independent system lacking any fundamentally important "connectedness" to the environment. (B) Ecological economics sees the economy as an open, growing, wholly dependent subsystem of a materially closed, nongrowing, finite ecosphere (Rees 1995; Daly 1992).

TABLE 8.1. Contrasting economic perspectives.

Neoliberal economics	Ecological economics
• Economic system is static, linear, deterministic	• Complex systems are dynamic, nonlinear, self-producing
• Economy separate from environment	• Economy a subsystem of ecosphere
• Models based on analytic mechanics	• Models recognize thermodynamics
• Substitutions are possible so there are . . .	• Complementarity dominates so there are . . .
• No limits to GDP growth	• Constraints on growth
• Analysis preoccupied with growth	• Analysis focused on development
• Efficiency orientated	• Equity oriented (intra- & inter-generational)
• Emphasis on production/consumption	• Emphasis on well-being (social capital)
• Short-term time frame	• Long-term horizon
• Favors monetary assessments	• Favors biophysical assessments

So it is that our scientific-industrial culture sees urbanization in the twentieth century mainly as a demographic transition driven by economics abetted by modern technology. We virtually ignore its ecological implications even though the mass "migration" of humans from all over the countryside into cities may well be the most ecologically significant phenomena since the emergence of *Homo sapiens* on the evolutionary scene.

This explains why I adopt an explicitly human ecological perspective—more precisely, an "ecological economics" perspective—in this chapter. Ecological economics was conceived, in part, to reconcile humankind with the rest of the natural world. It treats human beings not as outside nature, but rather as integral components of, and active participants in, the ecosystems that support them, i.e., as macroconsumers. From this point of view, much economic activity is really the material expression of human ecological relationships (Figure 8.1b). Ecological economics thus sees the human economy not as a separate system, but rather as an open, fully contained, and dependent growing subsystem of the materially closed, non-growing ecosphere (Daly 1992; Rees 1995). Table 8.1 compares the differing precepts and implications of ecological and neoliberal economics. The ecological economics framework better equips us to analyze cities—the economic engines of national economics—as ecological entities.

Ecological Economics and the "Second Law"

The second law of thermodynamics governs the unidirectional flow of energy through ecosystems. If we accept humans as ecological entities then we must also accept that human economic activity is governed by the second law (Georgescu-Roegen 1971; Daly 1991). This most fundamental of physical laws is central to ecological economics, but is virtually ignored by conventional economic models.

In its simplest form, the second law states that any isolated system will tend toward equilibrium; alternately, the "entropy" of any isolated system always increases. Available energy spontaneously dissipates, concentrations disperse, gradients disappear (Rees 1997a). An isolated system thus becomes increasingly unstructured in an inexorable slide toward thermodynamic equilibrium. This is a state of maximum entropy in which there is no structure and nothing can happen (Ayres 1994).

The second law was originally formulated for simple isolated systems close to equilibrium. We now recognize, however, that even open, far-from-equilibrium, self-organizing (autopoietic) systems are subject to the forces of entropic decay. It is clear, however, that not all such systems dissipate as expected. On the contrary, many biophysical systems, from individual fetuses to entire ecosystems actually gain in organizational complexity and mass over time (i.e., they *increase* their distance from equilibrium).

How can we reconcile this seemingly paradoxical behavior with the second law? The explanation lies in the functional relationships that develop among complex autopoietic systems in nature. Biophysical systems exist in loose, nested hierarchies, each component subsystem being contained by the next level up and itself comprising a chain of linked subsystems at lower levels. Kay, et al. (1999) define such complex hierarchic structures as "self-organizing holarchic open" systems. This arrangement enables each subsystem to maintain itself and to grow by importing available energy and material (essergy) from its host environment. Subsystems also export their degraded energy and material waste (entropy) back into their hosts. In effect, contemporary interpretations of the second law posit that all highly ordered self-producing systems develop and grow (increase their internal order) "at the [potential] expense of increasing disorder at higher levels in the systems hierarchy" (Schneider and Kay 1994). Because such systems maintain themselves by continuously degrading and dissipating available energy and matter, they are called "dissipative structures" (Prigogine 1997).

Economic "Production" as Consumption

The second law explains an important difference between ecological and economic perceptions of the economic process. Mainstream economists, who ignore thermodynamics, are preoccupied with increasing *output*, and see the economy mainly as a productive process. (Historically, they have considered pollution impacts as "external" to the market or pricing system.) By contrast, ecologists, who regard thermodynamics as fundamental, focus on resource *inputs* and waste outputs (potential pollution), and see the economy also as a consumptive process.

Indeed, ecologists would classify humans (along with all other mammals) as consumer organisms—macroconsumers to be precise. As noted, macroconsumers are large organisms, mainly animals, that consume other organisms (plants or animals) to satisfy their metabolic requirements (Odum 1971). There can be no dispute about this classification. Through population growth and technology, humans have inserted themselves as the dominant consumer organism in all the world's major ecosystem types, from grasslands and forests to rivers and the sea. (How many people conceive of humans as the most ecologically significant marine mammal?)

Humans, however, differ from other consumer organisms in important ways. Not only do we have a biological metabolism, we also have an industrial metabolism (Ayres and Simonis 1994). Like our bodies, all our toys and tools, factories and physical infrastructure, require continuous flows of energy and material from and to "the environment" for their production, operation, and maintenance. Thus, any complete material accounting for the human enterprise must factor in the "metabolic" demands and waste output of all our cultural artifacts. Many of these artifacts constitute what conventional economists refer to as "manufactured capital" or "the built means of production."

Of course, humans also are producers but there are significant differences between production in the economy and production in ecosystems. In nature, green plants are the "factories." They use an *extraterrestrial* source of relatively low-grade energy (light from the sun) and use it to assemble simple, dissipated chemicals (mainly water, carbon dioxide, and a few mineral nutrients) into the high-grade fats, carbohydrates, proteins, and nucleic acids upon which most other life forms depend. The assimilated solar energy is further degraded and dissipated in the process. Because they are "self-feeding" and use only dispersed (high entropy) substances for their growth and maintenance, green plants are called *primary* producers.

By contrast, humans are strictly *secondary* producers. To produce our bodies and all manner of economic goods and services, humans must extract large quantities of high-grade energy and material resources from ecosystems and other sources within the ecosphere. In short, all production by humans, from population growth to manufactured goods, services, and capital itself, requires the consumption of a larger quantity of energy and material *first produced by nature*. Thus, whereas the ecosphere develops and produces itself by dissipating solar energy, the economy grows by dissipating the ecosphere (Rees 1999a).

Are Cities Ecosystems?

We now have a framework from which to address the questions posed at the beginning of the chapter. Do anthills, feedlots—and cities—qualify as ecosystems? As previously noted, a universal structural feature of ecosystems is the coexistence of autotrophs and heterotrophs. Producer and consumer organisms (particularly micoconsumers) coexist in a mutually interdependent obligatory relationship, which ensures a cascade of energy and the continuous recycling of essential chemical nutrients through the ecosystem. Continuous energy and nutrient flows are essential if the system is to persist as a self-sustaining assemblage of components and relationships within a particular physical environment. In functional terms, autotrophs and heterotrophs are operationally inseparable from each other.

From this classical perspective, it is clear that neither anthills nor feedlots qualify as ecosystems. They may be systems at some level (e.g., as an economic unit in the case of the feedlot), but they are not ecosystems. Some of the defining parts (primary producers) are missing altogether and, certainly in the case of the feedlot, others (microconsumers) are insufficiently abundant for conditions in the artificial environment.

Let's stay with the feedlot for a moment. This system is dominated entirely by a single macro-consumer species, cattle destined for human consumption. The autotrophs that produce the feed for feedlot cattle (or pigs, or chickens, which are raised using even more constrained industrial methods) are located at great distance from the feedlot itself. Having separated the "operationally inseparable," industrial feedlotting usually short-circuits even the possibility of within-ecosystem organic decomposition or nutrient recycling. As a result, feedlots often accumulate vast quantities of manure, much of which is disposed of inappropriately, contaminating soils, surface, and subsurface waters. Such obvious ecological dysfunction is no small matter. In Canada, more nitrates are lost in manure that, because of distance, cannot be reapplied economically to crop and grazing land than is applied to crops in the form of artificial fertilizers (Canada 1991). Clearly, while industrial livestock operations may be economically viable for their operators, the unaccounted real economic and ecological costs born by society are considerable.

Cities, of course, are much more ecologically complex than feedlots. They certainly contain various ecosystems that, although greatly modified by human activity or inputs, include all the essential parts and function more or less normally. As already noted, such "urban ecosystems" are worthy objects of study because of adaptations their constituent species have made to the urban environment or because of their impacts on the quality of the urban environment for humans.

However, even though humans create cities, are the largest macroconsumer in the urban environment, and many consider cities to be the principal habitat for humankind for the foreseeable future, urban dwellers have an ambiguous relationship to most of the urban ecosystems referred to above. In-city ecosystems certainly serve as an aesthetic backdrop to human affairs and they may affect the quality of the urban physical environment either positively or negatively; however, people are arguably not a functional component of most urban ecosystems (backyard and community gardens excepted). In the main, modern cities to people are analogous to anthills for ants and feedlots for cattle. Humans may be the most obvious urban species, but the producer organisms that feed them, and the bulk of the microconsumers needed to complete the nutrient cycles of which they are a part, are located in other ecosystems at distance from the city. We can only conclude that even though urban ecology must consider humans, cities per se constitute only a small part of the total ecosystem complex needed to support urban human populations.

Ecological Footprint Analysis

How small a part can be shown using ecological footprint analysis. Eco-footprinting is an analytic tool designed to estimate the "load" imposed on the ecosphere by any specified human population. The metric used is the total area of productive land- and waterscape required to support that population (Rees 1992, 1996; Wackernagel and Rees 1996). Eco-footprinting is solidly based on the premises of ecological economics. For example, it: (1) recognizes that despite our technological wizardry, humans remain a part of nature and that the economic production/consumption process invariably appropriates the biophysical output of a finite area of terrestrial and aquatic ecosystems; (2) emphasizes biophysical (rather than monetary) measures of humankind-ecosystems relationships.

Eco-footprint analysis has helped to reopen the controversial issue of human "carrying capacity." For nonhuman species such as deer or cattle, carrying capacity (indicated by "K" in the logistic equation) is typically defined as the maximum population that can be supported indefinitely in a defined habitat without permanently damaging the habitat (Gever, et al. 1991; Meadows, et al. 1992). Because the physical environment is in constant flux, however, carrying capacity is often more theoretically than practically useful even for other animals, and economists have long rejected the concept as irrelevant to humans.

The economists argue that trade can usually relieve local resource shortages and, if that fails, technology can increase resource productivity or provide functional substitutes for specific goods and services of nature. Theoretically, these factors should remove any practical constraints on the growth of local populations or economic activity. According to some authors, human ingenuity has been so successful historically in pushing back the limits to growth that "the term 'carrying capacity' has by now no useful meaning" (Simon and Kahn 1984).

Eco-footprint analysis gets around the economists' argument simply by inverting the standard carrying capacity ratio: rather than asking how large population can live in a given area, eco-footprinting estimates how much area is needed to support a given population, wherever the relevant land is located. This approach recognizes that while trade enables increases in local populations, those populations are now dependent, in part, on the productivity of *distant* ecosystems. Thus, by shuffling resources around, trade increases total human load but *does not increase total carrying capacity*. Similarly, increasing technological sophistication has not decoupled the economy from the land. On the contrary, as we shall see, modern humans are arguably more land-dependent than ever. Carrying capacity is therefore still a valid concept and potentially limiting.

Methodological Premises

Eco-footprinting builds on traditional trophic ecology. We construct what is, in effect, an elaborate "food-web" for the study population by quantifying the material and energy flows supporting that population and identifying corresponding significant sources of resources and sinks for wastes. As previously noted, in compiling a human "food-web" we must account for the energy and material demands of not only our biological metabolism but also our industrial metabolism.

Eco-footprinting is further based on the fact that many material and energy flows (resource consumption and waste production) can be converted into land- and water-area equivalents. Thus, the ecological footprint of a specified population is the area of land and water ecosystems required on a continuous basis to produce the resources that the population consumes, and to assimilate the wastes that the population produces, wherever on Earth the relevant land or water is located (Rees 2001). A complete eco-footprint analysis would therefore include both the area the population "appropriates" through commodity trade and the area it needs to provide its share of certain free land- and water-based services of nature (e.g., the carbon sink function). Note that eco-footprint estimates are generally trade-corrected. For example, a population's consumption of wheat can be represented as follows: $\text{consumption}_{\text{wheat}} = \text{production}_{\text{wheat}} + \text{imports}_{\text{wheat}} - \text{exports}_{\text{wheat}}$.

Generally speaking, the ecological footprint in hectares (ha) for a specific consumption item is estimated by dividing total consumption of that item by the study population (kg) by the average yield of producing lands/waters ($\text{kg} \times \text{ha}^{-1}$). In the case of selected wastes, total output of a given waste is divided by average assimilation rates per hectare. The population's total eco-footprint is obtained by summing the footprints of all calculable consumption and waste items. We avoid double counting whenever it is recognized (e.g., leather is a by-product of beef production so is not separately estimated). Note that it is not generally possible to determine the actual locations of various ecosystems providing resources and services to a given study population. (Many wastes, for example, are generally spewed into the global commons. Carbon dioxide from Chicago is transported by the atmosphere and partially assimilated all over the world.) However, this affects neither the size of the eco-footprint estimate nor the reality that an equivalent aggregate area of real land or water is being used by that population, however widely it may be distributed across the planet.

A population's ecological footprint can be interpreted in "second law" terms. As noted, from the ecological perspective, humans and their economies are primary consumers. However, for sustainability, consumption by humans cannot persistently exceed production by the corresponding supportive ecosystems. Because biophysical production is essentially driven by photosynthesis (solar energy), a population's ecological footprint represents the area required continuously to generate photosynthetically the free energy and material (essergy) dissipated by that population's consumptive activities. In effect, it estimates the area of solar collector required to power the population's creative, maintenance, and growth-related activities.

Many factors bear on the ultimate area of a given population's eco-footprint, including the size of the population, the average material standard of living, the productivity of the land/water base, and the (technological) efficiency of resource harvesting, processing, and use. For practical reasons, only major categories of consumption (132 categories in recent calculations) and waste can be included in the analyses. Thus, most published eco-footprint calculations tend to be under- rather than overestimates. Fuller details of the method with examples can be found in Rees (1996); Rees and Wackernagel (1994); Wackernagel and Rees (1996); and Wackernagel, et al. (1999).

Urban Ecological Footprints

Recent eco-footprint analyses show that average citizens of North America, Europe, Japan, and Australia—the world's most intensely urban regions—require the biophysical output of 5–10 hectares (12–25 acres) of biophysically productive land and water *per capita* to support their consumer lifestyles (Wackernagel, et al. 1999). These findings should alter our perceptions about cities, urban ecosystems, and urban economies.

For example, in 1995 the author's home city of Vancouver, Canada, had about 472,000 residents living on a political footprint of 114 km^2 (11,400 ha). Assuming Vancouverites enjoy the average Canadian eco-footprint of 7.7 hectares *per capita* (Wackernagel, et al. 1990), we estimate that the aggregate eco-footprint of Vancouver is 3,634,400 ha, or 319 times its nominal area. Similarly, in 1996 Canada's largest city, the newly amalgamated Metro Toronto, had a population of approximately 2,385,000 and an area of 630 km^2. With a *per capita* footprint of about 7.6 hectares, the total ecological footprint of Toronto was about 181,260 km^2, or nearly 290 times larger than its political area (Onisto, et al. 1998). Finally, in an unusually comprehensive analysis, Folke, et al. (1997) estimated that the 29 largest cities of Baltic Europe appropriate for resource consumption and waste assimilation, an area of forest, agricultural, marine, and wetland ecosystems 565 to 1130 times larger than the areas of the cities themselves (depending on assumptions about waste assimilation). These studies show that the ecosystems *in* high-income cities constitute less than one percent, and as little as 0.1 percent, of the total ecosystem area required to support their human populations.

We have noted that traditional economics sees cities as the engines of the national economy—certainly the manufactured financial capital that produces most of the nation's money income is located in and around cities. However, ecological economics underscores the dependence of the urban economic machine on the 99+ percent of the supportive ecosystems that lie outside the city. Although cities produce the goods and services from which we earn our money incomes, stocks of natural capital in the countryside and the global commons produce the natural income—flows of resources and ecosystems services—upon which urban populations and the entire economic process feeds (see Box 8.1 for a fuller explanation of the natural capital/income concept).

Box 8.1: On Natural Capital and Natural Income

Natural capital refers to any stock of natural assets that yields a flow of valuable goods *or* services into the future. For example, a forest or a fish stock can provide a harvestable flow—timber and fish, respectively—that is potentially sustainable year after year. Similarly, natural capital provides such services as waste assimilation, erosion and flood control, and protection from ultraviolet radiation. (Thus, even the stratospheric ozone layer is a form of natural capital). The stocks that produce these flows are "natural capital" and the sustainable flows are "natural income." Because maintaining natural income often requires that the corresponding biophysical systems retain their capacity for "self-production," the integrity of such systems (particularly ecosystems) is an important attribute of natural capital.

There are three classes of natural capital. *Renewable* natural capital (e.g., living species and ecosystems) is self-producing and self-maintaining, and is generally derived from solar energy and photosynthesis. Renewable natural capital can yield marketable goods such as wood fiber, but may also provide unaccounted essential services when left in place (e.g., climate regulation). *Replenishable* forms of natural capital (e.g., ground water and the ozone layer) are nonliving, but, like living forms, are often ultimately dependent on the solar engine for renewal. Finally, *nonrenewable* kinds of natural capital (e.g., fossil fuel and minerals) are like inventories in that any use implies liquidating part of the stock.

Note that because adequate stocks of critical self-producing and replenishable natural capital are essential for life support and are generally nonsubstitutable, these kinds of natural capital are generally more important to sustainability than are nonrenewable forms.

Source: Liberally adapted from Costanza and Daly [1992] and Rees [1995].

The sheer scale of high-income eco-footprints produces some potentially troubling comparisons. A limited analysis for the International Institute for Economy and Development in London shows that the biophysical demands of London alone appropriate a productive area equivalent to all the ecologically productive land in Britain (IIED 1995). This result presages another key finding of ecological footprint analysis—most high-income countries have an ecological footprint several times larger than their national territories. In effect, these countries are running massive *ecological deficits* with the rest of the world (Rees 1996).

In fact, several studies suggest that urban industrial society is now running a *global* ecological deficit of 30 percent or more and is accumulating a nonrepayable ecological debt (Carley and Spapens 1998; Folke, et al. 1997; Wackernagel and Rees 1996; Wackernagel, et al. 1999). Even at current average levels of economic production and consumption, the human load exceeds the long-term carrying capacity of the Earth. Fisheries collapses, ozone depletion, greenhouse gas accumulation, and falling water tables are some of the better-known empirical trends showing that much of today's economic activity is derived from the unsustainable liquidation of natural capital.

As noted, many experts and ordinary people alike interpret the material abundance of the industrial world as evidence of humanity's growing *independence* of nature. According to a well-known growth advocate, the late Professor Julian Simon: "Technology exists now to produce in virtually inexhaustible quantities just about all the products made by nature." (Simon, cited in Bartlett 1996). Sometimes urbanization itself is taken as proof that people are throwing off the shackles of the land. These beliefs are illusion. Sustainability is unattainable from within the expansionist paradigm. The ecological reality is that productive croplands, pasture-lands, and forests everywhere are being used more intensely than ever to sustain the world's burgeoning urban populations. In fact, the average *per capita* and total loads imposed by people on the land have increased steadily with rising incomes since the beginning of the Industrial Revolution.

Reconsidering Urban Ecosystems: A Human Ecological Perspective

Great cities are planned and grow without any regard for the fact that they are parasites on the countryside, which must somehow supply food, water, air, and degrade huge quantities of wastes.—Eugene P. Odum (1971)

This chapter set out to explore the human element of "urban ecosystems" from an ecological economics perspective. What, then, have we learned? Several points flow from the preceding analysis. To summarize:

- While people create "urban ecosystems," and these systems provide the entire habitat for many nonhuman populations, urban ecosystems comprise only a tiny fraction (typically much less than one percent) of the total area of productive ecosystem required to support urban human populations.
- Urban dwellers play only a minor functional role within many "in city" urban ecosystems, but they are virtually the sole macroconsumers in vast areas of cropland, pasture, and forest outside the city scattered all over the world. Similarly, many wastes generated by people in the city are injected into the global commons—the atmosphere, rivers, and ultimately the oceans—for processing and possible recycling.
- This means that the ecosystems providing most of the biophysical life support for urban human populations are actually rural and other extra-urban ecosystems. Major cities (and many entire countries) are sustained largely on carrying capacity imported from the countryside and the global commons all over the world. Thus, people who live in cities are no less dependent on, and may be more demanding of, the productivity of rural landscapes than they would be if they were still living on the land.
- As is the case with livestock feedlots, cities lack key ecosystems components. The migration of people from the land to cities effectively disrupts human-dominated ecosystems by separating the "operationally inseparable" (primary production is spatially removed from consumption, and consumption from most subsequent decomposition). Most importantly, urbanization significantly alters natural biogeochemical cycles of vital nutrients and other chemical resources. It changes local, cyclically integrated ecological production systems into global, horizontally disintegrated, throughput systems (Rees and Wackernagel 1996). Thus, from the perspective of their main inhabitants, cities per se are not functional ecosystems.
- Cities are nodes of intense energy and material consumption and waste production embedded within a vastly larger global tapestry of productive land and water. In this singular sense, as Odum (1971) remarked, cities do resemble parasites in their relationship to the countryside. A parasite is an organism that gains its vitality at the expense of the vitality of its host. In thermodynamic terms cities are "dissipative structures" that grow and maintain their internal order using low-entropy energy and matter extracted from their host environments. They also discharge (dissipate) the resultant high-entropy waste back into their hosts. Because of their enormous energy and material throughput, cities generate most of the world's waste and their populations suffer the most heavily polluted environs on the planet.

Some critics might object to the emphasis here on cities' dependence (parasitism) on rural areas. One can argue, quite rightly, that the relationship is actually a two-way, mutualistic one. Certainly rural residents benefit from such things as urban markets, the products of urban factories, urban-based services, and technology transfers from urban areas; indeed, they freely trade the products of forest and field to obtain these benefits. It is also true, however, that while rural populations have always survived with or without cities, the ecological dependence of urbanites on rural environments is absolute. There can be no urban sustainability without rural sustainability.

That said, the above conclusions should not be construed as an argument against either cities or urbanization. Rather, the intent is to clarify the ecological role of cities in an increasingly human-dominated world. For example, knowing the extent to which cities currently undermine the very ecosystems that sustain them is essential to developing any strategy to reconcile cities with the rest of nature. Indeed, such knowledge is critical not only to the sustainable planning and development of cities per se, but also to developing overall strategies to reduce the human load on the ecosphere.

For example, the construction, operation, and maintenance of buildings presently accounts for 40 percent of the materials used by the world economy and for about a third of energy consumption (Worldwatch Institute 1995). This suggests that improvements in building technology, design, and maintenance alone could make a significant contribution to reducing the ecological footprints of cities (Rees 1999b). Moreover, the sheer concentration of population and consumption in cities gives them enormous leverage in dealing with the material dimensions of global sustainability. I call this the "urban sustainability multiplier" (Rees 1995). Some of the major advantages of cities include (Mitlan and Satterthwaite 1994; Rees and Wackernagel 1996; UNCHS 1996):

- Lower costs per capita of providing piped treated water, sewer systems, waste collection, and most other forms of infrastructure and public amenities;
- Greater possibilities for, and a greater range of options for, material recycling, reuse, re-manufacturing, and the specialized skills and enterprises needed to make these things happen;
- High population density which reduces the per capita demand for occupied land;
- Great potential through economies of scale, cogeneration, and the use of waste process heat from industry or power plants, to reduce the per capita use of fossil fuels for space-heating;
- Great potential for reducing (mostly fossil fuel) energy consumption by motor vehicles through walking, cycling, and public transit.

It is also important to recognize that ecological impacts that can be traced *to* cities are not necessarily the impacts *of* cities per se. Much of a city's ecological footprint is attributable to the consumption patterns of urban residents. What and how much we consume reflects prevailing social values as well as personal preferences and activities, and to a large extent these things are independent of where we live. Certainly if the fixed elements of an individual's footprint require the continuous output of two hectares of ecosystems scattered about the globe, it doesn't much matter where that individual resides (Rees and Wackernagel 1996). This means that while the ecological dysfunction caused by rural-urban spatial separation, and the sheer volume of waste production in cities are particular urban ecosystem problems that must be dealt with, modern industrial society itself may be inherently unsustainable.

This leaves unanswered the question of whether there is an ecologically optimal population distribution or settlement pattern. This is a complicated issue. For example, people often move to cities because of greater economic opportunities. To the extent that the higher average incomes result in increased average personal consumption (*net* of any savings resulting from urban agglomeration economies), the average urbanite's ecological footprint may well be greater than that of a typical rural resident in the same country. This would be true even though many categories of elevated urban consumption—higher clothing bills, cleaning costs, increased expenditures on security—do not contribute to net improvements in urbanites' welfare. These things nevertheless add to the individual's and therefore the city's total eco-footprint. Are today's cities inherently more or less sustainable than more dispersed settlement patterns? We cannot say with certainty on the basis of the mixed evidence to date.

We can argue, however, that urban economies should be much more efficient in their use of resources. Despite the absence of short-term market signals, there are two immediately compelling reasons for cities to strive to do more with less: the increasing probability of global climate change and the possibility of serious petroleum shortages in coming decades (Duncan and Youngquist 1999). Climate change may undermine agriculture and geopolitical stability, and therefore threatens the ecological and political security of urban populations dependent on reliable flows of resources from afar. Abundant cheap energy is a critical resource in itself and is the means by which humans obtain most other resources. Industrial society in general, and cities in particular, are the product of petroleum and may implode without it (Price 1995). Some energy analysts argue that given the continuing lack of suitable substitutes for myriad uses of petroleum as fuel and chemical feedstock, the life expectancy of our western techno-industrial society "is less than 100 years" counting from the 1930s! (Duncan 1993). Enhanced efficiency would both reduce carbon dioxide emissions (thus slowing climate change) and conserve oil supplies.

The prospect of global change and the needed efficiency revolution also argues for enhancing the self-reliance of urban regions (Rees 1997). With urbanization and an increasingly global market, many commodities and manufactured goods travel thousands of kilometers between the point of production and point of consumption. This involves a considerable and often unnecessary expenditure of energy and material on transportation and related infrastructure alone, and contributes significantly to urban ecological footprints. Increased regional self-reliance would produce immediate economic and ecological savings in the form of reduced transportation costs, lower carbon dioxide emissions, and fewer processing and storage facilities.

There are additional ecological advantages to local production for local consumption. The juxtaposition of production and consumption has the potential to restore, at least partially, the integrity of the total human-dominated ecosystem complex. For example, depositing urban organic compost on nearby farm and forestland would close the nutrient cycles broken by the spatial separation of rural ecosystem and urban populations today. This would make cities less destructive of their sustaining ecosystems and the money costs might be recovered by savings from lower artificial fertilizer bills and reduced pollution damage from both fertilizer and urban wastes.

All this suggests that we should even begin rethinking the city in a "whole-systems" framework. For completeness, should not the urban ecosystem as much as possible comprise both the consumption-based urban core and a production-oriented rural periphery? At the very least, there is an argument for rationalizing land uses within major urban regions to satisfy intra-regional urban demand. In this context, the bioregional philosophy of learning to live as much as possible "in place" has considerable appeal (Berg 1990; Sale 1985). The central idea of urban ecosystem planning would then be to reintegrate the geography of living and employment, of production and consumption, of city and hinterland. Such a transformed "homeplace, . . . rather than being merely the site of consumption, might, through its very design, produce some of its own food and energy, as well as become the locus of work for its residents." (Van der Ryn and Calthorpe 1986). In short, the urban ecosystem would finally become a complete ecosystem. Regrettably, the notion of regional self-reliance is anathema to those committed to the growth-based values of globalization and liberalized trade.

A Final Reflection: The Material Reality of Sustainability

As radical as such thinking may appear to today's liberal free-trading growth-bound mindset, it may be seen as excessively conservative in the near future. Up to one-half of the land on Earth has already been directly transformed by human action; more than half of the planet's accessible fresh water is being used by people; atmospheric carbon dioxide has increased by 30 percent in the industrial era; more atmospheric nitrogen is fixed and injected into terrestrial ecosystems by humans than by all natural terrestrial processes combined; two-thirds of the world's major fisheries are fully or overexploited; and biodiversity losses are accelerating (Lubchenco 1998; Vitousek, et al. 1997). The sheer scale of the human enterprise suggests the global economy is consuming the ecosphere from within. Contrary to the assumptions of expansionist thinking, so-called First World material lifestyles are not sustainably extendible to the entire world population in the foreseeable future.

Meanwhile, the present analysis has shown that human populations, however urbanized they may become, are increasingly in competition for the available load-bearing capacity of the planet. The wealthiest fifth of the human family appropriates the goods and life support services of five to ten hectares of productive land and water per capita to support their consumer lifestyles using prevailing technology; however, there are only about 2 ha of such ecosystems for each person on Earth (with no heed to the independent requirements of other consumer species). Carrying capacity studies by the Sustainable Europe Campaign suggest that to achieve sustainability, the world must lower its energy and material demands by approximately 50 percent in the coming decades (Carley and Spapens 1998). To clear sufficient "environmental space" to enable developing countries to grow, however, high-income countries must reduce their consumption of various nonrenewable resources by between 88 and 94 percent (Carley and Spapens 1998). This seemingly extraordinary challenge is supported by several other studies. For example, the United Nations Environment Program's recent "Global Environment Outlook 2000" report argues the need for a tenfold reduction in resource consumption by high-income countries (UNEP 1999). Even the Geneva-based World Business Council for Sustainable Development concurs that the industrial world must reduce material throughput and pollution by 90 percent by 2040 "to meet the needs of a growing world population fairly within the planet's ecological means" (Schmidheiny 1992).

The goal of sustainable economic growth in the Third World accompanied by rapidly declining material throughput in the First World is theoretically achievable through a massive increase in resource productivity (doing more with less) in combination with population control and an absolute reduction in material demand (i.e., the shift to simpler lifestyles). It is here that the ecological advantages of cities (the "sustainability multiplier") can really come to the fore. To achieve these subobjectives, however, unprecedented advances in technological efficiency, population policy, consumer values, material expectations, and governance are required. This in turn will require an unprecedented degree of moral courage, political leadership, and international cooperation. Powerful forces are raised in opposition to this endeavor, but success in achieving ecologically sustainable and equitable development would herald the final triumph of reason over technological hubris. It would also provide much-needed evidence that there is actually intelligent life on Earth.

References

Ayres, R.U., and U. Simonis. 1994. *Industrial Metabolism: Restructuring for Sustainable Development*. United Nations University Press, New York.

Ayres, R.U. 1994. *Information, Entropy and Progress: A New Evolutionary Paradigm*. AIP Press, Woodbury, NY.

Bartlett, A. 1996. The exponential function XI: The new flat Earth society. *The Physics Teacher* 34:342–343.

Berg, P. 1990. Growing a life-place politics. Pages 137–144 in V. Andruss, C. Plant, J. Plant, and S. Wright, eds. *Home! A bioregional reader*. New Society Publishers, Philadelphia, PA.

Canada. 1991. *The State of Canada's Environment*. Government of Canada (Supply and Services Canada), Ottawa.

Carley, M., and P. Spapens. 1998. *Sharing the world: sustainable living and global equity in the 21st century*. Earthscan Publications Ltd., London, UK.

Costanza, R., and H.E. Daly. 1992. Natural capital and sustainable development. *Conservation Biology* 6:37–46.

Daly, H.E. 1991a. The circular flow of exchange value and the linear throughput of matter-energy: A case of misplaced concreteness. Pages 195–210 in H.E. Daly, *Steady-State Economics*. Island Press, Washington, DC.

Daly, H.E. 1991b. *Steady-State Economics*. Island Press, Washington, D.C.

Daly, H.E. 1992. Steady-state economics: concepts, questions, policies. *Gaia* 6:333–338.

Duncan, R.C. 1993. The life-expectancy of industrial civilization: the decline to global equilibrium. *Population and Environment* 14:325–357.

Duncan, R.C., and W. Youngquist. 1999. Encircling the peak of world oil production. *Natural Resources Research* 8(3):219–232.

Georgescu-Roegen, N. 1971. *The Entropy Law and the Economic Process*. Harvard University Press, Cambridge, MA.

Gever, J., R. Kaufmann, D. Skloe, and C. Vörösmarty. 1991. *Beyond oil: the threat to food and fuel in the coming decades*. Third edition. Niwot, The University Press of Colorado.

Hayward, T., 1994. *Ecological thought*. Polity Press, Cambridge, UK.

IIED. 1995. *Citizen action to lighten Britain's ecological footprint*. (Report prepared for the UK Department of the Environment). International Institute for Environment and Development, London, UK.

Jacobs, J. 1984. *Cities and the wealth of nations*. Vintage Books, New York.

Kay J., H.A. Regier, M. Boyle, and G. Francis. 1999. An ecosystem approach for sustainability: addressing the challenge of complexity. *Futures* 31(7):721–742.

Lubchenco, J. 1998. Entering the century of the environment: a new social contract for science. *Science* 297:491–497.

Maturana, H. 1970. Biology of cognition. Reprinted in H. Maturana, and F. Varela. 1980. *Autopoiesis and cognition*. D. Reidel, Dortrecht, Netherlands.

Maturana, H., and F. Varela. 1980. *Autopoiesis and cognition*. D. Reidel, Dortrecht, Netherlands.

Maturana, H., and F. Varela. 1987. *The tree of knowledge*. Shambala, Boston.

Meadows, D.H., D.L. Meadows, and J. Randers. 1992. *Beyond the limits: global collapse or a sustainable future*. Earthscan Publications, London, UK.

Mitlin, D., and D. Satterthwaite. 1994. *Cities and sustainable development*. Background paper prepared for "Global forum '94", Manchester, 24–28 June 1994. International Institute for Environment and Development, London, UK.

Odum, E.P. 1971. *Fundamentals of Ecology*. Third edition. Saunders, Philadelphia.

Onisto, L.K., E. Krause, and M. Wackernagel. 1998. *How big is Toronto's ecological footprint?* Centre for Sustainability Studies and the City of Toronto, Toronto.

Price, D. 1995. Energy and human evolution. *Population and Environment* 16: 301–319.

Prigogine, I. 1997. *The end of certainty: time, chaos, and the new laws of nature*. The Free Press, New York.

Rees, W.E. 1992. Ecological footprints and appropriated carrying capacity: what urban economics leaves out. *Environment and Urbanization* 4(2):120–130.

Rees, W.E. 1995. Achieving sustainability: reform or transformation? *Journal of Planning Literature* 9:343–361.

Rees, W.E. 1996. Revisiting carrying capacity: Area-based indicators of sustainability. *Population and Environment* 17:195–215.

Rees, W.E. 1997a. Is "sustainable city" an oxymoron? *Local Environment* 2:303–310.

Rees, W.E. 1997b. Urban ecosystems: the human dimension. *Urban Ecosystems* 1(1): 63–75.

Rees, W.E. 1999a. Consuming the Earth: The Biophysics of Sustainability. *Ecological Economics* 29:23–27.

Rees, W.E. 1999b. The built environment and the ecosphere: A global perspective. *Building Research and Information* 27(4/5):206–220.

Rees, W.E. 2001. Ecological Footprint, Concept of *Encyclopedia of Biodiversity* (Simon Levin, Editor-in-Chief). Vol 2, pp. 229–244. Academic Press, San Diego.

Rees, W.E., and M. Wackernagel. 1994. Ecological footprints and appropriated carrying capacity: measuring the natural capital requirements of the human economy. Pages 362–390 in A.M. Jansson, M. Hammer, C. Folke, and R. Costanza, eds. *Investing in natural capital: the ecological economics approach to sustainability*. Island Press, Washington, DC.

Rees, W.E., and M. Wackernagel. 1996. Urban ecological footprints: why cities cannot be sustainable and why they are a key to sustainability. *Environ. Impact Assess. Rev*. 16:223–248.

Sale, K. 1985. *Dwellers in the land: the bioregional vision*. Sierra Club Books, San Francisco, CA.

Schneider, E., and J. Kay. 1994. Life as a manifestation of the second law of thermodynamics. *Mathematical and Computer Modeling* 19(6–8):25–48.

Simon, J., and H. Kahn, eds. 1984. *The resourceful earth: a response to global 2000*. Basil Blackwell, Oxford.

Schmidheiny, S. (with the Business Council for Sustainable Development). 1992. *Changing course: a global business perspective on development and the environment*. MIT Press, Cambridge, MA.

UN. 1995. *World urbanization prospects: the 1994 revision*. The United Nations, New York.

UNCHS (United Nations Centre for Human Settlements). *An urbanizing world: global report on human settlements 1996*. Oxford University Press, Oxford and New York.

UNEP. 1999. *Global Environmental Outlook 2000 (GEO-2000)*. United Nations Environmental Program, Nairobi.

UNDP. 1998. *Urban transition in developing countries*. United Nations Development Program, The United Nations, New York.
Van der Ryn, S., and P. Calthorpe. 1986. *Sustainable communities: a new synthesis for cities and towns*. Sierra Club Books, San Francisco, CA.
Vitousek, P.M., H.A. Mooney, J. Lubchenco, and J.M. Melillo. 1997. Human domination of earth's ecosystems. *Science* 277:494–499.
Wackernagel, M., and W.E. Rees. 1996. *Our ecological footprint: reducing human impact on the earth*. New Society Publishers, Gabriola Island, BC and Philadelphia, PA.
Wackernagel, M., L. Onisto, P. Bello, A.C. Linares, I.S.L. Falfán, J.M. Garcia, A.I.S. Guerrero, and M.G.S. Guerrero. 199.9. National natural capital accounting with the ecological footprint concept. *Ecological Economics* 29:375–390.
Worldwatch Institute. 1995. *State of the world, 1995*. Worldwatch Institute, Washington, DC.

9 Social Science Concepts and Frameworks for Understanding Urban Ecosystems

Carolyn Harrison and Jacquie Burgess

Introduction

Cities are a potent demonstration of humanity's domination of nature; they are also the source of a wide range of environmental problems that enmesh city residents in a process of globalisation capable of touching even the most remote and rural of communities. In the context of the agreement reached at the International Summit at Rio in 1992 that all nations should move in the direction of sustainable development, cities also have a critical role to play in determining the rate and nature of that change. For example, were city residents to adopt more pro-environmental lifestyles, then considerable progress would be made towards achieving sustainable development. Against this background, education and communication strategies which seek to promote understanding of the linkages between how people live their lives and the quality of our environment have a potentially important role to play in moving society in the direction of sustainable development.

The purpose of this chapter is to explore some of the concepts and frameworks social scientists use to understand how city residents make sense of their own attitudes, values, and behaviors toward the environment. It does this first by drawing on recent research in the social sciences that support contextualist approaches to society, and second by using the findings of a cross-cultural study undertaken in two European cities: Nottingham in the United Kingdom and Eindhoven in the Netherlands. This study was designed to compare how local residents and decision makers in each city discuss their responsibilities and behaviors toward the environment. By offering a cross-cultural comparison, the study serves to highlight the role that social, political, and cultural factors play in influencing people's willingness or reluctance to adopt more pro-environmental behaviors. It also serves to demonstrate how education strategies designed to promote public understanding of urban ecosystems can be informed by arguments individuals employ to challenge exhortations by governments and other agencies for citizens to "go for green."

Models of Sustainable Development

Many holistic models of sustainable development seek to emphasize links between society, economy, and environment in the manner of a natural system. Perhaps the most significant contribution these models have made is the extent to which they have encouraged decision makers and planners to identify and take account of the full costs to the environment of unsustainable development. Through the work of environmental economists in particular attempts have been made to assign costs to the losses and benefits of previously taken-for-granted environmental goods and services (see Costanza, et al. 1997); however, the social, political, and cultural constraints that prevent environmentally sustainable development from taking place have not been elucidated so clearly (Benton and Redclift 1994). Sustainable development is often framed as an *environmental* problem that can be solved by a scientific approach, thereby excluding (whether deliberately or not) debate about the wider sustainable development issues such as the North–South divide, social inequalities, debt burden, and the endless pursuit of consumption (Wynne 1994).

It is important to understand cities as natural systems and to adopt lifestyles consistent with prudent use of resources, such as decreasing dependency on the car, insulating buildings, and recycling and reclaiming materials. There is no guarantee, however, that individuals or institutions will respond to this logic. Because natural systems have no moral authority and environmental science claims about urban ecosystems are formed and transformed through a range of cultural, social, and political processes, strategies for environmental education and communication need to be informed by a range of intellectual and practical approaches. For example, exhorting the public to adopt more pro-environmental lifestyles involves issues of rights and responsibilities, and raises questions about the role that structures and norms in society play in governing how people engage with these concerns. Recent social models of sustainable development point to a range of approaches that can inform public education strategies about urban ecosystems and promoting pro-environmental lifestyles (Burgess, et al. 1999; O'Riordan and Voisey 1997).

Social Models of Sustainable Development

Throughout the last 30 years, new "contextualist" theories of society emerged in social sciences (e.g., psychology, anthropology, sociology, geography, and planning) (Giddens 1991; Giddens and Lash 1994). These contextualist theories have emerged in part to challenge the more traditional reductionist approaches in social science that posit society as an aggregation of individuals who behave rationally (i.e., in their own self-interest). In contrast, contemporary social theory sees individuals as social beings whose actions reflect their socially derived meanings, values, and knowledges. One of the leading theorists is Anthony Giddens, a British sociologist who has done much to explain how individual identity is an integral feature of the social structures that both shape, and are shaped by, individual actions (Giddens 1991). Contextualist theories suggest that how an individual behaves cannot be predicted as a logical outcome of cognitive processes alone. Instead behavior is seen as a more complex, reflexive process of active engagement that is contingent on many factors and circumstances. For example, what we might choose to do is contingent on people's experience with the past and with place, and also on the role structures and norms play in shaping behavior. "Structures" include institutions such as commerce, education, health care, and planning together with their rules and modes of organization, which literally structure social, economic, and political life. Formal and informal rules and regulations ensure that each society has "norms" and functions in a "proper" way. Viewed from a contextualist perspective, actions are interpreted as responses to feelings of emotional attachment and duty, questions of trust and authority, and to a sense of believing (or not) that individual actions can influence change. Given their emphasis on understanding behavior "in context," such approaches favor qualitative research methods where people are engaged in discussion, rather than experimental and questionnaire-based approaches characteristic of traditional social science.

Against the background of the present topic—understanding urban ecosystems—reductionist and contextualist approaches provide rather different perspectives on how scientific information about city environments is understood and acted on. For example, mass media campaigns designed to promote pro-environmental behavior tend to work with a reductionist cause-and-effect model in which the mind of each individual needs to be filled with new and "correct" information that will engender appropriate behavioral responses. On the other hand, contextualist models challenge this "stimulus-response" model by arguing that individuals engage critically with new information. In particular, information is always understood in the context of the social and cultural relations within which it is embedded. People already have well-developed ideas and opinions which are used reflexively to "interrogate" the authority, credibility, and legitimacy of new information. For example, several studies suggest that the social and cultural status of institutions has an important bearing on the extent to which the public trusts information (Wynne 1994; Irwin 1995). In the same way, questions of trust, authority, and legitimacy all influence public reception of communications seeking to promote an understanding of cities as ecosystems.

In this short chapter the reductionist and contextualist theories can only be treated schematically as "ideal types." This is what Figure 9.1 illustrates. Reductionist models anticipate that people will respond "rationally" to choices once successfully communicated; contextualist models suggest that any response is contingent on whether these choices have authority and

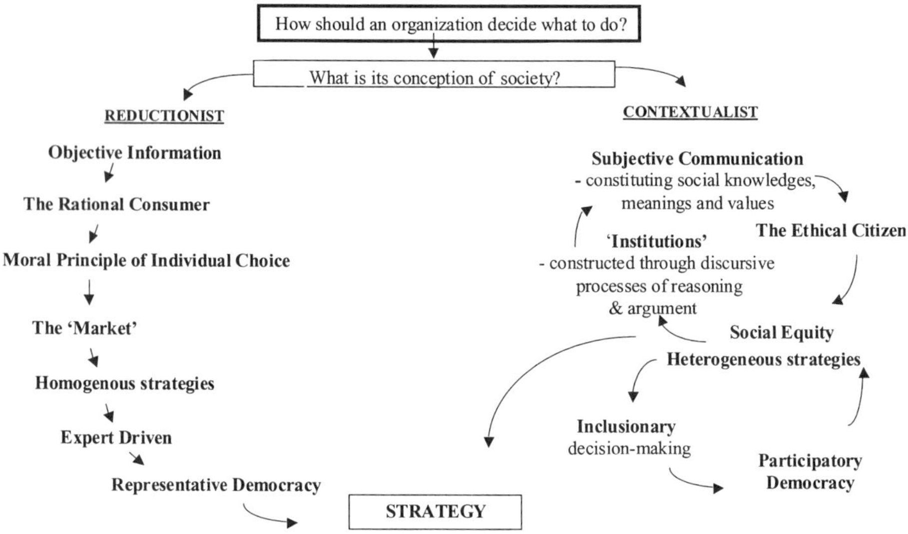

FIGURE 9.1. Approaches to sustainable development.

credibility in terms of social and cultural identification (or alienation) and not through any "assumed" or natural warrant. Scientific findings may not achieve authority with the public because the reception of information is shaped by a range of social, cultural, and political processes that change over time. Reductionist models construct individuals as "rational consumers" acting on their preferences, responding to market forces, and seeking to maximize their own self-interest; whereas contextualist theories construct them more as "ethical citizens" (see Figure 9.1). In the case of the ethical citizen, normative judgments figure prominently in decisions, especially when these decisions impact on communal resources such as the environment and the public domain of streets, parks, and plazas.

These two frameworks can also help illustrate different conceptions of how individuals engage with the political processes that determine the rules and norms of society. Reductionist models favor a dominant role for individual preferences as expressed through the market, for "experts" and "professionals" in decision making processes and are consistent with forms of "representative" democracy that perpetuate existing rights and power relations (Forester 1993). Contextualist approaches on the other hand favor a stakeholder approach in which anyone who has an interest in the outcome of decisions has a right to be involved. Consistent with a shift toward more equity in the allocation of rights and responsibilities, contextualist perspectives also favor more participative forms of democracy in which a wider range of knowledge is respected and given credence (see Bryant and Callewaert, Chapter 3 in this volume; Forester, 1989; O'Hara 1996; Irwin 1995). Integral to this process of greater participation is the reconstitution of social relations through a process of mutual learning and understanding. In this reflexive process individuals, structures, and norms may be redefined and reconstituted (represented by the feedback loop in Figure 9.1). The *transformative potential* of these deliberative and inclusionary processes of decision making contrasts with the reinforcement of existing power relations maintained by conventional, top-down and expert-driven processes of decision making (Fishkin 1991; Innes 1996; Healey 1997). In turn, communication strategies which seek to promote *new* understandings about the environment and society's relationship with it must provide opportunities for open and fair debate that can question existing understandings and social norms.

These ideal type alternatives serve to illustrate how new social concepts provide frameworks for thinking about sustainable development and cities as ecosystems. In what follows, elements of both are used critically to examine how more pro-environmental behavior can be encouraged among city residents.

A Cross-Cultural Study of Urban Residents' Commitments to Pro-Environmental Actions

The two-year study of residents' and decision-makers' attitudes to lifestyle changes required by global environmental change was undertaken in Nottingham (U.K.) and Eindhoven (the Netherlands) between 1993 and 1995. Both are medium-sized cities with a population of 274,000 and 195,000, respectively. Neither city had progressed environmental initiatives very far, although local authorities in both cities were sympathetic to developing integrated transport systems, recycling, and reclamation schemes. Nationally, the central government in the Netherlands had taken a more proactive approach to environmental planning than the U.K. government. Two National Environment Plans published in 1989 and 1993 set targets for all sectors of society to meet, and since 1990 the Dutch government has sponsored a mass media campaign to raise public awareness of how individual behavior could make a difference to global environmental problems. In the United Kingdom there were no such national plans, no sustained media campaign was undertaken, and the dominant approach gave priority to the operation of the market as the primary definer of both what were environmental problems and what their solutions might be. In the light of these national contexts, the overall purpose of the study was to determine whether citizens in Nottingham felt more or less empowered to assume responsibility and undertake pro-environmental behavior than citizens in Eindhoven, and if so to account for these differences.

Phase one of the study involved a questionnaire survey of 250 respondents in each city. The sample was generated randomly and the survey was conducted in comparable, suburban neighborhoods. Phase two involved conducting two in-depth discussion groups in each city—one with men and the other with women. The eight to ten participants in each group were recruited through the household survey and included a cross-section of the community as defined by age, income, and education. The groups met for 1.5 hours on each of five consecutive weeks. The household survey attempted to measure individual responses to questions about environmental awareness, attitudes, and behaviors, whereas the in-depth groups engaged discursively with a small number of people and gave participants time and opportunity to deliberate on the issues raised in the survey. The final stage of the research was to conduct a workshop with policy makers in each city to discuss the implications of the research for their environmental communication strategies (Burgess, et al. 1998). The findings of the household questionnaire and the in-depth groups are drawn on here to provide an understanding of how people rationalized their own environmental responsibilities (see Harrison, et al. 1996).

Anglo-Dutch Comparisons: Contrasts in Pro-Environmental Practices

First, we will briefly discuss the findings of the household survey as they relate to people's lifestyles and respondents' willingness to adopt more pro-environmental behavior. We will then move on to report on the findings of the in-depth discussion groups and focus on the reasons participants use to resist calls on them to "go for green."

One of the most intractable environmental issues facing cities is the demonstrable need to reduce traffic and to increase independent mobility without relying on the motorcar. In both cities local authorities had attempted to promote a number of measures designed to reduce car dependency, including car-sharing, promoting public transport, designating high-occupancy vehicle lanes on commuter routes and providing cycle routes. Overall, people in Nottingham exhibited a much higher dependency on the car than in Eindhoven. Car ownership was slightly higher in Nottingham (77 percent) than in Eindhoven (74 percent) but 69 percent of car owners in Nottingham reported using their cars 5 days a week or more compared with 41 percent of car owners in Eindhoven. In addition, there was a greater reluctance to change transport behavior in Nottingham. When asked if they had changed their transport behaviour in the last 5 years, only 37 percent of Nottingham respondents said that they had, compared with 60 percent in Eindhoven. Of this latter group, 35 percent said they now used their car less often compared with only 17 percent of the former. Alternative transport modes used most frequently involved walking and cycling in Eindhoven and using public transport in Nottingham. Only in Eindhoven did people mention that they had changed their behavior "for the sake of the environment" (13 percent). On this evidence, although the majority of people in both cities depended on the car, more people in Eindhoven reported that they had changed their behavior in favour of less-polluting transport modes, and for some people these changes had been made for "environmental" reasons.

When it came to addressing the wider issues raised by sustainable development, such as the need to reduce consumption and use natural resources in more prudent ways, a similar picture emerges. The level of pro-environmental behavior was much higher in Eindhoven than in Nottingham. For example, people purchased more green products, recycled more materials and shared these tasks among members of the household. In this sense the overall commitment to recycling in Eindhoven was much higher than in Nottingham, but it was not clear whether this pro-environmental behavior had become a matter of routine, signifying a change in lifestyle, or whether commitment was more pragmatic and ephemeral.

Analysis showed that respondents who were most "environmentally active" (excluding car use) lived in households that are better educated than average, had higher incomes, and held managerial and professional jobs, although members of all social classes participated in pro-environmental behavior. This is consistent with the findings of other surveys that suggest a marked shift in environmental behavior since the early 1980s (Witherspoon 1994). Certainly residents of Eindhoven seemed more environmentally committed than residents of Nottingham. Whether this was the result of access to more information associated with the mass media campaign, access to more recycling facilities, or a greater predisposition to a "collective" approach to solving problems could not be determined from the household survey. Detailed statistical analysis revealed very little consistency between pro-environmental behavior and gender, education, class, voting intention, or how active in the local community people report themselves to be. In other words no simple and coherent "green" view about how to address environmental problems existed among these city residents, and pro-environmental behavior could not be predicted with any confidence from recorded variables. One of the main purposes of the in-depth discussion groups was to explore the apparent ambiguities raised by the analysis of the questionnaires and to allow us to work with qualitative methods of inquiry that are more sensitive to contextualist accounts of society.

Resisting Calls to "Go for Green": Findings of the In-Depth Groups

The four groups (nine men and nine women in Nottingham and 10 men and 10 women in Eindhoven) met for 1.5 hours each week over 5 weeks. The groups followed a similar agenda. Topics included green consumerism, the impacts of technological and social changes on people's daily lives, their experiences of environmental changes, and ideas about sustainability. Through the discussions it became clear that assuming responsibility for addressing problems associated with global environmental change was a complex concept that involved a number of real and sometimes intangible constraints and benefits. Running through all the discussions was a powerful moral or normative dimension about what people *ought* to be doing, not only for the sake of the environment but also for the sake of society. For some people this sense of commitment came from deep personal conviction and was expressed with emotional force. Other people had a much weaker emotional commitment but wanted to engage altruistically in contributing to the collective good. Being able to exercise choice in what to buy and having the time to recycle was also important; however, people were concerned, too, about whether or not their actions were effective in achieving the goals they espoused and whether or not they could believe all the information they received about environmental problems and solutions.

The Role of Information

In both cities, the media played a particularly prominent role in discussions about these wider political and social concerns, in particular through their reporting of environmental issues, which served to expose the "contingent" nature of environmental "truths." In both Eindhoven groups, members felt overburdened by information that was often contradictory. In Nottingham, too, "media food scares" for example provoked a real sense of confusion for both men and women. Wanting the best for their families but being dependent on expert advice, and coping with the conflicting claims of different interests as represented through media reports, left everyone feeling very angry and confused. John felt very strongly about this: "We talked last week about aerosols. Why didn't they just ban them straight away if they're dangerous? And if they're not dangerous why scare us? I've actually lost confidence in, um, supposed "experts" on environmental issues. Because . . . then you get politicians coming in and they don't tell you the truth. . . . Suppose I had asked your advice about food, what food I should eat, or whether an aerosol is dangerous. I'd want to know the credibility, that . . . where you're coming from? What experience have you got, er, to make an opinion?" The men struggled to come to terms with their belief that experts such as scientists, politicians, and people in the media couldn't be trusted and how this affected their ability to make justifiable decisions. They all agreed with John when he said: "I live in a period of confusion."

The Dutch men talked about their response to the Dutch government's media campaign. This campaign used an image of a burning globe held in a hand that was accompanied by a message exhorting people to "act locally, think globally." One of the men said: "One person cannot blow out the candle to save the world—it's much more complex than that!" The Dutch women were equally cynical about the media as a purveyor of trustworthy information. One of them said: "Fifty percent cannot be believed, but it's difficult because I don't know which half!" In these circumstances "following your own instinct didn't help either, because there comes a point where that's very difficult if you are getting so much false or biased information. Often they say something in the morning and in the evening it's retracted."

Given this pervasive conviction that experts could not be trusted and a belief in the contingent nature of "truth" about environmental issues, it is not surprising to learn that people were unwilling to accept personal responsibility for the environment. This was not the passive response of an uncaring or ignorant public but rather an active resistance. It represented their own attempts to separate out environmental problems from the complex web of social, economic, political, and cultural practices people understood these problems to be embedded within. Much as Eden (1995) suggests, having a well-developed sense of "actionable responsibility" enables some committed environmentalists to adopt pro-environmental behavior even in the face of conflicting media reports of the efficacy of particular actions. Many more people when faced with the mixed messages promoted by the mass media feel impotent and do not know what to do for the best. In these circumstances they look to "others" to take the lead.

The Social Contract Between State and Its Citizens

Overall, what was impressive about the in-depth discussions held in the two cities was the extent to which the tenor of discussion in Eindhoven was much more optimistic and positive than in either of the two Nottingham groups. For example, with respect to recycling schemes, participants in Eindhoven linked recycling and reclamation of materials to possible improvements in the local economy. Such schemes were regarded as an industry requiring considerable investment but also as a source of potential new employment; however, although the Dutch groups looked to the national and local government to take these initiatives forward, they also believed that as individuals they had a social obligation to participate in the scheme. Introduction of a compulsory scheme locally had reinforced individual responsibility because as one man put it: "You can't hide from your responsibility at the local level." Organizing a compulsory local scheme seemed a good way of making abstract global problems "real." Although members of the Dutch men's group agreed when Jan said that governments "always promise more than they deliver," they also accepted his metaphor of environmental progress as the moves of a knight in a chess game. As Jan expressed it: "If the worst comes to the worst you go forward one and back two. But we must go on all out and try to keep going through it."

This positive attitude and willingness to accept some measure of self-ascribed responsibility for pro-environmental action, especially when national and local governments had taken the lead, contrasted with the seemingly more defeatist attitude that pervaded discussions in the Nottingham groups. Some members of the Nottingham groups had attempted to organize a recycling program in their local school, only to see it fail through lack of effort and the vagaries of the wastepaper market. Others had also tried to make use of whatever local facilities were provided, even though they were poorly run and serviced. These frustrating individual experiences led to a complex and often furious debate about where the responsibility for changing attitudes and practices resided. Was it with individuals, with government, or with commerce? In these discussions there was more disagreement about the nature of individual responsibility than in the Dutch groups, but there was a clear consensus that in the United Kingdom neither national nor local governments were setting an example for people to follow. The imposition of Value Added Tax on domestic fuel in 1993 for example, was interpreted in both Nottingham groups as a means of raising revenue dressed up as an environmental measure; as one man said, "That's how greedy this government is. It's not green. It points the way down the green road but doesn't go down it!"

Trying to separate issues of individual responsibility for the environment from broader changes in social values was difficult because these broader changes seemed to inhibit any real shift toward the kind of altruistic behavior that was required. For some of the Nottingham women it seemed that "People now are just so greedy and selfish. . . . It's like our country is selfish, we don't want to stop our people driving cars and stopping acid rain, because we want to drive our cars and we've got a right to do it. You know, we don't seem to have a moral conscience." Others felt that the free-market, individualistic ideology pursued by the national government was more to blame. Overall, however, they agreed that the absence of both personal and national commitment to the shared responsibility meant that there was no basis upon which a social contract between individuals and their neighbors and between the U.K. government and its citizens could be built. Under these circumstances the prospects for achieving sustainable development in the United Kingdom seemed more remote than in the Netherlands.

In Nottingham, people's frustration, alienation, anger, and in some cases despair, were all implicated in explanations for the contrast between the high level of environmental awareness reported in the questionnaire survey and the lower levels of reported pro-environmental behavior. In the Netherlands the discussion groups revealed a firmer basis to the social contract between the state and its citizens than was the case in the United Kingdom. Despite public scepticism about the effectiveness of the national government's mass media campaign designed to promote pro-environmental behavior, Dutch people were encouraged by the fact that state had taken the lead in acting responsibly towards the environment. By comparison, the ad hoc and laissez faire approach to promoting pro-environmental policy pursued by the U.K. government was often ridiculed by Nottingham residents. Taken together, such findings serve to highlight the multiple and pervasive influences that social, political, and cultural factors play in developing effective environmental communications—much as the contextualist conception of society suggests.

Conclusions

Summarizing and illustrating complex ideas in this brief way fails to do justice to the subtleties of both contextualist and reductionist conceptualizations of society. We suggest, however, that contextualist perspectives offer new insights about how individuals are engaged with society and how more effective strategies for environmental communication can be developed. Most obviously contextualist approaches ask natural scientists and policy makers to be more critical about their framing of who their publics are and what they will and will not do. In terms of developing a communication strategy for understanding urban ecosystems, educators and policy makers need to recognize the limitations of reductionist conceptions of society, which tend to assume a linear process of learning based on offering "the correct information." Numerous studies suggest that such an approach is not effective.

Working with contextualist conceptions of society means accepting that individuals are socially engaged actors whose environmental understanding and behavior is contingent on where they live, the history of events, their social networks, and social and moral norms. These approaches also recognize that the way society "works" depends upon a reflexive process of mutual trust through which individuals and structures (e.g., organizations, legal processes, rights and responsibilities) come to constitute each other. Gaining peoples' trust and support for education programs which seek to convert high public awareness of environmental problems into pro-environmental behavior, for example, is thus likely to require new ways of working. More participatory approaches to environmental communication and decision making that encourage face-to-face deliberation are capable of forging new social relations through a process that is based on mutual respect and trust. In this way, knowledge claims of experts such as educators, natural scientists, and politicians will add to, rather than displace, the legitimate knowledge claims of other groups in society.

In conclusion therefore, a contextualist approach to society suggests that effective education strategies which seek to promote a shared understanding of the inter relationships between lifestyles and environment will be:

- Heterogeneous in nature and content;
- Localized rather than universal in the scale of their delivery;
- Action-led rather than based on exhortation;
- Supportive of new public forums and arenas which encourage participatory democracy rather than reliant on existing structures and processes of representative democracy;

- Inclusive rather than exclusive in terms of the range of knowledges, experiences, and understandings they respect and accommodate.

Approached in this way, the task of understanding urban ecosystems is not simply one of information gathering and transfer, but one that also needs to acknowledge the influence of a range of other social, political, and cultural processes.

Acknowledgments. The cross-cultural study was funded by ESRC Research Grant L320253053 and the Dutch Institute of Forestry and Nature Research. The contextualist model of sustainable development was funded by The Environment Agency R&D Project Record E2/006/1.

References

Benton, T., and M. Redclift. 1994. *Social theory and the global environment.* Routledge, London.

Burgess, J., C. Harrison, and P. Filius. 1998. Environmental Communication and the Cultural politics of environmental citizenship. *Environment and Planning A* 30: 1445–1460.

Burgess, J., K. Collins, C. Harrison, R. Munton, and J. Murlis. 1999. *An analytical and descriptive model of sustainable development.* Sustainable Development Series 13. The Environment Agency, Bristol.

Costanza, R., R. d' Arge, R. de Groot, S. Farber, M. Grasso, B. Hannon, K. Limburg, S. Naeem, R.V. O'Neill, J. Paruelo, R.G. Raskin, P. Sutton, and M. van den Belt. 1997. The value of the world's ecosystem services and natural capital. *Nature* 387: 253–260.

Eden, S. 1995. Individual environmental responsibility and its role in public environmentalism. *Environment and Planning A* 25:1743–1758.

Fishkin, J.S. 1991. *Democracy and deliberation: new directions for democratic reform.* Yale University Press, London.

Forester, J. 1989. *Planning in the face of power.* University of California Press, Berkeley, CA.

Giddens, A. 1991. *Modernity and self-identity.* Polity Press, Cambridge.

Giddens, A., and S. Lash. 1994. *Reflexive modernization.* Polity Press, Cambridge.

Harrison, C.M., J. Burgess, and P. Filius. 1996. Rationalising Environmental responsibilities: a comparison of lay public in the UK and the Netherlands. *Global Environmental Change* 6:215–234.

Healey, P. 1997. *Collaborative planning: shaping plans in fragmented societies.* Macmillan, Basingstoke.

Innes, J.E. 1996. Planning through consensus-building: a new view of the comprehensive planning ideal. *Journal of the American Planning Association* 62:460–472.

Irwin, A. 1995. *Citizen science: a study of people, expertise and sustainable development.* Routledge, London.

O'Hara, S. 1996. Discursive ethics in ecosystem valuation and environmental policy. *Ecological Economics* 16:95–107.

O'Riordan, T., and H. Voisey. 1997. *Sustainable development in Western Europe*. Frank Cass, London.

Witherspoon, S. 1994. The greening of Britain: romance and rationality. Pages 107–139 in R. Jowell, ed. *British social attitudes: the eleventh report*. Dartmouth. SPCR.

Wynne, B. 1994. Scientific knowledge and the global environment. Pages 169–189 in T. Benton, and M. Redclift, ed. *Social theory and the global environment*. Routledge, London.

10
The Future of Urban Ecosystem Education from a Social Scientist's Perspective: The Value of Involving the People You Are Studying in Your Work

John B. Wolford

Introduction

In studying an ecosystem, is it even conceivable to exclude the human element? Likewise, is it conceivable to conduct a social scientific study of an ecosystem when minimizing the bio/physical/ecological dimensions? Perhaps the problem lies in the term *ecosystem* itself, which is defined in idiosyncratic ways, all of the definitions grounded within the discipline or field of the definer. The term inevitably must encompass the core elements of holism, mutualism, dialectic, inclusiveness, and—intrinsically—the concept of system. Although, in a romantic way, it is a comforting conceit to think of ecological environments in their "pure" biological habitats ("pure" meaning to many researchers habitats that exclude any human influence), all ecosystems do nonetheless involve humans necessarily. From a basic, phenomenological perspective, humans are involved in ecosystems simply because humans conceive of them—we think of them, and therefore they are. Operationally, however, humans are part of ecosystems because our biological, social, cultural, and personal effects are seen in all ecosystems on this planet. Our effect is inescapable. Our imprint is on all global ecosystems. In order to understand any ecosystem, then, a researcher necessarily must understand the importance of human involvement. This is true not only of urban ecosystems, which will be the focus of this paper, but of all ecosystems. In one sense, the urban environment may be the worst place to scrutinize human involvement, because such a study might unduly bias the researcher and any student to think that the urban environment is the only place people have an impact on the environment. On the other hand, urban ecosystems are perhaps the best place to engender an understanding of human impact, because the human involvement is so overwhelming and manifest.

Though my approach derives from an interdisciplinary, multifield perspective which involves the differential approaches exemplified by the social scientific perspective on the one hand and the humanistic perspective on the other, I do not represent all disciplines in their particulars. I no doubt do represent many in the broad strokes of my brush; however, even in my representation of what I know best, the anthropological and folkloristic approaches to urban ecosystems, my understanding exemplifies more the nascent development of the subfields of urban anthropology and urban folklore than it does standards within the larger disciplines. In that I will be addressing both social scientific and humanistic concerns, this paper serves as a platform to explore different threads that run through theoretical ecosystem models and determine whether some potential exists to interweave these threads into some wonderfully evocative tapestry of explanation.

Researchers often become so involved in their work that they become myopic, although whether they evince an objective or a subjective myopia tends to be idiosyncratic. Cultural anthropologists, for example, scrutinize culture and its effects on people, history, environment, and so forth, and might even come to the conclusion that culture is the touchstone for all understanding in this world. Of course, researchers in other fields recognize that position as patently absurd, because they know what they research is equally important in understanding the broad truths of life and universal mystery. Natural science, social science, humanities—all are human intellectual enterprises that home in on a particular aspect of what it is to be human. All disciplines develop canonical methods of inquiry; they all establish parameters of substantive investigation; and they all come up with mutable truths—principles—that result from their rigorous research. What they don't do, in general, is communicate very well with each other. What they don't address is the fact that each field has sequestered only an aspect of life and reality to explore, and has ignored—or dismissed as anomalous—the rest. All disciplines have forgotten that they must return to their beginning premise of isolated inquiry and attempt to integrate the obtained truths from the other fields of knowledge in order ultimately to construct a human enterprise of ontological teleology, or the inherent impetus toward fulfillment.

In this essay, I represent the position that knowledge and knowledge generation must be integrative, since life itself is fundamentally and necessarily an integrative process. Even those fields of study which appear pure—such as the application of natural science inquiry to the physically oriented field of ecology—*require* input from humanistic and social scientific fields of inquiry in order to be truly holistic, in order to reveal the "system" within the ecosystem. Further, all studies are simply part-truths, no study is ever complete, and single-field studies by implication will never accomplish completeness. Scholars, however, can approach holism by trying to integrate the different fields of knowledge in the research they do. Likewise, involving the very people who live and work and play in an area—the people who themselves are part of the ecosystem—creates a depth of understanding not only for outside researchers but also for the insiders. The final stage of developing a holistic study is by putting this multidimensional, multi-perspective understanding of an ecosystem to use—the understanding derived from differential academic and insider research. The ideal way to apply the research is through education, by involving learners (whether adult or child) in the research, in the dissemination of the research, and in the absorption of the knowledge. Ecosystem teaching, then, is process-oriented, one that emphasizes how to be inclusive in focus and in content in order for all to comprehend as much as possible. The very fact of an ecosystem study's inevitable lack of perfect completeness is itself, ironically, a motivator, since researchers and students will always be challenged. In this chapter I focus largely on anthropology, oral history, and folklore, but my work stands for all kinds of people-oriented research. The implications of this approach for urban ecosystem education are broad and inclusive, where not only students but also entire fields can benefit from the cross-fertilization of research and its resultant expansion of understanding.

Background for This Approach

Within the field of anthropology, holism is a fundamental principle for studying systems. By "holism" anthropologists mean documenting and analyzing in detail all of the cultural, social and personal behavior, thoughts, and artifacts that come to structure and provide nuance to the society under study. The holistic ideal logically leads to the study not only of human populations within their obvious human-constructed environments—their villages, nomadic habitats, horticultural residences, even cities—but also of the natural elements within and around human-centered environments. Julian Steward, a pioneer in ecological anthropology, devoted his long scholarly career (from the 1920s to the 1970s) to developing the concept of cultural ecology, the study of specific human cultures and societies interacting with and mutually affecting nature. He emphasized culture, and especially the technological aspects of culture, over any primacy of nature, his primary objective being to understand cultural processes. His contributions to the larger field of anthropology were (1) asserting that people live in a dynamic relationship with their natural environment and (2) maintaining that human technology is a determining factor in shaping the environment (Stewart 1956). In the 1950s and 1960s, Roy Rappaport reoriented Steward's by-then conventional ecological model in anthropology by presenting a normative functionalist model, which holds that the natural world and the physical objects created by a society (e.g., its buildings, its villages, its agricultural fields, its pottery traditions, etc.) are integrated within that society's cultural framework to create and balance the norms of the society. Labeled the "new ecology," his theory was critiqued as nonresponsive to the constants of dysfunction and non-normative elements within society. Rappaport responded to this criticism by modifying his model throughout his lifetime (1926–1997) to include more historical, political, and symbolic elements within the analytical equation. Out of his work grew other approaches to ecological studies of society—historical ecology, symbolic ecology, and political ecology, to name the most influential—labeled the "new ecologies." What all of these seek to create is a model that adequately addresses the dialectic involved (through human agency) between the natural environment, individual people, a people's structured society, and a people's cultural constructions (Biersack 1999; Biersack 1999; Rappaport 1968, 1969).

While the earliest efforts to address dynamic ecosystemic relationships focused narrowly on anthropological disciplinary concerns, the subsequent efforts have come closer to developing that interdisciplinary ideal of holistic understanding of the ecological system. As mentioned, Rappaport's "new ecology" shifted the focus from a dominantly cultural and technological one to a materialist one, where society could be interpreted to be, in effect, a product of the environment. The "new ecologies" that developed out of Rappaport's innovation—symbolic, historical, and political—attempted to diminish the determinist nature of the Rappaportian ecology by emphasizing the mutualistic aspect of ecological systems, whether foregrounding a cognitive, diachronic, or political primacy.

Aletta Biersack (1999) discusses the New Materialism, a synthesis in ecological anthropology that attempts to bridge the dichotomy between idealists and materialists. That is, rather than thinking about culture as explainable only in its own terms (idealism) or understanding culture simply as a product of nature (materialism), the New Materialism would dispense with such dualisms (nature/culture) and explain the world according to a strictly holistic model. It would think in terms of "an ecology of incommensurabilities," a holistic sensibility that views reality as a "life-world" of "an indivisible material/symbolic/political/social/historical reality." After Descola and Pálsson (1996) she calls this a monist ecological anthropology. Although analytically separable elements exist for scholarly scrutiny, in reality the life-world is all one, indivisible, mutualistic, and monistic. Although it seems anthropocentric, humans are in fact at the center of it all, in a liminal position—people are the mediators, the portals, the creators. They conceive and act in relation to and in affect to both nature and culture. The relationship is emphatically not dichotomous, but dialectical.

Local Case Studies

The Missouri Historical Society (MHS) reenvisioned its mission circa 1995. Determining that the metropolitan area of St. Louis should be our primary concern over the next few years, we isolated four central sociocultural themes that would serve as core concerns for our projects: transcendence, the importance of history, a people's social contract, and sustainability. A focus on these issues would lead us to produce cultural products—exhibits, research projects, outreach programs, educational programs—that people in the community could not only identify with but could engage in, simply because these are concerns that urban people hold in common. We understand all of these issues to be interrelated and mutually affecting, but the issue directly related to this paper, sustainability, contains within its rubric issues relating to the social, cultural, and natural environments, and implicitly deals directly with the concerns of urban ecosystemic balance.

Since the formation of the Core Values (as they were called), several initiatives have been realized or are in process at MHS. We have tripled our exhibit space with a new addition to the museum, and in this new space an exhibit, "Seeking St. Louis," documenting the history of that city, opened in 2000. A highlight of this exhibit, which exemplifies the Core Values, is an inlaid map of the Mississippi River that flows through the first floor and is a representation of the centrality of the river in the creation and ongoing life of St. Louis. Just as a city is dependent on its natural resources, the exhibit will be dependent on this symbol to transmit the idea of the interdependency—the systemic relation—of the people and the natural place in which they have thrived.

The oral history project I have worked on since 1994, *People and Place in Twentieth-Century St. Louis*, likewise takes into account the interchange between people and their natural environment, although it will do more so in the future. As originally conceived in 1994, this neighborhood-oriented project focused heavily on people and how they construct their built environments—their buildings, their fences, their gardens, their parks, their urban infrastructure (Figures 10.1 and 10.2). The emphasis was on the historic, cultural, and social elements involved in the human construction of place. What has become apparent as missing—or rather, implicit and thereby underrepresented—is the human/nature dialectic. To frame the human urban environment primarily in human terms and omit the integral role of nature, is to amplify *anomie*, the normlessness and the sense of isolation not only from one another in society but also from nature, a normlessness that typifies American culture.

Our original interview guide incorporated topics such as how people respond to their built environments—the significance of parks nearby; whether, how, and where people garden; whether and how people respond to the type and density of architecture in their neighborhoods; what kinds of outdoor activities people engage in and where; in short, how and why people relate to their surrounding space. What we have been asking has in one way or another elicited strong feelings about the neighborhood/environment dialectic that reflect in plain English Biersack's monist ecological anthropology. Greg Carter, for instance, a resident of and alderman for North Pointe, a primarily African American middle-class neighborhood, provides a deep description of how trees and especially the wood features

FIGURE 10.1. Stone fence with letter A.

of his house provide a deep, aesthetic satisfaction for him (Figure 10.3). [In these oral history transcriptions, the use of the three-dot ellipsis indicates a pause in the speaker's narrative. A four-dot ellipsis indicates narrator verbiage that is left out of the quote.]

> GC: Well, the outside of the house was, the layout was very nice. You know . . . I redid some landscaping on the front. They had some . . . I don't know . . . aluminum shutters. I went and had some real wood made and . . . I put those up myself . . . We did some tuckpointing. That's why it looks that smoother look. . . . You want your own look. We had a white birch in the front that died about 2 years ago, which was a beautiful tree, it was an absolutely gorgeous tree. In fact, it set the tone for this whole block, but some borers got in it and killed the tree so I had to take it down and we're thinking about putting it back up, but with the layout of the house, I'm thinking about just leaving it like it is. You know, because then my blue fir, my pine, and, you know, the landscaping for the bushes . . . It just sets the house to me. It's nice. Coming into the house, it's all wood. The floor, the rails, the stair casing, the trim in the ceiling is all wood, and that's cherry oak. . . . You're never finished with your house. (Carter, 11/07/1996, tape 2/2, side A:1–3)

Maxwell Culmore, another resident of the same large neighborhood, characterizes his neighbors in terms of whether they take care of their property

FIGURE 10.2. McPherson Community Gardens. Skinker-DeBaliviere neighborhood.

FIGURE 10.3. North Pointe residence.

by executing acceptable standards of landscaping, people he calls "grass cutters." Culmore is representing and validating a moral standard in terms of people's attention to their personal stake in the environment, their yards. In perfect synchronization with American cultural values, Culmore places the high moral ground on upkeep of personal property, which in terms of landscape means keeping it tidy, cut, neat, and in control. People do what they are "supposed to do," by cutting grass. Such a position highlights a key point in cultural ecology: the importance of people's day-to-day experience *in* the environment shaping their view *of* the environment.

JD: Now what is your impression of this neighborhood today?

MC: Well, it's middle-class working area. And basically everybody . . . I guess you have to go block by block. I can only tell you about this block. This block is grass cutters, everybody tries to fix up and do what a homeowner is supposed to do. (Culmore, 10/31/1996, tape 2/2, side A:6–7)

In a related way, residents in an older section of the city, Benton Park, consistently remark on the astounding collection of late nineteenth-century architecture that characterizes their neighborhood (Figure 10.4). Their aesthetic for brick and limestone residences along tree-lined avenues

FIGURE 10.4. Benton Park residences.

represents a worldview valuing the crafting and controlling of natural resources for social use—they find beauty in the societally-structured reformulation of clay and stone. Interestingly, one of the narrators even mentions the distraction that occurs when the natural forms of trees blocks her view of the larger urban environment. An element that enhances their appreciation for their neighborhood is the built-in worth, or potential worth, of their residence.

Susan: I think [our neighborhood is] the most gorgeous thing there is. . . . You know, we've got one of the most architecturally appealing, to me, you know, houses in the whole area. You'll see houses like this in Lafayette Square and stuff, but most of them are just brick front and stuff, you know. I like the limestone and the steps and, you know, I walk out the steps and I think, gee, this could almost be New York. . . .

And, when the trees aren't there, if you look out the window, you can see the city and the Arch. And, at night the Arch is lit up and, you know, these houses should be worth beaucoup bucks, but you know, they are not. . . .

Scott: That's one of our, that's one of my big gripes, you know. It's that if this were in New York or Philadelphia or Boston or something, I mean, this place would be worth a million dollars, but if here in St. Louis for some reason, people just think, well you should tear that old place down or something, you know. (Vogel, 5/21/1997, tape 2/3, side A:6–7)

Later, Susan Vogel summarizes this urban environmental aesthetic succinctly:

Susan: . . . [W]e just love to walk down the streets or drive down the streets and see the row of houses and we just go, "Isn't that beautiful?" (Vogel, 5/21/1997, tape 2/3, side A:9)

Through this project, we have gathered intimations of people's deep connection to the natural environment, but we have not explored it fully. We need to ask directly about how they use or ignore their personal landscape; why they prefer the materials they use for home, fences, and garden walls, and why they do not prefer others; what their lawn decorations mean to them; why they do or don't engage in gardening, and why they choose their particular gardening style over others; which smells they do or do not enjoy; what effect either wild or domesticated animals in the landscape have on them; whether air quality is an issue that keeps them in or out of an area. People's choices are microdocumentations of urban ecological values, and compilations of these microdocuments comprise strong statements that could and should affect public policy, urban planning, and educational initiatives around environmental concerns. That people want gardens, and green cities, and a sense of open space, all are factors that influence people to move out of closely built, asphalted, small-lotted urban areas into the more open suburban areas. Beyond sociocultural concerns, what social researchers need to do is form teams with ecological scientists to study an area in order to understand the biological, climatic, hydrologic, and other natural processes that are in interplay with the human macro- and micro-processes in the environment.

In this project's most recently covered neighborhood, the Skinker-DeBaliviere neighborhood, another aspect of human/nature monism emerged during the historical research. Skinker-DeBaliviere originally was a land grant to a prominent St. Louis family in the late eighteenth century, and through most of the nineteenth century it served as a farm and as a country estate for members of that family. Developed in anticipation of the 1904 Louisiana Purchase Exposition, this area's small meandering stream, the River Des Peres, proved to be a problem, because it ranged from drybed to flood status throughout any given year. Over time, it became a sewage conduit as well. For the Louisiana Purchase Exposition, the river was enclosed in its Forest Park reach. Up through the first half of the twentieth century, the River Des Peres remained exposed in the Skinker-DeBaliviere area; the city covered it eventually as a health concern, and situated Des Peres Avenue over its general underground course (Harleman, et al. 1973).

By 1920, the Skinker-DeBaliviere area was built up, with three story Arts-and-Crafts single-family residences mixed in with townhouses and multi-family residences on close-set lots along platted streets. All of these buildings are big brick buildings, typical of St. Louis residential architecture in general. Developers built heavy three story brick buildings along Des Peres Avenue as well, even though underneath and beside that street ran the still meandering, sometimes flooding sewage conduit that had been the River Des Peres. Over time, nearly all of those buildings along the River Des Peres developed cracks and severe structural damage from collapsing ground and water damage, and they were torn down, replaced by one story brick multifamily residences that are a complete anomaly within this neighborhood (Figures 10.5–10.8). This street and its built environment is a visual narrative, and it provides a testament to a society's initial disregard for the natural world and to its inevitable adaptation to it. Other researchers have noted the impact of human agency on the urban environment, notably Anne Whiston Spirn's (1984, 1998) research in Philadelphia, which deals precisely with this kind of discursive relationship between the landscape and human society. Understanding the history and the development of Des Peres Avenue within the context of a monist, holistic, integrative urban ecology allows for an enlightened recognition that we as individuals do not live alone, that we as a society do not function alone, and that we as natural creatures live within the constraints and rhythms of natural forces. On the flip side, using insights from historical ecology, we must comprehend that our actions inevitably craft shape, meaning and consequence out of the natural materials that form our environment. A reorientation of the Western worldview to become more accommodating and synchronous with nature rather than vanquishing and controlling—for example, to structure a community around a river rather than deny the river exists—would go far

FIGURE 10.5. Des Peres Avenue, looking south, Skinker-DeBaliviere neighborhood.

FIGURE 10.6. Des Peres Avenue and McPherson Avenue, Skinker-DeBaliviere neighborhood.

FIGURE 10.7. Typical Skinker-DeBaliviere single-family residence.

FIGURE 10.8. Typical Skinker-DeBaliviere single-family residence.

in preventing disasters, promoting social harmony, and protecting natural resources, human beings among them. People who live in the Skinker-DeBaliviere neighborhood in general are oblivious to the river that runs beneath their homes and yards, and most know little to nothing about its history or the impact it had in the formation of their community. Most do know of the River Des Peres, but know it as a sewage conduit that eventually flows into the Mississippi River miles away. They do not connect themselves to other communities along the river, nor does any other community along the river connect itself to others, because urban Americans in general do not think in environmental or natural-systems terms. Envisioning a pedagogical future that would integrate multifaceted understandings of the ecosystem is to envision a future that changes a people's self-concept to one where they realize that they exist not only within but as part of the flux of a natural environment.

Involving People in Research

Institutions have a way of thinking (by committee, often) that they know best how to run a project and how best to effect objectives. This is often wrong-headed, not necessarily because the people are misguided; rather, it is the nature of bureaucratic thinking to forget about humanity, even when people are the object and the subject of their projects. In conducting any project where people are even tacitly involved, their roles must be integral to the project, even when their views may not correspond with the project's goals. When discord arises (as it inevitably does), problems within the logic of the project must be addressed. Imposing linear thinking on a meandering stream does not work, as the River Des Peres has shown; and imposing categorical, bureaucratic thinking on chaotic, human-variable reality is just as unworkable.

By the 1960s and 1970s, after decades of European American anthropologists studying non-European American peoples, anthropologists came to realize, reflexively, that studies of a people are fuller, more resonant with meaning, if the people themselves are actively involved as partners in procuring, producing, processing, and interpreting the data. What is added is the insider perspective. In the same manner, the ideal in any applied urban community study is to involve community members as advisors and colleagues in the project. One program we are just now beginning at MHS, the Community Partners Gallery, has the goal of dedicating MHS space and personnel for an entire year at a time to a discrete community within St. Louis, to enable the community to present an exhibit and related programs that they conceive and fashion. Called "community scholars," the participating residents are able to express to the rest of the world how they perceive their community and how they would like to be perceived. In the 1980s and early 1990s, Betty Belanus of the Smithsonian Institution was a pioneer in this approach, that advocates public presentations partnering the community with institutions. The idea of "community scholars" has been embraced ever since by state folklorists, who have produced exhibits and programs that have significantly strengthened community involvement and educational outreach programs. Following from this model, urban ecological studies would benefit by involving not only the academic researchers who study the socionatural environment within an urban place but also the members of the community who comprise part of the operational study.

Although it is not an ecological study, Jacqueline K. Dace's oral history project, *Through the Eyes of a Child* (1996–2002) is a model community partners project that exemplifies what can be accomplished by involving the people researched within the organizational structure of a project. The project began by envisioning an African American history study. Dace coordinated focus groups of African American community leaders, academics, business people, and residents, who provided insight into the public history wishes of the community. From those groups and their recommendations, she formed two advisory committees, with which she and MHS staff worked to develop ideas for a project. What emerged was *Eyes*, which is a study of four long-term African American communities in the metro St. Louis area. Dace and other staff did all of the research and the interviewing and organizing, and made sure that the community was fully apprised of every step. She marshaled institutional and community resources to organize periodic community gatherings to celebrate the neighborhoods and the people involved. The project has resulted in an enormous amount of archival material, dozens of firsthand oral historical accounts, active oral history and community research by middle school students in their neighborhoods, an historic play using the actual words of neighborhood residents (*Through the Eyes of a Child: Coming Home*), and a video, *Through the Eyes of a Child* (which MHS distributes free to local middle schools) which won the Honorable Mention prize at the 2000 Hollywood Black Film Festival. Dace is following through on the promise of this project educationally by creating a website and a CD-ROM, which are to be used in conjunction with a proposed curricular program that educational students at the University of Missouri—St. Louis will be required to learn and implement within the local school districts. In 2003, MHS will open an exhibit based on this material. By making sure the people were intricately involved as primary content providers, advisors, organizers, and eventual recipients of pedagogical material, Dace shared ownership and control of the entire project with the people whose lives she was helping to document; by doing so, she crafted a project that was meaningful not only to MHS and external researchers, but to the community being documented.

An excellent example of an ecosystem project involving community participation is the Baltimore Ecosystem Study, a supremely structured multifield research project whose subtitle is "Human Settlements as Ecosystems" (Grove and Hinson 2002, Chapter 11 in this volume; Pickett, et al.

2002, Chapter 5 in this volume). Dr. Steward T.A. Pickett, the project director, has long been an advocate of recognizing the human element in ecosystem studies, and this project exemplifies his approach. Of the eight research categories involved in the project, five are scientifically oriented (soil, vegetation, meteorology, hydrology, and wildlife), two are social scientific (social and regional modeling [which draws mainly from economics]), and one is pedagogical (education). The social and regional modeling research addresses the human impact on the ecological system, whereas the educational initiative addresses the need to invoke participation by school children in environmental studies and to educate the public in more beneficial environmentally sustainable lifestyles. Certainly the structure that Dr. Pickett and his team have created is exemplary: it allows interdisciplinary teams to cover different aspects of the ecosystem; it implements a sophisticated plan to involve the local school communities in the project, thereby achieving the dual objectives of student education and involvement at different age levels; it operates under a long-term grant from the National Science Foundation that ensures it baseline funding for renewable 6-year periods; and it cooperates with different levels of the governmental, educational, academic, and civic leadership of an already involved metropolitan area. The multidisciplinary and especially the multifield team approach in this study is an outstanding model for future research in ecosystem studies. Although the social scientific and the pedagogical are given play within the model, it is an ongoing concern to provide balance between the academic approaches in studying all the different human and natural aspects that create an urban environment. The obvious solution, and the obvious problem, is to mediate differences and derive a balanced approach.

Conclusions

People are part of a natural environment not simply as biotic agents and actors; they also are primary in shaping the forms and affecting the viability of a natural environment. Cities, far from being distinct from studies in environmental ecology, are themselves natural environments, incredibly complex natural environments constructed from the internally byzantine processes inherent in human systems wedded to natural elements (including both flora and fauna). The most relevant fauna in any environment is *Homo sapiens*.

Only for the purpose of study does it make sense to divide the real world into analytical categories. The real world, however, itself is not so divided. It is integrated, systemic. Theorists talk about social ecology, human ecology, cultural ecology, political ecology, symbolic ecology, historical ecology (and all of these simply on the humanistic and social scientific side!); and researchers discuss these various ecologies for the purpose of understanding real-world processes and the products that incorporate them all. Each of these ecologies is one part of the elephant; no one of them sufficiently explains the totality of the elephant. As researchers interested in providing meaningful explanations, we must integrate our variable research, we must break the habit of being defenders of our disciplines, and we must try to explore the larger meanings inherent in the operational world. The objective for all researchers is to achieve understanding; the objective for all people is to achieve, ultimately, a sense of their place in the universe, to coordinate their worldview—a sociocultural construction of subjective reality—with the world, to mediate (in Rappaport's terms) the cognized with the operational world. In other words, researchers are not alone—*all* people are involved in trying to make sense of the world. We need to work together. One way to start is for academics, educators, and everyday people to partner as colleagues both in applied research and in educational initiatives.

References

Biersack, A., guest ed. 1999. Contemporary issues forum: ecologies for tomorrow: reading Rappaport today. *American Anthropologist* 101(1):5–112.

Biersack, A. Introduction: From the "New ecology" to the "new ecologies." *American Anthropologist* 101(1):5–18.

Carter, G. (J.K. Dace, interviewer). Oral history interview, 11/07/1996. Missouri Historical Society, *People and place in twentieth-century St. Louis* project, North Pointe neighborhood.

Culmore, M. (J.K. Dace, interviewer). Oral history interview, 10/31/1996. Missouri Historical Society, *People and place in twentieth-century St. Louis* project, North Pointe neighborhood.

Dace, J.K. *Through the Eyes of a Child* project. St. Louis: Missouri Historical Society, 1996–2003.

Descola, P., and G. Pálsson. 1996. *Nature and society: anthropological perspectives*. New York: Routledge 1996:1–21.

Grove, M., and K. Hinson. 2002. A social ecology approach to understanding urban ecosystems and landscapes. Chapter 11 in A. Berkowitz, K. Hollweg, and C. Nilon, eds. *Understanding urban ecosystems: a new frontier for science and education*. Springer-Verlag, New York.

Harleman, K., et al. 1973. *The neighborhood: a history of Skinker-DeBaliviere*. St. Louis: Skinker-DeBaliviere Community Council.

Pickett, S.T.A. 2002. Why is developing a broad understanding of urban ecosystems important to science and scientists? Chapter 5 in A. Berkowitz, K. Hollweg, and C. Nilon, eds. *Understanding urban ecosystems: a new frontier for science and education*. Springer-Verlag, New York.

Rappaport, R.A. 1967. "Ritual regulation of environmental relations among a New Guinea people." *Ethnology* 6:17–30.

Rappaport, R.A. 1968. *Pigs for the ancestors: ritual in the ecology of a New Guinea people*. Yale University Press, New Haven, CT.

Spirn, A.W. 1998. *The language of landscape*. Yale University Press, New Haven, CT.
Spirn, A.W. 1984. *The granite garden: urban nature and human design*. Basic Books, New York.
Steward, J. 1956. *Theory of culture change*. University of Illinois Press, Urbana.
Vogel, S., and S. Vogel. (J.K. Dace, interviewer). Oral history interview, 05/21/1997. Missouri Historical Society, *People and place in twentieth-century St. Louis* project, Benton Park neighborhood.

11
A Social Ecology Approach to Understanding Urban Ecosystems and Landscapes

J. Morgan Grove, Karen E. Hinson, and Robert J. Northrop

Introduction

If one's knowledge of one's whereabouts is insufficient, if one's judgment is unsound, then expert advice is of little use

Berry 1990

The shape and dynamics of cities are the result of physical, biological, and social forces. We include the term *dynamic* to emphasize that cities change over time and are the result of both idiosyncratic events and dominant trends. To begin to understand the patterns and processes of cities, we approach the idiosyncratic and dominant—whether it is physical, biological, or social—within an historical context.

To present our approach for understanding urban ecosystems, we separate this chapter into two sections. In Section I, we summarize our social ecology approach—its basis in different social sciences such as geography, history, sociology, and political science—and five concepts or types of analyses that we propose are useful for teaching about urban ecosystems. These five concepts are: (1) the human ecosystem; (2) units of organization and scale; (3) geographic imagination and spatial analyses; (4) linkages between scales and across geography; and (5) examples from everyday life—policies, plans, and management.

In Section II, we assert that a social ecology approach must meet four criteria in order for teachers to adopt it in their existing curriculum. A social ecology approach to the study of urban ecosystems must (1) relate to a teacher's subject matter; (2) be an integral component of their existing curriculum framework; (3) prepare students for achievement in district, state, and national assessments; and (4) be relevant to students' lives while producing significant and enduring learning. We use the activities of Karen Hinson, her students' eleventh grade Advanced Placement United States History course (1998–1999), and the Baltimore Ecosystem Study (BES) as an example to demonstrate the incorporation and application of a social ecology approach in an educational context. As part of the students' year-long project, and in partnership with the Forest Services of the Maryland Department of Natural Resources and the United States Department of Agriculture, the students were assigned to apply skills learned from their existing curriculum—both knowledge (information) and performance (analytical)—to understand the social history, current status, and future trends of the development of the City of Baltimore's drinking water supply.

Finally, although this course occurred in an academic setting, it is ultimately not an academic exercise. These students might have numerous land management roles in the future: as homeowners, concerned citizens, or professionals offering "expert" advice. In the Summary and Epilogue, therefore, we consider some key points from the course and important issues for the conservation and restoration of urban ecosystems.

A Social Ecology Approach to the Study of Urban Ecosystems

It is increasingly difficult to determine where biological ecology ends and social ecology begins (Golley 1993). Indeed, the distinction between the two has diminished through the convergence of related concepts, theories, and methods in the biological, behavioral, and social sciences. Social ecology may be thought of as a life science focusing on the ecological study of various social species such as ants, bees, wolves, dolphins, or orangutans. In this context, we may study *Homo sapiens*, or human ecology, as an individual social species or comparatively with the ecology of other social species. The subject matter of social ecology, like biological ecology, is of stochastic, historic, and hierarchical systems (Grove and Burch 1997). For instance, because living systems exhibit emergent properties, they cannot be reduced completely to laws of physics (Simpson 1964; Bailey and Mulcahy 1972; Mayr 1982). Further, living systems are not deterministic; they have an historical and contingent dimension that cannot be predicted from physical laws alone (Botkin 1990; Gould 1994).

The underlying basis for this life science approach to the study of human ecological systems depends upon three points (Grove and Burch 1997):

1. *Homo sapiens*, like all other species, are not exempt from physical, chemical, or biological processes. Human characteristics (biophysical and social) are shaped by evolution and, at the same time, shape the environment in which *Homo sapiens* live.
2. *Homo sapiens*, like some other species, exhibit social behavior *and* culture.
3. Social and cultural traits are involved fundamentally in the adaptation of social species to environmental conditions.

In this context, human ecology must address the challenge of reconciling social and biological facts to understand the behavior of *Homo sapiens* over time (Machlis, et al. 1997). Such a biosocial approach to human ecological systems (Burch 1988; Field and Burch 1988; Machlis, et al. 1997; Wilson 1975, 1998) stands in contrast to others who may adopt either a more traditional geographic or social approach (Hawley 1950, 1986; Catton 1994). This is not to say that social sciences such as psychology, geography, anthropology, sociology, economics, and political science are not important to social ecology. They are, as we will discuss further. Indeed, the most fundamental fact that distinguishes humans and their evolutionary history from other species—both social and nonsocial species—is that human social development has enabled the species to escape local ecosystems to such an extent that local ecosystems no longer regulate human population size, structure, or genetic diversity (Eldredge 1995). This is nowhere more apparent than in urban ecosystems.

The Human Ecosystem

Social scientists have focused on interactions between humans and their environments since the self-conscious origins of sociology, economics, geography, political science, psychology, and anthropology (Hawley 1950, 1986; Burch 1971; Burch, et al. 1972; Young 1974; Micklin and Choldin 1984; Field and Burch 1988; Lee, et al. 1990; Catton 1994; Agnew, et al. 1996). The explicit incorporation of the ecosystem concept within the social sciences dates to Duncan's (1961, 1964) journal articles "From Social System to Ecosystem" and "Social Organization and the Ecosystem." Recently, the social sciences have focused increasingly on the ecosystem concept because it has been proposed and used as an organizing approach for natural resource policy and management (Cortner and Moote 1999).

The ecosystem concept and its application to *Homo sapiens* is particularly important from a research and education perspective because of its utility as an analytical framework for integrating the physical, biological, and social sciences. The ecosystem concept owes its origin to Tansley (1935), in one of modern ecology's clearest yet most subtle founding documents. Tansley noted that ecosystems can be of any size, as long as the concern was with the interaction of organisms and their environment in a specified area. He noted further that the boundaries of an ecosystem are drawn to answer a particular question. Thus, there is no set scale or way to bound an ecosystem. Rather, the choice of scale and boundary for defining any ecosystem depends upon the question asked and is the choice of the investigator. Further, each investigator may place more or less emphasis on the chemical transformations and pools of materials drawn on or created by organisms; or on the flow, assimilation, and dissipation of biologically metabolizable energy; or on the role of individual species or groups of species in the flows and stocks of energy and matter. The fact that there is so much choice in the scales and boundaries of ecosystems, and how to study and relate the processes within them, indicates the profound degree

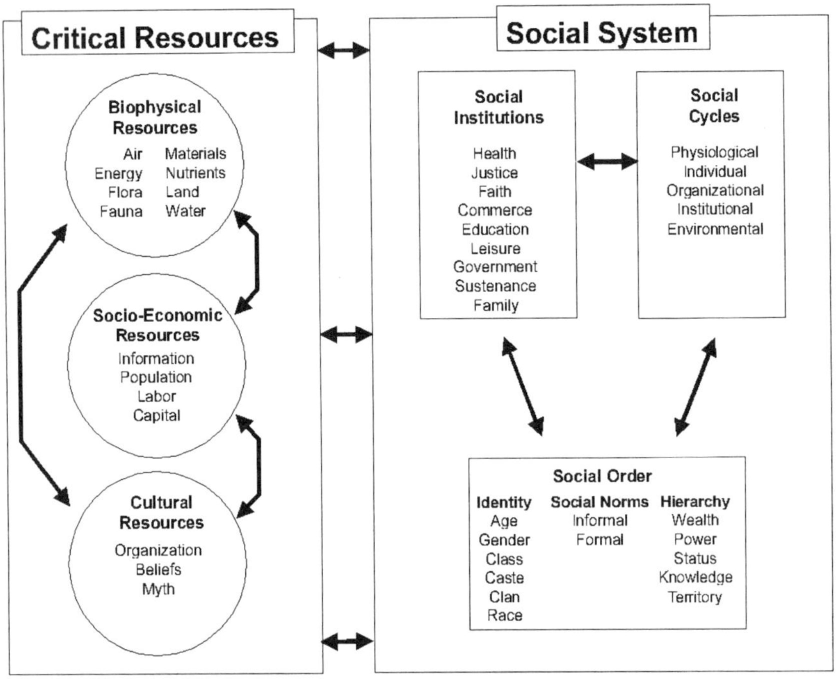

FIGURE 11.1. A framework for human ecological systems. Adapted from Machlis, et al. (1997).

to which the ecosystem represents a research approach rather than a fixed scale or type of analysis.

However, the application of an ecosystem approach to the study of anthropogenic ecosystems requires a revised analytical framework. The analytical framework or parts diagram (Figure 11.1) we identify here and have presented elsewhere (e.g., see Burch and DeLuca 1984; Machlis, et al. 1997; Pickett, et al. 1997) is not a theory in and of itself. As Machlis, et al. (1997) note,

> This human ecosystem model is neither an oversimplification nor caricature of the complexity underlying all types of human ecosystems in the world. Parts of the model are orthodox to specific disciplines and not new. Other portions of the model are less commonplace—myths as a cultural resource, justice as a critical institution. Yet we believe that this model is a reasonably coherent whole and a useful organizing concept for the study of human ecosystems as a life science.

We propose that there are several elements that are critical to the successful application of this framework. First, it is important to recognize that the primary drivers of anthropogenic ecosystem dynamics are both biophysical and social—there is no single, determining driver of anthropogenic ecosystems—and the relative significance of drivers may vary over time. Second, components of this framework need to be examined in the context of each other simultaneously (Machlis 1999). Finally, researchers need to examine how *dynamic biological and social allocation mechanisms*—ecological, exchange, authority, tradition, and knowledge—affect the distribution of *critical resources*—energy, materials, nutrients, population, genetic and non-genetic information, population, labor, capital, organizations, beliefs, and myths—within any human ecosystem (Parker and Burch 1992).

Some readers may examine this parts diagram and wonder if it is complicated with too many terms. However, consider Henderson's (1995) point that,

> . . . trying to run a complex society on a single indicator like the Gross National Product is literally like trying to fly a 747 with only one gauge on the instrument panel. There would be nothing to tell you whether the wing flaps were up or down, whether the fuel tank was full, or what the altitude was. In effect, you'd be flying blind. Or imagine if your doctor, when giving you a checkup, did no more than check your blood pressure!

Henderson's analogies can be taken further. The human ecosystem framework we show here has 45 concepts, which might be represented by one variable per concept, and each variable may be thought of as an "idiot" light or indicator. In contrast, a Cessna airplane may have approximately 50 "idiot" lights that inform the pilot about the status of critical functions of the plane, a Boeing 747 jet has approximately 500 lights, and a NASA Space Shuttle has about 5,000 lights. Given the importance to individuals, groups, and societies of having accurate, timely, and extensive data and understanding about the systems in which they inhabit and on which they depend for life, it may be reasonable some day to strive for at least a Boeing 747–level of "idiot lights" for most human ecosystems (Machlis 1999).

Units of Organization and Scale

Ecologists have a longstanding appreciation for the importance of scale and its associations with different processes. For instance, Urban, et al. (1987) identify different ecological processes/scales and order them by discrete space-time domains. Similarly, one might imagine that the social sciences would be predisposed to a multiscale approach, given that different levels of social organization such as individuals, families, communities, and societies correspond approximately to social science disciplines such as psychology, anthropology, sociology, and political science as well as multiscale disciplines such as geography and economics. Only a few social scientists, however, have explicitly applied scale as an analytical tool to the study of different levels of social organization, particularly as it relates to ecological systems.

Social scientists are increasingly beginning to realize the need for the explicit use of scale as an analytical tool, though it is occurring most frequently in an interdisciplinary context (Grimm, et al. 2000; Pickett, et al. 1999). For instance, the Baltimore Ecosystem Study (BES) has worked to articulate and understand the dynamics of different social scales over time based upon existing social theory. Some examples of issues studied in the BES include:

- Regional variations: urban–rural dynamics (Morrill 1974; Cronon 1991; Rusk 1993);
- Municipal variations: distribution and dynamics of land-use change (Burgess 1925; Hoyt 1939; Harris and Ullman 1945; Guest 1977);
- Neighborhood variations: power relationships between neighborhoods (Shevky and Bell 1955; Timms 1971; Johnston 1976; Agnew 1987; Logan and Molotch 1987; Harvey 1989);
- Household variations: household behavior within communities (Fortmann and Bruce 1988; Fox 1992; Grove and Hohmann 1992; Burch and Grove 1993; Grove 1995).

A Geographical Imagination and Spatial Analyses

All behaviors occur in space and are spatially dependent. For instance, the path an individual walks down a street may depend upon and respond to the path of an approaching person. Changes in one community may depend upon changes in another neighboring community, or the competitive advantage of one port versus another may depend upon their differential access to maritime and land routes for moving goods.

While behaviors occur in space and are spatially dependent, the spatial characteristics of areas frequently are variable or heterogeneous. For instance, the distribution of land uses within the Gwynns Falls watershed, in Baltimore, Maryland varies and has important implications for the economic structure, interactions among neighborhoods, and hydrology of the watershed (Figure 11.2). As we mentioned before, the location of these areas is important. If we took the same land use areas, or patches, and rearranged them like a new puzzle, it is likely that the economic, community, and hydrologic dynamics of the watershed would change.

Crucial to the study of spatial heterogeneity of urban ecosystems is an understanding of underlying processes. Spatial heterogeneity is the combined result of physical, biological, and social processes. An important, underlying driver of patterns of social heterogeneity is social hierarchies. Social hierarchies—or social differentiation—is a significant concept for understanding human ecosystems because it affects the *allocation or flow of critical resources* (re. Figure 11.1: biophysical, socioeconomic, and cultural) and, hence, ecosystem processes. In essence, social differentiation determines "who gets what, when, how, and why."

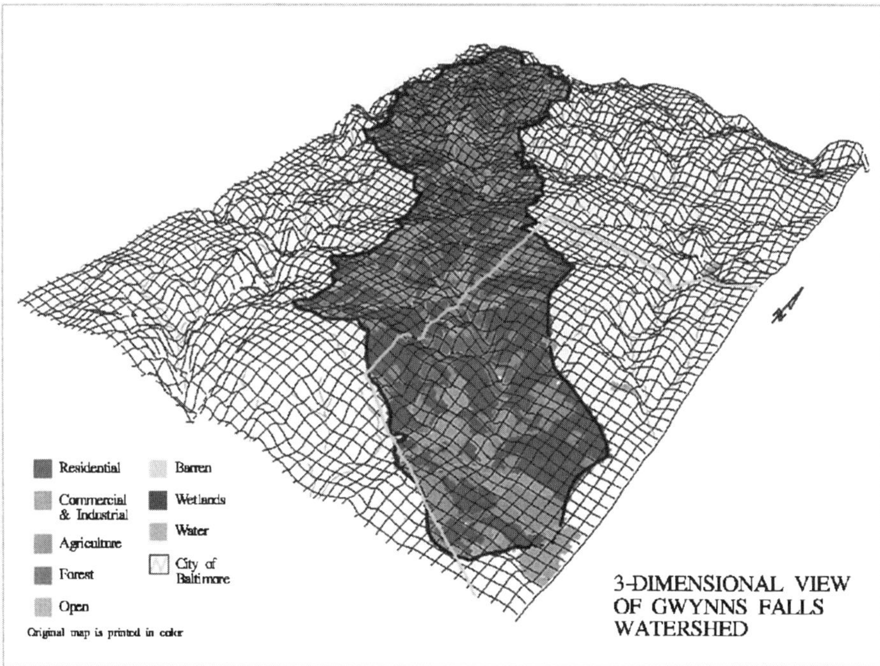

FIGURE 11.2. A three-dimensional view of the Gwynns Falls watershed. From Revitalizing Baltimore Program: Resource Information Systems Component 29 July, 1994.

The allocation of critical resources is rarely equitable and is frequently related to class, race, or ethnicity. As Machlis, et al. (1997) note, unequal access to and control over critical resources is a consistent fact within and between households, communities, regions, nations, and societies. Five types of sociocultural hierarchies are critical to patterns and processes of human ecological systems: wealth, power, status, knowledge, and territory (Burch and DeLuca 1984). *Wealth* is access to and control over material resources in the form of natural resources, capital (money), or credit. The unequal distribution of wealth is a central feature of human ecological systems. *Power* is the ability to alter others' behavior through explicit or implicit coercion (Mann 1984; Wrong 1988). The powerful, often elites with political or economic power, typically have access to resources that are denied the powerless. One example is politicians who make land-use decisions or provide services for specific constituents at the expense of others. *Status* is access to honor and prestige and the relative position of an individual (or group) in an informal hierarchy of social worth (Goode 1978; Lenski 1966). Status is distributed unequally, even within small communities, and high-status individuals may not necessarily have access to either wealth or power. For instance, a minister or an imam may be respected and influential in a community even though he or she is neither wealthy nor has the ability to alter coercively other people's behavior. *Knowledge* is access to or control over specialized types of information (technical, scientific, religious, and so forth). Not everyone within a social system has equal access to different types of information. Knowledge often provides advantages in terms of access to and control over the critical resources and services of social institutions. Finally, *territory* is access to and control over critical resources through formal and informal property rights (Bromley 1991; Burch, et al. 1972; Fortmann and Bruce 1988).

These processes of social differentiation of human ecosystems frequently have a spatial dimension that is usually characterized by patterns of territoriality and that lead to spatial heterogeneity on many scales. This spatial understanding of social differentiation in an ecological context enables an investigator to ask both "who gets what, when, how, why, and *where*?" and, subsequently, to ask about the relationships among spatial patterns and physical, biological, and social processes of a given area (Grove and Burch 1997).

Linkages Between Scales and Across Geography

Scales and spatial analyses can be linked within the context of hierarchy theory. Hierarchy theory attempts to describe the strong and weak linkages within and between scales of a system in order to understand the ways that components at different scales are related to one another. Thus, lower-level units interact to generate higher-level behaviors and higher-level units control those at lower levels. For instance, a hierarchical approach to urban ecosystems (Figure 11.3) may attempt to understand the ways that the interactions among households within a neighborhood affect the ability of a neighborhood to attract public and private investments in trees, parks, and policing, while the competition among neighborhoods in terms of relative political power subsequently affects the quality of government services that each household receives.

At each scale, there are associated endogenous (internal) and exogenous (external) processes. Examples of endogenous change within a neighborhood may include changes in population structure, housing conditions, or vegetation, whereas exogenous change to a neighborhood may include changes in financial markets, regional transportation, or climate. It is crucial to understand that endogenous change at one level appears as exogenous change to the next, lower level. Further, the links between one scale and another may have variable strength and be spatially-dependent. Understanding these linkages between scale, space, and hierarchy are fundamental to urban ecosystem studies.

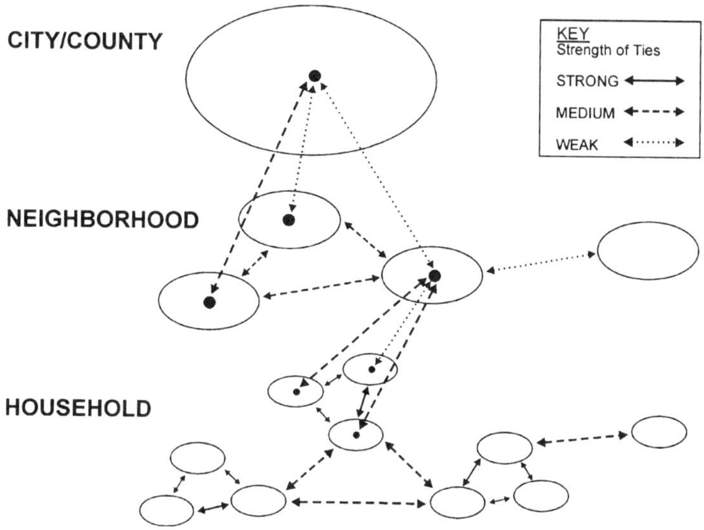

FIGURE 11.3. Example of an urban hierarchical system. Adapted from Urban, et al. (1987).

Learning from Everyday Life: Policies, Plans, and Management as Data

Urban ecosystems are messy systems where controlled experiment often is impossible. For this reason, everyday events—policies, plans, and management activities—represent significant interventions in and powerful means for understanding the dynamics of urban ecosystems. For instance, the Federal Highway Administration's programs of the 1950s significantly affected the settlement and commuting patterns of cities by making it easier for people to travel in and out of cities by car; while the Congressional Fair Housing Act of the 1960s affected the racial and religious composition of cities by removing formal, discriminatory barriers that determined where people could and could not live based only upon their race or religion. Other "everyday" examples include basic means by which cities support their populations with food, water, employment, and leisure.

A Study of Baltimore City's Protected Watershed Forest and Reservoirs

In this section, we use Baltimore City's water supply to illustrate the utility of a social ecology approach for understanding urban ecosystems in general and for education in particular. Our choice to use this example for the class is based upon three points. First, this case study is an appropriate application of the knowledge and skills that Karen Hinson expects her students to develop through their Advanced Placement U.S. History course (described later). Second, Rob Northrop and Morgan Grove are working on the development of forest management plans for the city's protected watershed forests and reservoirs and they are familiar with numerous resources that might be useful to the students. Third, the students' involvement in the project might provide them with an appreciation for the relevance of their studies to their day-to-day lives.

Course Description and the Appropriateness of a Social Ecology Approach

Western School of Technology and Environmental Science (Western) is a Baltimore County Public School (BCPS) dedicated to an integrated approach to instruction and 90 minute classes. The focus of the school is on career and technology programs and environmental science. Students choose from a wide range of science and technical courses and as seniors participate in research projects, internships, apprenticeships, and work experiences. Western prepares students for three options after graduation: college, the workplace, or both. Advanced Placement courses are offered to provide interested and motivated students with the opportunity to study materials in more depth.

Advanced Placement (AP) United States History is a course that is based upon the requirements and outline developed by the College Board and the curriculum developed by BCPS. The main purpose of this course is to prepare students for success on the AP United States History exam. The level of achievement required for success on the AP exam necessitates that students be able to analyze and synthesize data and developments in different areas, analyze themes, make comparisons, evaluate and assess verbal, graphic, and pictorial evidence, and demonstrate mastery of a wide range of knowledge (Advanced Placement Course Description, 1999).

The AP curriculum developed by BCPS is designed to increase student achievement throughout the academic year so that students attain a high level of performance on the AP exam. As a result, the BCPS AP United States History curriculum requires students to acquire, integrate, extend, and refine knowledge; use knowledge meaningfully; and think critically and creatively. Students in AP United States History therefore must be curious, independent learners; masters of content; and expressive writers. In turn, AP United States History teachers must use creative, active, dynamic methods of instruction to ensure a high level of success (AP US History Curriculum 1998).

A social ecology approach is one method of instruction that is available to teachers as they implement an AP program. In AP United States History, the Human Ecosystem Framework was one social ecology tool that was selected as the foundation of the research project. This approach provided a venue for exposing students to collegiate expectations and levels of performance while providing students with the opportunity to use the tools of historians and social scientists in order to find meaning and draw conclusions about the past by analyzing the effects of political, economic, social, and environmental factors throughout history. Social ecology and the Human Ecosystem Framework became the organizing idea around which students constructed meaning and extended their knowledge about the history of the United States.

Project Description and Its Relevance to the Course's Existing Curriculum

The infusion of social ecology and the Human Ecosystem Framework into the AP United States history course resulted in an integrated approach, which required students to investigate biological and human systems in order to identify interactions and patterns throughout history. Students utilized skills of inquiry and learning to answer student-generated questions about United States history. This approach did not require the students to investigate information outside of the realm of the curriculum. Rather it provided a unique opportunity to integrate students' prior learning in other disciplines with the new learnings of the AP United States History curriculum. In this instance students were able to transfer their knowledge from science to history. It is important to note that this project could have been just as successful if students were bringing their knowledge of history into an AP Environmental Science class. For instance, students could have applied their knowledge of civil war history to a discussion of late nineteenth century land use patterns around the Loch Raven Reservoir. As a result of such integration, students were able to process the AP United States history information on a variety of historical and theoretical levels, which prepared them for the higher-level thinking skills required by the AP exam. This process of integration is important because it allows students to apply knowledge meaningfully to address real-world issues, conflicts, and solutions.

Background

In conjunction with the Baltimore Ecosystem Study, the students were teamed with a diverse group of biological and social scientists and professionals in order to delve into Baltimore's and the United States' history and the links among the two. Students were able to explore the history of their local area as it relates to the history of the United States as a whole while at the same time exploring how the biological environment affected the development and needs of the human system. This project had two main objectives for students:

1. To apply primary and secondary research skills in order to determine the effects of political, economic, social, and environmental factors throughout history on the watersheds of Loch Raven Reservoir, Prettyboy Reservoir, and Liberty Reservoir.
2. To research and write a book on the land-use history of Loch Raven Reservoir, Prettyboy Reservoir, and Liberty Reservoir watersheds in order to explain and synthesize how the human ecological system has affected political, economic, social, and environmental history throughout the centuries in these locations.

Students were given the task of developing a historical narrative of the effects of political, economic, social, and environmental factors throughout the entire period of United States history in the study areas that became Loch Raven Reservoir, Prettyboy Reservoir, and Liberty Reservoir in order to provide a forester with the historical basis on which to develop a management plan for the forest ecosystems surrounding the three reservoirs. Students were instructed how to use field research as well as historical research in libraries and historical societies throughout the area in order to accomplish this task.

Approach

The research project was a multitiered approach designed to have the students learn from their own research, from each other, and from professionals in the field.

The first tier related to the organization of the research teams. Students were divided into three research teams, one for each reservoir in the Baltimore area: Loch Raven, Liberty, and Prettyboy. Each team had seven members, each of whom was responsible for researching a specific topic:

- Demographics
- Economic structure
- Political structure
- Transportation
- Class, race, and religion
- Public health
- Leisure and recreation

The topics selected are integral to a social ecology approach and the Human Ecosystem Framework. They also reflect the emphasis placed by the BCPS AP United States History curriculum and the AP United States History exam on "political institutions and behavior and public policy, social and economic change, diplomacy and international relations, and cultural and intellectual developments" (Advanced Placement Course Description 1999). By organizing students into three research teams, they were able to learn from each other about a wider range of topics, to compare results and synthesize data across topics, and thus to develop a more complete account of the land-use history of the reservoir.

The second tier focused on the learning of historical knowledge. As the class encountered each unit in AP United States History, the students' task was to collect data on her or his topic that corresponded to the unit. Each student was responsible for gathering and analyzing data regarding her or his topic and how the topic affected people's attitudes about and towards the environment throughout each historical period. As a result, students were constantly processing United States history at both a national and local level in regards to trend, events, and issues.

The third tier focused on the biological and social scientists and professionals that interacted with the students throughout their research project. The students' link to the scientists and professionals was a result of the connection between the teacher, Karen Hinson and a main contact, Morgan Grove. Dr. Grove was able to provide the students with the opportunity to meet, discuss, exchange e-mail with, and learn from scientists and professionals. Dr. Grove and Ms. Hinson facilitated the relationship between the students, scientists, and professionals by anticipating the needs and approaches of all those in the network.

Project Results and Preparation for Student Assessments

Many school systems throughout the United States are being confronted with county, state, and national assessments that gauge student achievement and performance in specific disciplines. When students are able to perform at high levels on performance assessments, they are better prepared for traditional assessments such as county, state, and national assessments. This research project was also a performance piece for these students. The performance assessment was the book that they provided to the forester in June 1999.

The ultimate assessment for these students is the AP United States History exam. The AP United States History exam requires students to analyze and synthesize data and developments in different areas, analyze themes, make comparisons, evaluate and assess verbal, graphic, and pictorial evidence, and demonstrate mastery of a wide range of knowledge. This project reinforced these skills by having students discover, assess, evaluate, and synthesize data they collected from primary sources such as journals, census records, letters, artwork, maps, and statistical data in order to make comparisons and analyze themes in the reservoir areas. The transfer of knowledge between the local history of the reservoirs to the national history of the United States allowed them to demonstrate proficiency and mastery of knowledge required for success on the AP United States History exam. In the end, they had to express their findings in a written form as they would on the AP United States History exam. The final result was a 22 percent increase in the students' AP scores from the previous year.

The students summed up the success of this project in the introduction to their book.

> Amongst the main objectives through the project we, as a class, hoped to gain knowledge of the correlation between people and their environment. These goals combined knowledge of research experience, observing how our own environment changed over time, to understand that historical trends, research methods, and analysis are important to many people in their daily work, to understand how history is relevant to decisions made about where we live, and to witness how our work will be used to protect and serve the reservoirs and people that are using the reservoir watershed. This was also the first time for many of us to complete a large-scale study that would represent our own work in a usable, real situation.

Finally, the students concluded that:

> There were successes, both personal and professional throughout our research exploration. Some personal successes included finding obscure data, completing a difficult portion of research, and working in a group to complete a project of which we are proud. Professional successes included making a reputation for our program, working with professionals throughout the study, and the use of such materials as the Human Ecosystem Framework to produce a complete and well-oriented discussion of our environment and habitat.

Relating to Students' Lives

When instruction relates to students' lives in a meaningful way, students become a part of the learning experience. In our course students felt connected to history and the learning became more enduring. By focusing on the communities in which the students live, they began to see history as a process that affects everyone as opposed to being a subject that is merely learned in high school. Students then began to see the relevance of history to their lives and the importance of their communities. They are then able to become capable and informed citizens and decision makers.

Students also began to see that the information they learn and collect is relevant to the public policy decisions relevant to the areas where they live. Public policy began to take on a new perspective for the students. Such decisions as where to locate a reservoir, which once seemed to be made in isolation, began to be seen as complex, multifaceted, historical decisions that can be influenced by informed citizens.

Students also began to see that for people to do their jobs successfully on a daily basis, knowing and understanding historical trends, research methods, and analysis are important. Students were surprised to hear biological scientists discuss history and social scientists discuss ecology. By interacting with a variety of biological and social scientists and professionals, students became exposed to job possibilities and careers of which they previously may have not been aware. They also became aware of the interdisciplinary nature of many jobs. Thus, the project became a career development experience as well as a research experience. Indeed, for several of these students the project became the impetus for senior research internships with the Baltimore Ecosystem Study.

Summary and Epilog

We have tried to provide what we consider to be fundamental concepts to a social ecology approach for teaching about urban ecosystems. Further, we have identified key criteria for a social ecology approach in existing curriculum. Finally, we illustrated this approach using our experience working with Karen Hinson's AP United States History course.

We would like to add another point from our experience during the 1998–1999 school year. We found that this project depended upon process as well as content and was successful because of the network of participants—students, teacher, main science contact, scientists, and professionals. When working with such a diverse population, it is important to:

- Ensure the project is appropriate to the curriculum.
- Make certain the project supports student assessments.
- Ensure the project relates to student lives.
- Anticipate the needs and approaches of all members of the network.

During the past 20 years the human population within the Chesapeake Bay ecosystem has increased from 12.5 million to 16 million. Most of the population growth occurred adjacent to metropolitan areas, where 75–100 percent increases are expected during the next century. This population growth and related land use change within the Chesapeake Bay ecosystem has been linked directly to declining water quality and living resources (US EPA 1983).

The city of Baltimore, Maryland, is representative of the problems and opportunities associated with these expanding metropolitan regions. The city and its sprawling suburbs now constitute a complex urban ecological system, characterized by widespread and substantial forms of human interactions with their biosphere (Pickett 1997). Implementation of conservation and restoration strategies for the sustainable development of these urban ecosystems, which now constitute a significant component of the greater Chesapeake ecosystem, has become an integral part of the interstate Chesapeake Bay restoration program.

Resource issues (e.g., those focused on by Karen Hinson's AP United States History course) ultimately concern human behavior and are similar to those faced by numerous metropolitan areas around the world. Initially developed in the rural hinterlands, Baltimore's three reservoirs now are embedded in rapidly urbanizing watersheds. Eutrophication and sedimentation now jeopardize the long-term ability of the reservoirs to store the quantity and quality of water needed by the 1.5 million residents of the region that depend on them. On top of the need for water, growing population, affluence, and accessibility have fueled an increased demand from residents throughout the region for public recreation opportunities and conservation of biological diversity.

The protection of important ecological processes such as hydrology, nutrient cycling, and maintenance of biodiversity across these heterogeneous landscapes requires the development of cooperative and collaborative working relationships among individuals, communities, businesses, and government agencies (Sample 1994). Forging and sustaining working partnerships will require a halt to mediating conflicts exclusively through technological fixes and expert advice. Resolving competing issues will need "transformative" processes (Bush and Folger 1994) leading to "a change or refinement in the conscience or character of individual(s) . . ." Such a change will require a willingness and ability to listen and learn and a more empathetic relationship with the environment, other members of the community, and other generations: current, past, and future.

Resource decisions will need to integrate the ecological and social dimensions of landscapes in order to provide a more realistic foundation for the selection of conservation strategies. This will require the ability to "reimagine" the biophysical components of Baltimore's watersheds within a social context that will affect the way we think about the reservoir lands and the questions we ask about them in assessing current conditions and future trends. To achieve this, natural resource managers and local citizens will need to understand how to incorporate social information and understanding into their decision making. They will need to incorporate methods committed to providing simple tools to facilitate the rapid collection and interpretation of relevant social data. Experts will need to make a demystified social science available as everyone's tool (Korten 1984).

The social ecology approach we have described provides a logical framework for interdisciplinary assessment, learning, and communication. It is a guide for identifying data needs and a tool for land managers to use in assessing the human dimensions of alternative conservation activities. It is a model for a curriculum that develops students' skills and insights needed for integrating scientific knowledge of ecological and social relationships. It is a significant initial step in our efforts to develop a unified, holistic approach to the conservation of our ecosphere and prompts us to continue our pursuit of the highest and best ideal of conservation: integrating people and their needs with their environment. Ultimately, we hope we have provided the basis for our students to develop the knowledge and judgment they will need as leaders for the conservation of urban ecosystems in the twenty-first century.

Acknowledgments. Support for this research was provided by the Burlington Laboratory (4454) and Global Change Program; Northeastern

Forest Research Station, USDA Forest Service; the National Science Foundation—NSF Grant #DEB–9714835; the Environmental Protection Agency—EPA Grant #R–825792-01-0; and Western School of Technology and Environmental Science, Baltimore County Public Schools, Baltimore County, Maryland. We would also like to thank the following individuals, who made presentations to the class: Mr. Mike Ratcliffe, Census Bureau, Department of Commerce; Mr. Peter D'Adamo, P.E., Johns Hopkins University; Mr. John C. Moore, P.E., Rummel, Klepper, and Kahl, L.L.P.; Dr. Gary Machlis, National Park Service, Department of the Interior; Mrs. Donna Williams, Maryland Historical Society; and Dr. William R. Burch and the students of his Ecosystem Management Course, School of Forestry & Environmental Studies, Yale University.

References

Advanced Placement Course Description. 1997. United States History. College Entrance Examination Board and Educational Testing Service, New York.

Advanced Placement United States History Curriculum. 1998. Baltimore County Public Schools, 6901 Charles Street, Towson, MD.

Agnew, J.A. 1987. *Place and politics: the geographical mediation of state and society*. Allen & Unwin, Boston.

Agnew, J., D.N. Livingstone, and A. Rogers, eds. 1996. *Human geography: an essential anthology*. Blackwell Publishers, Cornwall, Great Britain.

Bailey, K.D., and P. Mulcahy. 1972. Sociocultural versus neoclassical ecology: a contribution to the problem of scope in sociology. *The Sociological Quarterly* 13(Winter):37–48.

Berry, W. 1990. Word and flesh. *Whole Earth Review* Spring:68–71.

Bromley, D.W. 1991. *Environment and economy: property rights & public policy*. T.J. Press, Cornwall, Great Britain.

Burch, W.R., Jr. 1971. *Daydreams and nightmares: a sociological essay on the American Environment*. Harper & Row, New York.

Burch, W.R. 1988. Human ecology and environmental management. Pages 145–159 in J.K. Agee, and R.J. Darryll, eds. *Ecosystem management for parks and wilderness*. University of Washington Press, Seattle, WA.

Burch, W.R., Jr., N.H. Cheek Jr., and L. Taylor, eds. 1972. *Social behavior, natural resources, and the environment*. Harper & Row, New York.

Burch, W.R., Jr., and D.R. DeLuca. 1984. *Measuring the social impact of natural resource policies*. New Mexico University Press, Albuquerque, NM.

Burch, W.R., Jr., and J.M. Grove. 1993. People, trees and participation on the urban frontier. *Unasylva* 44(173):19–27.

Burgess, E.W. 1925. The Growth of the City: An introduction to a research project. Pages 47–62 in R.E. Park, E.W. Burgess, and R.D. McKenzie, eds. *The city*. University of Chicago Press, Chicago, IL.

Bush, R.A.B., and J.P. Folger, 1984. *The promise of mediation: responding to conflict through empowerment and recognition*. Jossey-Bass, San Francisco, CA.

Catton, W.R., Jr. 1994. Foundations of human ecology. *Sociological Perspectives* 37(1):75–95.

Cortner, H.J., and M.A. Moote. 1998. *The politics of ecosystem management*. Island Press, Washington, DC.

Cronon, W. 1991. *Nature's metropolis: Chicago and the great west*. W.W. Norton & Co., New York.

Delcourt, H.R., and P.A. Delcourt. 1988. Quaternary landscape ecology: relevant scales in space and time. *Landscape Ecology* 2:45–61.

Eldredge, N. 1995. *Dominion*. California University Press, Berkeley, CA.

Field, D.R., and W.R. Burch Jr., eds. 1988. *Rural sociology and the environment*. Social Ecology Press, Middleton, WI.

Fortmann, L., and J.W. Bruce, eds. 1988. *Whose trees? Proprietary dimensions of forestry*. Westview Press, Boulder, CO.

Fox, J. 1992. The problem of scale in community resource management. *Environmental Management* 16(3):289–297.

Golley, F.B. 1993. *A history of the ecosystem concept in ecology: more than the sum of the parts*. Yale University Press, New Haven, CT.

Goode, W.J. 1978. *The celebration of heroes: prestige as a social control system*. University of California Press, Berkeley, CA.

Gould, S.J. 1994. In the mind of the beholder. *Natural History* 2:14–23.

Grimm, N.B., J.M. Grove, S.T.A. Pickett, and C.L. Redman. 2000. Integrated approaches to long-term studies of urban ecological systems. *BioScience* 50:571–584.

Grove, J.M., and W.R. Burch, Jr. 1997. A social ecology approach to urban ecosystem and landscape analyses. *Journal of Urban Ecosystems* 1(4):259–275.

Grove, J.M., and M. Hohmann. 1992. GIS and social forestry. *Journal of Forestry* 90(12):10–15.

Guest, A.M. 1977. Residential segregation in urban areas. Pages 269–336 in K.P. Schwirian, ed. *Contemporary topics in urban sociology*. General Learning Press, Morristown, NJ.

Hagen, J.B. 1992. *An entangled bank: the origins of ecosystem ecology*. Rutgers University Press, New Brunswick, NJ.

Hansson, L., L. Fahrig, and G. Merriam, eds. 1995. *Mosaic landscapes and ecological processes*. Chapman Hall, New York.

Harris, C.D., and E.L. Ullman. 1945. The nature of cities. *Annals of the American Academy of Political and Social Science* 242:7–17.

Harvey, D. 1989. *The urban experience*. Johns Hopkins University Press, Baltimore, MD.

Hawley, A.H. 1950. *Human ecology: a theory of community structure*. Ronald Press, New York.

Hawley, A.H. 1986. *Human ecology: a theoretical essay*. University of Chicago Press, Chicago, IL.

Henderson, H. 1995. *Paradigms in progress: life beyond economics*. Brett-Koehler Publishers, San Francisco, CA.

Hewlett, J.D., and W.L. Nutter. 1969. *An outline of forest hydrology*. Revised edition, University of Georgia Press, Athens, GA.

Johnston, R.J. 1976. Residential area characteristics: research methods for identifying urban sub-areas—social area analysis and factorial ecology. Pages 193–235 in D.T. Herbert and R.J. Johnston, eds. *Spatial perspectives on problems and policies. Volume 2.* John Wiley & Sons, New York.

Korten, D.C. 1984. Rural development programming: the learning process approach. Pages 176–188 in D.C. Korten, and R. Klauss, eds. *People-centered development: contributions toward theory and planning frameworks*. Kumain Press, West Hartford, CT.

Lee, R.G., W.R. Burch, Jr., and D.R. Field. 1990. Conclusions: past accomplishments and future directions. Pages 277–289 in R.G. Lee, D.R. Field, and W.R. Burch, Jr., eds. *Community & forestry: continuities in the sociology of natural resources*. Westview Press, Boulder, CO.

Lenski, G.E. 1966. *Power and privilege: a theory of social stratification*. McGraw-Hill Book Company, New York.

Logan, J.R., and H.L. Molotch. 1987. *Urban fortunes: the political economy of place*. University of California Press, Los Angeles, CA.

Machlis, G.E., J.E. Force, and W.R. Burch, Jr. 1997. The human ecosystem as an organizing concept in ecosystem management. *Society & Natural Resources* 10: 347–367.

Mann, M. 1984. *The sources of social power: volume 1, a history of power from the beginning to A.D. 1760*. Cambridge University Press, New York.

Mayr, E. 1982. *The growth of biological thought: diversity, evolution, and inheritance*. Harvard University Press, Cambridge, MA.

McIntosh, R.P. 1985. *The background of ecology: concept and theory*. Cambridge University Press, New York.

Micklin, M., and H.M. Choldin. 1984. *Sociological human ecology: contemporary issues and applications*. Westview Press, Boulder, CO.

Morrill, R.L. 1974. *The spatial organization of society. Second edition*, Duxbury Press, Duxbury, MA.

Parker, J.K. 1997. Integrating biophysical and social science knowledge in developing countries: social ecology and its potential contributions to ecosystem management in the United States. Pages 61–71 in K. Cordell and J. Bergstrom, eds. *Integrating social sciences and ecosystem management*. Sagamore Press, Champaign, IL.

Parker, J.K., and W.R. Burch, Jr. 1992. Toward a social ecology for agroforestry in Asia. Pages 60–84 in W.R. Burch, Jr. and J.K. Parker, eds. *Social science applications in Asian Agroforestry*. IBH Publishing Co., New Delhi, India.

Pickett, S.T.A., W.R. Burch, Jr., S. Dalton, T. Foresman, J.M. Grove, and R. Rowntree. 1997. A conceptual framework for the study of human ecosystems in urban areas. *Journal of Urban Ecosystems* 1(4):185–199.

Pickett, S.T.A., W.R. Burch, Jr., and J.M. Grove. 1999. Interdisciplinary research: maintaining the constructive impulse in a culture of criticism. *Ecosystems* 2:302–307.

Rusk, D. 1993. *Cities without suburbs*. Woodrow Wilson Center Press, Washington, DC.

Sample, V.A. 1994. *Building partnerships for ecosystem management on mixed ownership landscapes*. Forest Policy Center of American Forests, Washington, DC.

Shevky, E., and W. Bell. 1955. *Social area analysis: theory, illustrative application and computational procedure*. Stanford University Press, Stanford, CA.

Simpson, G.G. 1964. *This view of life*. Harcourt, Brace, and World, New York.

Tansley, A.G. 1935. The use and abuse of vegetational concepts and terms. *Ecology* 16:284–307.

Timms, D. 1971. *The urban mosaic: towards a theory of residential differentiation. Volume 2.* Cambridge University Press, Cambridge, England.

Urban, D.L., R.V. O'Neill, and H.H. Shugart, Jr. 1987. Landscape ecology: a hierarchical perspective can help scientists understand spatial patterns. *Bioscience* 37: 119–127.

US EPA. 1983. *The state of the Chesapeake Bay*. Annapolis, MD.

Wilson, E.O. 1975. *Sociobiology*. Harvard University Press, Cambridge, MA.

Wilson, E.O. 1998. *Consilience: the unity of knowledge*. Alfred A. Knopf, Inc., New York.

Wrong, D.H. 1988. *Power: its forms, bases, and uses*. University of Chicago Press, Chicago, IL.

Young, G.L. 1974. Human ecology as an interdisciplinary concept: a critical inquiry. *Advances in Ecological Research* 8:1–105.

12
The Historical Dimension of Urban Ecology: Frameworks and Concepts

Martin V. Melosi

The historical evolution of cities, from an ecological perspective, requires a clear understanding of the place of the city in the physical world. In a discussion several years ago about the nature of cities, a long-time colleague admonished me by saying, "Cities are not trees." Despite the friendly remonstrance, the notion of cities as natural environments is worth exploring if for no other reason than it helps us to reflect upon the place cities occupy in the physical world. Comparisons between cities and anthills or beehives, for example, at the very least connect cities to the physical world rather than excluding them from it. In this sense, cities, like their human builders, are part of nature.

The differentiation between the natural and the human-made world is a persistent theme in environmental history. Nature is traditionally understood as nonhuman—what environmental historian Donald Worster (1988) has called "the world we have not in any primary sense created." In this definition he excluded the social environment (i.e., "the scene of humans interacting only with each other in the absence of nature"), and the built or artifactual environment (i.e., "the cluster of things that people have made and which can be so pervasive as to constitute a kind of 'second nature' around them"). Thus, he attempted to distinguish between the natural and the built environment "for it reminds us that there are different forces at work in the world and not all of them emanate from humans. . . ." Worster justified the exclusion of the built environment from his definition of the main contours of environmental history by arguing that "the built environment is wholly expressive of culture; its study is already well advanced in the history of architecture, technology, and the city" and concludes that "when we step beyond the self-reflecting world of humankind to encounter the nonhuman sphere, environmental history finds its main theme of study."

While the differentiation between the natural and the human-made is a long-standing motif, well before the current generation of environmental historians, such a view makes several unwarranted assumptions. First, it does not account for the generations-old debate about the nature of cities within an environmental context, initially expressed in the organic theory. For example, how can we justify as part of the main theme of environmental history the study of human intrusion in the natural world in the form of farming and husbandry, and not in the building of a town or city? Second, how can we understand "the role and place of nature in human life" if we create an artificial physical environment devoid of human communities—including cities? Humans have not simply encountered nature as individuals, but as parts of groups, and if not in cities then in towns and villages or as members of nomadic clans regularly setting up and breaking down camps. And third, while the built environment is expressive of culture, it is not wholly expressive of culture, since upon its creation it is part of the physical world, and interacts and sometimes blends with the natural world. From the vantage point of human history, isolating the "natural world" in such an unnatural way denies the powerful holistic quality that the discipline of history can bring to the study of urban ecology.

The articulation of organic theory vis-a-vis cities, an appreciation of the city as a modifier of the environment, the development of ecological theory by the Chicago School, and the emergence of systems theory are important underpinnings for grasping the historical dimension of urban ecology. These themes, by their nature multidisciplinary, can help to establish an effective context for teaching key concepts about the development and impact of urban ecosystems.

Organic Theory

While it never gained universal appeal, the idea of the city as a natural system inspired graphic metaphors. One notion—organic theory—related the structure and operation of the city to that of the human body. According to urbanist Graeme Davison (1983),

> few ideas have exercised as powerful an influence upon students of urban society as the organic or biological conception of the city. From Aristotle's *Politics* to the Chicago School and beyond, social theorists have likened cities to bodies or organisms; dissected them into constituent organs, such as 'heart,' 'lungs,' and 'arteries'; and charted their growth and decay. These metaphors reflect a longstanding conflict in western thought. On the one hand, cities were exalted as the intelligent creation of civilized man and were sharply distinguished from the products of unreflective nature. Yet they also manifested an astonishing order within their vast complexity, and demonstrated a capacity for growth and self-regulation that resembled the working of nature itself. Akin to nature, cities nevertheless stood apart from nature, and so reflected man's own ambiguous relationship to the natural order. From time to time, the balance between these ideas—the city as man-made; the city as natural—has shifted back and forth in response to changing experiences of urban life and changing assumptions about man and his place in nature.

Davison went on to suggest that the idea of the city as a natural system "became the dominant paradigm among the first generation of middle-class urban investigators" (in Great Britain at least) in the late eighteenth and early nineteenth centuries. On one level, it reinforced the theories of laissez faire economists and natural historians—"the chief ideologists of the commercial middle-class"; and on another it "endorsed the technocratic professionalism of sanitarians and other reformers" of a Malthusian bent (Davison 1983).

Such a theory had obvious flaws, most especially attempting to make a leap in logic from an integrated biological system subject to Darwinian processes, such as natural selection, to a less-integrated entity subject to very different modes of selection and development. Organic theory, in some cases, was unfairly put to the employ of certain class interests, who used their elite status to grant unjustifiable power to themselves. In asserting the notion that society was like the human body, they granted themselves the role of the head and certainly the clenched fist. Nevertheless, organic theory did elicit powerful images of community interdependency and the rational functioning of many components of towns and cities.

The organic theory also found its advocates in more modern times, among thinkers whose general philosophies vary substantially. Spenser W. Havlick (1974) argued that a city or town is "a transformed combination of resources (land, water, air, mineral, and human)" and that the major goal of urbanization is "to convert the resource base into cities." The result is the city as "a second order resource" that provides benefits to the urbanites, to the region, and to the nation.

Sociologist David Harvey (1973) agreed that an urban system is "a giant man-made resource system." Applying Marxist theory, he refined that concept by suggesting that "the growth of this man-made resource system involves the structuring and differentiation of space through the distribution of fixed capital investments." At the heart of Havlick's and Harvey's definitions is not so much an environment akin to natural systems but a construct dependent on reordering of natural resources to form a new order. In some sense it continues to embrace the organic nature of cities in form, if not in operation.

Like Harvey, city and regional planning professor Manuel Castells (1983) placed emphasis on human action in structuring cities, but also understood cities as dynamic rather than static: "Cities are living systems, made, transformed and experienced by people. Urban forms and functions are produced and managed by the interaction between space and society, that is, by the historical relationship between human consciousness, matter, energy and information."

Ascribing to the urbanization process a defining term normally limited to natural phenomena, geographers Thomas R. Detwyler and Melvin G. Marcus (1972) viewed the city as "a relatively new kind of ecosystem on the face of the earth." It is, however, an "open system"—not self-contained, not functioning independently or in isolation from the rest of the world. In this usage, "ecosystem" has some descriptive power without attempting to create a strict biological model.

The views of Havlick, Harvey, Castells, Detwyler, and Marcus are, in their ways, modifications of the organic theory, but still rooted in it. While the notion of a city as a human body analog is not altogether persuasive, the idea of the city as animate—if not "natural"—is essential for an understanding of urban growth and development. Cities are not static backdrops for human action, nor are they organic metaphors, but ever-mutating systems (Lynch 1960; Blumenfeld 1982).

Urban historians have recognized that point, at least in the broadest periodization of their field of study. It is common in American urban history, for example, to utilize an economic construct—preindustrial, industrial, postindustrial city—or even a spatial construct—walking city, networked city, metropolis—to identify city form from the colonial era to the present. In the former, the periodization owes considerably to the relationship between economic change and urban growth, which is understood as central to the process of city building. It depends largely on the assumption that changes in economic activity (e.g., from commercial to industrial development) resulted in requisite change in urban development. Few scholars would deny this connection, albeit with the reservation that waves of economic change were not geographically uniform and that economic forces were part of a more complex matrix of issues. In the latter case that focuses on spatial variables, factors such as transportation and service delivery are linked to outward and upward growth and expansion, complemented by changing residential and work patterns for the urban population.

David R. Goldfield and Blaine A. Brownell (1990) sought to combine the economic and spatial chronologies by formulating four urban periods: the Market Place, 1790–1870; the Radical Center, 1870–1920; the Vital Fringe, 1920–1970; and the Multicentered Metropolis, 1970–present. In such a chronology changes from commercial to industrial (and postindustrial) economic development were fused with urban growth patterns from the core out to the suburbs and beyond. Any of these periodization patterns, however, represents only the grossest attention to the relentless process of change taking place in cities from day to day, let alone from decade to decade.

From the perspective of some environmental historians, such as William Cronon (1991), cities are part of what he calls "Second Nature." In this context, cities are seen as human creations, not inhuman contrivances. They are places where people live and work, and virtually constitute our environment in many cases. In *Nature's Metropolis: Chicago and the Great West*, Cronon (1991) analyzes how the city acts as an agent for extracting resources from the hinterland, channeling them into the city to be converted into commodities of use to society. The focus is on grain, forests, and cattle, and how cities transformed them into flour, lumber, and meat. These commodity flows are utilized to demonstrate the connection between hinterland and city, rather than to establish a rigid model of how cities function spatially. Cronon's conception of Second Nature, however, keeps the built environment from being viewed as unnatural, and thus connects to organic theory in a general way.

City as Modifier

Cities also are major modifiers of the physical environment. "Their existence," geographer Ronald J. Johnston (1982) noted, "can influence the course of basic physical processes, such as the hydraulic cycle." Urbanization removes much of the filtering capacity of soil and rapidly channels precipitation into available watercourses, thus encouraging flooding. City building affects the atmosphere by increasing airborne pollutants and also creating heat islands where temperatures are greater than the surrounding area. Various urban activities produce huge volumes of waste products, which require complex disposal mechanisms (Johnston 1982). As Detwyler and Marcus (1972) concluded, "Unfortunately, the urban ecosystem seldom treats air and water resources by riparian standards; that is, they are not returned to the ecosphere in the same condition in which they were received." Such an observation reinforces the need to examine further how cities transform environmental conditions internally and externally over time. This has been one of the primary goals of urban environmental historians, who have dealt with air, water, and land pollution, city services, and public health throughout human history. The greatest number of books and articles, however, have focused on internal changes in environmental conditions in cities, rather than how cities impact their surroundings (Melosi 1993).

Alternatively, cities have the capacity—when properly designed—to use resources more efficiently than highly decentralized populations. Concentration can be an advantage in providing services, offering social and cultural opportunities, and producing and distributing goods. Such an observation has escaped the existing historical literature to a large extent. An ecological perspective on cities, however, raises the possibility of exploring such a potentially important topic.

Ecological Theory and the Chicago School

In an attempt to understand the broad features of the urban environment, sociologists and geographers in particular developed theories of urban ecology over the years. The origins of the ecological approach to spatial and social organization can be traced to nineteenth-century concepts and principles conceived by plant and animal ecologists. Urban sociology, however, was born at the University of Chicago during World War I with the work of Robert E. Park (Park, et al. 1925; Park 1952) and Ernest Burgess (1925) as noted by, among others, Berry and Kasarda (1977).

Some refer to the Chicago School as the "subsocial school," because, as Gideon Sjoberg (1965) stated, its members had been intent upon studying humans in their "temporal and spatial dimensions and explaining the resulting patterns in terms of subsocial variables." The fundamental subsocial variable was "impersonal competition," a concept borrowed from nineteenth-century Social Darwinism and classical economics, which emphasized laissez faire doctrine and the operation of the marketplace. Those committed to the ecological perspective of the Chicago School concentrated on factors determining urban spatial patterns and their social impacts. According to the proponents, spatial arrangement of cities was dependent on competitive economic and social forces. Variables such as family types and social status and problems such as crime and alcoholism, they argued, have spatial configurations within cities.

In the 1930s, the Chicago School's Roderick D. McKenzie introduced the theory of "ecological expansion" to the field of human ecology. According to Kasarda (1972), the theory stipulates that "population growth in peripheral areas of a system will be matched with an increase in organizational functions in its nucleus to insure integration and coordination of activities and relationships throughout the expanded system." For cities this implies that outward growth would occur without loss of contact with the core of settlement.

After its heyday in the 1930s and early 1940s, the ecological approach withered. But in 1950, Amos Hawley (1950) resurrected the ecological approach in the field of sociology. Building on the work of his mentor, Roderick McKenzie, Hawley attempted to explain the relationship between population size and urban organizational structure. According to the theory, population growth along the periphery of an urban system will be matched by an increase in organizational functions at the core to insure stability in the expanded system. This pattern of growth produced a core city and a series of dependent suburbs. This was a valuable starting point for thinking about peripheral growth of cities in general terms, but one that smacked of organic theory to some degree. Suburban development in American cities in particular, however, preceded Hawley's theory by many decades (certainly by the early nineteenth century) and manifested several variations. In the hands of historians, Hawley's work offered a broad guideline rather than a rigid model.

What began in sociology as an emphasis on social problems in central cities led to analyses of the relationships among communities within metropolitan areas and to comparative urban research. The theoretical focus also splintered into several distinct perspectives, among which an urban ecological approach appeared in various forms. Economic, technological, and socio-cultural variables received primacy in different theories. Otis Dudley Duncan (1960) and Leo Schnore (1965), however, employed the concept of the "ecological complex" with four basic components—environment, population, social organization, and technology—which they viewed as functionally interrelated, as noted in Sjoberg (1965) and Huth (1970). Contention over the key variable(s) in the spatial and social development of cities was a primary factor in splintering the adherents to urban ecology. For historians simply to embrace the most monocausal of those theories is inappropriate. But the notion of an "ecological complex" has merit precisely because it extends the study of urbanization beyond city walls, requiring the researcher to examine external as well as internal influences shaping growth and development.

The attempt by Eric E. Lampard (1983) to apply social science theory to the study of urban growth set an agenda for historians beginning in the early 1960s. Lampard perceived of the city as an ecological complex—somewhat like Duncan and Schnore—of population, the physical environment, technology, and social organization which could be employed to determined the "changing structure and organization" of communities. As he continued to assert throughout his career, "the fate of urbanized areas, like that of cities, is always determined in *interaction* with the world around." Even though Lampard contributed substantial basic research of his own, his major contribution was to conceptualize about the process of growth, coaxing others to do likewise (Lampard 1961, 1965, 1968).

Another point of contention in urban ecology has been the nature of the relationship between urbanization and social organization. The work of Lewis Mumford (1938, 1945, 1961, 1963) most immediately comes to mind. Rather than strictly an urban ecologist, Sjoberg (1965) treated Mumford as "more a moralizer than a scientist," whereas Mumford's biographer Donald L. Miller (1992) viewed Mumford as an "urban historian, urban visionary." Because, as Sjoberg (1965) perceptively noted, Mumford viewed the crucial problems of modern society as "products of an imbalance between nature and human culture," his works sharply condemned the modern metropolis for veering so far from the Athenian "polis." To Mary Jo Huth (1970), Mumford and some "traditional materialists" aligned with the Chicago School viewed the city in negative terms as "secular, impersonal, and segmental." From this perspective, urbanization was not conducive to social organization, but little attention was paid to quantifying such a general observation.

While Mumford's view of urbanization per se is not strictly negative—indeed his monumental work *The City in History* is a plea to bring the importance of the city into our consciousness—his critical appraisal of the "megalopolis" has influenced scores of scholars and commentators. On the other hand, Burgess's (1925) theories which linked social status with residential patterns tended to emphasize order rather than the social disorganization notions ascribed to Mumford and others. Burgess's "concentric zones" distributed population in a city according to economic and social status, where the inner rings of settlement were predominantly poor and the outer rings increasingly more affluent. Others discussed "sectors" which were not so much like ripples on a pond, but more like slices of a pie (Huth 1970). Burgess's concentric zones and related theories raised important questions about the capacity of cities for social organization or disorganization, especially the relationship between place and economic and social status and stratification; however, they have a relatively narrow role to play in sorting out an array of variables pertinent to the ecology of cities at large.

The 1950s also saw the ecological approach reemerging in urban geography, especially through the formulation of location theory, and through more extensive cross-disciplinary discourse with sociology. But the rejuvenated ecological approach was narrower in focus than its original incarnation, especially because it downplayed the interrelationships of human groups and devoted increased attention to the internal structure of cities and to land-use patterns through theories of city location.

Location theories, especially central place theory, almost began simultaneously with the formative years of the social sciences in the late nineteenth and early twentieth centuries. Central place theory, according to the *International Encyclopedia of the Social Sciences* (Sills 1968), "outlines the logic of systems of central places, focusing particularly upon the numbers, sizes, activities, and spatial distribution of such places and their associated regions." The theory has been discussed widely in social science literature (Beavon 1977; Carter 1981; Fales and Moses 1972; Berry 1974). The central place theory of German economic geographer Walter Christaller in the 1930s, however, most strongly influenced European and American scholars. Christaller was concerned with how cities served as central places for tributary regions particularly with respect to commerce and trade, manufacturing, service delivery, and administrative functions (Berry and Kasarda 1977). Central place theory, in particular, and location theory, in general, while not particularly useful as organizing principles for dealing with the urban environment as a whole, help to distinguish between development at the urban core as opposed to the periphery. Central place theory complemented the theory of agricultural production originally developed by J.H. von Thunen and the theory of industry location found in the work of Alfred Weber (Berry and Kasarda 1977; Berry and Horton 1970; Northam 1975).

Theories of ecological expansion and segmental growth help to focus on the relationship between technology and the organization of cities. Ecological expansion was meant to explain the relationship between population size and urban organizational structure. According to the theory, population growth along the periphery of an urban area will be matched with an increase in organizational functions at the core to insure stability in the expanded system. This pattern of growth, as stated earlier, produces a core city and a series of dependent suburbs. Technology, especially in the form of modern transportation, becomes a key variable in this theory because it reduces "the friction of space," which allows the core to retain control over its periphery. By contrast, the theory of segmental growth posits that the number of segmental units in an area increases as a function of population distribution, given that the friction of space is high, that is, no efficient means of transportation (or communications for that matter) exists. In this case, growth occurs primarily through increases in population density, not expansion (Webb 1974; Johnston 1982; LaGory 1979; Hawley 1950; Spates and Macionis 1987; Winsborough 1962; Steiss 1974; Walker 1978; Scargill 1979).

By the 1970s, as geographer Larry Bourne (1982) has noted, "a requiem had been written for most of the overly ambitious large-scale models of urban development that were popular in the 1960s." Where overarching models in urban ecology have tried to accomplish too much, efforts to define the city in environmental terms proclaims the value of thinking of cities as wholes, if not as intricately designed organisms. According to geographers Brian J.L. Berry and John Kasarda (1977), "the central problem of contemporary ecological inquiry is understanding how a population organizes itself in adapting to a constantly changing yet restricting environment." The four "reference variables" in addressing this problem are population, organization, environment, and technology. What needs to be made clear is how these variables lead to an understanding of how a population "organizes itself," and in what ways. It is necessary, of course, to look beyond spatial ordering toward more complex relationships.

Systems Theory

Systems theory has been in and out of fashion over the years, but it is useful because technological systems, for one example, provide an effective way to understand the growth of the city. Such systems include transportation, communication, energy sources, water supply, and waste disposal and may have an internal order that the city-building process as a whole lacks. The development of technical networks in the nineteenth century—and provisions for related services—were prime characteristics of the modern city. They sometimes offered an internal matrix for distributing services, controlling pollution, or providing access from one place to another. They also connect cities, as in the case of telephone or telegraph communication. Whereas industrialization remained local or regional for many years, new technological innovations were quickly diffused nationally. The technically networked city ushered in a remarkable period of modernization in the nineteenth century with a legacy extending well into the twentieth century. Several of the technical innovations of the nineteenth century—the automobile, electrical power networks, and the telephone—

were instrumental in transforming core cities into metropolises (Melosi 1990).

In 1964 geographer Brian J.L. Berry published an influential article entitled, "Cities as Systems within Systems of Cities." Among other things, he argued that "cities are systems susceptible of the same kinds of analysis as other systems and characterized by the same generalizations, constructs, and models."

"It is clear," he added, "that cities may be considered as systems—entities comprising interacting, interdependent parts. They may be studied at a variety of levels, structural, functional, and dynamic, and they may be partitioned into a variety of subsystems." Manual Castells (1983) changed Berry's focus somewhat by discussing "living systems." But that idea came very close to reintroducing an organic view to city-building that was not the intention in Berry's formulation.

As a way of applying an ecological approach to cities, the idea of a city as a system within a system of cities offered a powerful research approach for model building. But the systems models that became popular in the 1960s were criticized in the 1970s as "too formal and restrictive" (Bourne 1982). Since cities are strongly influenced by a range of external forces, it is best to think of them as "open systems" which departs from the kind of thinking that would make them insular or self-contained (Detwyler and Marcus 1972).

With some modification and rethinking, the systems approach to cities pioneered by scholars like Berry had a rebirth in the 1980s and has some obvious potential for evaluating the urbanization process. It is well to keep in mind historian Seymour Mandelbaum's (1985) caveat: "Systems thinking is formally holistic but it need not be catholic." I would take this to mean that systems theory is a valuable tool, but it need not become dogma. I also would add that cities may be regarded in the most generous terms as open systems, but within cities there are systems which can usefully be designated as "city" or "urban" and which offer insights about city-building, growth, and the urban environment in general.

Historian Thomas P. Hughes (1989) argued persuasively that systems "involve far more than the so-called hardware, devices, machines, and processes, and the transportation, communication, and information networks that interconnect them." He added, "Such systems consist also of people and organizations." As such, the systems—no matter how large or consolidated they are—do not become autonomous, but exist within limits imposed by the available technology, the hands of their operators, and the function dictated by their users.

Walking the tightrope between a rigid determinism and an organic view of systems is a substantial challenge. Yet, the elaboration of systems theory applied to the historical evolution of cities can be fruitful. The translation of systems theory into historical methodology has yet to reach its potential, however.

Conclusion

As the focus of this chapter suggests, I firmly believe that the historian's view of the city as ecosystem needs to be informed by a strong theoretical base, especially derived from the social sciences (insofar as they interpret and meld key concepts from the biological and physical sciences). Such a foundation provides a basis upon which more empirically based historical research can proceed.

Key to the definition of the city as an ecosystem are the concepts drawn from arenas explored earlier (Table 12.1) each should include the following ideas: (1) cities historically have had a metaphoric attachment to ecological concerns through ideas such as organic theory; (2) yet cities are indeed a part of (or at least an outgrowth of) the natural world; this may be further refined by the concept of "Second Nature"; (3) cities are also major modifiers of the physical world; (4) ecological theory of cities helps to understand patterns of growth and social ordering through urbanization; and (5) cities are open systems. Armed with these concepts, the city can be made more understandable from both an historical and an ecological perspective.

TABLE 12.1. Key factors from four areas of theory for defining the city as an ecosystem.

Organic Theory
- metaphor of society as a human body
- city as a natural system
- city as a transformed combination of resources
- city as "second nature"

City as Modifier
- impact of city on hydraulic cycle
- impact of city on atmosphere
- creation of "heat islands"
- city as waste producer

Ecological Theory
- humans in temporal and spatial dimensions
- urban spatial patterns and their social impacts
- population size and urban organizational structure
- concept of the "ecological complex"
- Mumford's social disorganization concept
- location theory and social organization
- theories of ecological expansion and segmental growth

Systems Theory
- internal and external impacts of technical systems
- "cities as systems within systems of cities"
- cities as "living systems"
- cities as "open systems"

Acknowledgments. Several portions of this chapter were adapted from sections of previous publications (Melosi 1997; 2000).

References

Beavon, K.S.O. 1977. *Central place theory: a reinterpretation.* Longman, London, UK.

Berry, B.J.L. 1974. Internal structure of the city. Pages 227–233 in K.P. Schwirian, ed. *Comparative urban structures: studies in the ecology of cities.* Heath, Lexington, MA.

Berry, B.J.L., and F.E. Horton. 1970. *Geographic perspectives on urban systems.* Greenwood, Englewood Cliffs, NJ.

Berry, B.J.L., and J.D. Kasarda. 1977. *Contemporary urban ecology.* Macmillan, New York.

Brian J.L., and B.J.L. Berry. 1964. Cities as systems within systems of cities. Regional Science Association Papers 13:147–163.

Burgess, E.W. 1925. "The Growth of the City: An Introduction to a Research Project." Chapter II (pages 47–51) in R.E. Park, E.W. Burgess, and R.D. McKenzie, eds. *The City.* University of Chicago Press, Chicago, IL.

Carter, H. 1981. *The study of urban geography.* Edward Arnold, London, UK.

Castells, M. 1983. *The city and the grassroots: a cross-cultural theory of urban social movements.* Edward Arnold, London, UK.

Cronon, W. 1991. *Nature's metropolis: Chicago and the great west.* Norton, New York.

Davison, G. 1983. The city as a natural system: theories of urban society in early nineteenth-century Britain. Pages 349–370 in D. Fraser, and A. Sutcliffe, eds. *The pursuit of urban history.* Edward Arnold, London, UK.

Detwyler, T.R., and M.G. Marcus, eds. 1972. *Urbanization and environment: the physical geography of the city.* Duxbury Press, Belmont, CA.

Duncan, O.D., W.R. Scott, and S. Liebers. 1960. *Metropolis and region.* Johns Hopkins University Press, Baltimore, MD.

Fales, R.L., and L.N. Moses. 1972. Land-use theory and the spatial structure of the nineteenth-century city. *Papers of the Regional Science Association* 28:49–80.

Goldfield, D.R., and B.A. Brownell. 1990. *Urban America: a history.* Houghton Mifflin Co., Boston.

Harvey, D. 1973. *Social justice and the city.* Johns Hopkins University Press, Baltimore, MD.

Havlick, S.W. 1974. *The urban organism: the city's natural resources from an environmental perspective.* Macmillan, New York.

Hawley, A. 1950. *Human ecology: a theory of community structure.* Ronald Press, New York.

Hughes, T.P. 1989. *American genesis: a century of invention and technological enthusiasm, 1870–1970.* Viking, New York.

Huth, M.J. 1970. *The urban habitat: past, present, and future.* Nelson-Hall, Chicago, IL.

Johnston, R.J. 1982. *The American urban system: a geographic perspective.* St. Martin's Press, New York.

Johnston, R.J., ed. 1968. *Roderick D. Mckenzie on human ecology: selected writings.* University of Chicago Press, Chicago, IL.

Kasarda, J.D. 1972. The theory of ecological expansion: an empirical test. *Social Forces* 51:165–175.
LaGory, M. 1979. Twentieth century urban growth: an ecological approach. *Sociological Focus* 12:187–202.
Lampard, E.E. 1961. American historians and the study of urbanization. *American Historical Review* 67:49–61.
Lampard, E.E. 1968. The evolving system of cities in the United States: urbanization and economic development. Pages 81–139 in H.S. Perloff, and L.W. Wingo, Jr., eds. *Issues in urban economics*. Johns Hopkins University Press, Baltimore, MD.
Lampard, E.E. 1965. Historical aspects of urbanization. Pages 519–54 in P.M. Hauser, and L.F. Schnore, eds. *The study of urbanization*. Wiley, New York.
Lampard, E.E. 1983. The nature of urbanization. Pages 3–51 in D. Fraser, and A. Sutcliffe, eds. *The pursuit of urban history*. Edward Arnold, London, UK.
Lynch, K. 1960. *The image of the city*. Technology Press, Cambridge, MA.
Mandelbaum, S.J. 1985. Thinking about cities as systems: reflections on the history of an idea. *Journal of Urban History* 11:139–150.
Melosi, M.V. 2000. *The sanitary city: urban infrastructure in America from colonial times to the present*. Johns Hopkins University Press, Baltimore, MD.
Melosi, M.V. 1990. Cities, technical systems and the environment. *Environmental History Review* 14:45–64.
Melosi, M.V. 1993. The place of the city in environmental history. *Environmental History Review* 17:1–23.
Miller, D.L. 1992. Lewis Mumford: urban historian, urban visionary. *Journal of Urban History* 18:280–307.
Mumford, L. 1945. *City development: studies in disintegration and renewal*. Harcourt, Brace, New York.
Mumford, L. 1961. *The city in history: its origins, its transformations, and its prospects*. Harcourt, Brace, New York.
Mumford, L. 1938. *The culture of cities*. Harcourt, Brace, New York.
Mumford, L. 1963. *The highway and the city*. New American Library, New York.
Northam, R.M. 1975. *Urban geography*. Wiley, New York.
Park, R.E. 1952. *Human communities: the city and human ecology*. Free Press, Glencoe, IL.
Park, R.E., E.W. Burgess, and R.D. McKenzie. 1925. *The City*. University of Chicago Press, Chicago.
Scargill, D.I. 1979. *The form of cities*. St. Martin's Press, New York.
Schnore, L. 1965. *The urban scene: human ecology and demography*. Free Press, New York.
Sills, D.L. ed. 1968. *International encyclopedia of the social sciences*, Vol. 2. Macmillan, New York.
Sjoberg, G. 1965. Theory and research in urban sociology. Pages 157–189 in P.M. Hauser, and L.F. Schnore, eds. *The study of urbanization*. Wiley, New York.
Spates, J.L., and J.J. Macionis. 1987. *The sociology of cities*. Wadsworth, Belmont, CA.
Steiss, A.W. 1974. *Urban systems dynamics*. Heath, Lexington, MA.
Walker, R.A. 1978. The transformation of urban structure in the nineteenth century and the beginnings of suburbanization. Pages 165–212 in K.R. Cox, ed. *Urbanization and conflict in market societies*. Maaroufa Press, Chicago, IL.

Webb, S.D. 1974. Segmental urban growth: some cross-national evidence. *Sociology and Social Research* 58:387–391.
Winsborough, H.H. 1962. City growth and city structure. *Journal of Regional Science* 4:48.
Worster, D. 1988. Doing environmental history. Pages 289–307 in D. Worster, ed. *The ends of the earth: perspectives on modern environmental history*. Cambridge University Press, New York.

13 Urban Ecosystems, City Planning, and Environmental Education: Literature, Precedents, Key Concepts, and Prospects

Anne Whiston Spirn

The ecosystem concept provides a powerful tool for understanding the urban environment: it furnishes a framework for perceiving the effect of human activities and their interrelationships; it facilitates weighing the relative costs and benefits of alternative actions; it encompasses all urban organisms, the city's physical structure, and the processes that flow within it; and it is appropriate in examining all levels of life, from an urban pond to a metropolitan region. Regarding the city as an ecosystem permits every individual to perceive his or her cumulative impact on the city, and the designer of every building and park to perceive its place within the whole. It also permits the planner of a transportation network or regional park system to trace the effect of comprehensive change on smaller parts of the system. A knowledge of the system's dynamics yields a different appreciation for boundaries in space and time than is normally permitted in day-to-day pursuits and highlights the shortcomings of designing solely within political boundaries and time spans of less than several human generations (Spirn 1984).

The Granite Garden: Urban Nature and Human Design (Spirn 1984) describes the urban natural environment and the natural processes that shape it, and demonstrates how cities can be planned and designed in concert with natural processes rather than in conflict. The book brings together and applies knowledge from many disciplines and includes a lengthy bibliography and a review of the relevant literature in urban climatology, geology, hydrology, soil science, ecology, horticulture, forestry, wildlife management, civil and environmental engineering, urban and environmental history, landscape architecture, urban planning, and design. Research for the book led to several discoveries: (1) much is known about the urban natural environment; (2) there exist many successful models of urban ecological planning; (3) most of these examples are not known to the public, or to scientists, or even to planners; and (4) in their ignorance of existing knowledge and precedents, researchers and planners have repeatedly "reinvented the wheel."

Studying urban ecosystems is not a new frontier for science or for planning, though it may be for education. Then how does one account for the failure to build on past work? Part of the answer lies in strong disciplinary barriers, which are reflected in the literature, and by the fact that interest in and funding for urban ecosystem research has been intermittent. Even more important is the deep-seated resistance to seeing cities as part of the natural world. Ideas of nature and what is "natural" stem from strongly-held feelings and beliefs. If we are to better understand and manage urban ecosystems, environmental education will play an important role. People should learn to read urban landscapes as scenes of dynamic processes connecting air, earth, water, and life and as habitats humans share with other organisms.

Urban Ecosystems and Planning: Literature and Precedents

It is important to recognize the scope of the existing literature on urban ecosystems, especially that which is directly relevant to planning, and the precedents for an ecological approach to city planning. Since the classic compilation *Man's Role in Changing the Face of the Earth* in 1956 (Thomas Jr. 1956), the literature on the urban natural environment has grown within many disciplines. Climatologists have described how the form of cities and the activities that take place there modify climate at macro- and microscales, and how this affects energy flows and the concentration and dispersion of air pollutants (e.g., Chandler 1976; Spirn 1987). Geologists have shown how urban development processes can accelerate or retard earth processes such as erosion, mass wasting, and subsidence (e.g., Leveson 1980). Hydrologists and civil engineers have described how urban form and infrastructure alter the hydrologic cycle at scales both of regional and local watersheds and how this affects fluvial processes and patterns of groundwater movement and water pollution (e.g., Dunne and Leopold 1978). Soil scientists have documented processes of urban soil formation, how they influence plant growth and reproduction, and how to reconstruct and manage urban soils to enhance plant survival (e.g., Craul 1999). Zoologists have described populations, behaviors, and habitats of certain animal species, mainly birds and mammals, and how they can be managed to enhance human health, safety, and recreation (e.g., Gill and Bonnett 1973). Plant ecologists have studied processes of succession on abandoned land. Although much is known about urban ecosystems, this knowledge is published in journals of many disciplines, and referencing (and reading?) across disciplines is rare. The literature on urban climate provides a striking example of this phenomenon; studies by climatologists and biometeorologists contain no references across these fields (e.g., Chandler

1976; Herrington 1980). Authors also tend to cite the most recent work in their area and to ignore work published during past cycles of interest in urban ecosystems, which seem to come in distinct waves as federal funding for studying urban natural environments has surged and receded: early 1950s (e.g., Thomas 1956), 1970s (e.g., Stearns and Montag 1974), and 1990s (e.g., Platt, et al. 1994).

Though an ecological approach is not part of mainstream city planning and design, it has a strong historic tradition, a foundation of knowledge to support it, and projects that demonstrate its benefits. The roots of this tradition are deep (Spirn 1985): Hippocrates's treatise *Airs, Waters, Places* from the fifth century B.C.; Alberti's *Ten Books on Architecture* of 1485; Evelyn's *Fumifugium* of 1661; Marsh's *Man and Nature* of 1864; Olmsted's papers and reports; and, in the twentieth century, Geddes's *Cities in Evolution* of 1915; McHarg's *Design with Nature* of 1969; in 1984 my own *Granite Garden* (Spirn 1984) and Hough's *City Form and Natural Process* (Hough 1984); and Lyle's *Design for Human Ecosystems* (Lyle 1985) and *Regenerative Design for Sustainable Development* (Lyle 1994), among others.

There are many successful examples of urban ecological planning. Most were initiated in response to a single environmental problem specific to that city: air or water pollution, flooding or landslides. Most were inspired by fear of destruction, health hazard, or litigation. A series of catastrophic floods prompted the formulation of Denver's urban storm drainage strategy, for example. There, downtown rooftops and plazas, and a regional system of parks are part of a comprehensive infrastructure for flood control and storm drainage. The Fens and Riverway in Boston are an earlier precedent, designed in the 1880s and 1890s by landscape architect Frederick Law Olmsted as a landscape system to accommodate the movement of people, the flow of water, and the removal of wastes (Spirn 1995). The Fens, which originally functioned as a stormwater detention basin before it was altered in the early twentieth century, was the first attempt anywhere, so far as I know, to *construct* a wetland. Werribee Farm, outside Melbourne, Australia, is a 27,000 acre reservation that treats 60 percent of the sewage from a metropolitan area of nearly three million people. Sewage treatment lagoons include wetlands designed and built as bird habitat, and the site is open to bird watchers. These are a few landmarks among many examples; like similar models elsewhere, their multifaceted functions are not well known among residents of the cities in which they occur.

Urban Ecosystems and Planning: Some Key Concepts

Although plans must be tailored to local conditions—social, cultural, political, economic, and environmental—it is possible to summarize some general concepts and strategies.

Humans Are Part of the Urban Ecosystem; Environmental History Reveals How

Reconstructing the environmental history of an urban landscape provides a window into the interaction of natural and social processes through time (Spirn 1998). This kind of history can be gleaned from historical documents, especially maps and photographs, from verbal descriptions, and from the landscapes themselves. Studying environmental change over time helps foster an understanding of urban landscapes as dynamic. It shows how natural processes are significant agents in urban development, and how social and cultural processes are active ingredients of urban ecosystems.

Take the example of abandoned lands in American inner cities, which are not scattered evenly, but tend to form significant spatial and temporal patterns. In Boston's Dudley Street neighborhood, in the 1980s, few vacant lots were on hilltops and hillsides, whereas 90 percent of the lowest parts of the valley were vacant, sites of community gardens and successional meadows and woodlands. One reason was water flowing down from hillsides into the valley, saturating the soil, and puddling after rain. Old maps show a stream, the boundary between the towns of Roxbury and Dorchester; buried in a sewer in the late nineteenth century, the stream now carries the community's wastes and rainfall. Successive atlases from 1876 to 1886, 1892, 1903, 1910, 1922, 1934, 1948, and 1964 trace landscape change from farm and field to streetcar suburb to inner-city neighborhood. Houses in the valley were built later than houses on the hills, many as apartments for poorer people. The first vacant lots in the valley were recorded in the early twentieth century, within twenty years after construction there. By 1964, large areas in the valley bottom had been abandoned. Water flowing underground and ground subsiding over buried floodplains are only part of the story. Abandonment in the Dudley Street neighborhood was also the product of social and political processes: industrial decline, unemployment, red-lining (denial of insurance and loans for homes and businesses in areas with older homes and immigrant or "nonwhite" populations), and arson. City planners commonly recognize the last set of problems without appreciating the existence and contributing effects of buried streams. Corresponding patterns exist in many other American cities, such as in West Philadelphia's Mill Creek neighborhood (Spirn 1986 and 1998).

Natural Processes Shape and Structure Urban Landscapes at All Scales; Urban Planning Must Take This into Account

There is a common tendency to focus on natural features (e.g., rivers and trees) rather than the processes that shape and structure them (e.g., flow of air, water, and materials; plant reproduction and growth). Ignoring natural processes leads to harmful consequences, including the failure of planners to accommodate dynamic change, their failure to make connections among seemingly unrelated issues and phenomena, and their inability to realize opportunities. Take the example of how flowing water shapes and structures rivers, floodplains, watersheds and their topographies and boundaries. Water flow is not constant, but ever changing, and floodplains are dynamic zones, places where water rises and falls, pools and seeps. Changes in ground surface and permeability in one place within a watershed alter floodplain boundaries and the form and size of streambeds and banks in other places. Burying a stream in a sewer and filling in the floodplain alters, but does not eliminate, many of the floodplain's characteristic qualities. Seen from the perspective of hydrological processes, the drainage system of neighborhood, city, or region consists not only in channels officially designated for stormwater flow, but also in all the surfaces and water reservoirs within a watershed: in roofs, roads, and parking lots; gardens, parks, and forests; in soil, plants, and valley bottoms as well as in ponds.

Environmental Issues Must Be Addressed Within Appropriate Boundaries at Appropriate Scales

Natural processes form territories, pathways, and boundaries, which may coincide with political and cultural boundaries or cut across them. Environmental problems felt in one place may be caused by activities that take place elsewhere. Solving an environmental problem may require taking action in a different location than where the problem is felt. Without broad understanding of urban ecosystems among both planners and the public, it is difficult to take the measures required.

Integrate Planning and Design at All Scales to Achieve Desired Goals

Policies shape the physical form of the city; urban form, in turn, can support or undermine the goals of policy. Urban planning often focuses upon policy formulation, regulation of human activities, and the spatial allocation of land uses without explicitly addressing the shape and structure of the city—its buildings, streets, parks, neighborhoods, and overall form. Land-use planning is most effective in allocating space for relatively large areas of land use, for preserving natural resources, and avoiding environmental hazards, but planning alone cannot achieve environmental goals. Thousands of decisions are made daily about the physical shape and structure of individual buildings, plazas, parks, and streets. These decisions are usually made without concern for the city's larger environmental context. Collectively, these independent decisions have an enormous influence upon environmental quality—local, regional, and even global.

An understanding of urban ecosystems should underlie all aspects of the physical planning and design of the city. In particular, the integration of all open urban land into a unified plan could extend the traditionally accepted aesthetic and recreational value of open space to a crucial role in health, safety, and welfare. Parks and plazas, water bodies and streams, floodplains and marshy lowlands, steep hillsides and rocky outcrops, and even parking lots and highway corridors could be part of a cohesive open space system to improve air quality and climate; to reduce flooding and improve water quality; to limit the impact of geological hazards such as earthquakes, subsidence, and landslides; to provide a diverse community of plants and animals within the city; to conserve energy, water, and mineral resources; and to enhance the safe assimilation of the city's wastes (Spirn 1984).

Plan for Both Comprehensive, Large-Scale Change and Incremental Changes That Address Local Conditions

Planning should combine the regional-scale overview and the local view. Comprehensive and advocacy (or grassroots) approaches to planning each have strengths; when pursued independently they may have disastrous results. The comprehensive approach can take regional phenomena into account and produce fast results, but it may also lead to unanticipated consequences that are difficult to reverse. An incremental, grassroots approach may address local needs and permit the assessment of successes and failures and the refinement of subsequent interventions, but it may be rendered dysfunctional by failure to account for the larger context of city and region. These two approaches should be integrated. Establish a comprehensive framework for large-scale investment that addresses citywide needs and that encourages and accommodates smaller-scale projects tailored to the needs of specific people in particular places. This framework should be one within which many diverse agents of change can play a role, including not only government agencies and private businesses, but also institutions, small organizations, and even individuals.

Define Multipurpose Solutions to Comprehensively Defined Problems

The problems facing contemporary cities loom large and seem to dwarf the resources available to address them effectively. Social and economic issues, on the one hand, and environmental and aesthetic issues, on the other, compete for attention and scarce funds. Given limited resources and growing challenges, cities can no longer afford single-purpose solutions to narrowly defined problems. Such an approach wastes resources and causes unanticipated consequences. Planners should seek integrated solutions to social, economic, cultural, and environmental problems. One strategy for promoting such integration is for a city to start with its most pressing problem for which there is widespread public support, and then find ways to address other concerns as well. An important environmental problem, such as air pollution, water pollution, or flooding may be the central, organizing issue within which social, economic, aesthetic, and other environmental problems are addressed. Alternatively, a social or economic problem, such as unemployment, may serve as the focus, and ways found to incorporate solutions to environmental problems.

Cities Can Be Planned, Built, and Managed to Require Less Energy and Resources to Build and Sustain

The city, its infrastructure, and every building, garden, and park within it, should be designed as much as possible as closed systems, which consume fewer resources and produce fewer wastes. Cities that can adapt to changing needs without requiring large-scale destruction will also conserve materials and energy. Lynch (1958) outlined strategies for achieving environmental flexibility in urban form. An additive structure (e.g., a grid layout) can accommodate growth or decline at the periphery without major change to the overall structure at the center of a neighborhood or city. Using communication systems to accommodate changing needs may reduce the need for radical alteration of the city's physical structure. Urban form that is too narrowly specialized (e.g., districts that consist entirely of one land use), like monocultures, are inherently unstable; they may not be adaptable to change.

Every City Has a Distinctive Ecosystem with Characteristic Climate, Geology and Topography, Water Regime, and Plant Communities, Which Affords Resources and Poses Hazards for Human Settlement

While urbanization radically changes the surface of the urban landscape, there is a more enduring structure, with distinctive rhythms. This enduring context or "deep" structure expresses the fundamental climatic, geomorphic, and biotic processes in a particular place. The deep structure of Boston and Denver and the rhythms they engender, for example, are strikingly different: maritime, temperate climate, inundated, glaciated basin, and deciduous forest; continental, semiarid climate, mountain and plain, grassland. Enduring context is the product of processes operating and interacting across vast scales of time at the scale of the larger metropolitan region and at the microscale (Spirn 1993). The physical shape and structure of a city—its infrastructure of roads and sewers, its buildings and parks—should be consonant with the deep structure of the local natural environment. When urban structure obscures or opposes this deep local structure, by burying floodplains, for example, it requires additional energy and materials to sustain. Urban design that reveals and responds to deep structure is likely to be more economical and more easily sustained over time.

Landscape Literacy: A Basis for Environmental Education, Civic Action, Urban Planning, and Community Development

If knowledge of urban ecosystems is to be applied to the planning, design, and management of cities, we will require more than additional information and new ideas; it demands a transformation of perception. Most people, including ecologists and city planners, see cities as existing outside nature. They are not persuaded that natural processes of air, earth, water, and life shape cities just as social, economic, and political processes do. Changing these attitudes must begin with early education; it is too late to wait until the professional education of ecologists and planners. For planning to be effective, the public should be aware of how natural processes shape the places they live. Environmental education should present ecological principle, as they are manifest in local landscapes, not just as abstract ideas or as concepts that apply to distant places. Landscape literacy should be a fundamental part of environmental education and a basis for community development (Spirn 1998).

Take the example of West Philadelphia's Mill Creek neighborhood, one of the poorest in Philadelphia and the site of the West Philadelphia Empowerment Zone, where I have been working on issues of vacant land, flooding, subsidence, environmental education, and community development since 1987 (Spirn 1998). Blocks of tumbled buildings, slumped streets, and vacant land mar the Mill Creek neighborhood, but few recognize the pattern this devastation forms. A linear zone meanders through the lowest parts of the landscape, the former floodplain of Mill Creek, a buried river whose sewered channel still drains half of West Philadelphia, carrying sewage and stormwater. After large rainstorms, the sewer overflows into the Schuylkill River. For more than 60 years, the ground has fallen in along the line of the sewer. In several places on the buried floodplain, entire city blocks are vacant. Young woodlands of alianthus, sumac, and ash have grown up on older lots; urban meadows on lots recently vacated. Mill Creek—the stream, its watershed, and the community—provide rich material for studying urban ecosystems and have been a laboratory for testing and refining the concepts outlined in the previous section.

In the late 1980s, I identified low-lying vacant lots that are prone to flooding and proposed that the city use these sites to detain stormwater in order to eliminate combined sewer overflows (CSOs). My students and I developed comprehensive watershed management strategies to reduce local flooding and CSOs and designed demonstration projects integrating stormwater management, parks, and environmental study areas. But city officials were unconvinced that there is a correlation between vacant land, derelict buildings, and the old floodplain of the Mill Creek and that this understanding must be incorporated into plans for the neighborhood. Despite its persistent, devastating effects on the community over the years and its continued contribution to degraded regional water quality, they largely ignored the existence and significance of Mill Creek until recently.

Community residents and local teachers were more receptive to these ideas. Since 1995, my students and I, together with teachers and students at Sulzberger Middle School, have been reading this landscape and have developed a new curriculum organized around Mill Creek and its urban watershed; the Mill Creek Project combines learning, community development, and water resource management. Like most residents, Sulzberger students are African American. The school is next to the Mill Creek's former floodplain; the cafeteria and gym flood after heavy rains. Students are bussed, occasionally, to suburban environmental study areas in order to see "nature," yet all around the school are successional meadows on vacant lots where former houses and businesses in the old Mill Creek floodplain caved in or were demolished, affording ample opportunity for studying the processes of nature and politics at work.

Since 1996, dozens of university students and hundreds of sixth-, seventh, and eighth-graders at Sulzberger have learned to read the neighborhood's landscape, to trace its past, to understand its present, and envision its future. Armed with the knowledge of the creek's past gleaned from the contemporary landscape and from old maps and photographs, middle school and university students write and draw their visions for the future and publish them on the West Philadelphia Landscape Project web site (*www.upenn.edu/wplp*). In 1997, Sulzberger introduced an Environment Small Learning Community (10 teachers and approximately 240 students each year in grades 7 and 8) where themes of regional watershed and local community introduced into classes in science, math, social studies, computing, language, and art in order to study problems and identify solutions to actual problems of sustainable development in the community, and to bring these problems and potential solutions to public attention. As a result of our work, the Philadelphia Water Department Combined Sewer Overflow Program made the Mill Creek Watershed a special study area in spring 1997, and in 1999 decided to build a demonstration project on vacant land in Mill Creek, which will combine a storm water detention facility to reduce combined sewer overflows and an environmental study area for Sulzberger Middle School. In 2001, The City of Philadelphia received a federal grant to extend this project.

Key concepts of urban ecosystems and planning underlie both the middle-school curriculum and my university courses. By studying the environmental history of Mill Creek across 400 years, both sets of students have gained an appreciation for how human activities interact with natural processes to shape and structure urban ecosystems at a range of scales: from moments to centuries, from household to region. They formulated plans that take this interaction into account. Tracing water movement and use through their home, neighborhood, city, and region, students explored environmental issues within a succession of watershed boundaries and discussed what actions by individuals and groups might be taken at these diverse scales in order to conserve water and improve the quality of rivers and wetlands. Middle school and university students have devised designs for an outdoor classroom on a vacant lot near the school, which would also be a wetland, water garden, and stormwater detention basin. This project represents both a proposal for incremental change that addresses local conditions and a comprehensive approach to improve regional water quality; it is a multipurpose solution to comprehensively defined problems. For the university students, the project serves as a demonstration of how cities can be planned, built, and managed to require less energy and resources to build and sustain. It is an example of how every city has a distinctive ecosystem with characteristic climate, geology and topography, water regime, and plant communities, which affords resources and poses hazards for human settlement.

Sulzberger teachers have reported that their students' performance in all subjects improved, in some cases dramatically, and attributed this to the Mill Creek Project. Learning to read the significance in patterns of open spaces, streets, and plants on vacant lots and in a community garden nearby enthralled the middle-school and college students in ways that their more general history and science texts do not. Studying the area's natural and man-made features brought the place alive. It also introduced them to broader environmental, social, and political issues. Deciphering the stories in old documents and the landscape itself not only increased the middle-school students' self-confidence; it also seemed to give them a sense of orientation. My undergraduate students in liberal arts and engineering and graduate students in landscape architecture and urban planning apply theories they learned in the classroom to investigate, understand, solve, and reflect upon difficult real-world problems. Their experiences and observations corroborated some theories and called others into question; this developed confidence in their own reasoning and gave them a sense of where they fit in intellectual and professional debates.

Landscape literacy, which is the ability to read landscapes; and landscape fluency, the capacity for expression; are empowering. Learning to perceive environmental phenomena in local urban landscapes, to derive information from those observations, to frame questions, and find and test possible answers develops a basis for informed public decisions about urban ecosystems. Landscape literacy should be a fundamental goal of environmental education, a basis for civic action, and a prerequisite for urban planning and community development.

References

Alberti, L. 1485. *On the art of building in ten books*. J. Rykwert, N. Leach, and R. Tavenor, translators. 1988. MIT Press, Cambridge, MA.

Chandler, T.J. 1976. *Urban climatology and its relevance to urban design*. Technical Note 149. World Meteorological Association, Geneva, Switzerland.

Craul, P. 1999. *Urban soils: applications and practices*. Wiley, New York.

Dunne, T., and L. Leopold. 1978. *Water and environmental planning*. W.H. Freeman, San Francisco, CA.

Evelyn, J. 1661. *Fumifugium: or the inconvenience of the aer and smoake of London dissipated*. Reprinted in 1930. Old Ashmolean Reprint, Oxford, UK.

Geddes, P. 1915. *Cities in evolution*. Williams and Norgate, London, UK.

Gill, D., and P. Bonnett. 1973. *Nature in the urban landscape: a study of urban ecosystems*. York Press, Baltimore, MD.

Herrington, L.P. 1980. "Urban Vegetation and Microclimate." in G. Hopkins, ed. *Proceedings: national urban forestry conference*. College of Environmental Science and Forestry, State University of New York, Syracuse, NY.

Hippocrates. ca. Fifth century B.C. Airs, waters, places. Volume 1, T.E. Page, ed. 1962. *Hippocrates*. Loeb Classical Library, Harvard University Press, Cambridge, MA.

Hough, M. 1984. *City form and natural process*. Van Nostrand Reinhold, New York.

Leveson, D. 1980. *Geology and the urban environment*. Oxford University Press, New York.

Lyle, J.T. 1985. *Design for human ecosystems*. Van Nostrand Reinhold, New York.

Lyle, J.T. 1994. *Regenerative design for sustainable development*. Wiley, New York.

Lynch, K. 1958. Environmental adaptability. *Journal of the American Institute of Planners* 24:16–24.

Marsh, G.P. 1864. *Man and nature*. Harvard University Press, Cambridge, MA.

McHarg, I.L. 1969. *Design with nature*. Natural History Press, New York.

Olmsted, F.L. 1886. *The problem and its solution*. Speech to the Boston Society of Architects. Olmsted Papers. Library of Congress, Washington, DC.

Olmsted, F.L., and J.B. Harrison. 1889. "Observations on the treatment of public plantations, more especially related to the use of the axe." Pages 362–75 in F.L. Olmsted Jr., and T. Kimball, eds. 1973. *Forty years of landscape architecture: Central Park*. MIT Press, Cambridge, MA.

Platt, R.H., R.A. Rowntree, and P.C. Muick. 1994. *The ecological city: preserving and restoring urban biodiversity*. University of Massachusetts Press, Amherst, MA.

Spirn, A.W. 1984. *The granite garden: urban nature and human design*. Basic Books, New York.

Spirn, A.W. 1985. Urban nature and human design: renewing the great tradition. *Journal of Planning Education and Research* 5(1):39–51.

Spirn, A.W. 1986. Landscape planning and the city. *Landscape and Urban Planning* 13:433–441.

Spirn, A.W. 1987. Better air quality at street level: strategies for urban design. Pages 310–320 in A. Vernez-Moudon, ed. *Public streets for public use*. Van Nostrand Reinhold, New York.

Spirn, A.W. 1993. Deep Structure: On process, form, and design in the urban landscape. Pages 9–16 in T.M. Kristensen, S.E. Larsen, P.G. Moller, and S.E. Petersen, eds. *City and nature*. Odense University Press, Odense, DK.

Spirn, A.W. 1995. Constructing nature: the legacy of Frederick Law Olmsted. Pages 91–113 in W. Cronon, ed. *Uncommon ground: rethinking the human place in nature*. W.W. Norton, New York.

Spirn, A.W. 1988. *The language of landscape*. Yale University Press, New Haven, CT.

Stearns, F., and T. Montag, eds. 1974. *The urban ecosystem: a holistic approach*. Dowden, Hutchinson & Ross, Stroudsberg, PA.

Thomas, W.L., Jr., ed. 1956. *Man's role in changing the face of the earth*. University of Illinois Press, Chicago, IL.

14
A Human Ecology Model for the Tianjin Urban Ecosystem: Integrating Human Ecology, Ecosystem Science, and Philosophical Views into an Urban Eco-Complex Study

Rusong Wang and Zhiyun Ouyang

The Framework of an Urban Eco-Complex Model of Tianjin, China

China is experiencing rapid growth in urbanization and industrial transition. The pace, depth, and magnitude of these changes, accompanied by the persistence of past practices, are threatening to the regional and global environment and human well-being. Sustainability can only be assured with a human ecological understanding of the complex interactions among environmental, economic, political, and social/cultural factors and with careful planning and management grounded in ecological principles. Taking Tianjin, the third largest city in China as an example, this chapter tries to integrate human ecology, ecosystem science, and philosophical views of cities into an urban eco-complex model. This model combines a mechanism model, a planning model, and a regulation model. Urban ecological relationships are modeled through identification of the city's key factors, feedback, and function; and partial simulation of its problems, processes, and alternative policies. Some integrative strategies for regulating its technological, institutional, and behavioral aspects are put forward for helping local people to help themselves.

The ecosystem concept, one of the key concepts in ecology, has been used for more than half a century since Tansley defined it as "the biome considered together with all the effective inorganic factors of its environment" (Tansley 1935). People are used to calling a biome and its physical environment in a specific area an ecosystem and pay more attention to its inner physical and biological entities. Though this may be appropriate for studying the relatively isolated biome such as a lake or an island, it is difficult to deal with an urban ecosystem, which has more material, energy, and information exchange and biological migration with outside ecosystems than do other natural ecosystems. In his newest edition of the text *Ecology*, E.P. Odum defines *ecology* as the "separate discipline that integrates the study of organisms, the physical environment, and human society," a "discipline that emphasizes a holistic study of both parts and wholes" (Odum 1997). Urban ecology is just this kind of new discipline.

Unlike biological communities, a city is a kind of artificial ecosystem dominated by human activities, sustained by natural life support systems, and vitalized by ecological processes. It was named by Shijun Ma a "social–economic–natural complex ecosystem" (Ma and Wang 1984). Its structure is expressed as an eco-complex between human beings and their working and living settlements (including geographical, biological, and artificial environs), its regional environment (including sources for material and energy, sinks for products and wastes, pools for buffering and maintaining), its social networks (including culture, institution, technology), and its economic networks (industries and infrastructural services). Its natural system consists of physical and and biological factors, in terms of the Chinese traditional five elements: metal (minerals), wood (living organisms), water (sources and sinks), fire (energy), and soil (nutrients and land) (Figure 14.1). Its functions include production, consumption, supply, assimilation, steering, and buffering, which play key roles in sustaining the complicated network of human ecological relationships (Figure 14.2).

With a population greater than 8.4 million, Tianjin is the third largest industrial city in China, a major port, and historically the gateway to Beijing,

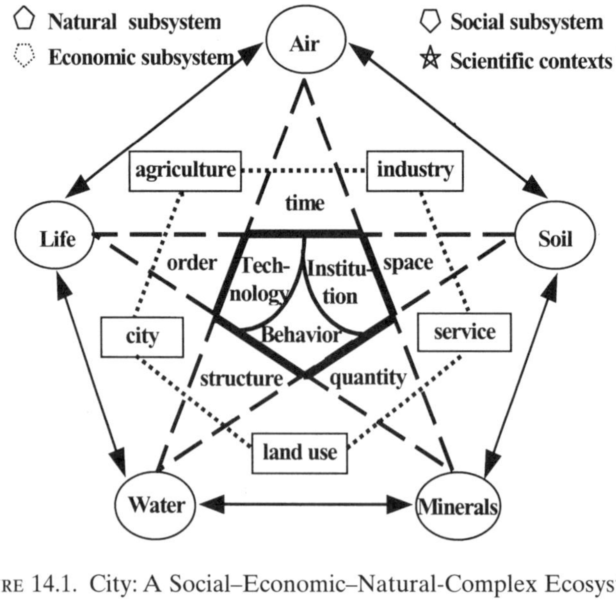

FIGURE 14.1. City: A Social–Economic–Natural-Complex Ecosystem.

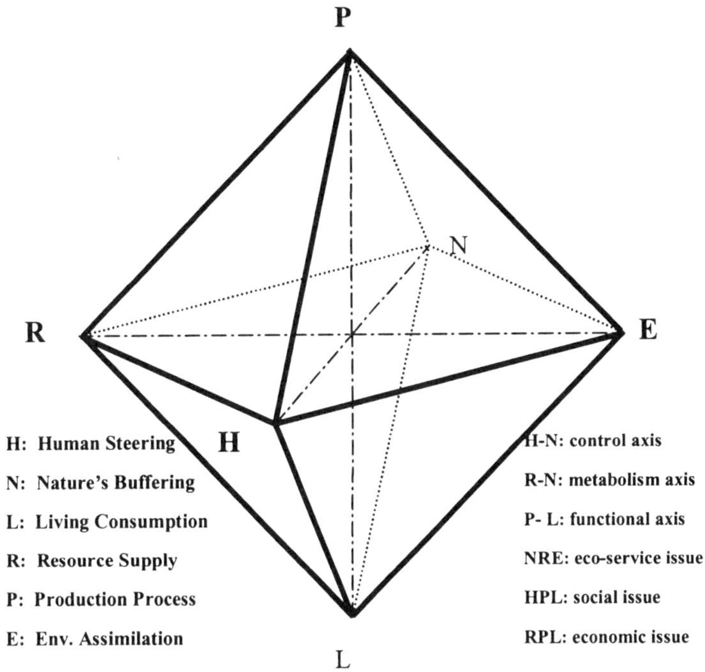

FIGURE 14.2. Functions and their interactions in an urban ecosystem.

the capital. The rapid growth of Tianjin in recent years is expected to continue, along with consequential problems of urban development such as water pollution, air pollution, housing shortages, sewage disposal, and traffic congestion.

In this chapter, a human ecology concept model is developed as a framework to identify the structure and function of the urban ecosystem through key factor analysis. We then explore ways to simulate the dynamics and cybernetics of urban systems through problem diagnosing, process tracing, and policy testing, all in search of strategies for urban ecological regulation (Figure 14.3). The components of each of the three models comprising the overall eco-complex model are shown in Table 14.1 (CERPG 1995).

The Eco-Mechanism Model: Understanding Urban Ecological Dynamics and Cybernetics

Traditional views of urban development pay attention to only two kinds of functions—economic production and social consumption. The often neglected and most important one is the sustaining or service function. There

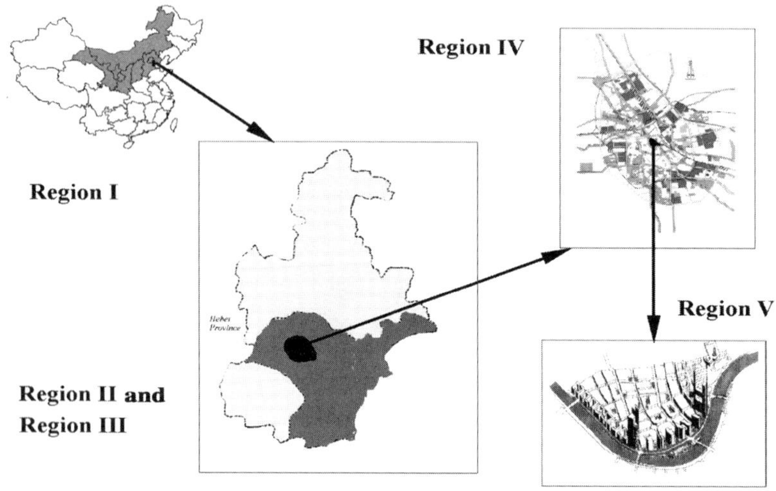

FIGURE 14.3. Different spatial scales or regions of the Tianjin urban ecosystem.

are two kinds of service functions. One is for inner-city services that include the functions of resource supply, products and wastes acceptance, transportation and communication, spatial containing, safety guarding, and cultural sustaining. The other service function is for regional services which

TABLE 14.1. The components of the three main models comprising the eco-complex model of Tianjin, China.

The Eco-Mechanism Model
- A **dynamics model** for understanding the main driving forces and main metabolic processes.
- A **cybernetics model** for understanding the main positive and negative feedbacks, and main risks and opportunities.
- A **contexts model** dealing with temporal evolution, spatial pattern, metabolic balance, institutional coupling and functional order.

The Eco-Planning Model
- A **key factor identification model** to identify boundaries, key limiting, promoting, buffering, and critical factors, key dominating and compensating components, and key negative and positive feedbacks.
- A **partial simulation model** for problem diagnosing, process tracing, policy testing.
- An **adaptive optimization model**, such as pan-objective ecological programming.

The Eco-Regulation Model
- An **eco-industry model** for technological innovation to incubate totally functioning technologies.
- An **ecopolis model** of institutional reform to cultivate systematically responsible institutions.
- An **eco-cultural model** of behavioral inducement to encourage an ecologically vivid culture.

include pollution purification, climate buffering, soil cultivation, biodiversity conservation, water conservation, and landscape harmonization. It is through just this type of specific function that the economy, society, and physical environment can interact with one another to sustain a harmonized urban ecosystem. According to Lao Dan, a famous ancient Chinese philosopher, this sustaining function is "such a thing which seems to spring forth from nowhere, and yet it penetrates everywhere. It is formless, shapeless, vague, indefinite, imperceptible and indescribable, always changing, and reverting to the state of nothingness" (Yang 1968).

An urban ecosystem is driven by two kinds of agents. One is the physical force originated from various kinds of solar energy. The flow of energy results in various physical, chemical, and biological processes in the city and causes the urban ecosystem's succession. Another comes from three social forces including money, the lever of economic incentives; power, the lever of social integration; and spirit, the lever of cultural regulation. Energy drives metabolism, money stimulates competition, power promotes symbiosis, and spirit induces self-reliance. These interact with each other and formulate the ecological force driving the functional flow of material, energy, people, capital, and information; maintaining the structural network; and pushing the functional succession (Wang 1988).

For thousands of years, Chinese philosophers have investigated the harmonious relationship among Tian (heaven or universe), Di (earth or resource), and Ren (people or society), forming the bases of Chinese human ecological thoughts. The most fruitful period was Cun-qiu (Spring and Autumn to Warring States, 720–221 B.C.), when various schools, including Confucianism, Taoism, Legalism, Yin-Yang, and Logicianism flourished (Wang 1991). The result is a systematic set of principles for managing the relationships between man and his or her environment, including Dao-Li (natural relationship with the universe, geography, climate, etc.), Shi-Li (planning and management of human activities, such as agriculture, warfare, politics, family, and others), and Qing-Li (ecological ethics, psychological feelings, and motives and values towards the environment). The Yin and Yang theory (negative and positive forces play upon each other and formulate all ecological relationships), Wuxing theory (five fundamental elements and movements within any ecosystem promoted and restrained with each other), Zhong Yong (things should not go to their extremes but keep equal distance from them or take a moderate way), and Feng-Shui theory (Wind–Water theory expressing the geographical and ecological relationship between human settlements and their natural environment) are some of these principles (Wang and Qi 1991).

Based on ancient human ecological philosophy in China, and through observation and study of the dynamics of natural and human ecosystems, the Tianjin human ecology model was focused on the interactions between human behaviors (the decision behaviors, life attitudes, and economic activities) and urban ecosystem structure and processes. The city is understood

TABLE 14.2. Some of the dynamic characteristics of structural and functional coupling in the eco-mechanism model of the urban eco-complex.

Structural Coupling

- *Hierarchy and networking.* The urban eco-complex is organized in an ecological order by both vertical and horizontal connections, forming units at different scales from the individual to the community and the ecosystem.
- *Dominance and diversity.* Having its dominant and diversified components, the city is able to drive and maintain its productive and sustainable development.
- *Openness and independence.* Being open to the outside lets the city make full use of external resources, and being independent from the outside enables the city to be more self-reliant and keep away from outside risks.
- *Robustness and flexibility.* Robust structure enhances the city's fertile productivity, whereas flexible structure enables the city to adapt to the changing environment.

Functional Coupling

- *Exploitation and adaptation.* Severe environmental stress leads to adaptation as the city changes its eco-niche and adapts to alternative resources, making the most efficient use of the environment.
- *Competition and symbiosis.* All urban sectors survive through competition for resources and fertile production as well as symbiosis for maintaining sustainability. Competition stimulates high efficiency of resource use and symbiosis encourages sustainability of the ecosystem.
- *Proliferation and compensation.* When the function of an ecosystem is disturbed, some of its components might take the chance to expand or proliferate unusually so as to dominate the system, whereas other components might compensate so as to maintain the original function of the system. An ecosystem may benefit or suffer from these proliferation and compensation mechanisms. To stabilize a city, the compensation mechanism should be encouraged, whereas to raise its productivity, proliferation may play a key role.
- *Exhaustion and stagnation.* Due to resource exploitation, when outputs from a city are much higher than inputs and inputs fall short of the minimum cost for restoring depleted functions, ecosystem exhaustion will occur. On the other hand, when inputs into the city are much higher than outputs of final products, much material and energy will leak into the environment and an ecological stagnation may occur. In a totally functioning ecosystem, the input/output (I/O) ratio is appropriately one (Wang, et al. 1990).

as a structurally and functionally coupled eco-complex with dynamics outlined in Table 14.2.

The Eco-Planning Model: Probing Urban Ecosystems' Complexity and Sustainability

An urban ecosystem is completely different from a mechanical or a natural ecosystem. Its attributes and features usually are fuzzy, rough, incomplete, and often subject to rapid change, and corresponding parameters and data often fail to meet the standard requirements for modeling. We therefore have to find an alternative way to plan and regulate it. Urban modeling should be more than a mathematical optimization process under some simplified conditions and subjective hypotheses. The method developed here is to encourage a modeling revolution: from quantification to qualification, from optimization to adaptation, and from physical control to ecological learning. It is rather an interactive and evolutionary learning process consisting of partial learning of the key determinants, a process-oriented evolutionary optimization where participants are involved in integrative understanding and planning.

Key Factor Identification Model

Though there are nearly an infinite number of eco-units and connections between them in an urban ecosystem, the number of key factors, components, and connections that directly relate to the target problems and determine the dominating dynamics of the system usually is finite. In practice, we need only to identify the system's partial cybernetic relationships and to simulate its dominating dynamics. In the following sections, we present the results of seven facets of a key factor analysis of the Tianjin ecosystem.

Task-Oriented Flexible Boundary Definition

It is difficult to define the boundary of an urban ecosystem because its functional flows usually are not confined within a specified area. In our research, the boundary we identified differs depending on the different problems, processes, and policies we are concerned with. The urban ecosystem boundary may be discontinuous in space or vary during the course of a study according to different goals. In the Tianjin project, 5 different boundaries were identified according to different purposes (Figure 14.3 and Table 14.3).

TABLE 14.3. Five different boundaries and the corresponding purpose in the Tianjin Project (see also Figure 14.3).

- *Region I.* The eco-economic region of the Haihe River watershed and economic hinterland in 6 provinces of northern China, with a population about 120 million in nearly 1 million km^2, was taken as the target for studying regional, ecological, and economic impacts and development strategies.
- *Region II.* An administrative region with a population of nearly 8 million and an area of 11,660 km^2, identified for the study of urban–rural ecosystem interactions.
- *Region III.* The urbanized area that includes 6 urban districts, 4 close suburban districts, and 3 coastal districts with a population of 4.86 million in an area of 3,137 km^2, identified for investigating industrialization and urbanization strategies.
- *Region IV.* The built-up area with a population of 3.8 million living in an area of 330 km^2 identified for investigating land use–induced ecological problems and strategies for old town rehabilitation.
- *Region V.* The Guangfudao neighborhood within the downtown area with an area of 1.06 km^2 and 37,300 residents, the site for demonstration projects of urban renewal planning.

Key Factor Identification

For any ecosystem, there exists some combination of favorable environmental factors which promotes ecosystem development. Likewise, there are limiting or buffering factors that inhibit or stabilize an ecosystem's development, and potentially these same or other critical factors that dominate the dynamics of the ecosystem. In Tianjin, abundant land, a well developed transportation network, experienced technicians, and the city's geographical position are its key promoting factors; the water resource shortage and pollution, and some institutional limitations are key limiting factors; the widespread wetlands and conservative tradition are key buffering factors; and the relationship between Tianjin and Beijing, which share the same watershed and socioeconomic region, is the critical factor having both promoting and limiting effects on its development.

Key Component Identification

All ecosystems have dominant species or components that determine the main dynamics and govern the main processes within the system, while there often are compensating components that can spontaneously enhance or compensate for weakening functions in the system. In Tianjin city, the chemical industry, the coastal development district, and the strong administrative power are the key dominating components influencing the city's development.

Key Metabolism Flow Identification

Water, energy, material, information, people, and capital flows in the urban ecosystem have been investigated through spatial and time series analysis in Tianjin. Twenty-two main categories of material flow through the city have been investigated. The main streams of each material and the main stagnation points in the flows within each production sector have been ascertained. The efficiency in 15 different industrial sectors from 1949 to 1988 was compared. The results show that while production increased significantly during the last 40 years, the material-consuming coefficients have risen simultaneously, which is not an ecologically sound way of development.

Key Feedback Analysis

The relationships between population and urban construction, and between education level and production are the key loops of positive feedback within the Tianjin system. Key negative feedback loops include the relationship between land-use density and quality of life, and between industrial development and environmental quality. The main risks are regional ecosystem deterioration and falling behind the other three main coastal economic centers in China: Guangzhou, Shanghai, and Dalian. The potential for collaboration with Beijing and the environmental capacity for industrial development (available land and atmosphere for pollution dissipation) are the main opportunities identified in this analysis.

Leading Function Assessment

Economic production, social well-being, and natural buffering were investigated through comparative study of the 24 largest Chinese cities, and time series analysis of 40 years of data. These studies show that though Tianjin's urban economy increased more than twofold and its urban infrastructure improved dramatically since 1978, the city's economic development was slower than in the three other largest coastal cities: Guangzhou, Shanghai, and Dalian, even though remained more stable than its competitors. This shows that both the opportunities and the risks are relatively low in Tianjin compared to the three other coastal megacities.

By investigating 40 years of data on 25 natural and social indicators which influence citizens' quality of life, we found that Tianjin's environmental quality has remained at the same level as 8 years ago, while the quality of the social environment is quite improved. Some quantitative indicators such as the per capita housing area, the total length of roads within the city, and the gas and electricity supply have increased by about 50 percent. But generally speaking, in comparing Tianjin to 12 other large cities, the total quality of life is lower than that of other middle-size cities in China.

Although the natural buffering function of the region is relatively strong, which allows assimilation of all of the sewage discharged from both Beijing *and* Tianjin (the second and third largest industrial cities in China), the aquatic carrying capacity of the region is significantly overloaded, especially in some rivers and suburb areas. The green area coverage ratio ranks the lowest among the 25 largest cities in China. It is an urgent task for Tianjin city to enhance and restore its ecosystem vitality.

Dominating Dynamics Analyses

A number of dynamic processes can dominate urban ecosystem development, and these can be identified through key risk and opportunity analysis while varying space, time, quantity, layout, and order. Natural disasters (geological and hydrological), economic fluctuations, social turbulence, transportation, and socioeconomic edge effects during the last 2000 years all had major influences on Tianjin's development. Urban development stagnated before 1978, and has been thriving since then. The urban infrastructure has been greatly improved during last two decades. A counter-urbanization dynamic has taken place in the outskirts of the city rather than in the coastal area or in the four satellite towns that were planned originally by local decision makers. This occurred because of the attractiveness of the area's production and living niche rather than due to administrative orders. Tianjin's former position as the third largest economic center in

China is being surpassed by Guangzhou and Dalian or even one of the other new industrial star cities like Shenzhen and Suzhou. The reasons are: the strong constraining role played by Beijing, the capital; the lower diversity and openness of the Tianjin urban system; the more rigid natural and hydrological conditions; inactive institutions; the unreasonable town distribution in the region; and the loss of the hinterlands. This challenging dynamic could be changed by switching the Tianjin/Beijing relationship from one of competition to one of symbiosis.

Partial Simulation Model

Though it is difficult to simulate a whole urban ecosystem, we can do a partial simulation in three stages: (1) problem diagnosing, (2) process tracing, and (3) policy testing. In the Tianjin project we chose the problems of housing shortage and irrational land use, the processes of water flow from its source to sink, expansion of built-up area, and the policies of old town renewal and green space development as our starting points. Partial simulation led to some useful results through collaboration with local decision makers.

Problem Diagnosing

The high density of human activity and low quality of citizens' life in the downtown area caused many urban problems, including poor housing, traffic, noise, and lack of green space. Starting from diagnosing the old town rehabilitation and the housing problem at the Hedong District, we first ascertained which objective and subjective factors caused or influenced the problems, then simulated their system interrelationships to find key factors. In this way we identified directions and alternatives for improvement, and tested the effects and feasibility of these scenarios. In an analysis of housing problems in the Hedong District, for example, we identified 110 different indicators in 18 neighborhoods through the use of 350 questionnaires. Key problems included significant gaps in housing conditions among different neighborhoods. Potential solutions were identified through eco-niche and eco-potential evaluation, and ecological–economics assessment.

Process Tracing

All urban problems are connected, to some extent, with inappropriate ecological process. Through tracing, quantitatively and qualitatively, the spatial and temporal processes of water and energy flow from source to sink using different simulation techniques, we found some integrative strategies for mitigation of water and air pollution, for tapping alternative water and energy resources, and for reasonable production distribution and planning. The interaction pattern of Tianjin water flow was simulated through 33 alternative strategies and some technological, institutional, and management recommendations were put forward to local decision makers.

Policy Testing

The common shortcoming in traditional policy making is its failure to adopt a cybernetic point of view. Decision makers are used to cause-and-effect "monothinking," which has caused many severe side problems in Tianjin. Taking the economic development strategy of the Tanggu coastal district, the third largest port in China, and the Greenbelt construction policy analysis as examples, we developed a policy analysis method for local decision makers. The policy testing scheme allows decision makers to evaluate alternatives, comparing their positive and negative, current and potential, and local and total effects and impacts, taking into consideration some objective and subjective standards, quantitative and qualitative measures, and theoretical and empirical knowledge. After investigating 15 main industrial sectors and 28 key enterprises together with the local environmental protection and government policy agencies, 16 development measures were tested by checking their social, economic, and natural benefits and costs, through pan-objective ecological programming.

Adaptive Optimization Model

The essence of traditional mathematical modeling is to turn complex reality into a simpler mathematical framework and to optimize it according to some fixed rules. It is in fact a projection of some simplified parameters to an optimum result. Though a good method for well-defined physical systems, it is hardly suitable for an urban ecosystem. In fact, man can neither fully understand nor optimally control the whole man-dominated urban ecosystem. People usually make their decisions not through optimization but rather through a simple trial-and-error method, with the result chosen from out of a large heap of schemes. This often is more satisfactory and feasible than decision making from a strict mathematical programming perspective. It reminds us that the most important task in studying urban systems is not to find an optimum control scheme but rather to trace the ecosystem's processes and search for a healthy course of development that is locally satisfying and systematically responsible. In the Tianjin project, a new method called pan-objective ecological programming was developed, which is aimed at relearning and rearranging the development process by using eco-cybernetics and system analysis techniques and interacting with local policy makers. The core of this technique is to let local decision makers and managers find the system's key factors, key dynamics, and key opportunities; to adjust continuously its efficiencies, its harmoniousness, and vitality; and to search for an environmentally sound, economically productive, behaviorally feasible, and systematically responsible process (Wang, et al. 1991).

Sustainability Evaluation of Tianjin Urban Ecosystem

Building on the analyses of the structure, function, and flows of the whole city described above, we did a comprehensive evaluation of the city's sustainability using six groups of indicators (Table 14.4). From 1978 to 1991, life quality indicators increased most rapidly while indicators of production efficiency and institutional structure improved more slowly. Compared with Beijing and Shanghai, social order indicators in Tianjin are higher, whereas the other five indicators are inferior; though compared with 27 other provin-

TABLE 14.4. Indicators of sustainability used in an analysis of change from 1978 to 1991 in Tianjin, China.

Category of indicators	Subindicators	Change 1978–1991
(1) *Production efficiency*	• Growth rate of economy • Productivity (per capita GNP, profits, and taxes, etc.) • Resource use efficiency (water, energy, main raw material, and capital) • Waste emission and regeneration (air, sewage, solid waste)	2.5%
(2) *Life quality*	People's satisfaction with: • Income • Supply of housing • Traffic • Food • Education • Recreation • Other basic conditions and facilities • Life expectancy and health conditions.	8.3%
(3) *Institutional harmony*	Compromise between: • Dominance of the leading industry and products and diversity of various alternative opportunities • Self-reliance and openness to the outside system • The social regulation ability and individual or sectorial creativity ability	2.6%
(4) *Capability of people*	The capability of: • Decision makers (policy appropriateness, sensitivity of information feedback, ecological responsibility) • Entrepreneurs (creativity and vitality) • Citizens (literacy, values, and attitudes)	4.0%
(5) *Human ecological order*	• Social order (social mode, security, and morality) • Economic order (sustainable resource supply, inflation rate, unemployment, etc.)	2.9%
(6) *Natural order*	• Landscape • Water body • Atmosphere • Biodiversity	3.5%

cial capital cities, Tianjin's comprehensive score ranks third. Thus, Tianjin is a somewhat conservative city with a relatively stable social life and economic activity. It has had less social turbulence than other cities during the past 15 years, and its economic growth has not been as rapid as that of some other coastal cities, which, from a long-term point of view for sustainable development, might not be a bad thing. The lower indicators of the people's capabilities and institutional structure, however, suggest important challenges for the sustainable development of the city in the future.

The Eco-Regulation Model: Integrating Hardware, Software, and Mindware into Urban Ecological Development

Based on the cybernetic and planning models, a more practically oriented instrument—an eco-regulation model—was developed to help work toward a totally functioning technology, systematically responsible institutional structure, and ecologically vivid culture to benefit society while sustaining nature. Here the key is integration of "hardware" (technological innovation and integrative design), "software" (institutional reform and system optimization), and "mindware" (behavioral inducement and capacity building) (Wang, et al. 1998). This eco-regulation model includes technological innovation through ecological engineering, institutional reform through ecological management, and behavioral inducement through capacity building.

Hardware Regulation: Urban Ecological Engineering

Hardware regulation involves innovation of traditional production technologies in a search for alternative means of local resource exploitation and waste regeneration. Strategies include eco-industry incubation and technological innovation; eco-building development to encourage "greening," solar and bio-gas systems, eco-sanitation, and the use of eco-building materials; eco-infrastructure development in the areas of water, energy, traffic systems, and waste regeneration; and ecological landscape and vegetation restoration. The result can be considered a Totally Functioning Technology (Table 14.5). The Tianjin project resulted in a number of ecological engineering recommendations for technological innovation that were provided to decision makers and some of these have been put into implementation.

Software Regulation: Urban Ecological Management

City administrators are accustomed to short-term, small-scale, cause-and-effect reasoning and monoobjective management methods, and attempt to manage a complicated ecological network through a simple chain-linked

TABLE 14.5. Characteristics of a totally functioning technology (TFT) developed through urban ecological engineering.

(1) *Comprehensive integration of technology*
- High technology and traditional low technology
- Trans-industrial, inter-sectional, and interdisciplinary technology
- Environmental technology, cleaner technology, and ecosystem technology

(2) *Ecologically adaptive technology*
- Turning the open loop of material metabolism into a closed loop at the ecosystem scale
- Enhancing the structural and functional flexibility of production and consumption to have high adaptability to the changing environment

(3) *Economically efficient technology*
- Encouraging deeper processing and higher efficiency
- Using local resources
- Decentralized and cost-effective methods of waste treatment to turn negative environmental impacts into positive economic benefits

institution. They don't or can't understand and manage the effects of time lags, regional impacts, and information feedback that characterize a dynamic, diversified, and large-scale ecosystem. Furthermore, they often remain in a given position for only a few years and have only a limited scope of responsibility. To reform and adjust the unreasonable institutions such as the structure of production and products, spatial layout, administrative and management systems, and external connections, we encourage the establishment of a systematically responsible institutional structure in the city.

A Decision Support System for Urban Ecosystem Regulation (DSSUER) has been developed that includes a database, graphic and mapping base, methods base, knowledge base, and a management interface. The database consists of social, economic, and environmental data for the city, forming the basis for analyzing past and current conditions in the urban ecosystem and identifying problems. The graphic base can show trends in different factors, their relationships, and feedback loops, giving decision makers a more visual impression of the dynamic features of the system. The methods base incorporates various analysis tools—quantitative and qualitative, deterministic and stochastic, fuzzy and artificial intelligence, classic and innovative systems analysis—for users to choose among. The artificial intelligence technique, for instance, is ideally suited for exploring dual relationships in the system. Through modifications in parameters of the model, corresponding dynamic features of the system are simulated, and strategies for dealing with the resultant changes are analyzed. The knowledge base includes information from related disciplines as well as the knowledge and experiences of experts and decision makers with different backgrounds.

Mindware Regulation: Capacity Building for Eco-Culture Cultivation

The consciousness, creativity, and capability of policy makers, entrepreneurs, and the public are the keys for implementation of ecological development. Ecologically speaking, the "modern" city of Tianjin is a somewhat inefficient, immoral, unhealthy, counterintelligent, and less ecologically viable habitat. Its efficiency of resource use is much lower than that of a natural ecosystem. It exploits resources through degradation of hinterland ecosystems and imposes environmental impacts on its surroundings. Its people are estranged and competitive rather than intimate and cooperative. Its artificial living and working environments are far away from the real needs of human health. People more and more rely on electricity, water, cars, and chemicals to survive, more and more separate from nature.

In order to jump out of this ecologically decaying culture, a refinement of people's concepts, thoughts, values, manners, emotions, tastes, customs, and habits should be encouraged. Only when the lifestyle is harmonious with nature in metabolic process, structural pattern, and functional development, and human activities enhance rather than deplete the life-supporting system, can sustainable development be expected to be realized (Wang, et al. 1996). Recommendations for behavioral inducements were provided to and partly accepted by the local government through the Tianjin project.

Conclusions

The main characteristics of our urban ecology model are to apply the principles of ecological cybernetics to the identification and simulation of an urban ecosystem, and to help local people to understand their own city and find better strategies for its sustainable development. Its main procedures are the determination of system boundary, identification of its key factors, analyses of its interaction patterns, simulation of its dynamics and cybernetics, and interpretation of the modeling results. It includes three models: a cybernetic model, a planning model, and a regulation model through a methodological revolution from quantification, optimization, and computerization to Ecologically Intelligent Integration. The project emphasized interdisciplinary collaboration between natural and social scientists and local decision makers, and trained young scientists with interdisciplinary knowledge. Some of the research results and recommendations have been accepted by the local government and have shown significant benefits for the city's development.

References

Cooperative Ecological Research Project Group. 1995. *Towards sustainable city: methods of urban ecological planning and its application in Tianjin, China*. Urban Systems Consult, GmbH, Berlin.

Ma, S.J., and R.S. Wang. 1984. Social—Economic—Natural Complex Ecosystem. *Acta Ecologica Sinica* 4(1):1–9.

Odum, E.P. 1997. *Ecology: a bridge between science and society*. Sinauer Associates, Inc, Sunderland.

Tansley, A.G. 1935. The use and abuse of vegetational concepts and terms. *Ecology* 16:284–307.

Wang, R.S., and J.S. Yan. 1998. Integrating hardware, software and mindware: ecological engineering in China. *Journal of Ecological Engineering* 11:277–290.

Wang, R.S., J.Z. Zhao, and Z.Y. Ouyang. 1996. *Wealth, health and faith,—sustainability studies in China*. China's Science and Technology Press, Beijing.

Wang, R.S. 1991. Probing the nothingness—human ecological relationship analysis. Pages 1–6 in R.S. Wang, ed. *Human systems ecology* (English version). China Science and Technology Press, Beijing.

Wang, R.S., and Y. Qi. 1991. Human ecology in China: its past, present and prospect. Pages 183–200 in S. Suzuki, ed. *Human ecology coming of age: an international overview*. Free University Brussels Press, Brussels.

Wang, R.S., B.J. Yang, and Y.L. Lu. 1991. "Pan-objective ecological programming and its application to ecological research." Pages 321–330 in P. Korhonen, A. Lewandowski, and J. Wallenius, eds. Multiple criteria decision support, lecture notes in economics and mathematical systems, Vol. 356. Springer-Verlag, Berlin.

Wang, R.S., Z.Y. Ouyang, and Q.T. Zhao. 1990. On ecological construction. *Journal of Environmental Science (China)* 12(3):1–12.

Wang, R.S., J.Z. Zhao, and X.L. Dai 1989. *Human ecology in China*. China Science and Technology Press, Beijing.

Wang, R.S. 1988. *High efficiency and harmonious relationship: the principles and methodology of urban ecological regulation*. Hunan Educational Press, Changsha.

Yang, J.L. 1968. *Truth and nature*. International Publishing Company, Hong Kong.

Section III
Foundations and Frontiers from Education Theory and Practice: Themes

KAREN S. HOLLWEG, ALAN R. BERKOWITZ, and CHARLES H. NILON

Natural and social scientists who study cities have identified numerous concepts that they find important for understanding urban ecosystems. For most people, however, the term *ecosystem* applies to the tropical rainforests and other "natural" areas featured in spectacular TV footage. Most students learn about decomposition and materials cycling as things that occur on forest floors where rotting leaves and logs put nutrients back into the soil and promote new growth, and not as things that take place in cities. Citizens find out from the news daily that planning and other decisions about their city have been made in the halls of government, without their or scientists' input.

There is a gulf between the scientists' perspective and the perspective of the residents of cities and citizens everywhere. Further, if, as we have claimed in previous sections of this book, people must understand the basic concepts of ecosystem dynamics, boundaries, functions, history, and spatial patterns to be able to understand urban ecosystems and make decisions based on these concepts, then we must bridge that gulf. This section examines the role that education can play in doing that by focusing on two central questions: What is known about how people learn the kinds of concepts, skills, dispositions, and motivations identified as important by those who study and know much about the ecological functioning of our cities? What is being done and what is still needed to provide urban residents and people everywhere with learning experiences that will enable them to understand the urban ecosystems in which they live or with which they interact, so they can make the kinds of decisions that will lead to a healthy, sustainable future?

Contemporary ideas about teaching and learning are built on research conducted over the last 30 years. The National Research Council's (NRC) report *How People Learn* (1999) summarizes that research and highlights the need to focus on *developing understanding* and *learning to apply knowledge* (e.g., see Figure 1.2)—as opposed to learning facts by memorizing them, gaining skills through drill, and "covering" content, as we used to. The report reviews numerous studies and describes the rationale for beginning the process of teaching by discovering learners' existing notions, engaging learners in exploring new ideas, and eventually developing the conceptual frameworks for organizing factual knowledge and skills—frameworks that are key for using that knowledge in solving problems. It also describes studies about context and applicability that indicate that students must also learn *when* it is relevant to use new ideas and the situations in which it is appropriate to apply them. This emphasis on context parallels that used by scientists themselves in defining key concepts for urban ecosystem understanding (see the themes introduction and the chapters in Section II). The NRC study reminds us how important school-community connections are by pointing out that students only spend 14 percent of their time in school and 53 percent of their time in their homes and communities. (The other 33 percent is spent sleeping.) To make the most of in-school learning, it is important to acknowledge and build upon students' nonschool experiences, to enable students to learn in a way that will enable them to apply what they learn in out-of-school settings.

David Orr (1991) and others (for examples, see Simmons, Chapter 17; Keiny, et al., Chapter 19; Fialkowski, Chapter 21; and Grant, Chapter 22 in this section) argue for an integrative approach to education that teaches people more than "how to think" by addressing the interrelatedness of the sciences, social studies, and humanities; valuing harmony between human and natural systems; and equipping students to take part in reshaping our existing patterns of unsustainable resource use and harmful resource flows. This approach emphasizes the knowledge and skills needed to meet individual local needs and rebuild local communities.

The chapters in this section look at ways people gain an understanding of urban ecosystems and apply those understandings in their lives. While some chapters look in depth at the acquisition of skills and concepts (e.g., Roseman and Stern, Chapter 16), others focus on attitudes, motivations, and habits of mind or a combination of factors (Hogan and Weathers, Chapter 15; Chawla and Salvador, Chapter 18). Authors address what research tells us about how students develop conceptual understandings of complex systems and affective notions about the ecosystem in which they live. Since the body of research regarding learning about and in urban ecosystems is still rather limited, the authors draw on the larger body of literature concerning other types of education—as well as drawing on urban ecosystem education work that has been and is being done in schools (e.g., Roseman and Stern, Chapter 16; Smith, Chapter 20), colleges (e.g., Grant, Chapter 22; Campa, et al., Chapter 23), and communities (e.g., Fialkowski, Chapter 21), some of it published here for the first time. They describe a variety of ways in which learning opportunities are organized and approached, and reveal the degree to which the achievement of educational objectives has been documented. The learners described here range from children at play outside their homes gaining a sense of their place in the city (Chawla and Salvador, Chapter 18), to policy makers in governmental centers learning from constituents as they attempt to plan for a sustainable future (Roberts, Chapter 24).

The authors all are actively involved in improving education in and about urban ecosystems. They are working to solve problems they confront in their own specific fields, regions, or locales, presenting what they have learned and done so far and posing questions that remain. Their chapters describe the education frontier on which we find ourselves.

References

National Research Council (NRC). 1999. *How people learn. Brain, mind, experience and school.* National Academy Press, Washington, DC.

Orr, D.W. 1991. *Ecological literacy. Education and the transition to a postmodern World. SUNY series in constructive postmodern thought.* State University of New York Press, Albany, NY.

15 Psychological and Ecological Perspectives on the Development of Systems Thinking

Kathleen Hogan and Kathleen C. Weathers

Studying urban ecosystems is a venture in analyzing complexity (Douglas 1983; Pickett, et al. 1997a). Introducing urban ecosystem studies into schools thus presents a prime opportunity for students not only to learn principles of human ecology, but also to develop skills and habits for grappling with complexity. Connections between the study of ecosystems such as cities and the development of young peoples' thinking capacities is the theme of this chapter.

The growing interdependence of world economies, expansion of global media coverage, ease of travel, and use of telecommunications flood us with information about complex societal and environmental interconnections. Some of these factors and events directly affect our daily lives, and some do not, but all offer potential to heighten our awareness that simple analyses of situations and simple solutions to problems do not suffice. People who are able to recognize and analyze complexity will have the tools for participating as thinking members of a global community, and perhaps be better positioned to be effective problem solvers and decision makers in personal and local spheres. This is the premise that motivates our focus on educating young people to become systems thinkers, which we define as individuals who habitually look at things within the context of the environments that affect them, consider multiple cause-and-effect relationships, anticipate the long-term consequences and possible side effects of present actions, and understand the nature of change.

Natural and social scientists who strive to understand dynamic systems are expanding their capacities for systems thinking. They are adopting new approaches to interdisciplinary collaboration (Finholt and Olson 1997), new methods of analysis and modeling (Zimmer 1999), and perhaps most fundamentally, new attitudes and world views (Goldenfeld and Kadanoff 1999). A Newtonian, deterministic view of the universe as a machine that is ruled by linear cause and effect is giving way to a systems view that focuses on relationships, contextuality, and integration (e.g., Mann 1988), and recognizes the probabilistic nature of complex phenomena (Ulanowicz 1997).

Given the growing prevalence of systems analyses in the development of scientific knowledge, as well as the increasing complexity of the systems within which we live and work, it should be incumbent upon schools to work toward developing students' systems thinking capacities. We propose that in order for educators to foster students' abilities to engage in systems thinking and to understand the systems (e.g., cities) of which they are integral components, it is necessary to conceive of thinking itself as a complex system. We will develop this idea in the pages that follow by reviewing how systems thinking is currently taught, taking a retrospective look at the development of ecologists' systems thinking, presenting an expanded view of systems thinking, and providing examples of how urban ecology education can foster such thinking.[1] We conclude with brief reviews of areas of research in psychology and education that shed some light on various facets of systems thinking, and then we suggest directions for future research that will help us continue to refine our approaches to teaching and learning systems thinking.

How Is Systems Thinking Currently Defined and Taught by Educators?

Systems thinking comprises skills that allow a person to analyze open systems (i.e., those that exchange matter and energy with a surrounding environment) by recognizing how multiple factors interact, and by seeing and predicting patterns of change over time. One goal of teaching systems thinking is to help students think about whole systems in terms of their component parts, and to understand how the parts relate to one another and to the whole (American Association for the Advancement of Science [AAAS] 1993).

Educators who explicitly teach systems thinking[2] use a set of concepts and analytic tools derived from general systems theory (von Bertalanffy

---

[1] Throughout we use Sukopp's (1990) definition of urban ecology: "Urban ecology investigates the biosphere in towns and cities using ecological methods." We consider an urban ecosystem to be a defined area in space that includes the built environment and interactions among biological (including humans and their social systems), chemical, and physical worlds.

[2] The American Association for the Advancement of Science (1993) argues against explicit introduction of systems vocabulary and abstractions as a starting point for learning about systems. Rather, they say, the goal should be to enable students to attend to whole systems, such as when troubleshooting why a device is not working by looking at inputs, outputs, and interactions, whether or not they can label them as such. Although it is likely that students need concrete experiences with particular systems in order for systems principles to make sense to them, the role and timing of introduction to general systems terminology in systems-related learning and performance is an open question for research.

1968). As early as the 1940s, systems theorists suggested using concepts such as nonlinear causality and feedback to aid in analyzing and solving complex practical and academic problems in a diversity of fields. In fact, a classic application of systems analyses and modeling in the social sciences was in the field of urban policy studies (Forrester 1969). Studies such as this spawned a set of vocabulary and computer modeling tools that are used today by educators who seek to develop their students' systems thinking abilities.

Teachers within schools in the Catalina Foothills School District in Arizona, for instance, teach systems thinking by introducing students to a set of concepts (e.g., circular causality) and analytic tools (e.g., software programs for building models) that they apply across disciplines and grade levels (Yates 1998). Students in these schools are as likely to use behavior-over-time diagrams (Lannon 1994) to analyze story plots and characters in an elementary school language arts class as they are to use them to plot rates of flow in a high school physics class. The point of using a common systems vocabulary and tools across disciplines is not just for students to come away understanding particular systems, but also for them to become familiar with general templates and approaches for analyzing complex systems. Some of the general principles about dynamic systems that educators in the Arizona school district and elsewhere target include (e.g., VanderVen 1997): processes and things comprise systems; the properties of whole systems are usually different from those of its component parts; systems have emergent properties that arise from interactions of systems components; systems have boundaries that delineate external systems and internal subsystems; systems and subsystems interact via flows of inputs and outputs; systems can have feedbacks in which the output from one part of a system becomes the input to other parts; and effects can arise from complex interactions of multiple causal factors, so that it may not always be possible to predict accurately the result of changing a single system component or connection.

A popular approach to teaching students about systems and developing their systems analysis skills is to use computer-based models and simulations that make it possible to track how interacting components of a dynamic system change over time. Some of these software tools are role-playing simulations with a strong storyline that reflects real-life scenarios. For instance, Simons (1993) describes a computer-based simulation of urban planning in which the user takes on the role of a city mayor who must respond to a request from a city council member to divert money from public housing projects to schools. As mayor, the student views maps of the city and consults figures and trends in urban development over time in relation to population swings, unemployment, and other factors. The mayor can then run models to observe potential consequences of different decisions about public housing. In a similar story-based computer simulation, called Fish Banks (Meadows 1995), groups of students become managers of a fishing company who try to maximize their assets while considering economic competition and the dynamics of a renewable natural resource. One systems concept that this simulation conveys is that seemingly rational actions at local levels can have unintended and even detrimental consequences for a system as a whole (e.g., the depletion of a fishery).

Another category of computer-based tools for fostering systems thinking is generic modeling tools that allow teachers and students to create their own models. Mandinach and Cline (1994) report on a project that developed high school teachers' abilities to use modeling software called STELLA (High Performance Systems 1997) within their science and social studies courses in a Vermont school district. The teachers used the software to enhance their teaching about dynamic phenomena (e.g., oxygen production and population dynamics in biology, reaction rates in chemistry, acceleration in physics, war and race riots in social studies) by emphasizing systems principles such as causality, feedback, variation, and interaction. College biology educators also use STELLA to predict and model food web dynamics (Rueter and Perrin 1999).

A less complex, semiquantitative modeling tool called Model-It (Jackson, et al. 1996) is being used with middle school students for projects such as modeling stream ecosystems using water quality and other data collected from a river that passes through a city (Stratford, et al. 1997). Another new modeling tool developed for educational purposes is called StarLogoT (Resnick 1994). Whereas when using traditional modeling tools such as STELLA modelers manipulate aggregate properties (e.g., an entire population of an animal), in StarLogoT modelers manipulate individual entities, such as the behavior of a single ant in an anthill or individual rabbits in a predator/prey scenario. This allows students to see how complex, unpredicted behaviors of a system as a whole emerge from a collection of individual behaviors, and how random events can lead to predictable patterns (Resnick and Wilensky 1998).

These current methods of teaching systems thinking help students develop analytic skills and concepts that are useful for understanding complex systems. For instance, through using systems-oriented modeling software to study the factors affecting the quality of an urban stream, ninth graders gained competency in using a variety of cognitive strategies, such as analyzing system components, synthesizing information, testing ideas with models, and building explanations for observed phenomena (Stratford, et al. 1997). Also, students who use systems modeling software learn to use multiple approaches to solving complex, realistic problems, and express and examine their conceptions about the systems they are studying (Doerr 1996).

Are understanding systems concepts, however, and being able to use such systems analysis tools as computer models all that is entailed in being a competent systems thinker? Within the field of cognitive studies, comparisons of how experts and novices in a domain (physics, for instance) approach problem solving has yielded productive insights for the education of students who are learning to solve problems in that domain. Thus, to explore the possibility of expanding our conceptions of systems thinking for educational purposes, we turn now to examining the development of knowledge about complex systems as practiced by experts in the domain of ecology.

How Is Systems Thinking Practiced by Professional Ecologists?

We set out to define the characteristics and components of expertise in systems thinking by examining how scientists have developed successful ecosystem models.[3] We traced the development of three ecosystem models. The first is the universal model of energy flow (e.g., Odum 1959), which we chose because it was one of the first templates that ecosystem ecologists used to understand the controls on and complexity of energy flow in ecosystems, including urban ecosystems (e.g., Douglas 1983), and because it appeared in one of the first ecology texts that explicitly addressed ecosystem science. The text in general, and this model in particular, were the underpinnings for a significant body of ecosystem research during the 1960s and 1970s. E.P. Odum's continued awareness of the importance of the human influence on ecological systems also played a role in encouraging research on urban ecosystems (Douglas 1983). The second model we studied is the Hubbard Brook small watershed model of nutrient cycling, which has been—and continues to be—extremely influential in the development of ecosystem thinking for ecologists (e.g., Bormann and Likens 1967; Likens, et al. 1977; McIntosh 1985; Golley 1993). We chose this model primarily because of its importance as the classic empirically-derived watershed model of inputs, outputs, and major pathways and accumulations of nutrients in forest ecosystems. We note in particular that the small watershed approach is central to urban ecosystem research in the newly initiated Baltimore Ecosystem Study (Pickett, et al. 1997b). The third paradigm is the "Mice, Mast, Moths" model (e.g., Ostfeld, et al. 1996; Jones, et al. 1998), which is an example of a more recent model of ecological complexity. This model not only portrays the key interactions and drivers among popula-

[3] Although we are aware that systems ecology and ecosystem ecology are not necessarily synonymous (McIntosh 1985), here we use ecosystem models as an example of systems thinking. In addition, because urban ecosystem study is an emerging field, we drew on the development of models of systems that have been within the traditional purview of ecologists. The first two models, in fact, have served, and are serving, as a foundation for urban ecosystem research (e.g., Douglas 1983; Pickett, et al. 1997b). Ecologists' general processes of systems thinking are similar despite differences in the systems they study.

tion, community, and ecosystem levels, but also incorporates humans as components of ecosystems through generating predictions that have important implications for human health. All three models represent ecosystem complexity, and have had an important influence on the field of ecology. Taken together, they provide snapshots of systems thinking in the discipline of ecology from the 1960s through the present.

We used written accounts—autobiographical (e.g., Likens 1992; Bormann 1996; Likens 1998), historical (e.g., McIntosh 1985; Hagen 1992; Golley 1993), and empirical (see previous references)—to trace how the models were developed. This approach gave us a macro perspective on how the models evolved over time, not a micropsychological perspective of the online thinking of particular scientists. Nonetheless, we felt that these readings could help us uncover some common factors in the development of ecosystem models that might be instructive for teaching systems thinking. We also interviewed some of the key players in the development of the Hubbard Brook and Mice, Mast, Moths models. The interviews were extraordinarily important in helping us to understand subtle aspects of the development of ecologists' systems thinking, and so we draw on them for examples within the following descriptions of the two broad categories of ingredients of systems thinking that emerged in our analyses—cognitive and contextual components.

Cognitive Components of Systems Thinking in Ecology

Certainly, in-depth disciplinary knowledge was a necessary component of the systems thinking of ecologists who developed the ecosystem models we examined. In addition to a basic knowledge of one's own field, personal knowledge of ancillary fields was often necessary. The scientists developed disciplinary knowledge through formal academic training, but also through informal discussions with other practitioners throughout their careers and intensive reading of related literature while developing their own ideas and research focus. Such activities also helped scientists develop a firm grasp of previous research, such as the history, development, and limitations of important concepts in the field, as well as ideas for methodological approaches. For instance, one of the developers of the Hubbard Brook model (G. E. Likens) was exposed to the study of ecosystems through whole-lake manipulations while in graduate school at University of Wisconsin, Madison. This exposure later influenced his ideas for experimental manipulations of whole watersheds in New Hampshire.

Another key cognitive aspect in the development of the ecosystem models we examined was intuition that grew from the scientists' firsthand, intimate familiarity with elements of the natural world comprising the systems that they were seeking to understand. All of the models were empirically derived, so individuals in each research team were the primary collectors of data, or they worked with long-term databases they had initiated and maintained. This familiarity allowed the scientists to develop not just knowledge, but also intuition about the systems they studied.

The development of the ecosystem models also required cognitive skills such as the ability to identify and delineate (sometimes abstract) boundaries; use imagery and analogies; construct conceptual, empirical, and mathematical models; and in some cases, extend beyond one's own data to hypothesize about other systems. In all of the cases we examined, creative and flexible use of these cognitive skills were central to the development of the models. Also, somewhat unquantifiable characteristics, such as a passion for understanding complexity and a willingness to push the envelope beyond current modes of thinking in the field emerged as central to development of successful models. Scientists also held certain epistemological commitments, for instance a commitment to parsimony (attempting to "capture complexity simply," which Likens described as the motivation behind the development of his team's first conceptual model), or to post-Newtonian views of causality as being decentralized and probabilistic, which influenced how they perceived and described the systems they studied.

Finally, the collaborative skills and inclinations of the members of the research teams emerged as one of the most important enablers of systems thinking (e.g., McIntosh 1985; Bormann 1996). Cross-disciplinary interactions were essential to the development of all of the models. For instance, the Mice, Mast, Moths model emerged through a collaboration among chemical, forest, plant, and vertebrate ecologists which made it possible to uncover the impact of mast years (a year in which there is a synchronous production of fruits within a particular tree species across a wide region) in oak trees on mouse populations (i.e., that food production, more specifically the *timing* of that food production, drives mouse populations); the intimate link between mouse population dynamics and the presence, distribution, and Lyme-disease infection rates of deer ticks; and the effect of mouse populations on gypsy moth outbreaks. In order for collaborations that yield such knowledge to work, scientists must be open to the approaches of other disciplines or subdisciplines and points of view. In fact, based on his experiences working in multidisciplinary teams for many years, one of the developers of the Hubbard Brook model has stated that successful team-building, which depends in part on the collaborative skills of individual team members, is essential for the future productivity of ecosystem science (Likens 1998).

Contextual Components of Systems Thinking in Ecology

The development of new scientific knowledge in general, and of systems understanding in particular, is necessarily accomplished within social contexts that range from small work groups to the society at large. Interactions within these social and cultural contexts have an important influence on the development of intellectual products through peer encouragement and feedback, the ability to assume or be offered leadership within working groups, and access to funding to support the research.

As was noted earlier, the development of ecosystem models resulted from distributed expertise and cross-disciplinary interactions. To illustrate, the Mice, Mast, Moths research resulted in 13 scientific papers between 1994 and 1999. Of those papers, 11 required input and expertise from several different subdisciplines, five integrated among levels of organization and approaches, while only two could be considered narrower, subdisciplinary papers. The researchers we interviewed also noted that having a balance of temperament and personality types in a research team was as important as having a blend of disciplinary expertise.

Social structures facilitated the development of individuals in the research teams, which fed back to the development of their science. Personal connections, shared passions about topics or goals, and friendships among individuals were often the foundations upon which more formal collaborations were built. Events such as invitations to present one's research via seminars and at symposia afforded opportunities to put forth new ideas and get feedback from peers. In fact, the business of science in general, and of creating successful systems models in particular, involves the ongoing persuasion of peers through the exchange of ideas. Feedback within the scientific social structure often results in invitations to sit on panels (e.g., to review grant proposals) or advisory boards (e.g., to guide the direction of a new project), which present additional opportunities to learn what is "hot" as well as how to promote one's own ideas. These types of strong networks have the additional advantage of attracting good students and collaborators.

The task of writing and rewriting research proposals—a necessarily iterative process—was useful not only for obtaining funding, but for convincing collaborators within the research groups, as well as outside reviewers, that the research ideas were sound. This feedback and input was critical to honing ideas. The feedback came not only through reviews of proposals and papers, but from giving talks and seminars (for instance, Likens still recalls a thought-provoking question about his long-term data from an audience member at one of his talks over thirty years ago, who asked "Are you really sure that this is how the system will work in the future?"), and from the presence of graduate students and postdoctoral associates, who often served to facilitate or expand research into a new intellectual arena.

Larger societal contexts have also played an important role in ecologists' systems thinking. For instance, the increased incidence of Lyme disease in humans created a very practical societal need that partly motivated the scientists' quest to understand the ecology of the disease. The Hubbard Brook studies yielded the first descriptions of acid rain in North America, which triggered a cascade of societal concern and reaction that has resulted in ongoing feedback between the research and environmental policy (e.g., Likens 1992).

Summary

We found from our retrospective analyses of the development of three ecosystem models that grappling with complexity via systems thinking requires a suite of cognitive and interpersonal attributes, as well as sensitivity to contextual constraints and opportunities. Thus, ecologists' systems thinking has both cognitive and contextual components (Figure 15.1). The Cognitive Dimension box of Figure 15.1 contains small boxes that represent a multitude of psychological attributes of individuals. These include deep knowledge of a scientific discipline and natural phenomena; habits of mind, such as an inclination to expand beyond the boundaries of one's own discipline; ambition; vision; collaborative skills and inclinations; and skills such as analytic and analogical thinking abilities entailed in modeling. The Interpersonal and Societal/Cultural Dimension boxes represent the contextual factors outside of any single individual that allow systems ideas to emerge. Interpersonal factors include access to networks of colleagues and opportunities to collaborate, share work, and get feedback. Finally, larger trends in science and society can create a nurturing seedbed for ecosystem

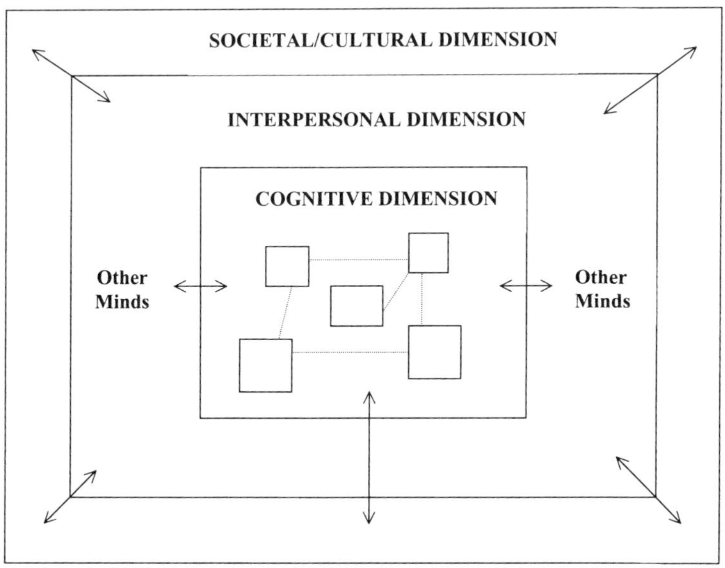

FIGURE 15.1. Representation of the interplay of cognitive and contextual factors in scientific thinking and the development of knowledge products such as ecosystem models.

science. The arrows connecting the boxes illustrate the interplay of cognitive and contextual factors in systems thinking, and indeed, in the development of all kinds of knowledge.

An Expanded Definition of Systems Thinking as a Suite of Competencies

Our portrait of the multifaceted and contextualized nature of ecological scientists' systems thinking is similar to portrayals of expertise that have emerged through cognitive and educational research. Building on his studies of expert performance in a number of domains, the cognitive scientist Carl Bereiter (1990) postulated that (1) complex thinking and learning always occur in response to a particular performance or problem context, and (2) expertise in performance is characterized by the activation of a whole suite of personal attributes that are finely adapted for use in that context. He calls that suite of attributes a "contextual module" to reflect that they function as a cohesive unit when activated by a particular context that requires sophisticated thinking or performance. For instance, studying for a test is an activity context that is very familiar to students, and one for which they might activate a "test-preparation module"—a whole set of knowledge, skills, goals, and attitudes that are activated together in an attempt to meet the demands of that task.

We argue that making sense of complex systems (such as urban ecosystems) is a type of task context that should activate a "systems thinking module." What types of ingredients would coalesce to comprise a systems thinking module? The following discussion contains the seven core elements of contextual modules that Bereiter proposed; however, we have elaborated on each element with suggested applications to systems thinking and the pursuit of understanding urban ecosystems.

Knowledge

Knowledge that is relevant to systems thinking includes knowledge about systems properties (e.g., feedback, emergent properties, interactions, multiple agents, and random and probabilistic events); the components and characteristics of specific systems (e.g., people, institutions, and natural resources in city ecosystems); and related scientific, sociological, economic, and other disciplinary subconcepts (e.g., principles of thermodynamics, units of social organization). Another type of knowledge that can support systems thinking and analyses is understanding the nature and goals of knowledge building within disciplines such as science (e.g., that models are intellectual tools for building, testing, and even generating new ideas and hypotheses, and why one goal of science is to build predictive models).

Reasoning Skills

The cognitive skills that are perhaps most central to systems thinking are those that enable a person to reason dynamically, flexibly, and reflectively. Specific types of reasoning that are entailed in the analysis of complex systems include:

(a) analytical reasoning—the ability to examine the parts of a whole (e.g., the factors leading to the outbreak of a disease in a densely populated area);
(b) analogical reasoning—the ability to discern similarities between two or more things (e.g., seeing how material and energy inputs and outputs of a city are similar to and different from a plumbing system);
(c) dialogical reasoning—the ability to think within multiple frames of reference (e.g., seeing the issue of publicly provided safe drinking water from the perspectives of citizens of various socioeconomic groups, government officials, youth, parents, and the elderly);
(d) inferential reasoning—inductive and deductive thinking that link general principles or conclusions to specific evidence or instances (e.g., linking a widespread commitment to recycling in some European cities to a scarcity of natural resources and consequent regulatory policies);
(e) evaluative reasoning—judging the worth or quality of something according to some external criteria (e.g., forming an opinion on the protection of open space in cities); and
(f) integrative reasoning—bringing together different pieces of information into a unified whole (e.g., creating an energy budget for an urban school building).

Goal Structures and Motivations

Since systems thinking can be mentally taxing because it demands keeping track of multiple pieces of information, people need to have some sense of why it is worth engaging in such efforts. Goals and motivations for engaging in systems thinking will be different for different people and will vary for any one person from situation to situation. They might include intrinsic goals related to attaining personal intellectual satisfaction, or extrinsic goals related to rewards (e.g., grades, professional advancement) received for successfully accomplishing an analysis of a complex system. Other goals and motivations for systems thinking could be related to personal interests, sociopolitical commitments, or opportunities for interpersonal interaction.

Identity and Self-Concept

A view of oneself as an active and capable thinker could be an important element that supports engagement in systems thinking. Students' self-referenced perspectives include not only their self-efficacy for systems thinking, but also their personal identification with broader social and cultural groups that they may or may not interpret as being affiliated with intellectual or activist pursuits related to systems thinking.

Problem Models

Through engaging in a variety of systems analysis tasks, people are likely to develop a repertoire of generic problem types or phenomena that are conducive to systems analysis. Internalizing a set of common characteristics of such situations makes it possible to recognize readily an opportunity for systems thinking, understand what doing a systems analysis of the situation will entail, and then mobilize other supporting components of the systems thinking module.

Affect

Emotions are associated with just about everything we do. A systems thinking context might evoke emotions ranging from anxiety to elation at being challenged to stretch one's thinking capacities in a solo or group context. Particular emotional associations with engaging in systems thinking will likely shift with time and situation.

Codes of Conduct

Particularly when systems thinking occurs in a social context, there will be a variety of roles, expectations, and obligations of the various participants. Becoming proficient in the practices of a discipline (e.g., natural science, social science) entails understanding the norms that characterize the cultural practices of that knowledge-building community. When urban ecology is studied within a science class for instance, then the norms of science (e.g., respect for evidence) might establish a basis for certain codes of conduct among participants. Students also have their own social norms that will permeate their intellectual collaborations for systems analyses.

Although Bereiter proposed contextual modules more as an heuristic framework for broadening the focus of educational practices than as a formal cognitive theory, there is increasing evidence of the interdependence of cognitive elements, such as in studies relating knowledge to reasoning (Schauble 1996) and motivation to cognition (Pintrich, et al. 1993). The contextual modules framework also integrates context with cognition, which reflects a growing awareness among cognitive researchers that much of our cognitive activity is socially situated (Greeno 1997).

By conceptualizing systems thinking as entailing a suite of competencies anchored in a particular context, we move toward defining systems thinking as an holistic way of relating to real-world situations that demand systems analysis. Contextual modularity provides a psychological framework that supports what environmental educators have advocated for years: that learning about and acting on behalf of the environment requires development of cognition, skills, affect, and behaviors. The modularity perspective emphasizes that these are mutually supportive and interdependent elements of skilled performance, so that changes in one element bootstrap or feed back to the development of another element. The whole amounts to more than the parts in isolation. Thus, thinking itself can be regarded as a dynamic system.

Developing Students' Systems Thinking Capacities Through the Study of Urban Ecosystems

When scientists and other experts exercise and apply their suite of systems thinking capacities, they do so in the context of complex, compelling problems. Indeed, cognitive theorists propose that demanding performance contexts are necessary for activating sophisticated thinking modules. This leads us to recommend that educators immerse students in challenging problem contexts to foster their development of systems thinking capacities. These need not just be culminating events that allow students to apply knowledge they already possess, but rather can be used to motivate the construction of knowledge as it is needed for solving problems (Greeno 1997).

It could be argued that deriving educational approaches from analyses of the practices of experts such as scientists is misguided because the jobs, life contexts, and maturity levels of adult experts and young students are so different. For instance, scientists focus on the unknown, while science education concentrates on the known. Ecosystem science is rarely an individual effort—research is the combined work of players representing multiple perspectives and disciplines that need to be combined to achieve advanced understanding. In classrooms, however, there typically is one expert (the teacher) and many novices, and it is individuals' accomplishments that ultimately are valued and assessed. Perhaps most fundamentally, students and scientists have different objectives. Whereas scientists work to create what the philosopher of science Sir Karl Popper called World 3 knowledge—products of thought that exist independently of any single mind, such as theories and other ideational objects that are the currency of scholarly exchange, students in school aim to produce World 2 knowledge, or personal, in-the-head understanding of World 1 (the physical world) phenomena (Bereiter 1994).

What if students, however, were to undertake work that mirrors the activities of scholarly communities by dedicating themselves to creating knowledge products that have a useful life of their own, and are open for exchange and debate, rather than solely to developing personal knowledge structures? Educators are finding that such experiences result in learning that is deeper, more dynamic, and more motivated (McGilly 1996) than typical schoolwork experiences. In designing urban ecology learning experiences, then, it could be beneficial to consider what types of projects would give students an opportunity to produce viable ideas and tangible products within a community of fellow inquirers, while also developing their own suite of interdependent capacities for systems thinking.

Imagine a project-based learning experience centered around the theme of invasive exotic plants in urban woodlands.[4] Students from elementary through high school levels could be given an initial challenge, such as to (1) map the distribution of exotics in a plot of woodland and then present the information to a management agency; (2) design, write, and produce interpretive signs or brochures that educate the public about native and nonnative species; (3) make and execute plans for removing invasive exotics from an area of urban forest; (4) raise native plants that could be used for habitat restoration; (5) design and administer a survey to poll neighborhood residents about their perspectives on urban forests, invasive exotic plants, and related ecological issues; or (6) debate whether invasive exotics should be considered a "problem," and, if so, to suggest management policies that could stem the introduction or spread of invasive exotics.

Once a project context that captures students' interest is established, there will be innumerable opportunities to stress systems themes. Principles of systems dynamics could be introduced to students through discussion of factors—biological, physical, social—that interact to determine the structure and function of urban woodlands. Students could draw causal loop diagrams (Lannon 1994) and other conceptual maps to develop ideas about relations among factors such as landscape structure, human settlement patterns, plant community dynamics, and human values, behaviors, and management tactics. Advanced students could use modeling software to build dynamic models of interacting factors, or to run pre-built models to make and test predictions about changes in woodlands given different management or perturbation scenarios. By creating, sharing, and critiquing their own and their peers' models and other knowledge products, students would generate the kinds of World 3 knowledge that experts produce, in addition to gaining personal understanding of the interrelations of humans' knowledge, values, and behaviors with the characteristics of the landscapes they inhabit.

Thus, it is possible to see how urban ecology projects have the potential to teach students about complex systems by challenging them to create practical products as well as academic knowledge. These activities certainly can develop students' systems-related knowledge and reasoning

---

[4] The City Parks Foundation in New York City runs a project-based, after-school program in urban ecology in which students have studied invasive exotic plants in forested urban parks. We thank the education director of that program, Mary Leou, for conversations about project-based learning of urban ecology.

skills, but what about the other components of their systems thinking modules?

Affect, identity, motivations, problem models, and codes of conduct will automatically be present as students engage in projects such as those described for the urban woodlands study, given cognitive theory that specifies that these are integral components of human thinking and performance. For instance, the experience of conducting investigations in teams is likely to elicit a variety of emotions about interpersonal interactions. Whereas some of the personal attributes listed here may be well adapted to the systems-analysis tasks at hand, others may be counterproductive. These elements of systems thinking, therefore, need to be consciously refined and honed, just as knowledge and reasoning skills need to be deliberately developed.

We propose that fostering development of these additional components of a systems thinking module can be accomplished with one central tactic: reflection on learning. Although the mutually supportive components of a thinking module are activated together without deliberate reflection, reflection is necessary to make the elements accessible to conscious modification (Bereiter 1990).

In practice, this would entail raising students' awareness of the full suite of elements that coalesce to support competent systems thinking, and giving them tools for reflection. For instance, along with introducing students to a conceptual framework (e.g., plant reproduction and distribution strategies) and vocabulary (e.g., feedback loops) that facilitate urban forest systems analysis and assessment, teachers could introduce a framework that describes the elements of systems thinking. This could then become the basis for self-analysis and assessment. Throughout a learning experience, students would be given prompts to reflect in writing or discussion on how their affect, identity, motivation, and the like support or inhibit their learning and performance in a systems analysis context. Structured self-assessment rubrics for assessing collaborative processes (see Hogan 1994 for an example) could help students become aware of productive and counterproductive codes of conduct as they collaborate to develop knowledge about urban woodlands. Developing self-awareness of one's own and one's peer group's assets and liabilities as thinkers builds the basis for future regulation of performance. Feedback from a teacher assists students to develop strategies to improve.

In summary, the general approach we suggest for developing systems thinking abilities is immersion in a compelling project, introduction to systems analysis frameworks and vocabulary, and use of an assessment framework and tools that support self-evaluation and development. Just as using templates for systems analysis has become habitual for students who attend schools that emphasize systems themes across grade levels and academic disciplines, so, too, could reflection on the components of a self-system that supports systems thinking become second nature for students who are regularly asked to reflect on their own performance. Through these experiences, students should be able to see how their learning experiences relate to the world beyond school, to their identities as self-directed learners, and to their growth toward mature participation in valued social practices.

Challenges to Teaching and Learning Systems Thinking: Insights from Research

What do teachers need to know in order to be effective facilitators of students' systems thinking? Certainly they need a good conceptual understanding of systems principles and in-depth knowledge about the particular system about which they want to teach, such as an urban ecosystem. But they should also understand what learners are likely to find problematic and how they tend to reason so that they can plan and sequence learning experiences to meet their learners' needs. Teachers create effective learning experiences by integrating knowledge about learners with knowledge about the discipline to be taught.

The following sections highlight some areas of educational research that provide useful background information for educators who want to develop systems thinking and urban ecosystems learning experiences. Given the scope of this chapter, these sections cannot provide thorough reviews of each topic. Rather, they provide brief summaries of some key findings along with pointers to relevant literature.

Misconceptions and Naive Notions

In order to engage in powerful relational and systemic reasoning about a topic such as urban ecosystems, students need a detailed, integrated body of relevant conceptual knowledge (Bishoff and Anderson 1998). A whole host of subconcepts are entailed in developing an understanding of urban ecosystems. We focus in this section on science concepts, but acknowledge that social science concepts play an equally important role in understanding urban ecosystems. Even just the science concepts embedded within the multidisciplinary topic of urban ecosystems, however, are numerous and complex, including concepts relating to thermodynamics and energy flow, populations and food web dynamics, the cycling and conservation of matter, photosynthesis and respiration, and so on. Although these science concepts may not relate to the most pressing or personally relevant things that students will want or need to know about cities, they are components of the required science curriculum in all schools, so teachers could make the concepts more meaningful to students, especially to those living in cities, by teaching them within the context of urban ecosystem studies.

What do educators need to know in order to help students develop scientific knowledge that can serve as a partial foundation for understanding and analyzing urban ecosystems? A guiding tenet of science education practice over the past several decades is captured in the following words of a cognitive psychologist (Ausubel 1963): "The single most important thing you can know is what a learner already knows; ascertain this and teach him accordingly." Thousands of studies (Pfundt and Duit 1998) have documented that students often have strong naive preconceptions about natural phenomena[5] that can interfere with their ability to develop more accurate scientific explanations of how the world works. Science educators have thus come to think of their job as precipitating conceptual change, which requires students to confront their prior ideas in order to construct canonical knowledge (Posner, et al. 1982).

One strand of studies on students' prior conceptions that is relevant to teaching about urban ecosystems concerns the naive ideas that children ages 5–16 in Europe and the United States display about a variety of ecological concepts (Griffiths and Grant 1985; Webb and Boltt 1990; Leach, et al. 1996a, 1996b; Hogan and Fisherkeller 1996; Hellden 1998). Although these studies did not survey students' ideas about urban ecology in particular, they did address topics that are integral to understanding key processes in urban ecosystems such as the sources, flows, and fate of matter and energy into and out of a city.

One robust finding across studies is that students of all ages find it hard to accept that plant biomass comes from gas and water, rather than from solid materials such as soil. This makes it hard for even most 16-year olds to interrelate photosynthesis, respiration, and decay into a comprehensive view of matter cycling, and thus they would have difficulty seeing how these processes contribute to the overall matter and energy budget of a city, for instance. Students' knowledge tends to be compartmentalized, so for instance if older students know something about the role of oxygen in the respiration of food, they do not relate this easily to the flow of matter and energy in ecosystems.

Students also harbor misconceptions, and have missing conceptions, about food web relationships. Primary school–aged children think of organisms only as individuals, rather than a members of populations, and do not think about food as being scarce and competed for by individuals within populations. Precollege students have difficulty explaining population dynamics in terms of abstract resources such as energy, and relating principles of the conservation of matter to feeding relationships and decomposition. Students also have difficulties tracing relationships among trophic levels. Young students, for instance, apply teleological reasoning to conclude

[5] For an overview of students' preconceptions about social phenomena and institutions, see Cheek 1993.

that organisms in lower trophic levels are there to satisfy those at higher levels. Once students are able to describe population dynamics within single food chains, difficulties tracing population dynamics among multiple food chains that are connected within food webs persist, even among first-year college students.

Students also exhibit misconceptions about issues concerning human-accelerated environmental change, such as global warming and the ozone hole (Fisher 1998; Rye, et al. 1997; Boyes and Stanisstreet 1991). Two prevalent misconceptions are that ozone layer depletion is a major cause of global warming, and that carbon dioxide destroys the ozone layer. These and related misconceptions persist from ages 10 to 16, and indeed are commonly held by adults (Kempton, et al. 1997).

Finally, young students may have difficulties understanding the systems concept itself. For instance, they may assume that each ingredient of a system has properties that are identical to those of the system as a whole (AAAS 1993). One example of this type of thinking is when children assume that molecules that comprise liquids are themselves fluid, and that those that comprise solids are hard.

An intriguing proposal is that some of students' difficulties in understanding ecological concepts could be due to a common root problem: that they mentally categorize dynamic processes as things (Chi, et al. 1994), such as by believing that heat and energy are physical substances. In this case, conceptual change requires changing students'ideas about the basic ontological category to which phenomena belong. Because being able to differentiate matter and energy is necessary for understanding basic ecosystem processes, it may be well worth considering an educational approach that explicitly addresses students' ontological categories.

One promising sign that emerged in several of the studies cited earlier is that with increasing age young people showed more awareness that there were weaknesses and gaps in their explanations of natural phenomena. They were also able to reflect on how their ideas originated in their experiences in everyday life and school. Such self-awareness, and the corresponding ability to critically examine and become dissatisfied with one's own ideas, facilitates the conceptual change process.

Causal Reasoning

One focus of cognitive research in recent years has been on trying to understand the relative importance of a firm knowledge base on the one hand, and content-independent reasoning skills on the other, in the development and execution of higher-order thinking processes (e.g., problem solving, critical thinking, decision making). Most researchers agree that these two facets of cognition are intimately related (e.g., Mintzes, et al. 1998; Schauble 1996).

Therefore, understanding how students reason is as important as understanding the nature of their conceptual knowledge when planning an educational program that aims to foster students' systems thinking capacities through the study of urban ecology. An increasing number of studies provide encouraging evidence that even elementary school students can acquire the thinking skills necessary for analyzing complex systems.

Causal reasoning is one type of reasoning that is central to systems thinking. There are several types of causal patterns that occur within complex systems, including linear causal chains (with and without single or multiple branches of effects from a common cause); cyclic, spiral, and reciprocal cause and effects; and probabilistic causal relations (Grotzer 1997; Grotzer and Bell 1999). After finding that students readily apply linear, but not more complex causal reasoning to analyses of systems interactions, Grotzer and her colleagues (Grotzer and Perkins 1999) designed interventions to expose students to graphical templates representing various patterns of causality. The intent was to increase students' sensitivity to the variety of possible causal patterns, and to increase the likelihood that they would recognize forms of causality embedded in particular science concepts. The researchers expected that this would decrease students' tendency to impose linear causal models where they did not fit. Preliminary results of their studies with elementary and middle school students learning about biological and physical systems reveal that enhancing science learning activities with presentations of the causal templates, plus explicit conversation about different causal patterns, leads to increased gains in conceptual understanding of complex systems (Grotzer and Perkins 1999). Similar findings at the college level demonstrate that when students receive explicit information about principles of emergent causality (e.g., emergent systems cannot be explained in terms of simple linear cause-and-effect relations) they learn about systems that display these characteristics (e.g., electrical circuits) better than students who are exposed only to traditional textbook information about the systems (Slotta and Chi, in press).

Although there is not yet a large literature on teaching students to reason about complex causality, there are indications that direct exposure to models of causality is one promising approach. These findings support a general model of cognition that represents thinking process as proceeding on two levels: the object level, where cognitive operations are performed; and the meta level, where a person holds information, known as metacognitive knowledge, about the object level (Winne 1997). Monitoring and regulation processes control the flow of information between the object level and the meta level when people engage in reflective thinking (Metcalfe and Shimamura 1994). Generic models of causality are an example of a type of meta knowledge that can guide systems analyses occurring on the object level.

There is some evidence that the reasoning skills necessary for understanding properties of dynamic systems such as feedback and emergence develop in parallel with cognitive structures that underlie general capacities for abstraction and propositional reasoning (Chandler and Boutilier 1992), typically around the beginning of adolescence. Based on empirical findings and theory development, however, educational researchers are recommending that it is not necessary to delay instruction in abstract or inferential thinking until adolescence as is typically done in schools (Metz 1996). In fact, young children are quite capable of using sophisticated causal reasoning skills within the context of problems that they care about and have a good knowledge base on which to draw (Koslowski 1996). Therefore, some of the observed shortcomings in children's thinking may be due more to their limited experience with thinking about complex causality within the school curriculum than to inherent developmental constraints.

Emotions

Feelings and values are intertwined with knowledge and reasoning skills in young people's cognition about nature and environmental issues. Children's feelings of affiliation with nature are evident even in impoverished urban communities where opportunities to develop such affiliations could be constrained by lack of access and freedom of movement (Kahn 1997; see also Chawla and Salvador, Chapter 18 in this volume). Yet urban children can also be more fearful and anxious in natural areas than rural children (Cohen and Horm-Wingerd 1993), presumably reflecting their differing levels of familiarity with such areas, as well as associations that urban children have with wooded or swampy areas within cities as sites of crime (Kahn and Friedman 1995). Although urban children like trees, animals, open space, and water, they prefer parklike to wilderness settings, but also worry about experiencing physical discomfort or harm in such settings (Simmons 1994).

In addition to basic likes and dislikes for different kinds of outdoor environments, there are also moral dimensions to children's environmental perspectives. One study of black children between the ages of 7 and 11 living in an economically impoverished inner-city community (Kahn and Friedman 1995) found that they regarded pollution as a violation of humans' moral obligation to nature. Their knowledge that animals can be harmed by pollution was linked with the value of caring about the harm that people inflict on animals. However, young people also tend to exhibit more anthropocentric rather than biocentric rationales for protecting the environment or preserving species, reasoning that nature is important to human welfare rather than that nature has intrinsic rights regardless of its benefits to humans (Kahn and Friedman 1995; Kahn 1997; Palmer 1997).

One interesting finding about city children's environmental concerns is that when asked about what environmental issues they discuss with parents at home, children living in poor inner-city communities name topics such as drugs and violence as the top environmental issues that concern their families (Kahn 1997). Although these children's interpretation of the term "environmental" may have differed from the researcher's definition of the term, their responses point out that issues of socioeconomic and ecological viability are inseparable.

In summary, children integrate scientific principles with value judgments when they reason about nature and make decisions about socio-scientific issues (Ratcliffe 1997). Their environmental knowledge, sensitivities, and commitments are woven within the larger fabric of their cultural context and life-world domains (Kahn 1997). Given the interdependencies of cognition and affect, emotions must be recognized as an important dimension that students bring to the learning of a topic such as urban ecology, and to related development of systems thinking skills.

Intellectual Collaboration

Understanding complex, dynamic phenomena is aided by teamwork, as was evident in our earlier profile of scientists' systems thinking. Cooperative learning has been a popular pedagogical strategy over the past two decades for promoting teamwork in classrooms (see Webb and Palincsar 1996 for a review). Traditional cooperative learning techniques, however, tend to stress prosocial rather than intellectual behaviors. Dividing up responsibilities for a school task by assigning a different role to each member of a group, for instance, does not necessarily help students build on one another's ideas and strengths to attain a synergistic combination of contributions.

Table 15.1 lists the strengths and pitfalls of the collaborative reasoning of eighth graders who worked in small groups to construct conceptual models of the nature of matter (Hogan 1999a). Their task was in essence to engage in collaborative systems thinking to analyze a dynamic microsystem—the relations among atoms, molecules, and environmental conditions.

Sharing findings such as these directly with students is a potential means of helping them develop a more sophisticated mental model of effective intellectual collaboration than simply holding the idea that it is necessary to get along with your partners. Helping students become aware of what effective intellectual collaboration entails is a step toward enabling them productively regulate their own collaborations.

Directions for Educational Research on the Development of Systems Thinking

Although there have been many studies that address one or two components of systems thinking, it is difficult to examine empirically how all of the multiple components of cognition function together. There are research

TABLE 15.1. Factors that can promote and constrain collaborative reasoning.

Factor	Promote collaborative reasoning	Constrain collaborative reasoning
Metacognitive knowledge	• Accurate knowledge about the task	• Wrong or inadequate knowledge about the task • Knowledge about self as a thinker that contains negative evaluations
Metacognitive regulation	• Negative evaluation of and dissatisfaction with current ideas, and subsequent regulation of their improvement	• No evaluation of ideas; positive evaluation of weak ideas; or failure to regulate weak ideas
Conceptual knowledge	• Pieces of relevant prior knowledge available	• Very little prior knowledge available
Intellectual curiosity	• Sharing of queries and "why" questions • Expressing curiosity about others' ideas	• Few queries expressed • No expression of genuine curiosity about others' ideas
Discourse style	• Connected discourse that acknowledges, builds, and elaborates on others' ideas	• Disjointed discourse with little uptake of one another's ideas
Social relations and orientations	• Expression of enjoyment of working together • Skillful mediation of conflicts	• Expression of dislike between group members; verbal bullying • Expression of individualistic rather than group orientations
Engagement	• Active involvement in the task • Productive use of teacher hints and guidance	• Passive approach to the task • Failure to use teacher guidance productively

programs within science education that seek to understand the multidimensional and contextualized nature of cognition, including research on personal orientations to science learning (Shapiro 1994), contexts of personal meaning (Bloom 1992), personal frameworks for science learning (Hogan 1999b), and worldview theory (Cobern 1996). However, these programs have not investigated systems thinking per se.

A conundrum of research is that as a greater number of interacting factors and contexts are taken into account, conclusions typically become less generalizable (Bereiter 1990). Also, although finding correlations among factors and empirically accounting for variance might reveal which factors are important in learning, such analyses do not specify how the factors operate. Thus, some psychologists are creating computational models of cognition and human development by applying nonlinear dynamics and systems analysis methods in attempts to understand the processes and not just the ingredients of learning (Thelan and Smith 1994; van Geert 1998). An ongoing challenge is to scale up these models from representing atomistic units of behavior such as motor skills development to representing cognition on scales of real educational significance.

A systems perspective thus has potential not only to transform what and how we teach, but also how we do research in education and psychology. Continuing to explore multiple methods, from qualitative anthropological accounts of how students' systems thinking operates in context to quantitative computational models of cognition, will expand our capacities for making non-reductionist, integrative analyses of complex human behaviors such as systems thinking.

Conclusions

Systems thinking is a valuable educational goal because it can enrich the lives of learners in ways that will be meaningful to them as they tackle all manner of complex problems, not solely because it is good for society to have citizens who can think systemically. Based on analyses of systems thinking expertise, we proposed that becoming a competent systems thinker requires more than the types of knowledge and skills that schools explicitly foster. We have tried to broaden what "understanding" urban ecosystems means to include not only academic knowledge, but also identity development and capacities for mindful action and ongoing learning. We suggested that developing these capacities can be facilitated by immersion in challenging, real-world problem contexts; exposure to systems analysis language, tools, and procedures; and internalization of a multidimensional framework that can be used as a basis for reflection on one's own development as a systems thinker.

We seek a conception of systems thinking that applies to everyday thinking, learning, and acting in one's local environment. Refining this vision and creating practical applications of it will continue to challenge us as educators and researchers to eschew linear and compartmentalized thinking about the multiple components of students' development in favor of a systems view of thinking and learning.

Acknowledgments. We thank the book's editors for their comments on a draft of this chapter, Bill Carlson and Rick Ostfeld for their feedback on our early ideas, and Gene Likens and Rick Ostfeld for sharing with us their views on the development of the Hubbard Brook and the Mice, Mast, Moths models.

References

American Association for the Advancement of Science. 1993. *Benchmarks for science literacy*. Oxford University Press, New York.

Ausubel, D.P. 1963. *Educational Psychology: A cognitive view*. Holt, Rinehart and Winston, New York.

Bereiter, C. 1990. Aspects of an educational learning theory. *Review of Educational Research* 60:603–624.

Bereiter, C. 1994. Constructivism, socioculturalism, and Popper's World 3. *Educational Researcher* 23:21–23.

Bishoff, P.J., and O.R. Anderson. 1998. A case study analysis of the development of knowledge schema, ideational networks, and higher cognitive operations among high school students who studied ecology. *School Science and Mathematics* 98: 228–237.

Bloom, J.W. 1992. Contexts of meaning and conceptual integration: How children understand and learn. Pages 177–194 in R.A. Duschl and R.J. Hamilton, eds. *Philosophy of science, cognitive psychology, and educational theory and practice*. State University of New York Press, Albany, NY.

Bormann, F.H. 1996. Ecology: A personal history. *Annual Review of Energy and Environment* 21:1–29.

Bormann, F.H., and G.E. Likens. 1967. Nutrient Cycling. *Science* 155:424–429.

Boyes, E., and M. Stanisstreet. 1991. Misconceptions in first-year undergraduate science students about energy sources for living organisms. *Journal of Biological Education* 25:209–213.

Chandler, M.J., and R.G. Boutilier. 1992. The development of dynamic system reasoning. *Human Development* 35:121–137.

Cheek, D.W. 1993. Students and learning: Student conceptions of the social sciences and social institutions. Pages 97–102 in R.E. Yager, ed. *The science, technology, society movement*. National Science Teachers Association, Washington, DC.

Chi, M.T.H., J.D. Slotta, and N.A. deLeeuw. 1994. From things to processes: Toward a theory of conceptual change. *Learning and Instruction* 4:27–43.

Cobern, W.W. 1996. Worldview theory and conceptual change in science education. *Science Education* 80:579–610.

Cohen, S., and D. Horm-Wingerd. 1993. Children and the environment: Ecological awareness among preschool children. *Environment and Behavior* 25:103–120.

Doerr, H.M. 1996. STELLA Ten years later: A review of the literature. *International Journal of Computers for Mathematical Learning* 1:201–224.

Douglas, I. 1983. *The Urban Environment*. Edward Arnold, Baltimore, MD.

Finholt, T.A., and G.M. Olson 1997. From laboratories to collaboratories: A new organizational form for scientific collaboration. *Psychological Science* 8: 28–36.

Fisher, B.W. 1998. There's a hole in my greenhouse effect. *School Science Review* 79: 93–99.

Forrester, J.W. 1969. *Urban dynamics*. MIT Press, Cambridge, MA.
Goldenfeld, N., and L.P. Kadanoff. 1999. Simple lessons from complexity. *Science* 284:87–89.
Golley, F.B. 1993. *A History of the Ecosystem Concept in Ecology*. Yale University Press, New Haven, CT.
Greeno, J.G. 1997. On claims that answer the wrong questions. *Educational Researcher* 26:5–17.
Griffiths, A.K., and B.A.C. Grant. 1985. High school students' understanding of food webs: Identification of a learning hierarchy and related misconceptions. *Journal of Research in Science Teaching* 22:421–436.
Grotzer, T.A. 1997. "Children's understanding of complex causality in natural systems." Paper presented at the annual meeting of the American Educational Research Association, March, Chicago.
Grotzer, T.A., and B. Bell. 1999. Negotiating the funnel: Guiding students toward understanding elusive generative concepts. Pages 1–19 in L. Hetland and S. Veenema, eds. *The project zero classroom: views on understanding*. Fellows and Trustees of Harvard College, Cambridge, MA.
Grotzer, T.A., and D.N. Perkins. 1999. *Understandings of consequence project: progress report no. 1*. Unpublished Technical Paper, UCP1, Harvard Project Zero, Harvard University, Cambridge, MA.
Hagen, J.B. 1992. *An entangled bank: the origins of ecosystem ecology*. Rutgers University Press, New Brunswick, NJ.
Hellden, G. 1998. "A longitudinal study of students' conceptualization of ecological processes." Paper presented at the Annual Meeting of the National Association for Research in Science Teaching, San Diego, CA.
High Performance Systems. 1997. *STELLA technical documentation*. Author, Hanover, NH.
Hogan, K. 1994. *Eco-Inquiry: A guide to ecological learning experiences for the upper elementary/middle grades*. Kendall-Hunt, Dubuque, IA.
Hogan, K. 1999a. Thinking aloud together: A test of an intervention to foster students' collaborative scientific reasoning. *Journal of Research in Science Teaching* 36:1085–1109.
Hogan, K. 1999b. Relating students' personal frameworks for science learning to their cognition in collaborative contexts. *Science Education* 83:1–31.
Hogan, K., and J. Fisherkeller. 1996. Representing students' thinking about nutrient cycling in ecosystems: Bidimensional coding of a complex topic. *Journal of Research in Science Teaching* 33:941–970.
Jackson, S., S.J. Stratford, J.S. Krajcik, and E. Soloway. 1996. Making system dynamics modeling accessible to pre-college science students. *Interactive Learning Environments* 4:233–257.
Jones, C.G., R.S. Ostfeld, M.P. Richard, E.M. Schauber, and J.O. Wolff. 1998. Chain reactions linking acorns to gypsy moth outbreaks and Lyme disease risk. *Science* 279:1023–1026.
Kahn, P.H., Jr. 1997. Children's moral and ecological reasoning about the Prince William Sound oil spill. *Developmental Psychology* 33:1091–1096.
Kahn, P.H., Jr., and B. Friedman. 1995. Environmental views and values of children in an inner-city Black community. *Child Development* 66:1402–1427.
Kempton, W., J.S. Boster, and J.A. Hartley. 1997. *Environmental values in American culture*. MIT Press, Cambridge, MA.

Koslowski, B. 1996. *Theory and evidence: The development of scientific reasoning*. MIT Press, Cambridge, MA.

Lannon, C.P., ed. 1994. *A beginner's guide to systems thinking*. Pegasus Communications, Waltham, MA.

Leach, J., R. Driver, P. Scott, and C. Wood-Robinson. 1996a. Children's ideas about ecology 2: Ideas found in children aged 5–16 about the cycling of matter. *International Journal of Science Education* 18:19–34.

Leach, J., R. Driver, P. Scott, and C. Wood-Robinson. 1996b. Children's ideas about ecology 3: Ideas found in children aged 5–16 about the interdependency of organisms. *International Journal of Science Education* 18:129–141.

Likens, G.E., F.H. Bormann, R.S. Pierce, J.S. Eaton, and N.M. Johnson. 1977. *Biogeochemistry of a forested ecosystem*. Springer-Verlag, New York.

Likens, G.E. 1992. *The ecosystem approach: Its use and abuse*. Ecology Institute, Oldendorf/Luhe, Germany.

Likens, G.E. 1998. Limitations to intellectual progress in ecosystem science. Pages 247–271 in M.L. Pace and P.M. Groffman, eds. *Successes, limitations and frontiers in ecosystem science*. Springer-Verlag, New York.

Mandinach, E.B., and H.F. Cline. 1994. *Classroom dynamics: Implementing a technology-based learning environment*. Lawrence Erlbaum Associates, Hillsdale, NJ.

Mann, K.H. 1988. Towards predictive models for coastal marine ecosystems. Pages 291–316 in L.R. Pomeroy and J.J. Alberts, eds., concepts of ecosystem ecology. Springer-Verlag, New York, NY.

McGilly, K., ed. 1996. *Classroom lessons: Integrating cognitive theory and classroom practice*. MIT Press, Cambridge, MA.

McIntosh, R.P. 1985. *The background of ecology*. Cambridge University Press, Cambridge, MA.

Meadows, D. 1995. *Fish Banks, LTD*. Fourth edition. Institute for Policy and Social Science Research, University of New Hampshire, Durham, NH.

Metcalfe, J., and A. Shimamura. 1994. *Metacognition: knowing about knowing*. Bradford Books, Cambridge, MA.

Metz, K.E. 1996. Reassessment of developmental constraints on children's science instruction. *Review of Educational Research* 65:93–127.

Mintzes, J.J., J.H. Wandersee, and J.D. Novak, eds. 1998. *Teaching science for understanding: a human constructivist view*. Academic Press, San Diego, CA.

Odum, E.P. 1959. *Fundamentals of ecology*. Second edition. W.B. Saunders Co., Philadelphia, PA.

Ostfeld, R.S., C.G. Jones, and J.O. Wolff. 1996. Of mice and mast: ecological connections in eastern deciduous forests. *BioScience* 46:323–330.

Palmer, D.H. 1997. Students' application of the concept of interdependence to the issue of preservation of species: Observations of the ability to generalize. *Journal of Research in Science Teaching* 34:837–850.

Pfundt, H., and R. Duit. 1998. *Bibliography: Students' alternative frameworks and science education*. Fourth edition (distributed electronically). Institute for Science Education at the University of Kiel, Kiel, Germany.

Pickett, S.T.A., W.R. Burch, Jr., S.E. Dalton, and T.W. Foresman. 1997a. Integrated urban ecosystem research: themes, needs, and applications. *Urban Ecosystems* 1:183–184.

Pickett, S.T.A., W.R. Burch, Jr., S.E. Dalton, and T.W. Foresman. 1997b. A conceptual framework for the study of human ecosystems in urban areas. *Urban Ecosystems* 1:185–199.

Pintrich, P.R., R.W. Marx, and R.A. Boyle. 1993. Beyond cold conceptual change: The role of motivational beliefs and classroom contextual factors in the process of conceptual change. *Review of Educational Research* 63:167–199.

Posner, G.J., K.A. Strike, P.W. Hewson, and W.A. Gertzog. 1982. Accommodation of a scientific conception: Toward a theory of conceptual change. *Science Education* 66:211–228.

Ratcliffe, M. 1997. Pupil decision-making about socio-scientific issues within the science curriculum. *International Journal of Science Education* 19:167–182.

Resnick, M. 1994. *Turtles, termites and traffic jams: explorations in massively parallel microworlds*. MIT Press, Cambridge, MA.

Resnick, M., and U. Wilensky. 1998. Diving into complexity: Developing probabilistic decentralized thinking through role-playing activities. *Journal of the Learning Sciences* 7:153–171.

Rueter, J.G., and N.A. Perrin. 1999. Using a simulation to teach food web dynamics. *The American Biology Teacher* 61:116–123.

Rye, J.A., P.A. Rubba, and R.L. Wiesenmayer. 1997. An investigation of middle school students' alternative conceptions of global warming. *International Journal of Science Education* 19:527–551.

Schauble, L. 1996. The development of scientific reasoning in knowledge-rich contexts. *Developmental Psychology* 32:102–119.

Shapiro, B.L. 1994. *What children bring to light: a constructivist perspective on children's learning in science*. Teachers College Press, New York.

Simmons, D.A. 1994. Urban children's preferences for nature: lessons for environmental education. *Children's Environments* 11:194–203.

Simons, K.L. 1993. New technologies in simulation games. *Systems Dynamics Review* 9:135–152.

Slotta, J.D., and M.T.H. Chi. (In press). Understanding constraint-based processes: A precursor to conceptual change in physics. *Cognitive Science*.

Stratford, S.J., J. Krajcik, and E. Soloway. 1997. *Secondary students' dynamic modeling processes: analyzing, reasoning about, synthesizing, and testing models of stream ecosystems*. Paper presented at the annual meeting of the American Educational Research Association, Chicago.

Sukopp, H. 1990. Urban ecology and its application in Europe, in H. Sukopp, S. Hejny, eds., and I. Kowarik, co-ed. *Urban ecology: plants and plant communities in urban environments*. Academic Publishing, The Hague, The Netherlands.

Thelan, E., and L.B. Smith. 1994. *A dynamic systems approach to the development of cognition and action*. Bradford Books with MIT Press, Cambridge, MA.

Ulanowicz, R.E. 1997. *Ecology, the ascendant perspective*. Columbia University Press, New York.

VanderVen, K. 1997. Chaos/complexity theory, constructivism, interdisciplinarity and early childhood teacher education. *Journal of Early Childhood Teacher Education* 18:43–48.

van Geert, P. 1998. A dynamic systems model of basic developmental mechanisms: Piaget, Vygotsky, and Beyond. *Psychological Review* 105:634–677.

von Bertalanffy, L. 1968. *General system theory: foundations, development, applications*. George Braziller, New York.

Webb, N.M., and A.S. Palincsar. 1996. Group processes in the classroom. Pages 841–873 in D.C. Berliner, and R.C. Calfee, eds. *Handbook of educational psychology*. Simon and Schuster Macmillan, New York.

Webb, P., and G. Boltt. 1990. Food chain to food web: a natural progression? *Journal of Biological Education* 24:187–190.

Winne, P.H. 1997. Experimenting to bootstrap self-regulated learning. *Journal of Educational Psychology* 89:397–410.

Yates, J. 1998. *Starting early: how K-12 students are becoming systems thinkers* (audiotape). Pegasus Communications, Waltham, MA.

Zimmer, C. 1999. Life after chaos. *Science* 284:83–86.

16 Toward Ecology Literacy: Contributions from Project 2061 Science Literacy Reform Tools

Jo Ellen Roseman and Luli Stern

By current measures, today's students are not ecology literate. They fail to grasp some basic concepts essential for understanding ecosystems, such as the role of plants in removing carbon dioxide from the air and using it to build carbohydrates that both the plants themselves and animals, including humans, can use (Leach, et al. 1992; Anderson, et al. 1990; Smith and Anderson 1986). The extent of this illiteracy is vividly captured in a videotape made by the Harvard-Smithsonian Center for Astrophysics in which MIT graduates are asked, "Where does the weight of a log come from?" None of them know, even though it is apparent from their answers that they have studied photosynthesis in school. When asked what they would say if told the mass comes from the carbon dioxide in the air, most of their responses range from shock to disbelief. One articulate student, who is able to recall the phrase "photosynthetic version of the electron transport chain," nonetheless asserts that "from what [he] knows about organic chemistry, carbon isn't much of a building block."

If students are to progress toward ecology literacy, which includes a basic understanding of urban ecosystems, it is essential to (1) specify the knowledge and skills that constitute adult ecology literacy, (2) articulate discrete and appropriate cognitive steps toward adult literacy, (3) describe connections among the steps and when students can appreciate them, and (4) identify and test learning experiences for their effectiveness in promoting it. Project 2061 of the American Association for the Advancement of Science (AAAS) has defined adult literacy in ecology, articulated steps toward that goal, and mapped out connections among the ideas and skills that make up ecology literacy. Each of these efforts has resulted in a publication for K–12 educators, university faculty, and others to use in their local education reform efforts. Part of Project 2061's current work seeks to analyze curriculum materials to identify learning experiences likely to help students make progress toward ecology literacy.

Research on student learning plays a major role in the development of Project 2061 tools for improving science education. The tools, in turn, raise new questions for research to address. This chapter describes the role of research into students' understanding of science in Project 2061's R&D efforts and suggests how tools based on research can be used by reformers intent on helping students learn important ideas in ecology.

Defining Adult Literacy in Ecology

Working with panels of scientists, mathematicians, and technologists, Project 2061 set out to identify the basic knowledge and skills adults should have in five subject areas: biological and health sciences; mathematics; physical and information sciences and engineering; social and behavioral sciences; and technology. *Science for All Americans* (AAAS 1989) integrated these reports into a single document that provides a coherent set of recommendations for what all high school graduates should know and be able to do in science, mathematics, and technology.

During this time other groups and publications were also recommending changes in our educational efforts. *A Nation at Risk*, the 1983 report of the National Commission on Excellence in Education, severely criticized the American education system and called for fundamental reform. State governments were starting to support national education standards, and in 1989 the National Governors Association and President George Bush, Sr., endorsed national education goals at the nation's first education summit. That same year, the National Council of Teachers of Mathematics published *Curriculum and Evaluation Standards for School Mathematics*. Stigler and Hiebert (1999) recently cited the clear commitment to common goals as the real reason for optimism about educational improvement:

> In a field where fads have ruled, we are seeing something new: a growing commitment to the idea that clear and shared goals for student learning must provide a foundation on which to improve education and achievement. Without clear goals, we cannot succeed, for we cannot know in which direction to move.

Science for All Americans speaks directly to many of the concepts people need to grasp to be able to understand urban ecosystems. Many of these important ideas are found in Chapter 5: The Living Environment. For example, its section Flow of Matter and Energy begins by summarizing the link between living and physical systems:

> However complex the workings of living organisms, they share with all other natural systems the same physical principles of the conservation and transformation of matter and energy. Over long spans of time, matter and energy are transformed among living things, and between them and the physical environment. In these grand-scale cycles, the total amount of matter and energy remains constant, even though their form and location undergo continual change.

The many scientists involved in drafting the adult science literacy recommendations considered these ideas to be among the most important for making sense out of the world and to serve as a lasting foundation on which to build more knowledge over a lifetime.

To allow time for understanding the most important ideas and their interconnections, *Science for All Americans* limited the total number of ideas to be learned. Hence, in life science (i.e., physical and social science, mathematics, and technology) it left out several topics that are typically included in textbooks. For example, details of plant anatomy (e.g., xylem, phloem, petioles, grana, stroma, palisade cells, and guard cells) and metabolic steps of photosynthesis and respiration (e.g., photosystems I and II, light and dark reactions, glycolysis, and Kreb's cycle) were not considered essential for important tasks such as making sense of everyday phenomena or making social and personal decisions about matters involving science, mathematics, and technology. The fact that the hundreds of scientists involved in drafting and reviewing the *National Science Education Standards* (National Research Council 1996) came to similar conclusions about what constitutes core knowledge in science for K–12 students lends support for these omissions.

Steps Toward Ecology Literacy

Science for All Americans served as the basis for *Benchmarks for Science Literacy* (AAAS 1993), which described appropriate cognitive steps—benchmarks—towards science literacy for students completing grades 2, 5, 8, and 12. These publications in turn provided the groundwork for the K–4, 5–8, and 9–12 grade-level recommendations—fundamental concepts—for student learning in the *National Science Education Standards* (National Research Council 1996). The substance and grade level placement of both benchmarks and fundamental concepts were informed by research on how students learn science ideas, which is summarized in *Benchmarks*, chapter 15. Driver, et al. (1994) published a more detailed summary of research on selected science topics, including plant nutrition, photosynthesis, nutrition and energy flow, and matter cycling in ecosystems.

Most research studies report that many students have difficulties, some of which persist into college and beyond, in understanding science concepts. For example, research on student understanding of ecosystems reveals that students view matter as being created or destroyed, rather than as being transformed (Smith and Anderson 1986). Students who do see matter as being transformed view it as being transformed into energy rather than into simpler substances. Also, students view plants as taking in food from the environment, rather than as taking in raw materials that they convert to food (Bell and Brook 1984; Roth and Anderson 1987; Anderson, et al. 1990). In writing *Benchmarks*, reports that students have difficulty understanding particular ideas led to (1) stating less sophisticated precursors of a benchmark idea; (2) adding prerequisite ideas from different subject areas, for example, adding the idea that air is a substance; and (3) moving benchmarks to higher grade levels than where they are currently taught. It is important to note that there is little research into the effects of carefully focused instruction on students' understanding of scientific ideas. Some studies have shown positive results from this kind of instruction and have led to a slightly earlier placement of a few benchmarks (Lee, et al. 1993). As more such studies are carried out, it should be possible to determine whether students' conceptual difficulties are developmental or instructional in nature so that benchmarks can be positioned more precisely.

Conceptual Sequence

Drawing on this research base, *Benchmarks'* placement of learning goals in specific grade ranges suggests a sequence of learning that is consistent with what is known about how and when students are able to understand a particular idea or skill. The development of ideas about the flow of matter and energy in the living environment—ideas that are essential to understanding urban ecosystems—starts in grades K–2 with the idea that living things have specific needs—food, water, and light. In grades 3–5, energy is added to the list of needs and food, coming originally from plants, is now viewed as the source of energy and materials for growth and repair. In grades 6–8 the emphasis is on the transformation of matter and energy in living systems. Transformation of matter is cast mainly in terms of substances (e.g., carbon dioxide and water into sugar, sugars into storage forms, storage forms back to sugars and then back to carbon dioxide and water). Energy transformation is cast mainly in terms of tracking changes from one form to another (without an underlying mechanism): light energy into food energy, food energy into heat and (more importantly) into energy for survival (e.g., maintenance, growth, and so forth). The ideas culminate in grades 9–12 with an integrated molecular interpretation of the flow of matter and energy among living things and between them and the physical environment.

The same principles of conservation of matter that apply to physical systems also contribute to understanding conservation of matter in living systems. Understanding conservation can start in grades K–2 with the idea that people can see where things come from and where they go. In grades 3–5, understanding can develop further with the idea that things never appear out of nowhere and never just disappear. By grades 6–8, students can understand the law of conservation of mass in chemical and physical changes and can explain what they observe in these changes in terms of the rearrangement of a fixed number of atoms. This will make it possible for students to see food webs and cycles in terms of the breakdown and reassembly of invisible units (rather than as the creation and destruction of matter as most students do).

Ideas about matter and energy in the living environment also are intimately related to ideas about matter and energy in physical systems. Ideas in *Benchmarks* about energy that contribute to these understandings include the grades 3–5 idea that energy is needed to make anything go, run, or happen, and the grades 6–8 ideas that energy appears in different forms and that energy can be changed from one form to another. Energy transfer and transformation can be made concrete more easily for students in the context of physical systems rather than in biological systems. For example, students can observe that sunlight warms water and that falling water can make a wheel turn. But storing energy in molecular configurations is difficult to show, even with models. Starting with energy transformation in physical systems and using that understanding plus the idea that energy (like matter) is neither created nor destroyed can pave the way to understanding that light energy is transformed into chemical energy (during photosynthesis), which can be stored or transformed into mechanical energy and heat.

Representing Important Connections Among Ideas

As described above, ideas contributing to an understanding of the transformation and flow of matter and energy in ecosystems come from different disciplines and may be taught in different parts of the curriculum. To aid curriculum and instructional materials developers, Project 2061 has been developing "strand maps" to display the progression of benchmark ideas from kindergarten through grade 12 and the connections among them. *Atlas of Science Literacy* (AAAS 2001) presents a collection of strand maps that show how students' understanding of the ideas and skills that lead to science literacy in science, mathematics, and technology might grow over time.

The Strand maps on the flow of matter and energy separately show benchmarks on the flow of matter and flow of energy in ecosystems and the closely related physical science concepts of conservation of matter and energy transformation (*Atlas*, pp. 77 and 79).

Figure 16.1a shows a single strand from the Flow of Matter in Ecosystems map to illustrate basic map features. Each box on the map contains the text of specific learning goals derived from *Benchmarks*. Arrows between boxes show the contributions of one learning goal (at the tail of the arrow) to another (at the arrow head). Learning goals appropriate for kindergarten through grade 2 are shown at the bottom; those for grades 9–12 at the top. The result is a graphic depiction of how specific learning goals in *Benchmarks* support one another as they build toward adult ecology literacy.

The actual Flow of Matter in Ecosystems map illustrates how ideas about the conservation of matter connect to ideas about the flow of matter in ecosystems, though not until the 6–8 and 9–12 grade ranges (*Atlas*, p. 77).

FIGURE 16.1a

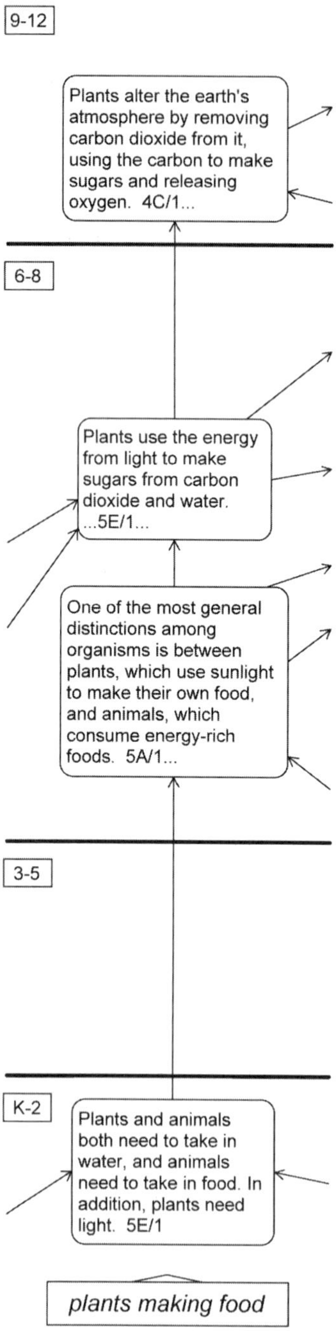
9-12
Plants alter the earth's atmosphere by removing carbon dioxide from it, using the carbon to make sugars and releasing oxygen. 4C/1...
6-8
Plants use the energy from light to make sugars from carbon dioxide and water. ...5E/1...
One of the most general distinctions among organisms is between plants, which use sunlight to make their own food, and animals, which consume energy-rich foods. 5A/1...
3-5
K-2
Plants and animals both need to take in water, and animals need to take in food. In addition, plants need light. 5E/1
plants making food

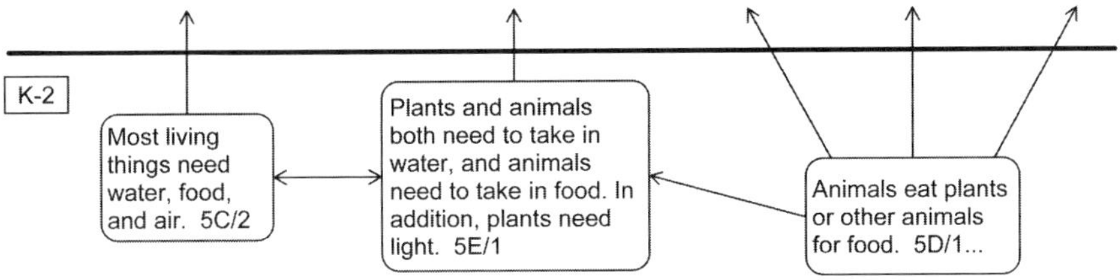

Fig. 16.1b

In these later grades, connections span disciplines, whereas in grades K–2 and 3–5 connections are more local. For example, K–2 ideas about the flow of matter in ecosystems (shown in Figure 16.1b) are connected to one another but not to K–2 *or* 3–5 ideas about the conservation of matter, which appear in the Conservation of Matter strand map (*Atlas*, p. 57) and are shown in Figure 16.1c. Recycling of matter in living things will make little sense until students appreciate that substances (including water) never appear out of nowhere and never just disappear, and that air is a substance (to be learned in grades 3–5).

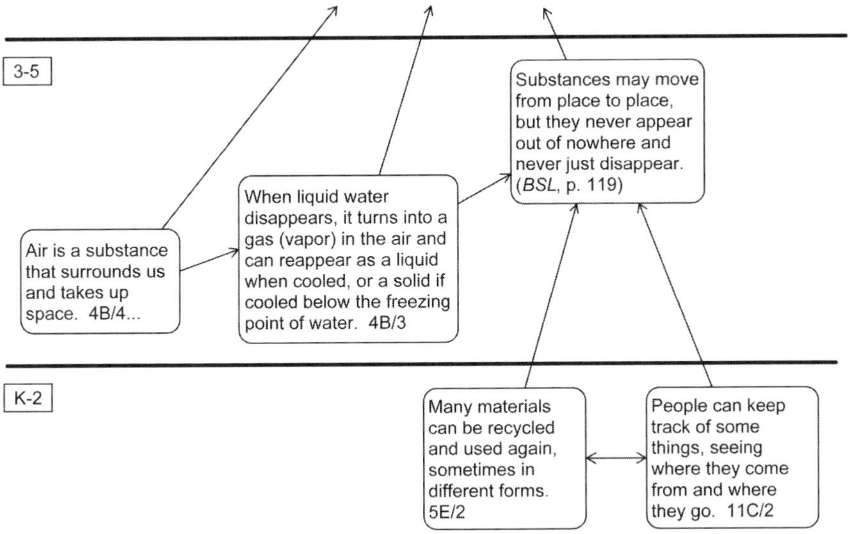

Fig. 16.1c

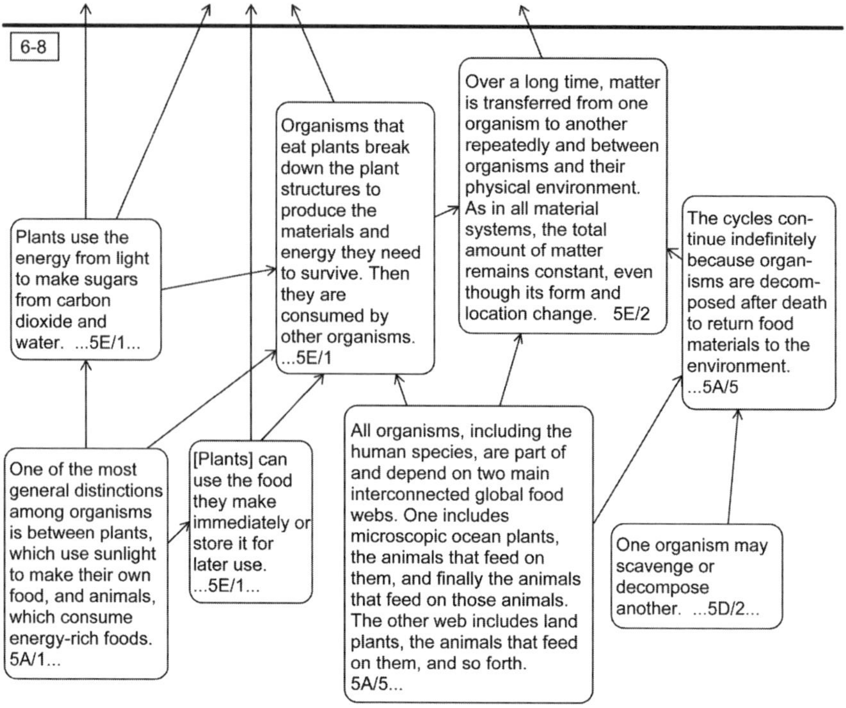

Fig. 16.1d

In the Flow of Matter in Ecosystems map, grades 6–8 emphasizes understanding of food chains and processes like food making, breakdown, and decomposition in organisms in terms of substances being transformed into other substances (Figure 16.1d). In grades 9–12, armed with an understanding of atoms and molecules, these same processes can be understood in terms of the combination and recombination of atoms (Figure 16.1e). At

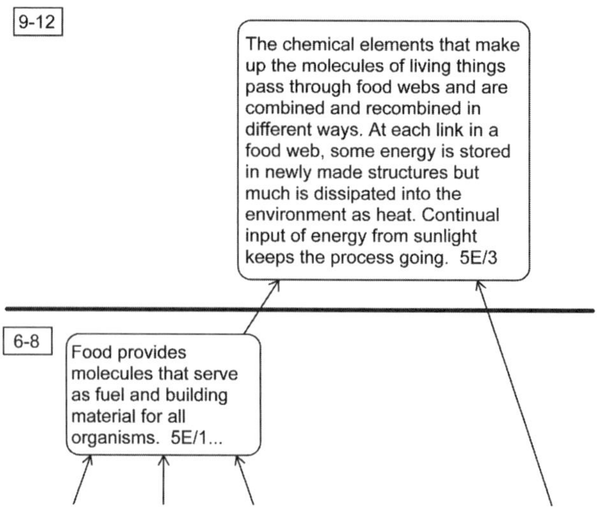

Fig. 16.1e

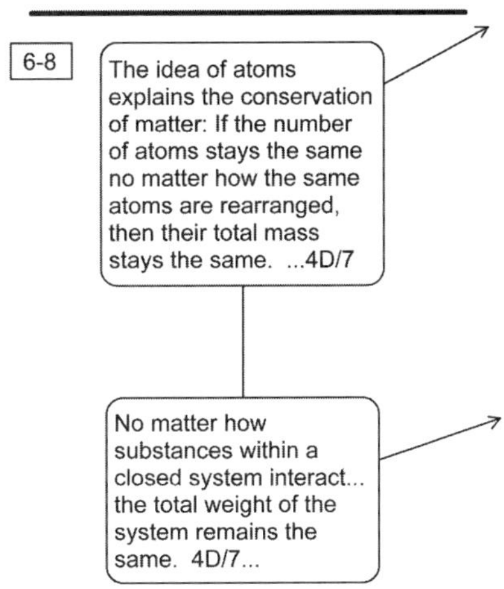

Fig. 16.1f

the same time, the parallel development of the idea of conservation begins with the observed mass conservation in closed physical systems and then with its explanation in terms of the combination and recombination of atoms (Figure 16.1f). Within these same grade ranges, students are also developing their understanding of the connection between matter flow in ecosystems and the conservation of matter (Figure 16.1g), again first in terms of substances and then in terms of atoms. In addition to moving from substances to a molecular interpretation, the developmental progression represented in the strand map also moves from understanding matter conservation in physical systems to understanding it in (less readily observable) living systems.

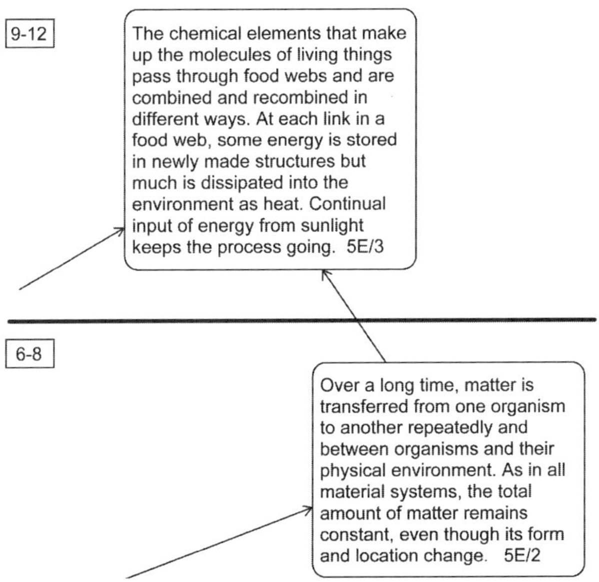

Fig. 16.1g

Implications for Instruction

By showing conceptual connections to be made within a grade range and potential prerequisites to understanding particular ideas, strand maps can guide the development of learning experiences to help K–12 students move toward understanding urban ecosystems. For example, when students are learning the idea that all matter is made up of atoms (which they will then use to revisit such processes as food synthesis and decomposition in terms of atoms and molecules), it is important to be sure they recognize that air is a substance. If students don't appreciate that air is made of atoms and molecules that, like the molecules of water, can be used to make larger molecules, then students will have difficulty seeing how the mass of a log could have come from carbon dioxide in the air.

It is important to note that the existing research is insufficient for precisely specifying an instructional sequence. Hence the maps do not (and cannot) lay out an instructional plan. Within a map, some ideas that appear at an earlier grade level may be understood initially in a rudimentary way and then more fully after students have learned later ideas. For example, ideas involving transformation at the substance level are lower on the map than ideas about a molecular explanation of such phenomena, that is, the conversion of carbon dioxide and water to sugars or the breakdown of starch to sugars all deal with transformation of substances before a molecular explanation is given. However, research indicates that without a molecular explanation the transformation of substances may remain mysterious for students, and they may even see it as alchemy (Driver, et al. 1994). Based on these findings, it is probably inappropriate to expect students to explain these interactions until they understand them in terms of molecules.

Curriculum Materials to Support Ecology Literacy

Specifying clear learning goals and suggesting connections among them is an essential first step, but educators need more than that to change instruction in a meaningful way. In most U.S. secondary schools, textbooks largely determine what is taught (Schmidt, et al. 1997; Anderson 1991). Although most science teachers claim that they do not simply follow the textbook, they rarely do more than add, omit, or replace one or two activities in the chapter they choose to teach. Given the textbook's important role in determining what is taught and how, the selection of textbooks ought to be based on demonstrated evidence of student learning. Unfortunately, such evidence is scarce, particularly for the widely used, commercially published textbooks. The few published studies on student learning that exist involve small, virtually unknown research units (e.g., Lee, et al. 1993).

Using the student learning data for these research materials, a reasonable first step toward evaluating other materials would be to predict the likelihood of student learning from them based on the extent to which they resemble those materials for which learning has been demonstrated. Indeed, Project 2061 has developed an analytical tool for evaluating the quality of a material's instructional support for student learning that uses criteria based on student learning theory and specific features found in these research-based teaching units. Over the past 4 years, Project 2061 staff collaborated with education researchers, materials developers, teachers, and scientists in identifying and field testing a set of content and instructional criteria that, taken together, describe high-quality instructional materials (Roseman, et al. 1996; Kesidou and Roseman 2002). The rationale for this approach is that materials that meet these criteria are much more likely to promote student learning than materials that do not. It is hoped that the application of these criteria to a variety of instructional materials will stimulate empirical studies of student learning that will lead to refinements of the analysis criteria themselves and to more precise and useful evaluations of materials.

The analysis requires that a coherent set of ideas be chosen to serve as its basis. The analysis then carefully examines how well a material's content aligns with those specific ideas and how well the instructional strategies in the text and teacher's guide can support students in learning them.

Content Alignment

Alignment means fidelity to the substance and sophistication of the specific ideas chosen for the analysis, not just to their general topic. For example, an activity in which students separate plant pigments through paper chromatography may fit with the general topic of photosynthesis but does not align with the substance of either of the following ideas about matter and energy transformation or any other ideas on the "Flow of Matter and Energy in Ecosystems" strand map (*Atlas*, p. 77).

- Plants use the energy from light to make sugars from carbon dioxide and water.
- A chlorophyll molecule can be excited to a higher-energy configuration by sunlight.

The pigment chromatography activity could be used to explain the basis for the color change of leaves in the fall because it shows that even green leaves contain the pigments for their fall colors. However, the activity won't be useful for explaining the very important ideas regarding energy flow stated above and, hence, would not be judged to align with them.

Other activities might appear to be more relevant to the substance of an idea but are aimed at either too low or too high a level of sophistication. For example, neither an activity in which plants are shown to grow toward the light nor an activity in which students read about and discuss the light-capturing step in photosynthesis addresses the sophistication of the idea that plants use light (grades 6–8 in the strand map). The former activity addresses the less sophisticated idea that plants need light (grades 3–5), and the latter addresses the more sophisticated idea that a chlorophyll molecule can be excited to a higher-energy configuration by sunlight (grades 9–12). Unfortunately, both of these activities are typically found in middle grades life science textbooks, taking up time and space in the curriculum that could be better used to focus on the most important ideas. For a closer match to the substance and intended sophistication of the idea in grades 6–8 idea, for example, students might benefit from activities that enable them to observe that in the absence of light, there is no sugar production. (Whether evaluating or designing instructional materials, the principle is the same: select the most important ideas to teach and tenaciously stick with instruction that is aimed at them. Activities that are merely fun or easy to do must fall by the wayside.)

Instructional Support

As noted above, Project 2061's curriculum-materials evaluation procedure uses a research-based set of criteria to make judgments about instructional quality. The criteria are based on available learning research (Anderson and Roth 1989; Berkeimer, et al. 1990; Ericsson, et al. 1993; Glaser 1994; Lee, et al. 1993; Smith, et al. 1993) and, where research is lacking, on the consensus of experienced educators and researchers. The criteria have been extensively field-tested by a variety of evaluators and on a variety of curriculum materials to maximize their validity and reliability. For each criterion, there are written explanations that clarify it, indicators that specify what it means to meet the criterion, and a range of examples from curriculum materials that illustrate a "poor," a "satisfactory," and an "excellent" score.

When teams of reviewers trained in the use of the Project 2061 procedure examine materials independently, the reliability achieved is more than 90 percent in math (Kulm and Grier 1998) and more than 80 percent in science (see Appendix A for the complete list of criteria). The 25 specific criteria are organized into seven categories: providing a sense of purpose for students, taking account of student ideas, engaging students with relevant phenomena, developing and using scientific ideas, promoting student thinking, assessing progress, and enhancing the learning environment. The following examples provide a more detailed look at how the criteria in three of these categories can be applied to analyze the instructional support provided by material that is targeting one of the grades 6–8 ideas found in Figure 16.1d: "Plants use the energy from light to make sugars from carbon dioxide and water."

Criteria within the category "Taking Account of Student Ideas" examine how well the material (1) develops and builds on prerequisite ideas (such as the idea in Figure 16.1c that air is a substance and other prerequisite

TABLE 16.1. Alerting to commonly held student ideas.

Scientific Idea	Commonly Held Idea
Plants make their own food using carbon dioxide, water, and sun.	Plants take in their food from the soil.
Food provides the molecules that provide energy and serve as building material for all provide energy and organisms.	Food is needed to keep plants and animals alive and to help them grow.
Energy can change from one form to another in living things. Animals get energy from oxidizing their food, releasing some of its energy as heat. Almost all food energy comes originally from the sun.	Plants take the sun's energy and turn it into matter.

ideas represented on the partial strand map); (2) alerts teachers to commonly held student ideas reported in the research literature (such as the misconceptions that plant food comes mainly from the soil and that matter and energy are interconvertible in living organisms); (3) provides specific questions and tasks to identify ideas unique to the teachers' particular students (which can be adapted from questions used in research studies); and (4) attempts to address students' commonly held ideas (challenging, for example, the misconception that plant food comes from the soil by having students predict what would be the results of van Helmont's famous experiment and then explaining these results, or by having students observe that plants thrive when grown in aquaculture).

The following examples, drawn from material being developed at Michigan State University to teach students ideas about food making in terms of transformation of substances, illustrates an effective way to meet two criteria in the category "Taking Account of Student Ideas." Table 16.1 shows how the material alerts teachers to commonly held student ideas—by contrasting them with the scientific idea to be learned. Figure 16.2 shows some of the questions provided to assist teachers in identifying their own students' ideas (so that they will be able to select appropriate activities in the unit and to modify or supplement them as necessary to best address these ideas).

Criteria within the category "Engaging Students with Relevant Phenomena" probe whether materials provide vivid experiences with phenomena that can make the scientific ideas plausible and are explicitly linked to them. For example, the following phenomena could be used to make plausible the ideas that plants make sugars from carbon dioxide and water and that these sugars provide building materials for all organisms:

- Sugar can be detected in the leaves of a plant kept in the presence of carbon dioxide but not in its absence.
- Air, water, and minerals are the only substances given to a plant, yet it grows and produces structures that look different from these

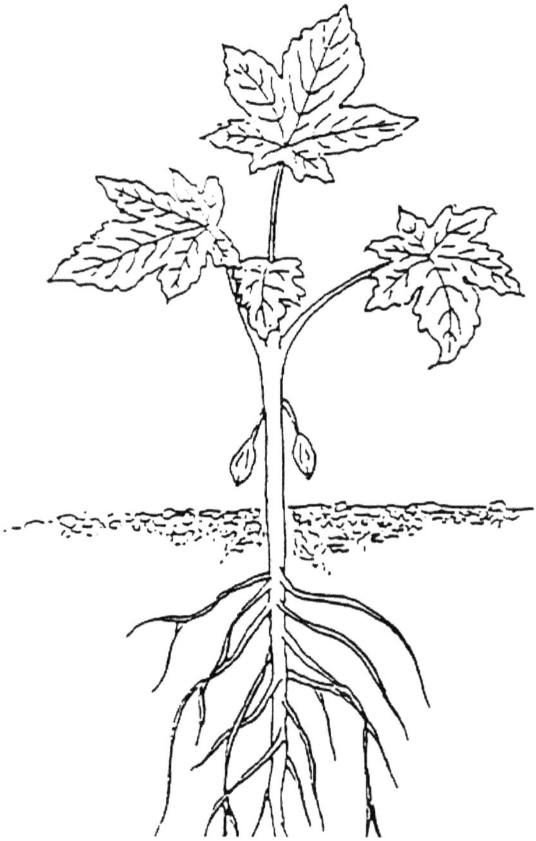

FIGURE 16.2. Assisting teachers in identifying their own students' ideas. (1) Do green plants need **food** to live and grow? Why or why not? (2) Do green plants need **light** to live and grow? Why or why not? (3)Draw arrow to show how **water** moves in a green plant. Explain why water needs to move like that in the plant. (4) Draw arrows to show how **food** moves in a green plant. Explain why the food needs to move this way. (5) A maple seed is small enough to hold in your hand. It grows into a tree that is so large that you can barely hold a log from it. Where does the "stuff" of the log come from?

inputs. Furthermore, the plant weighs more than the water it uses in photosynthesis (though, admittedly, this would be difficult for students to measure).

- The size of a carrot correlates with the size of its leaves, the site of food production.
- If leaves are cut off of flowering bulbs in the spring, the bulbs will be much smaller when dug up in the fall than bulbs where the leaves

were left on. The smaller bulbs will usually not produce flowers the next season.
- During years of severe drought, tree rings are narrower than in years where water is plentiful.

Naturally, students would need help in seeing how ideas about matter and energy transformation can be used to explain the above phenomena and even more help in deducing the ideas from the phenomena in the first place. For example, it would be easier for students to learn how to use the idea that plants make sugars from carbon dioxide and water to explain the observation that sugar is present in leaves of a plant kept in the presence of carbon dioxide but not in its absence, than to learn how to infer the idea from the phenomenon. The phenomenon itself simply shows that the plant needs the carbon dioxide to make the sugar. Convincing students that carbon dioxide is incorporated into the sugar would require additional data (e.g., data showing that the amount of sugar produced correlated with the amount of carbon dioxide supplied or that if the plant was kept in the presence of C^{14}-labeled carbon dioxide, then the sugar would be C^{14}-labeled as well).

Understanding such experiments would require students to know more sophisticated ideas about atoms and molecules and the role of evidence and logic in scientific investigations. And even explaining some phenomena, however intriguing they might be to students, could take considerable effort. For example, to understand why tree rings are narrow in years of drought, students would also need to appreciate that the lack of water exerts its effect indirectly (by leading to stomatal closure and, hence, to reduced CO_2 uptake). This sophisticated idea is not included as a middle grades learning goal by either *Benchmarks* or *National Science Education Standards*.

The Project 2061 procedure also includes criteria to examine whether a material helps students to see the utility of abstract ideas in explaining phenomena. For example, if students are first given an idea and then shown some phenomena, then materials should demonstrate how the idea can be used to explain one or two of the phenomena before giving students a chance to practice explaining some on their own. If the demonstration itself is complex, then models or other types of representations—either as part of the text itself or included in instructions for teacher-led discussions—will be helpful. On the other hand, if students are expected to *infer* an idea from a phenomenon, then the material should include tasks and/or question sequences to guide student interpretation and reasoning about the phenomenon. For example, to guide student reasoning about the phenomenon that plant leaves contain more sugar or starch if they have been grown with added amounts of carbon dioxide, students could use molecular models to represent the transformation taking place in the two situations (disassembling the starting models to produce product molecules), count the number of molecules being transformed in the two situations, and then respond to questions like these:

- For the reaction in which you did not add extra carbon dioxide, how many atoms of carbon are there in the molecules of the reactants? How many atoms of oxygen are there in the molecules of the reactants? What about hydrogen?
- For the same reaction, how many carbon, oxygen, and hydrogen atoms are present in the products?
- What do you notice about the numbers of the atoms in the starting substances and in the ending substances in the reaction in which you did not add extra carbon dioxide?
- Respond to the previous questions for the reaction in which you added extra carbon dioxide. (*Hint*: If you use more carbon dioxide molecules to start with, do you also not need to use more water molecules to start with?)
- How does the number of molecules of sugar formed in the first situation compare to the number of molecules of sugar formed in the second situation? Is this what you observed in your experiments?
- Summarize how ideas about conservation and transformation of matter have been used to explain the observations.

A similar strategy is used to guide students to interpret observations about mass conservation in chemical reactions in terms of atoms and molecules in the unit *Chemistry That Applies* (Michigan Science Education Resources Project 1993). Indeed, students would have a greater appreciation of the utility of ideas of matter and energy transformation in living systems if they could be helped to see their applicability in physical systems as well, or vice versa. Connecting these ideas to systems thinking more generally would also be useful. Ecosystems could be compared to other biological or physical systems to help students to think about food webs in terms of inputs, outputs, and control. The importance of defining the boundaries around systems could help students to see that air, and not only the soil, is part of a tree's system and provides it with inputs. These connections can help students appreciate that science is not made of a collection of concepts but of a coherent network of ideas.

All Students

Both the adult science literacy recommendations and the steps toward achieving them are intended for all students. Project 2061 does not think that the core learning goals for students who live in cities should be different from those for students who live in suburban or rural areas. While the *experiences* that help urban students make sense of ideas about matter and energy might be different from those for rural students, the fundamental *ideas* to be remembered and used should be the same for all students. For example, all students should know that decomposers transform dead organisms into simpler, reusable substances. Students who encounter decomposing compost piles in their everyday experience may only need to be helped to interpret their observations in terms of the transformation of matter and energy. On the other hand, students without ready access to such experiences in their communities will need to have such experiences provided for them, either in classrooms or informal education settings.

Conclusions

Project 2061's tools have significant implications for efforts to help students achieve ecology literacy. *Science for All Americans*, *Benchmarks*, and *Atlas* strand maps can guide curriculum designers in selecting core learning goals, illuminating student misconceptions that can impede students in achieving them, and identifying prerequisites and other conceptual connections to foster learning them. The Project 2061 curriculum-materials evaluation procedure can provide specifications for designing new materials and improving teaching. What is more, all of the tools suggest productive new directions for research on cognition, teaching methods, curriculum development, and assessment.

For example, in the context of evaluating middle grades life science materials, we have identified a number of phenomena (some are described in this chapter and others were presented at the National Association for Research in Science Teaching 2000 Annual Meeting) that could be used to help students make sense of the ideas of energy transformation, matter transformation, and matter conservation. Some of these phenomena come from the middle grades textbooks we have examined, but most have been identified in trade books or invented by staff and reviewers. Most of these have not been tested for their effectiveness in helping a variety of students learn benchmark ideas. Little is known about what it would take to make the phenomena vivid for different students, nor whether students will be able to follow the arguments needed to link the phenomena to benchmark ideas. These are important and badly needed areas for new research.

Not only phenomena need to be tried out with students, however. As argued before, there is ample evidence that today's students are not ecology literate (let alone literate about urban ecosystems). The MIT graduates who did not know where the weight of a log comes from clearly did not receive appropriate instruction to help them understand the relevant ideas, and their misconceptions have never been challenged. Representations, questions, and other curricular interventions that seem plausible in helping students learn these difficult but important ideas will all need to be tried out with students.

References

American Association for the Advancement of Science. 1989. *Science for all Americans*. Oxford University Press, New York.

American Association for the Advancement of Science. 1993. *Benchmarks for science literacy*. Oxford University Press, New York.

American Association for the Advancement of Science. 1997. *Resources for science literacy: professional development*. Oxford University Press, New York.

American Association for the Advancement of Science. 2001. *Atlas of science literacy*. Oxford University Press, New York.

American Association for the Advancement of Science. (In press). *Resources for science literacy: curriculum materials evaluation*. Oxford University Press, New York.

Anderson, C.W. 1991. Policy implications of research on science teaching and teachers' knowledge. Pages 5–30 in M. Kennedy, ed. *Teaching academic subjects to diverse learners*. Teachers College Press, New York.

Anderson, C.W., and K.J. Roth. 1989. Teaching for meaningful and self-regulated learning of science. Pages 265–310 in J. Brophy, ed. *Advances in research on teaching*. Volume 1. JAI Press, Greenwich, CT.

Anderson, C., T. Sheldon, and J. Dubay. 1990. The effects of instruction on college nonmajors' conceptions of respiration and photosynthesis. *Journal of Research in Science Teaching* 27:761–776.

Bell, B., and A. Brook. 1984. *Aspects of secondary students understanding of plant nutrition*. University of Leeds, Centre for Studies in Science and Mathematics Education, Leeds, UK.

Berkheimer, G.D., C.W. Anderson, and T.D. Blakeslee. 1990. *Using a new model of curriculum development to write a matter and molecules teaching unit* (Research series No. 196). Michigan State University, Institute for Research on Teaching. East Lansing, MI.

Driver, R. 1983. *The pupil as a scientist?* Open University Press, Milton Keynes, UK.

Driver, R., A. Squires, P. Rushworth, and V. Wood-Robinson. 1994. *Making sense of secondary science: Research into children's ideas*. Routledge, New York.

Ericsson, K.A., R.T. Krampe, and C. Tesche-Romer. 1993. The role of deliberate practice in the acquisition of expert performance. *Psychological Review* 100:363–406.

Glaser, R. 1994. *Application and theory: Learning theory and the design of teaching environments*. Learning Research and Development Center, Pittsburgh, PA.

Harvard Smithsonian Center for Astrophysics. 1987. *A private universe* [videotape]. Available from: Annenberg/CPB Math and Science Collection, P.O. Box 2345, S. Burlington, Vermont, 05407–2345.

Kesidou, Sofia, and Jo Ellen Roseman. 2000. How well do middle school science programs measure up? findings from Project 2061's Curriculum Review. *Journal of Research in Science Teaching* (in press).

Kulm, G., and L. Grier. 1998. *Math curriculum materials reliability study*. Project 2061, American Association for the Advancement of Science, Washington, DC.

Leach, J., R. Driver, P. Scott, and V. Wood-Robinson. 1992. *Progression in understanding of ecological concepts by pupils aged 5 to 16*. The University of Leeds, Centre for Studies in Science and Mathematics Education, Leeds, U.K.

Lee, O., D. Eichinger, C.W. Anderson, G.D. Berkheimer, and T.D. Blakeslee. 1993. Changing middle school students' conceptions of matter and molecules. *Journal of Research in Science Teaching* 30:249–270.

McDermott, L.C., and P.S. Shaffer. 1992. Research as a guide for curriculum development: an example from introductory electricity. Part I: Investigation of student understanding. *American Journal of Physics* 60:994–1003.

Michigan Science Education Resources Project. 1993. *Chemistry that applies*. Michigan Department of Education, Lansing, MI.

National Commission on Excellence in Education. 1983. *A nation at risk: the imperative for educational reform*. U.S. Department of Education, Washington, DC.

National Council of Teachers of Mathematics. 1989. *Curriculum and evaluation standards for school mathematics*. National Council of Teachers of Mathematics, Reston, VA.

National Research Council. 1996. *National science education standards*. National Academy Press, Washington, DC.

Roseman, J., L. Stern, A. Caldwell, and L. Kurth. 2000. *Can middle and high school science textbooks help students learn important ideas?* Project 2061's curriculum evaluation study. Symposium presented at the annual meeting of the National Association for Research in Science Teaching, New Orleans, LA.

Roth, K.J. 1984. Using classroom observations to improve science teaching and curriculum materials. Pages 77–102 in C.W. Anderson, ed., *Observing science classrooms: Perspectives from research and practice*. 1984 yearbook of the Association for the Education of Teachers in Science. ERIC/SMEAC, Columbus, OH.

Roth, K., and C. Anderson. 1987. *The power plant: teacher's guide to photosynthesis*. Occasional paper no. 112. Institute for Research on Teaching, Michigan State University, East Lansing, MI.

Schmidt, W., C. McKnight, and S. Raizen. 1997. Executive summary. *A splintered vision: an investigation of U.S. science and mathematics education*. U.S. National Research Center, Michigan State University, MI.

Scott, P.H., H.M. Asoko, and R.H. Driver. 1992. Teaching for conceptual change: A review of strategies. Pages 310–329 in R. Duit, F. Goldberg, and H. Niedderer, eds. *Research in physics learning: theoretical issues and empirical studies*. Institute for Science Education at the University of Kiel, Kiel, Germany.

Shaffer, P.S., and L.C. McDermott. 1992. Research as a guide for curriculum development: an example from introductory electricity. Part II: Design of instructional strategies. *American Journal of Physics* 60:1003–1013.

Smith, E., and C. Anderson. 1986. *Alternative conceptions of matter cycling in ecosystems*. Paper presented at the annual meeting of the National Association for Research in Science Teaching, San Francisco, CA.

Smith, E.L., and C.W. Anderson 1984. Plants as producers: a case study of elementary school science teaching. *Journal of Research in Science Teaching* 21:685–695.

Smith, E.L., T.D. Blakeslee, and C.W. Anderson. 1993. Teaching strategies associated with conceptual change learning in science. *Journal of Research in Science Teaching* 30:111–126.

Stigler, J.W., and J. Hiebert. 1999. *The teaching gap*. The Free Press, New York.

Appendix A: AAAS Project 2061 Middle Grades Science Textbooks Evaluaton

Criteria for Evaluating the Quality of Instructional Support

Category I. Providing a Sense of Purpose
Conveying unit purpose. Does the material convey an overall sense of purpose and direction that is understandable and motivating to students?
Conveying lesson purpose. Does the material convey the purpose of each lesson and its relationship to others?
Justifying activity sequence. Does the material involve students in a logical or strategic sequence of activities (versus just a collection of activities)?

Category II. Taking Account of Student Ideas
Attending to prerequisite knowledge and skills. Does the material specify prerequisite knowledge/skills that are necessary to the learning of the benchmark(s)?
Alerting teacher to commonly held student ideas. Does the material alert teachers to commonly held student ideas (both troublesome and helpful) such as those described in *Benchmarks*, Chapter 15: The Research Base?
Assisting teacher in identifying own students' ideas. Does the material include suggestions for teachers to find out what *their* students think about familiar phenomena related to a benchmark before the scientific ideas are introduced?
Addressing commonly held ideas. Does the material attempt to address commonly held student ideas?

Category III. Engaging Students with Relevant Phenomena
Providing variety of phenomena. Does the material provide multiple and varied phenomena to support the benchmark idea(s)?
Providing vivid experiences. Does the material include activities that provide firsthand experiences with phenomena when practical or provide students with a vicarious sense of the phenomena when not practical?

Category IV. Developing and Using Scientific Ideas
Introducing terms meaningfully. Does the material introduce technical terms only in conjunction with experience with the idea or process and only as needed to facilitate thinking and promote effective communication?
Representing ideas effectively. Does the material include accurate and comprehensible representations of scientific ideas?
Demonstrating use of knowledge. Does the material demonstrate/model or include suggestions for teachers on how to demonstrate/model skills or the *use* of knowledge?

Providing practice. Does the material provide tasks/questions for students to practice skills or *using* knowledge in a variety of situations?

Category V. Promoting Student Thinking About Phenomena, Experiences, and Knowledge
Encouraging students to examine their ideas. Does the material routinely include suggestions for having each student express, clarify, justify, and represent his/her ideas? Are suggestions made for when and how students will get feedback from peers and the teacher?
Guiding student interpretation and reasoning. Does the material include tasks and/or question sequences to guide student interpretation and reasoning about experiences with phenomena and readings?
Encouraging students to think about what they've learned. Does the material suggest ways to have students check their own progress?

Category VI. Assessing Progress
Aligning assessment to goals. Assuming a content match between the curriculum material and this benchmark, are assessment items included that match the same benchmark?
Testing for understanding. Does the material include assessment tasks that require application of ideas and avoid allowing students a trivial way out, like using a formula or repeating a memorized term without understanding?
Using assessment to inform instruction. Are some assessments embedded in the curriculum along the way, with advice to teachers as to how they might use the results to choose or modify activities?

Category VII. Enhancing the Science Learning Environment
Providing teacher content support. Would the material help teachers improve their understanding of science, mathematics, and technology necessary for teaching the material?
Encouraging curiosity and questioning. Does the material help teachers to create a classroom environment that welcomes student curiosity, rewards creativity, encourages a spirit of healthy questioning, and avoids dogmatism?
Supporting all students. Does the material help teachers to create a classroom community that encourages high expectations for all students, that enables all students to experience success, and that provides all students a feeling of belonging in the science classroom?

17
An Interdisciplinary Approach to Urban Ecosystems

Bora Simmons

As has been so eloquently expressed throughout this volume, the study of ecosystems is predicated on a consideration of interrelatedness. To study one aspect of an ecosystem in isolation limits our understanding of the whole. The issue is only exasperated when one considers that within nature there are systems within systems. To add to this complexity, in urban areas especially, a variety of systems have been created (e.g., transportation, housing, water and sewer, education).

Although each of these systems is, by its very nature, a subsystem of the urban ecosystem, there is a tendency to consider urban areas as wholly human endeavors, disconnected from the natural world. To make matters worse, the subsystem itself is rarely fully considered. It is easy to discuss transportation within a metropolitan area, for example, in terms of peak traffic capacity, ridership on a metro line, or the capital expenditures of infrastructure improvements. A piece by piece examination of an urban transportation system, however, provides unsatisfactory answers to the quest for understanding the whole, and may well leave out essential connections such as the relationships among transportation options, social equity, and quality of the environment. The fundamental assumption underlying the concept of an urban ecosystem is the understanding that ecosystems and urban systems are not mutually exclusive and are, in fact, inexorably bound together.

The physical structures that humans build and how they use them interact directly with the rest of the ecosystem we inhabit. From the heat island effect to whether open space is available to water quality to biodiversity to lead poisoning, a strong argument can and should be made that we can no longer think of urban areas and ecosystems as something set apart from one another. How we address issues concerning urban ecosystems relates directly to our long-term quality of life and the life of the planet.

As we begin to shift the paradigm by building a stronger understanding of urban ecosystems, we must also examine the nature of education about the environment within an urban ecosystem frame. Educational programs that enable students to gain a deep understanding of urban ecosystems and the skills necessary to make informed, forward-thinking decisions about urban issues must be developed and sustained. If the goal of environmental education is a society in which citizens make choices about critical issues with creativity, responsibility, and democratic participation, urban ecosystem education must play an integral role throughout our educational systems. This means, however, that we must make significant changes in how and what we teach.

Creating Synergy: Linking the Curriculum and the Environment

Learning about urban ecosystems can be achieved through studies within the traditional school disciplines, and these can effectively and efficiently facilitate the learning of specific concepts and process skills. It is all too easy, however, to fall into the trap of considering urban systems element by element—a unit on the water cycle in science, an historical investigation of the founding of the city, or the creation of a map of landforms in geography. By considering the urban ecosystem within the structure of a disciplinary approach, we are in danger of thinking the systems apart and failing to understand the interrelationships.

Urban ecosystem education calls for a truly interdisciplinary approach to teaching and learning. If we are to teach about interrelated systems, we must be able to create learning environments where understandings that cross-disciplinary boundaries are intertwined. Thus, the challenge is to develop urban ecosystem education opportunities that not only support complex understandings of the urban environment, but also address a commonly held criticism of schooling: The curriculum is treated as a collection of discrete content areas in which teachers move from one topic to another in lockstep fashion. As a result, lessons are often developed in isolation from one another and fail to help students relate their new learnings to what they already know (Applebee, et al. 1989).

Perhaps the explicit focus of urban ecosystem education on the integration of knowledge and skills is one of the primary distinguishing factors between it and a traditional view of curricular disciplines. Because the urban ecosystem is itself made up of interrelated natural and human-produced systems, it can provide an often missed opportunity for synthesis, connecting learnings to create a whole. Disinger (1993) suggests:

> An implicit assumption of disciplinary philosophies is that students will be able to perform their own synthesis when it becomes necessary to do so, by drawing as needed on their learnings from separate content areas. But rarely do students receive instruction or engage in guided practice in developing syntheses and drawing generalizations . . . Environmental education can provide a convenient and challenging mechanism for overcoming the shortcomings of monodisciplinary edu-

cation, by using the interdisciplinary entity that is the environment as a focus for teaching and learning.

To be effective, those who design urban ecosystem education programs must recognize that the development of integrated learnings as well as the synthesis and application of those learnings are deliberate and essential outcomes.

Although taking an integrated approach to urban ecosystem education can be argued on its own merits, the educational efficacy of interdisciplinary study is also well established. Educational theorists suggest that interdisciplinary teaching helps learners develop enduring patterns and connections that make learning meaningful. Because concepts and skills are interwoven throughout a series of lessons using a variety of methods and materials, learning is reinforced, varying learning styles are supported, and the needs of culturally diverse learners are more fully met (Caine and Caine 1994; Conley 1993; Loundsbury 1995; Seely 1995).

Urban ecosystem education must, however, go beyond connecting disciplines through the use of a particular theme. The urban ecosystem is not only the subject to be studied, but the place where learning occurs. Drake (1993) suggests: "When we set curriculum in the context of human experience, it begins to assume a new relevance. Higher-order thinking skills become a necessity as students begin to grapple with real issues and problems that transcend the boundaries of disciplines." Learners must be engaged in direct experiences through hands-on, minds-on investigations that provide them with ample opportunities to construct their own understandings of their environment. This type of education can help to address what Conley (1993) argues is education's past failure to teach process skills, which "led to the inevitable fragmentation of knowledge into 'infobits,' and to graduates who appeared unable to apply much of what they had learned to real-world situations." It is only through real-world, authentic experiences that urban ecosystem education will be able to help students think critically about the world around them and develop the necessary problem-solving skills to help them identify and tackle the environmental issues that face urban areas. Then, it will fulfill its potential ". . . as an exemplary vehicle for what many believe all of education should consider its primary function: furthering the development of higher-order skills—critical thinking, creative thinking, integrative thinking, problem solving" (Disinger 1993).

As an example, the students of Summit School in Seattle built their own school yard garden with flower, herb, and vegetable plots, and a wetland area. Through this process, the students talked with nearby residents to learn about the neighborhood (social studies), invited residents to join their effort (communication skills and cooperation), investigated the habitat needs of the animals they might attract (science), learned about how to propagate plants (science), took measurements of the space (mathematics) in order to design the gardens (problem solving and art), and kept written records of their activities (writing). This project provided an opportunity for the students to confront real-world problems and to use them as a mechanism for learning. In building their own garden and wetlands, the students learned about specific concepts such as the water cycle, plant propagation, and habitats. Importantly, they began to understand more fully the *relationships among* various parts: people, water, soil, habitat, plant growth, species diversity, and food production. Because members of the community were involved, they learned that their school yard habitat was not an island, but in reality, physically and socially linked to the neighborhood that surrounded them. They learned how others viewed natural areas, gardens, and gardening. The efficacy of this approach is, perhaps, best expressed by one the students, "When I was in fourth grade and we learned how a plant grows, we could have used a place like this to have a better example—instead of little seeds inside foam cups" (Williams and Agyeman 1999).

Facing the Challenges

It is one thing to argue that effective urban ecosystem education must be based on real-life experiences, integrated learnings that cross the disciplines, and the development of higher-order thinking skills. It is something entirely different to suggest that urban ecosystem education programs are easy to implement and currently in wide practice. Developing an urban ecosystem education program requires overcoming a number of significant challenges.

By its very nature, the study of an urban ecosystem involves great complexity. Discerning any one particular system (e.g., water, transportation) requires a vast array of understandings and skills. Connecting these systems to the whole, and comprehending that systems are nested within systems, dictates an even higher level of sophistication.

Teaching about urban ecosystems means that the teachers themselves must have some understanding of the whole and its intricacies. For the most part, however, teachers have gained their knowledge through traditional discipline-based approaches and may never have had the opportunity to explore the concept of urban areas as functioning ecosystems. In turn, these same teachers must have the skills to be able to translate the necessary systems understandings into a curriculum that is age appropriate and locally relevant for their students. When dealing with this complexity, the teacher must find the balance between overly simple and overly detailed or abstract explanations. With simplification, there is a danger that the texture and richness of the systems may be lost. However, with overly detailed explanations, the elegance of the system may be overshadowed in the drive to accumulate knowledge about each individual element.

Although teacher preparation represents a challenge, teacher perceptions may actually pose a more significant barrier to the implementation of an urban ecosystem education program. Research suggests that urban teachers express a variety of concerns when asked about taking students to urban areas for study. Most often, teachers identify funding/cost, time, lack of knowledge and skills, safety, and lack of resources as barriers (Young and Simmons 1992; Simmons 1993).

Teachers may not perceive the full array of teaching opportunities provided by urban areas that could support an integrated approach to urban ecosystem education. In one study, urban teachers were shown black-and-white photographs of various urban settings and asked what they might do with their students if they took them to these places. The teachers perceived distinct differences among the urban settings. The photographs of urban rivers, ponds, and streams elicited suggestions of science- and recreation-related activities. They associated wooded scenes with tree identification and wildlife study. Pictures of urban parks, open fields, and school sites were associated with providing various recreational, but few educational, opportunities. Finally, scenes of the city center (Chicago) invoked responses focused primarily on the use of nearby museums and the study of architecture (Simmons 1993; Simmons 1996). In addition, setting-specific training seems to play a significant role in whether or not teachers see a particular urban area as an educational resource (Simmons 1999). Interviews conducted with urban teachers suggested that there was no statistical relationship between participation in some form of general environmental education training and willingness to take students to specific types of urban settings. On the other hand, 100 percent of those teachers who had received training at a particular type of setting reported willingness to take their students to that type of setting for school-related activities. As might be expected, those teachers who had participated in setting-specific training were able to identify a greater number of possible education activities for their students, compared to those teachers who had not had this training. This study suggests that setting-specific training seems to provide teachers with a greater repertoire of skills and a heightened ability to recognize urban educational potentials.

Finally, teachers may well hold very different (often more negative) perceptions of urban areas than do their students. Interviews conducted with urban teachers and their students showed that they view urban natural settings quite distinctly (Simmons 1994). Both groups were asked to rate seven urban settings (e.g., wooded areas, school sites, parks, open space, cityscape) for preference and comfort level. A comparison of the rank ordering of the mean ratings demonstrated that the groups' preferences were opposite to one another. The school site was the most highly preferred by the children and least preferred by the teachers. Water-related settings were most highly preferred by the teachers and least preferred by the children. The places children preferred most and felt most comfortable in were those that the teachers preferred least and felt least comfortable in.

Urban ecosystem education must also face a number of structural challenges furthered by the educational system itself. On the whole, schooling is rooted within studies of the disciplines. From textbooks and resource materials to the school day schedule to the state learning standards and their associated tests, learning is boxed into and measured by the traditional disciplines. Integrated studies are difficult to accomplish in a 50-minute period. An English-language arts teacher may feel reluctant to tread into areas of study that encompass science or social studies. Teachers may be hard pressed to find texts or other materials that are not monodisciplinary in focus. And, when they do find materials that support interdisciplinary studies and provide needed background and lesson ideas, these materials are most likely produced at the national level, leaving teachers on their own to create locally relevant learning experiences.

Toward Urban Ecosystem Education

For urban ecosystem education to be truly integrated in a meaningful way throughout the curriculum, a comprehensive, cohesive program needs to be developed. To that end, *Excellence in Environmental Education—Guidelines for Learning (K–12)*, published by the North American Association for Environmental Education (1999), provides guidance for the development of locally relevant, effective programs. *Excellence in Environmental Education* sets expectations for performance and achievement at the end of the fourth, eighth, and twelfth grades, suggests a framework for effective environmental education programs, offers a vision of environmental education that makes sense within the education system, and promotes progress toward sustaining a healthy environment and quality of life. Because *Excellence in Environmental Education* is founded on a set of essential underpinnings (see Box 17.1) that promote systems thinking, provide real-world contexts and issues from which concepts and skills can be learned, and foster an understanding of the importance of where one lives, it can provide the scaffolding necessary to support the development of meaningful urban ecosystem education programs. These essential underpinnings furnish both a philosophical and a pedagogical foundation for urban ecosystem education.

As has been discussed, an understanding of systems and the interdependence of humans within those systems, especially within urban areas, is fundamental. *Excellence in Environmental Education* offers a conceptual framework that facilitates the teaching of these understandings (see Box 17.2). Although the framework was designed for environmental education in general, the concepts and skills presented are equally applicable to urban ecosystem education. In particular, strand no. 2 asks that students develop

Box 17.1: Essential Underpinnings of Environmental Education (NAAEE 1999)

Environmental education builds from a core of key principles that inform its approach to education in general and urban ecosystem education specifically. Some of these important underpinnings are:

Systems—Systems help make sense of a large and complex world. A system is made up of parts that can be understood separately. The whole, however, is understood only by understanding the relationships among the parts. The human body can be understood as a system; so can cities. Organizations, individual cells, communities of plants, animals and people, families, or municipal water supplies can all be understood as systems. And systems can be nested within other systems.

Interdependence—Human well-being is inextricably bound with environmental quality. Humans are a part of the natural order. We and the systems, we create—our societies, political systems, economies, religions, cultures, technologies—have a large impact on the systems and cycles of the rest of nature. Since we are a part of nature rather than outside it, we are challenged to recognize the ramifications of our interdependence.

The importance of where one lives—Beginning close to home, learners forge connections with, explore, and understand their immediate surroundings—their homes, schools, neighborhoods, and cities. The sensitivity, knowledge, and skills needed for this local connection provide a base for moving out into larger systems, broader issues, and an expanding understanding of causes, connections, and consequences.

Integration and infusion—Urban ecosystem education must span the natural sciences, social sciences, and humanities as they connect through the medium of the environment and environmental issues. Urban ecosystem education offers opportunities for integration and when infused throughout becomes a seamless part of the curriculum.

Roots in the real world—Learners develop knowledge and skills through direct experience with the environment, environmental issues, and society. Investigation, analysis, and problem solving are essential activities.

Lifelong learning—Critical and creative thinking, decision making, communication, and collaborative learning are emphasized. Engaging in individual and group work helps learners develop these capacities independently and in collaborative situations that anticipate the ways in which environmental problem solving happens in the community, on the job, and in the family. A strong emphasis on developing communication skills helps learners demonstrate and disseminate their knowledge. These skills are essential for active and meaningful learning, both in school and over a lifetime.

Box 17.2: Excellence in Environmental Education—Guidelines for Learning (K–12) (NAAEE 1999)

Excellence in Environmental Education provides students, parents, educators, home schoolers, administrators, policy makers, and the public a set of common, voluntary guidelines for environmental education. The guidelines are organized into four strands, each of which represents broad outcomes of environmental education and its goal of environmental literacy. These four strands can be easily adapted to relate explicitly to urban ecosystem education:

Strand #1: Questioning and analysis skills
An urban ecosystem education program develops learners' ability to ask questions, speculate, and hypothesize about the world around them, seek information, and develop answers to their questions about their urban environment. Learners must be familiar with inquiry, master fundamental skills for gathering and organizing information, and interpret and synthesize information to develop and communicate explanations.

Strand #2: Knowledge of environmental processes and systems
An important component of urban ecosystem education is an understanding of the processes and systems that comprise the environment, including human systems and their influences. That understanding is based on knowledge synthesized across the traditional disciplines. The guidelines in this strand are grouped into four subcategories: the Earth as a physical system; the living environment; humans and their societies; and environment and society.

Strand #3: Skills for understanding and addressing environmental issues
Skills and knowledge are refined and applied in the context of urban environmental issues. These environmental issues are real-life dramas where differing viewpoints about the underlying environmental problems and their potential solutions are played out. Environmental literacy includes the abilities to define, learn about, evaluate, and act on environmental issues. In this section, the guidelines are grouped into two sub-categories: skills for analyzing and investigating environmental issues and decision-making and citizenship skills.

Strand #4: Personal and civic responsibility
Environmentally literate citizens are willing and able to act on their own conclusions about what should be done to ensure environmental quality. As learners develop and apply concept-based learnings and skills for inquiry, analysis, and action, they also understand that what they do individually and in groups can make a difference.

knowledge of environmental processes and systems (e.g., the Earth as a physical system; the living environment; humans and their societies; and environment and society).

Using the guidelines as a map for curriculum development, students at the eighth-grade level might engage in a series of activities that trace the flow of energy in their city using examples that encompass several different transfers and transformations. They might construct food webs that identify relationships among organisms found in their own neighborhood, and describe how energy, which enters their ecosystem as sunlight, changes form and is transferred in the exchanges (production, consumption, and decomposition) that comprise these food webs. Widening their investigation, they might consider the sources of energy (e.g., fossil fuels, hydroelectric power) commonly used for heating, lighting, transportation, manufacturing, and other activities in their local community. They might perform a personal energy audit, detailing the energy they use on a daily or weekly basis in their home, at school, and as they move about their community. In addition to tracing the path of energy through their community, they might investigate the potential impacts of the by-products of fossil fuel use on their airshed or analyze the social, economic, political, and environmental implications of siting power generation plants or fuel transfer stations in or near their community. By the end of their investigations, the students should hold a firm grasp on the nature of energy, how it flows through the various systems within their city, their personal use of energy, and some of the potential implications of energy use.

No matter what the topic, urban ecosystem education should use what has been called a concentric circle approach to program development. "Start the learner in his or her most immediate environment: the home, school, or community environmental center. This appropriately focuses attention on the environment that the learner can most effectively influence. Then expand the learner's experiences out . . . in ever widening circles" (Frank and Zamm 1994). Experiencing and observing the local environment is an essential part of urban ecosystem education. Understanding their surroundings helps learners build a strong foundation of skills and knowledge for reaching out further into the world and deeper into the conceptual understandings that overall environmental literacy demands. Direct experience in the environment also helps foster the awareness and appreciation that motivate learners to further questioning, better understanding, and appropriate concern and action.

Excellence in Environmental Education suggests that young students might explore and learn about their local ecosystem by tracing the source of their drinking water and where it goes after it is used; keeping a journal of resident animal species and those that pass through on migratory routes; or mapping basic types of habitats within a short walk of their school. Middle school students might classify local ecosystems and create food webs that show the interaction of specific plant and animal species; monitor changes in water or air quality at various locations in their city; or see how their city is linked to other regions by creating a map that shows where food that is consumed locally comes from. High school students might document changes in land use and their environmental effect; evaluate sources of non–point source pollution; calculate the potential for generating wind or solar power at sites within their city; or trace human population trends for their urban area and make projections, based on research findings, for the future.

To facilitate this type of learning, the real environmental questions and issues that students experience in their daily lives become meaningful educational vehicles. As an example, while on an after-school, service learning, community clean-up walk, fourth graders from a school in New Mexico traced a puddle of dirty oil to the dumpster behind an auto lubrication service (Hampton 1998). The students talked to the owner, who assured them this was not normal procedure, and showed them how they collect motor oil for recycling.

A follow-up class discussion generated a lot of questions about oil pollution. Many students were particularly concerned about a recent oil spill in their area, which prompted an Oil Spill Clean Up Contest. Allowed to work independently or in groups, the students were challenged to clean a tablespoon of gear lube oil from a beaker of water. Using methods of science inquiry, they were given three days to conduct research and plan their approach and each team was allowed to bring from home one shoebox full of equipment. To ensure safety, plans had to be approved by the teacher. Students tested their techniques, recording the time required to complete their process. Using mathematics and observation skills the students then rated the cleanliness of each beaker and entered their findings into a database later used to examine the advantages and disadvantages of each method.

Using their geographic research results, students also mapped the size and location of the world's largest spills and explored actual methods of cleaning oil spills. Finally, students devised their own assessments to show what they had learned, and still wanted to learn, about oil spills and oil use in their own community. Assessments included books created for third graders, a computerized presentation, a comic book, and illustrated essays.

These New Mexico students studied science, mathematics, social science, art, and English-language arts during their investigation into oil spills. They learned about cycles, how a human-made substance (motor oil) can contaminate soil and water if not properly recycled, the impacts of oil contamination on the ecosystem, and the technologies that can be used to clean-up contaminated water. In doing so, they also identified their own questions about an environmental issue that was close to home, developed hypotheses, designed their own environmental investigations, collected information from a variety of sources, evaluated the results of their investigations, considered ways in which humans interact with their environment, and communicated the results of their investigations to others in their school.

When based in real-world experiences, urban ecosystem education is an integrative undertaking that draws from and blends a broad content knowledge of the natural and social sciences, arts, mathematics, and humanities. Constructing an educational program that addresses both traditional areas of the curriculum (e.g., science, social studies, communication skills) and the goals of urban ecosystem education is a complex task. Seeing how an urban ecosystem education program relates to a standards-based curriculum, and making the direct linkages between integrated learnings and specific disciplines, may not be obvious. Again, *Excellence in Environmental Education* can provide some guidance for the teacher who wants to connect learnings about the urban ecosystem to the traditional curriculum. Each of the four strands outlined in *Excellence in Environmental Education* (i.e., questioning and analysis skills, knowledge of environmental processes and systems, skills for understanding and addressing environmental issues, and personal and civic responsibility) is further defined by a series of guidelines that suggest general goals for learner achievement. Where appropriate these guidelines are correlated to the national standards (e.g., science, mathematics, social studies, fine arts, English-language arts, economics) set by professional organizations of the academic disciplines. Consequently, by using the conceptual framework offered by *Excellence in Environmental Education* in designing an urban ecosystem education program, teachers can consciously and deliberately design a standards-based curriculum that makes sense.

Greater Than Its Basic Parts

By taking an holistic approach to the teaching of urban ecosystem education, synergy among its basic "parts" can be created. As students analyze and evaluate the complexities of an urban environmental issue, they begin to understand the intricacies of the connections that they could not have discovered if the information was presented fact by fact and subject by subject outside the context of the urban ecosystem as a whole. As we develop programs for students, we must alter our thinking and begin to not only consider the study of ecosystems as a part of the content of the curriculum, but as a metaphor for urban education itself. Only the development of a comprehensive urban ecosystem education program insures that learnings about the urban ecosystem will not be marginalized or fragmented. "Urban ecosystem education" that is simply dropped into the curriculum as an interesting activity to do on a Friday afternoon or as a field trip "to the city" is not true urban ecosystem education, and in the long run, will not promote environmental literacy and responsible citizenship. To be effective, urban ecosystem education programs must be constructed and cannot be left to chance. They must be founded on an understanding of the knowledge and skills that lead to environmental literacy, and a wider vision of environmental education's place within the school curriculum.

References

Applebee, A., J. Langer, and I. Mullis. 1989. *Crossroads in American education.* National Assessment for Educational Progress and Educational Testing, Princeton, NJ.

Caine, R.N., and G. Caine. 1994. *Making connections: teaching and the human brain.* Addison-Wesley, Menlo Park, CA.

Conley, D.T. 1993. *Roadmap to restructuring: Policies, practices, and emerging visions of school.* ERIC Clearinghouse on Educational Management, Eugene, OR.

Disinger, J. 1993. Environmental education in the K–12 curriculum: An overview. Pages 23–43 in R. Wilke, ed. *Environmental Education Teacher Resource Handbook.* Kraus International Publishing, Milwood, NY.

Drake, S.M. 1993. *Planning integrated curriculum: The call to adventure.* Association for Supervision and Curriculum Development, Alexandria, VA.

Frank, J., and M. Zamm 1994. *Urban environmental education.* Kendall/Hunt Publishing Company, Dubuque, IA.

Hampton, E. 1998. *Personal communication.*

Loundsbury, J.H. 1995. *Connecting the curriculum through interdisciplinary instruction.* National Middle School Association, Columbus, OH.

North American Association for Environmental Education. 1999. *Excellence in environmental education—Guidelines for learning (K–12).* North American Association for Environmental Education, Rock Springs, GA.

Seely, A.E. 1995. *Integrated thematic units.* Teacher Created Materials, Inc., Westminster, CA.

Simmons, D. 1993. Facilitating teachers' use of natural areas: perceptions of environmental education. *Journal of Environmental Education* 24:8–16.

Simmons, D. 1994. A comparison of urban children's and adults' preferences and comfort levels for natural areas. *The International Journal of Environmental Education and Information* 13:399–414.

Simmons, D. 1996. Teaching in natural areas: what urban teachers feel is most appropriate. *Environmental Education Research* 2:149–157.

Simmons, D. 1999. Training and urban teachers' willingness to teach environmental education. *The International Journal of Environmental Education and Information* 18:29–38.

Williams, E., and J. Agyeman. 1999. Educating for a more livable urban environment. *Educator* Spring:26–30.

Young, C., and D. Simmons. 1992. Urban teachers' perspectives on teaching natural resources. *Women in Natural Resources* 13:39–43.

18 Children for Cities and Cities for Children: Learning to Know and Care About Urban Ecosystems

Louise Chawla with Ilaria Salvadori

Children in Cities: The International Framework

Urban ecosystems are people places. Among the roughly 3 billion people who inhabit urban areas around the world, a significant proportion are children under the age of 18. In the high-income nations of the North, young people in this age group constitute about 30 percent of the population. In most low- and middle-income nations, they constitute 40–50 percent (United Nations Population Division 1998). Urban ecosystems, therefore, are people places in which children play an important part. The city is not a subject for children to learn about as something separate from themselves, but a subject that includes them as an important component.

Children are a significant part of urban ecosystems not only because of their numbers but also because of their role. The last decades of the twentieth century outlined the great challenge of the twenty-first: to craft a sustainable society as can preserve the biological integrity of the planet at the same time as it provides for social equity, through ecological and social systems that meet the basic needs of the present without squandering the ability of future generations to meet their needs (World Commission on Environment and Development 1987). Children are central to this definition of sustainability and its realization. They are humanity's bridge from the present to the future: those who must carry the concept and practice of sustainable development forward through time.

Cities, too, must be central to the practice of sustainable development. There is a progressive trend toward the urbanization of world societies, and through more efficient land and resource use and service provision, urban areas offer advantages as well as risks for the implementation of sustainable practices. These potential advantages are heightened when we consider that 70–90 percent or more of the population live in urban areas in high- and middle-income nations in Latin America, North America, Western Europe, and the Pacific Rim, while the proportion of the population that lives in urban areas in other Asian nations and Africa is growing. Therefore urban areas and their residents, including children, must be central to planning for sustainability.

The importance of children in cities has been formalized by international law. In 1989, the United Nations ratified the Convention on the Rights of the Child, which places children's relationship to their societies and cities on new ground. The Convention recognized that children have basic rights as persons, in harmony with the Universal Declaration of Human Rights, but in addition, they have special needs for protection and provision, given their vulnerability and requirements for healthy development. The articles of the Convention, therefore, can be divided into three groups: those to ensure children's *protection*, *provision* for their basic needs, and *participation* in decisions that affect their lives, to the level of their ability. Guidelines for the implementation of the Convention explicitly spell out that these rights include participation in decisions that affect their living environment (Hodgkin and Newell 1998). Building on the Convention, *Agenda 21* (the Program of Action from the United Nations Conference on Environment and Development in 1992) recognizes children and youth as major actors who must be involved in participatory processes of environmental management and decision making. In a further step, the *Habitat Agenda* (the Program of Action from the Second United Nations Conference on Human Settlements in 1996) identifies children and youth as critical resources for the creation of sustainable settlements. In the words of the Preamble to the *Habitat Agenda*: "Special attention needs to be paid to the participatory processes dealing with the shaping of cities, towns, and neighborhoods; this is in order to secure the living conditions of children and of youth and to make use of their insight, creativity and thoughts on the environment" (UNCHS 1997, paragraph 13).

These landmarks in the definition of children's roles in their environment are the framework of this chapter because it is assumed that this book's goals of furthering an understanding of urban ecosystems, and applying this understanding to education, are for the larger end of creating sustainable cities and societies. The Convention on the Rights of the Child, *Agenda 21* and the *Habitat Agenda* indicate children's potential in contributing to this end.[1]

In the context of these goals, this chapter focuses on how children learn to know and care about their cities. It will review research on how children learn to understand large-scale environments like cities, and the limited

[1] The fact that the United States and Somalia, alone among the member nations of the United Nations, have not ratified the Convention does not detract from the importance of this international framework. Somalia, in the absence of an officially recognized government, lacks authority for ratification. The United States has signed *Agenda 21* and the *Habitat Agenda*, which integrate the substance of the Convention; and the Convention has achieved such wide recognition that it has already gained the force of customary law.

research that exists regarding childhood experiences that may motivate people to protect urban quality. It will discuss forces that are currently working against young people's acquisition of environmental knowledge and concern, and it will propose programs, policies, and design approaches to ensure that young people have the positive urban experiences that they need. It will do so with an emphasis on informal learning: what children learn by direct contact with the cities beyond their school walls as they move through urban places in play, travel, and everyday routines.

This emphasis on out-of-school learning is based on two theoretical principles. One is that children's learning is *constructivist* (Kahn 1999; Piaget 1963). This means that their conceptions of physical and social worlds reflect a continuous exchange of assimilation, accommodation, and equilibration in which they take in what they are ready to absorb according to current levels of understanding; while at the same time they progressively adjust their understanding in response to new and unexpected experiences. They are not passive learners, but active producers of knowledge with an innate drive to explore and learn about the environment. Classrooms at their best encourage learning of this kind. The principle of constructivism, however, implies that children are learning all the time, in more or less enabling ways depending on the quality of the experiences available to them, out of school as well as in. Our society tends to equate learning about complex subjects like ecosystems with formal in-class experiences; but urban ecosystems are, in fact, urban children's worlds, and therefore children will be best equipped to understand this complex world if formal curricula in their school, informal programs in their city and community, and opportunities for unstructured exploration in their everyday environment enhance each other. This principle implies that to construct an integrated understanding of urban ecosystems, children need a foundation of engaging experiences with different system components in the city around them. Therefore this often-neglected "third dimension" for learning, beyond the school and out-of-school programs, will be the focus of this chapter.

A second theoretical principle that guides this chapter is that urban environments can be evaluated in terms of *affordances* (Lewin 1936), or opportunities for action and learning which the environment offers children. Closely allied to this idea of affordances is the concept of *behavior settings* (Barker and Wright 1955; Schoggen 1989), or places and routines where people display regularly occurring patterns of behavior. Through their transactions with behavior settings, children have opportunities to cultivate a variety of skills, roles and understandings (Fuhrer 1998). This chapter focuses on what Burch and Carrera (this volume) have described as "the intimate ecology of experience" in everyday out-of-school environments, because this is where children have many of the experiences they need in order to construct their understanding of ecosystems, depending upon the variety and quality of the behavior settings that the urban environment affords.

Learning Large-Scale Environments

Whether or not children have opportunities to learn about their cities through a variety of experiences and interactions depends upon a number of factors, but central among them are children's range of movement and their understanding of the city as a geographic entity through which they can navigate, and the elements of which are meaningful to them. This section will review basic research on the development of children's spatial cognition and, closely associated with it, their range of mobility.

Spatial Cognition

Spatial cognition refers to knowledge about routes and landmarks, the configuration of places, place-related events and activities, and their social, cultural, and ecological meanings. Therefore it is a fundamental dimension of the understanding of urban ecosystems. It is often used in connection with the concept of a "cognitive map"—a term first introduced by Tolman (1948) to explain the behavior of rats in a maze. The term does not mean an actual map image in the head, and even less a free recall map that a person draws on request, but it refers to the processes people use—whatever they may be—as "a means of structuring, making sense of, and coping with the complexities of environments external to the mind" (Golledge 1987; Matthews 1992).

Notwithstanding these distinctions, much of the research to explore spatial cognition has focused on the analysis of children's free recall maps or models of their environments. In an early synthesis of this research, Hart and Moore (1973) proposed that children pass through three main stages of map drawing or modeling, and it is inferred, through comparable stages in their understanding of large-scale environments. In the first, egocentric stage, maps and models are primarily composed of landmarks defined by the child's own actions and interests (Figure 18.1). These place views are largely uncoordinated pictures of personally significant objects in the environment. In the second stage, the child develops a fixed system of reference structured around familiar locations such as home, school and friends' houses, which are pieced together into routes. At this stage, environmental understanding shows differentiated, partially coordinated clusters of elements. In the third stage, children are able to use a reference system based on abstract geometric coordinates and cardinal directions (Figure 18.2). At this point, children are able to integrate their knowledge of geographical configurations, routes, and landmarks into an abstract, hierarchical, and metric system of space.

Hart and Moore (1973) related this model of development to Piaget and Inhelder's (1967) research on environmental cognition, in which the first stage is roughly comparable to the Piagetian sensorimotor and preop-

FIGURE 18.1. A drawing of a child's neighborhood, in which personally significant objects are loosely placed around the child's house in the center.

erational stages from approximately 2–7 years, the second to the stage of concrete operations from approximately ages 7 through 11, and the third to the stage of formal operations which begins about age 12. These are rough generalizations, however. A large body of research that has reviewed Piagetian theory has noted that although it may be generally true that children's thinking proceeds from the egocentric to the multiperspectival and from the concrete to the abstract, and that children show increasingly integrated and hierarchic mental structures with age, this development is highly dependent on motivation, experience, and the conditions of measurement (Scholnick, et al. 1999). In fact, this conclusion that environmental understanding is a function of environmental experience and interest as much as age and general cognitive capacity is consonant with Piaget's (1963) principle that children learn primarily through their own active and interactive exploration of the world. This constructivist principle predicts that children's spatial cognition of large-scale environments will be highly dependent on their opportunities to encounter the environment and assimilate new experiences according to their individual levels of understanding and interest.

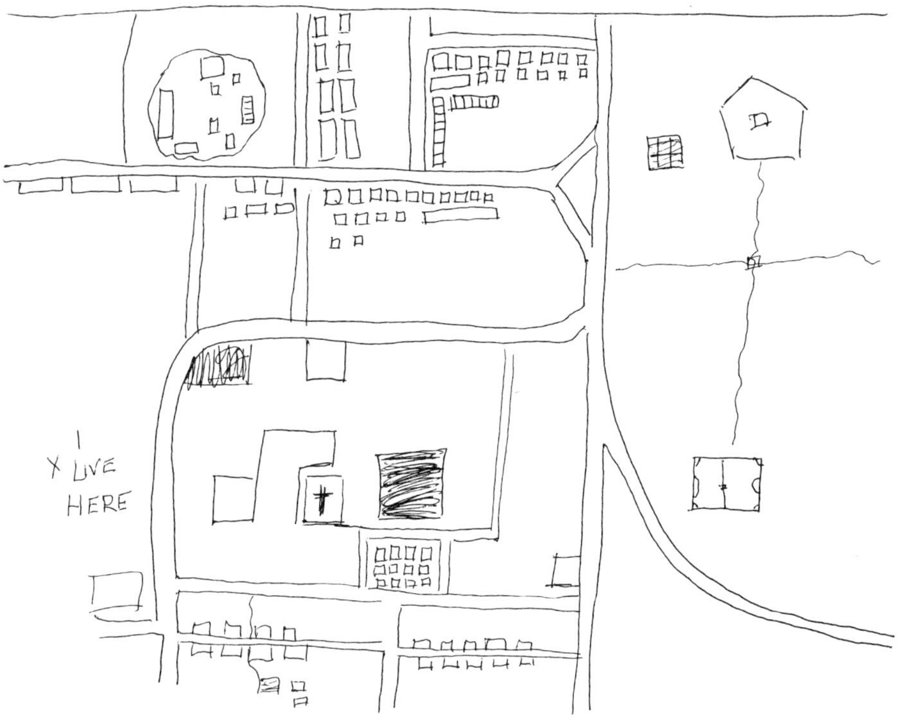

FIGURE 18.2. An advanced drawing of a child's neighborhood, which integrates routes and landmarks within an abstract system of cardinal directions and relative distances.

Other theories of spatial cognition emphasize the processes through which learning is acquired. According to Siegel and White (1975), children first notice and remember landmarks, which they gradually connect via paths and routes. These landmarks and routes are formed into clusters, which are at first uncoordinated with each other, but over time they are integrated into an increasingly accurate and coordinated frame of reference. According to Golledge (1978), spatial cognition develops around anchor points of primary, secondary, and tertiary nodes and the paths that link them. Primary nodes are intimately known places within the immediate locality, such as the home. These serve as anchor points from which the rest of the spatial hierarchy develops. Over time, other places become fixed around the primary nodes, forming secondary and tertiary nodes. Environmental understanding develops through interactions among these nodes and the paths between them, until a highly integrated view of place forms. Gibson (1979) and Heft (1983) have emphasized the interactive nature of spatial perception, as people move through landscapes and coordinate the information gained through the perception of successive vistas over time.

As Matthews (1992) notes in his review of these different explanations of spatial learning, all of them are highly dependent on people's activity, familiarity, and amount of mobility in the environment, their mode of movement, motivation, and the complexity of the environment itself. Therefore this literature on spatial learning needs to be considered in connection with the study of children's range of movement.

Range of Movement

One of the most extensively studied aspects of environmental learning is children's range of movement and experience, which has been investigated through interviews, questionnaires, place-use diaries, and by asking children to trace their routes of movement on maps or aerial photographs. This tradition of research has accumulated to the point where historical observations can now be made regarding children's changing mobility in their towns and cities since the beginning of the twentieth century.

Whiting and Edward's (1988) study of children's socialization in six cultures showed roughly similar transitions from the *lap* or *back* child who is kept to the closely bounded space of the mother or other caretakers until sometime around age two; the *knee* child who begins to move about in larger areas that can still be monitored by caretakers; the *yard* child who by age four or five gains independent access to the entire house and yard and often neighboring yards; and the *community* child who by the age of six or seven begins to move out of sight and sound of the home to go to school, do chores, or play with friends. To what degree, however, a child is allowed the freedom to become a wide-ranging community child varies dramatically from place to place and culture and culture. As a rule, suburban children travel farther than city children (Anderson and Tindall 1972; Moore and Young 1978; Van Vliet 1983), and boys have larger ranges of movement than girls, although girls may know their smaller range of places more intimately (Hart and Saegert 1978; Moore and Young 1978; Hart 1979; Van Vliet 1983; Webley 1981). Some studies have suggested that range increases with social class and majority versus minority racial or ethnic status (Maurer and Baxter 1972; Van Vliet 1983). Rules for movement vary depending on whether children travel alone, with friends, or in the company of adults (Moore and Young 1978; Hart 1979).

According to Hart (1981), range of movement through the environment and freedom to explore "are the most important forces influencing the quality, as well as the extent, of children's ability to map the spatial relations of places in the large-scale environment" (p. 207). In a Vermont town and in Coventry, England, respectively, Hart (1979) and Matthews (1992) found strong relationships between children's ability to map and model their territory and interpret local maps, and the extent of their range of environmental experience. Those who traveled most widely showed the broadest representations of space. In the city center of Coventry, Matthews found that 11–18-year-olds' knowledge of the city's configuration became more complete and integrated as their trips to the center became more frequent and self-directed. In metropolitan Toronto, Van Vliet (1983) found that teenagers who traveled farther from home participated in a larger number of different activities and scored higher in their knowledge about urban resources. Webley (1981) found that boys' superior mapping ability compared to girls' (a frequently observed difference) could be explained by their greater range of movement and familiarity with their area.

Both the amount of time that children spend in places and the variety of places that they experience are important dimensions of their learning. Moore and Young (1978) defined habitual, frequented, and occasional levels of experience. Van Vliet (1983) distinguished spatial range, or the distance traveled from home, from behavioral range, or the variety of activities accessible within this distance. There is no advantage to wide-ranging travel, by itself, if the environment is empty. In a four-nation study of urban experience, for example, Lynch (1977) noted that young adolescents in Melbourne, Australia had a wider range of movement than their peers in Cracow, Warsaw, Mexico City, or the provincial Argentinian capital of Salta, but they seemed driven by boredom to seek out widely scattered points of interest in a barren landscape.

It is also important *how* people travel. In large-scale environments in particular, where it is necessary to coordinate multiple perspectives of space, children who walk or bicycle show a more integrated understanding of the environment compared to those who view the environment passively (Cohen and Cohen 1985; Hart 1981; Matthews 1992). Working with 7–14-year-olds in Houston, Maurer and Baxter (1972) found that those who walked to school drew more natural features like trees, grass, sky, sun, and animals, and were generally better at mapping tasks, compared to others who traveled to school by car or bus. A bicycle allows children to gather extensive independent experience (Southworth 1970; Hart 1979); whereas subway riding does not contribute to mapping accuracy (Southworth 1970).

All of these variables in mobility relate to Moore and Young's (1978) concepts of range extension and range development. Moore and Young argue that it is not only important that children have opportunities to travel to a widening radius of places with increasing age (range extension), but also that they have opportunities to independently explore, manipulate, and in some cases transform their territory. At their best, with each new visit, children discover new possibilities for continuing involvement, so that places grow in meaning and value as children grow (range development).

All of these range factors are important dimensions of learning about urban ecosystems. Without direct experience of different parts of a city's geography, the concept of a city itself as an integrated identity with many districts and elements will remain an abstraction at best. Unless children can travel through a city via different modes of transportation, walk by its riversides, play in its parks, and visit its diverse neighborhoods, they will not have a vivid personal knowledge of its transportation system, watersheds, natural ecology, or cultures. Although movement, by itself, does not ensure knowledge, involvement, or affiliation, it is a precondition for the development of these means of relationship to place.

Recent research on children's urban ranges, however, suggests that opportunities for direct experience are becoming more and more restricted. Gaster (1991), in a study of children's range of free play in the Inwood district of Manhattan between 1910 and 1980, found a decrease since the 1940s in the number of play places visited, and a steady increase in the age at which children were first allowed out alone, the number of self-perceived barriers to the environment, and the amount of adult-supervised play. Hart, Mora, and Iltus (1991), attempting to study children's home range in the South Bronx, found it to be almost nonexistent. Blakely (1994), in a study of a stable working-class neighborhood in New York, found most children's free range of movement to be less than 400 feet. When Hillman and others (1997) replicated a 1971 British study of children's range of free movement, they found a steep decline in children's opportunities to leave their home and do things independently. For the first time, Valentine (1997) found that parents in Britain were beginning to keep their boys as close to home as girls out of fear for their safety. A study of low-income 10–15-year-olds in nine neighborhoods around the world (Chawla 2002) found that even those in the upper age range usually remained in their immediate neighborhood, for various reasons such as fear of violence, heavy traffic tightly programmed schedules, or a lack of money to travel farther. These findings have disturbing implications with regard to the understanding of urban ecosystems, as the research on spatial cognition indicates that the more that children move about freely, the better they understand their city's content, configuration, and manifold meanings. Conversely, the more constrained their experience, the less they know and the less prepared they are to adventure out and learn more.

Learning to Care

It is not enough to know about urban ecosystems. The chapters in this volume carry with them values about wanting to protect or restore natural ecosystems in cities and to create more sustainable cities with lighter ecological footprints on their surrounding regions. Therefore in addition to understanding how children cognitively construct cities as systems, it is important to understand how they achieve a positive identification with their city and an investment in its maintenance and protection. There is much less research on this emotional dimension of environmental experience; however, the limited research that exists is suggestive. Drawing on a meta-analysis of more than 100 studies of environmental attitudes and behavior by Hines (1986/87), Hungerford and Volk (1990) proposed a model of the genesis of responsible environmental citizenship in which people are motivated by certain "entry-level variables" to feel a personal sense of ownership of certain environmental issues, and as they acquire an in-depth knowledge and personal investment in these issues, become empowered to take action for the sake of the environment. Chief among these entry-level variables is "environmental sensitivity," which has been defined as a predisposition to take an interest in the environment and feel concern for it on the basis of formative experiences (Chawla 1998). In a growing body of literature, many of the formative experiences that people talk about go back to childhood and adolescence (Tanner 1998). In a review of seven of these autobiographical studies, Chawla (1998) found that people repeatedly recalled five primary types of experiences: positive encounters with the environment itself, influential family role models, teachers and education, environmental organizations, and witnessing environmental destruction (see Table 18.1).

Most of these studies have investigated people's reasons for protecting wilderness and other natural areas. In people's memories, however, the "wild" area they remember fondly may have been as small as a nearby grove of trees or riverbank: places that may exist in cities and suburbs as well as rural areas. In one study, which included activists who worked to improve urban quality (Chawla 1999), people described happy associations with gardens, bikeways, and accessible streets in the metropolitan areas of their childhood and youth.

If positive childhood experiences of the environment are important precursors of care for its protection, then it becomes important to understand how children evaluate urban quality. According to an international study of 10–15-year-olds in nine low-income urban communities around the world (Chawla 2000), important positive qualities are social as well as physical, and include a sense of social acceptance and integration, safety, a cohesive

TABLE 18.1. Significant life experiences that people consider sources of their concern and care for ironment.

Childhood Experiences	Percentage of Respondents in Seven Studies
Positive experiences in the type of environment one seeks to protect	75–90
Family role models, especially parents	50–75
Teachers, education	30–60
Environmentally oriented organizations	20–55
Witnessing environmental destruction	20–40

Source: Based upon seven studies reviewed by Chawla (1998).

community identity, peer gathering places, a variety of interesting activity settings, freedom of movement, and access to basic services and—where they are available—green areas. Both Hungerford and Volk's model and the literature of significant life experiences described in Table 18.1, however, suggest that in addition to positive environmental experiences, people need role models in the home, school, or neighborhood who demonstrate the environment's value, as well as opportunities to practice the skills needed to translate care into action.

An Oak Park Case Study

All of the preceding topics—spatial cognition, range, motivations for protecting urban quality—can be brought to life through a case study of a group of immigrant 9–14-year-olds in Oakland, California. In the spring of 1997, this chapter's co-author, Ilaria Salvadori, began field work at this location as a part of the international Growing Up in Cities project that aims to progress from research with children and adolescents that engages them in documenting how they use and evaluate their urban environment, to participatory interventions and programs that apply the children's ideas and energies to local environmental improvements. (For a full report of this work, see Salvadori 2002; Chawla 2002.)

Oakland embodies the character of the United States as a nation of immigrants at a time when, globally, more people are on the move than at any previous era in human history through residential mobility, refugeeism, and internal and transnational migration. In Oakland, the public schools contain children from more than 30 different language groups, and less than half of the school population speaks only English at home. To explore the city through these children's eyes, Salvadori worked with 28 children from families of transient workers from Mexico and political refugees from Cambodia who lived in a private housing complex in the low-income neighborhood of Fruitvale.

The Oak Park housing complex turns inward, built around a courtyard enclosed by two story–high blocks of apartments and surrounded by a parking lot and streets. This enclosure is mirrored in the children's perceptions of their home and the surrounding city. When the children were asked to draw their "neighborhood," the majority drew Oak Park itself. When they were asked to identify photographs of different parts of Oakland, not one of them was able to identify any city landmark. In their drawings and interviews, not one represented routes of movement around the city. When they were asked to tell about all the places that they knew outside Oak Park, "school," the "parking lot," and some stores that they visited with their parents accounted for almost all of the answers. When they were asked to tell about the farthest place that they had been in Oakland, "don't know" was the most frequent answer, followed by a few Oakland locations, and

TABLE 18.2. Answers given by 9–14-year-olds living in the Fruitvale neighborhood of Oakland, asked about the furthest place that they had been in Oakland.

What is the furthest place that you have been in Oakland?

Don't Know

Lake Tahoe, Nevada
San Leandro, California
Oakland Zoo
Church
Jack London Square

Pittsburgh, California, Nevada, Lucky Store, San Francisco, Boston, Canada, Washington, Reno, 98th Avenue, Bay Bridge, Toys R Us, Cambodian store, Alameda, a store we drive to

$N = 28$, 13 M / 15 F.
In ascending order of size, each font represents answers given by one, two, or three children, respectively.

answers which suggested that a number of the children did not know what "Oakland" meant. Their answers to this question are shown in Table 18.2, which indicates that their concepts of "Oakland" included Lake Tahoe, Reno, San Francisco, Canada, and New York.

Beyond the boundaries of their apartments, their building staircases, terraces, and courtyard—which were locations of intense social interaction and play—the children described geographies of fear. The streets were dangerous and belonged to older kids, drunks, and the police. In local stores, they feared that they were at risk of being robbed or kidnapped. If they walked to the local creek, they might slip on the banks and drown in the water. When the children were asked who forbid them to go to these places, they said, "Mom," "Grandmother," "My teacher." When they were asked whether something bad had ever actually happened in these places, most of them answered, "No" or "I don't know."

These fears are typical of those harbored by many children in the contemporary United States (Buss 1995; Louv 1990); but these projections of fears and judgments onto the physical environment distort the real dimension of problems and prevent children from considering the environment as a realm over which they can have control. As a first line of action against this powerlessness, the children in Oak Park were asked to participate in a design process that made their courtyard a site that they could reimagine and transform. Through focus group discussions, a collage exercise that overlapped new images on black-and-white pictures of the existing space, field trips to green spaces around the city, and pictures of playgrounds around the world, the children discovered that the ground is not always paved with concrete and trees are not always fenced. In the end, the children built three-dimensional models of their ideas, agreed on one plan, and presented it to their parents.

A small community garden, which was part of the plan, now grows in Oak Park, and fences no longer surround the trees—removed during an afternoon of play by some children who were participants in the project. Was this a play experiment, or a statement about their new relationship to their environment? Creating the models and the vision for Oak Park probably has not enhanced these children's mobility or reduced their fear of the outside world, but it has taught them a new language that they can use in dialogues with adults, and enabled them to see new possibilities for understanding and reshaping the places where they live.

Accessible Cities

The Oak Park case study illustrates many of the forces that are keeping children in the United States and other countries to narrowly restricted spheres of urban experience—among them geographic segregation by race and class, fear of crime, fear of traffic and natural hazards, and parents' prohibitions—that are often based on the rule of television and a resulting perception of a "mean world" beyond the dwelling walls, and a lack of money to travel. To this list could be added, in many places, tightly programmed child and family schedules (Wilhjelm, 2002) and adult fears of teenagers that prompt policemen and others to force this age group to "move on" and vacate public spaces (Malone and Hasluck 2002; Percy-Smith 2002). For children who live in suburbs on the periphery of cities, urban sprawl has thinned opportunities for experience. As noted before, recent research indicates that many contemporary urban children have narrowly limited spatial and behavioral ranges.

This phenomenon has disturbing implications for learning about urban ecosystems. As the section on learning large-scale environments has shown, the more widely that children are able to travel through cities, through their own independent exploration and wayfinding, the more detailed and accurately they are able to map and model their city and identify locations of interest. The more dense their behavioral range, the more behavior settings they experience, and with them opportunities for exploring and learning about the environment and social and cultural roles. Without this vivid personal experience, a city will remain an abstraction, and as the Oak Park case study has shown, often a confused one at that. Add to these implications the research reviewed above which suggests that positive place experiences and adult mentors who affirm a place's value may be critical motivations for a sense of responsibility and care for a place, and this impoverishment of many children's urban experience is a reason for concern.

In response to this concern, this conclusion will suggest a variety of strategies to make cities more accessible and inviting for children. It builds on the principle advocated by Burch and Carrera (2002, Chapter 26 in this volume) that:

> Enhancement of a holistic ecology as experienced by actual human beings within their environment of buildings, lots, streets, parks and other infrastructure, with their associated plants, animals, air and waters, is seen as the proper place for capturing learning for adulthood. . . . Yet this venue . . . is seldom given the dignity and systematization of a learning environment.

Direct experience of city elements, when children explore their world at their own pace, is not the only important aspect of informal out-of-school learning. More accessible and visible buildings, lots, streets, parks, and other infrastructure elements can also create many opportunities for urban environmental education through nonformal community-based programs and formal school-based programs. This section will review a number of approaches to enhancing this primary environment for learning.

Child-Friendly Networks and Plans

Efforts to make cities more open and accessible for children will be most effective if they are coordinated. *Cities for Children* (Bartlett, et al. 1999), a UNICEF-sponsored handbook for municipal authorities and other city leaders interested in children's well-being, recommends that one of the first steps needs to be the creation of a local plan of action for children that brings municipal plans into compliance with the Convention on the Rights of the Child. Such a plan will require attention to the three dimensions of the Convention: protection from harm, provision for basic needs, and opportunities for children's participation in relevant local decision making and in their societies' public places and cultural life. As the convention stipulates, preparation for participation requires freedom to gain and share information. Certainly information about urban ecosystems is a necessary precondition for participation in the creation of sustainable cities, and therefore an implied part of preparing children for responsible citizenship. This information should not be conceptualized only as facts learned secondhand through books, television, videos, and computer programs, but also as direct experience with different parts of the city system.

A number of books give detailed examples of how children can be involved in drafting local action plans, including *Children's Participation* by Hart (1997), *Changing Places* by Adams and Ingham (1998), and *Creating Better Cities for Children and Youth* by Driskell (2002). Structural ways to ensure continuing attention to children and their needs include the appointment of a child ombudsman or advocate within city government, advisory councils composed of children and youth, and the creation of an advisory board to the mayor composed of influential people with an interest in children, representing different sectors such as government, nonprofit organizations, community-based organizations, private businesses, and large city institutions such as universities and hospitals. These strategies provide for integrated approaches to planning for children in cities that can address the responsibility of urban societies and institutions to children and their basic needs, on the one hand, as well as create opportunities for children to learn and care about their cities, on the other. (Chawla and Malone 2002)

Nature in the City

Until the late twentieth century, in all but the densest downtown areas, nature was an evident part of the urban fabric: not only in parks but also in backyard gardens, overgrown lots, fields to graze delivery horses, farms on the city's edge, harbors, and riverbanks (Chawla 1995; Gaster 1991). Much can be done to ensure that elements of natural ecosystems remain accessible to children. The outdoor classroom or school yard habitat movement is based on the recognition that the school yard is the main access to the outdoors for many urban children, as well as an immediately available laboratory for incorporating activities with nature into all parts of the school curriculum (Billimore, et al. 1990; Rivkin 1995). Beginning with the principle that nature is everywhere, even in the most built-up environments, the VINE program develops children's skills to learn from their existing school yard and neighborhood (Hollweg 1995). Evaluations of this program have shown that children involved in these activities express a significantly greater interest in science, and specifically the life sciences, than those not involved (Hollweg 1997). Efforts to enrich available experiences range from planting more trees and establishing gardens, ponds, bird feeders, and weather observation instruments on school grounds, to ambitious transformations of the school site to recreate the ecology of different regional biomes (Moore and Wong 1997). When these efforts are combined with the mission of making the school yard a year-round community resource, they can profoundly enrich children's opportunities to learn about the natural world. Harvey (1989/90) has shown that increases in the number of plant species and nature activity centers in school yards can significantly increase children's botanical knowledge and their inclination to value and protect the natural environment.

People have also tried to ensure that children have opportunities to learn about natural cycles through city farms and urban gardens. The city farm movement preserves or recreates small farms within urban residential districts, as a field trip location for schools and as a green space for local residents (Perez, et al. 1981/82). There children can help in caring for domestic animals, and can witness animal births, growth, and deaths. Some farms make a deliberate effort to involve children in cycles such as growing and cutting hay, feeding animals, clearing stables, composting, and spreading compost on the fields. Many urban gardens have programs for children that also seek to instruct as well as engage (Moore 1995).

Much can be done in urban design and landscaping to create "nearby nature" in urban residential neighborhoods (Francis 1984/85; Kaplan, et al. 1998). On the scale of the housing site, some areas can be allowed to grow wild, where children can play with earth, water, and vegetation and observe insect and other animal life. Rainwater can be allowed to run off through channels of rocks and plantings rather than being immediately culverted into drainage pipes. Green spaces of different sizes can be linked through greenways that provide routes for urban wildlife as well as people (Flink and Searns 1993). Research by Taylor and others (1998) has shown that children in more vegetated inner city areas have significantly greater access to adults and to play, including creative play.

Transportation Pathways

In the most livable cities, people can travel by foot, bicycle, skateboard, or roller blades as well as by car, and these alternative pathways are kept separate from motor routes. Safe and well-used pathways of this kind are particularly important in providing children with independent mobility to different urban points of interest. These pathways are sometimes combined with markers that relate the natural, architectural, and social history of the adjacent area. In a low-income community of Melbourne, Australia, characterized by typical levels of fear of crime in the street, creating a recreational trail of this kind was a high priority for young adolescents as a way of increasing pedestrian activity and "eyes on the street" (Malone and Hasluck 2002). Van Andel (1984/85) has mapped increases in children's use of local streets when various design techniques are used to slow traffic in residential areas.

Southworth (1990), who proposed that every city should have an Urban Service for Children, in the sense of a citywide program of travel services, argued that it is necessary to strip away cognitive as well as physical barriers between the city and children. This includes making urban places and events known to children and helping children travel to them. His ideas include: street corner information kiosks with computer terminals or pictorial maps that indicate where people are and nearby places of interest; illustrated trailblazer maps or computer programs that lead people through significant points of history, ecology, geology, architecture, industry, and commerce; a child-designed travel guide or newspaper; a "passport" with child and youth discounts for entry into points of interest; an "AAA for kids" with counters at key points around the city; part-time jobs for high school and college students as members of an "urban service corps" who can give instructions on public transportation to significant places and take younger children on urban field trips; and special buses to take children on weekend excursions around the city.

Urban Studies and Ecology Centers

The preceding ideas are ways to increase children's mobility in cities and their access to points of interest, the natural environment of their region, and cycles of nature. Carr and Lynch (1968) and Wurman (1971) have given examples of numerous ways to make built systems more visible as well. Wurman's suggestions for two-dimensional and three-dimensional ways to illustrate urban elements such as transportation systems, land patterns, and public buildings are effective visual techniques that can be incorporated into geographic information systems; and anticipating the era of the personal computer, he recommended that each city should have an "urban data center" that links small systems, such as school-based centers, into a collective data system that creates an "urban observatory" for monitoring urban conditions.

Wurman's suggestion combines the contemporary capabilities of geographic information systems with the facilities of urban studies centers and ecology centers. As Bishop, Kean, and Adams (1992) have noted in their history of urban studies centers and urban ecology centers, it is a well-grounded tradition that is ready for revival. With a base in a school, university, museum, nonprofit organization, or other city institution, these centers bring together advocacy planners who are seeking ways to reach out to communities, with teachers, artists, community environmental leaders, and others who are also seeking to understand and improve urban conditions. Through a central staffed location, the center serves as a storehouse of ideas, materials and resource networks for scholastic and out-of-school programs. These centers serve as "incubators" to collect, disseminate, and facilitate ideas to improve understanding of natural and built elements of the urban ecosystem, and to bring people together across disciplines to engage children and other community members in not only learning about but also improving the local environment.

Wurman (1971) argued that rather than just studying cities from within schools, the city itself should be thought of as a schoolhouse and environment for learning. "If we can make our urban environment comprehensible, observable, and understandable," he wrote, "we will have classrooms with unlimited windows on the world." In that case, he added, "We should create the kind of confidence in a student that will enable him to judge and develop criteria that might be used in the evaluation and creation of his own environments." When children learn to judge and act on criteria that create cities that conserve and restore natural resources and provide institutions and services for the well-being of all inhabitants, this, ultimately, is what an understanding of urban ecosystems is for.

References

Adams, E., and S. Ingham. 1998. *Changing places*. The Children's Society, London.

Anderson, J., and M. Tindall. 1972. The concept of home range, in W.J. Mitchell, ed. *Environmental design: research and practice*. Proceedings of EDRA 3. Environmental Design Research Association, Washington, DC.

Barker, R.G., and H.F. Wright. 1955. *Midwest and its children*. Harper and Row, New York.

Bartlett, S., R. Hart, D. Satterthwaite, X. de la Barra, and A. Missair. 1999. *Cities for children*. Earthscan Publications/UNICEF London, UK.

Bishop, J., J. Kean, and E. Adams. 1992. Children, environment and education. *Children's Environments* 9(1):49–67.

Billimore, B., J. Brooke, R. Booth, and K. Funnell. 1990. *The outdoor classroom*. Building Bulletin 7. Her Majesty's Stationery Office, London, UK.

Blakely, K. 1994. Parents' conceptions of social dangers to children in the urban environment. *Children's Environments* 11(1):16–25.

Burch, W., and J. Carrera. 2002. Chapter 26 in A. Berkowitz, K. Hollweg, and C. Nilon, eds. *Understanding urban ecosystems*. Springer-Verlag, New York.

Buss, S. 1995. Urban Los Angeles from young people's angle of vision. *Children's Environments* 12(3):340–351.

Carr, S., and K. Lynch. 1968. Where learning happens. *Daedalus* 97:1277–1291.

Chawla, L. 1995. Revisioning childhood, nature, and the city. Pages 101–108 in K. Noschis, ed. *Children and the city*. Comportements, Lausanne.

Chawla, L. 1998. Significant life experiences revisited. *Journal of Environmental Education* 29(3):11–21.

Chawla, L. 1999. Life paths into effective environmental action. *Journal of Environmental Education* 31(1):15–26.

Chawla, L., ed. 2002. *Growing up in an urbanising world*. Earthscan Publications/ UNESCO, London, UK.

Chawla, L., and K. Malone. 2002. Neighbourhood quality in children's eyes. Pages 118–141 in P. Christensen and M. O'Brien, eds., *Children in the City*, Routledge Falmer, London, UK.

Cohen, S., and R. Cohen. 1985. The role of activity in spatial cognition, in J. Wohlwill and W.S. Van Vliet, eds. *Habitats for children*. Lawrence Erlbaum, Hillsdale, NJ.

Driskell, D. 2002. *Creating better cities with children and youth: a manual for participation*. Earthscan Publications, London, UK.

Flink, C.A., and R.M. Searns. 1993. *Greenways*. Island Press, Washington, DC.

Francis, M. 1984/85. Children's use of open space in Village Homes. *Children's Environments Quarterly* 1(4):36–38.

Fuhrer, U. 1998. Behavior settings as vehicles of children's cultivation. Pages 411–434 in D. Gorlitz, H.J. Harloff, G. Mey, and J. Valsiner, eds. *Children, cities and psychological theories*. Walter de Gruyter, Berlin.

Gaster, S. 1991. Urban children's access to their neighborhood. *Environment and Behavior* 23(1):70–85.

Gibson, J.J. 1979. *The ecological approach to visual perception.* Houghton-Mifflin, Boston.

Golledge, R.G. 1978. Learning about urban environments, in T. Carlstein, D. Parkes, and N. Thrift, eds. *Timing space and spacing time.* Edward Arnold, London, UK.

Golledge, R.G. 1987. Environmental cognition, in D. Stokols and I. Altman, eds. *Handbook of environmental psychology.* Wiley, New York.

Hart, R. 1979. *Children's experience of place.* Irvington Press, New York.

Hart, R. 1981. Children's spatial representations of the landscape. Pages 195–233 in L.S. Liben, A.H. Patterson, and N. Newcombe, eds. *Spatial representation and behavior across the life span.* Academic Press, New York.

Hart, R. 1997. *Children's participation.* Earthscan Publications, London, UK.

Hart, R., and G.T. Moore. 1973. The development of spatial cognition, in R.M. Downs and D. Stea, eds. *Image and environment.* Aldine, Chicago.

Hart, R., and S. Saegert. 1978. The development of sex differences in the environmental competence of children, in M. Salter, ed. *Play: anthropological perspectives.* Leisure Press, Cornwall, New York.

Hart, R., S. Iltus, and R. Mora. 1991. *Safe play for West Farms.* Children's Environments Research Group, City University of New York Graduate Center, New York.

Harvey, M. 1989/90. The relationship between children's experiences with vegetation on school grounds and their environmental attitudes. *Journal of Environmental Education* 21(2):9–15.

Heft, H. 1983. Wayfinding as the perception of information over time. *Population and Environment* 6:133–150.

Hillman, M. 1997. Children, transport and the quality of urban life. Pages 11–23 in R. Camstra, ed. *Growing up in a changing urban landscape.* Van Gorcum, Assen.

Hines, J.M., H.R. Hungerford, and A.N. Tomera. (1986/87). Analysis and synthesis of research on responsible environmental behavior. *Journal of Environmental Education* 18(2):1–8.

Hodgkin, R., and P. Newell. 1998. *Implementation handbook on the Convention of the Rights of the Child.* UNICEF, New York.

Hollweg, K. 1995. *Volunteers teaching children.* North American Association for Environmental Education, Washington, DC.

Hollweg, K. 1997. *Are we making a difference?* North American Association for Environmental Education, Troy, OH.

Hungerford, H.R., and T.L. Volk. 1990. Changing learner behavior through environmental education. *Journal of Environmental Education* 21(3):8–21.

Kahn, P.H., Jr. 1999. *The human relationship with nature.* MIT Press, Cambridge, MA.

Kaplan, R., S. Kaplan, and R.L. Ryan. 1998. *With people in mind.* Island Press, Covelo, CA.

Lewin, K. 1936. *Principles of topological psychology.* Harper, New York.

Louv, R. 1990. *Childhood's future.* Houghton Mifflin, Boston.

Lynch, K., ed. 1977. *Growing up in cities.* MIT Press, Cambridge, MA.

Malone, K., and L. Hasluck. 2002. Australian youths, in L. Chawla, ed. *Growing up in an urbanising world.* Earthscan Publications/UNESCO, London, UK.

Matthews, H. 1992. *Making sense of place.* Barnes & Noble Books, Savage, MD.

Maurer, R., and J.C. Baxter. 1972. Images of neighborhood among Black, Anglo- and Mexican-American children. *Environment and Behavior* 4:351–358.

Moore, R.C. 1995. Children gardening. *Children's Environments* 12(2):222–233.

Moore, R.C., and D. Young. 1978. Childhood outdoors. Pages 83–130 in I. Altman and J. Wohlwill, eds. *Children and the environment*. Plenum Press, New York.

Moore, R.C., and H. Wong. 1997. *Natural learning*. MIG Communications, Berkeley, CA.

Percy-Smith, B. 2002. Contested worlds. Pages 57–80 in L. Chawla, ed. *Growing up in an urbanising world*. Earthscan Publications/UNESCO London, UK.

Perez, C., R. Hart, L. Rivlin, S. Jeffers, and L. Chawla, eds. 1981/82. *City farms*. Special issue of Childhood City Newsletter, No. 26. Center for Human Environments, City of New York Graduate Center, New York.

Piaget, J. 1963. *The origins of intelligence in children*. W.W. Norton, New York.

Piaget, J., and B. Inhelder. 1967. *The child's conceptions of space*. Routledge and Kegan Paul, London, UK.

Rivkin, M. 1995. *The great outdoors*. National Association for the Education of Young Children, Washington, DC.

Salvadori, I. 2002. Between fences. Pages 183–200 in L. Chawla, ed. *Growing up in an urbanising world*. Earthscan Publications/UNESCO, London, UK.

Schoggen, P. 1989. *Behavior settings*. Stanford University Press, Stanford, CA.

Scholnick, E.K., K. Nelson, S.A. Gelman, and P.H. Miller. 1999. *Conceptual development: Piaget's legacy*. Lawrence Erlbaum Associates, Mahwah, NJ.

Siegel, A.W., and S.H. White. 1975. *The development of spatial representations of large-scale environments*, in H.W. Reese, ed. Advances in child development and behavior, Vol. 10. Academic Press, New York.

Southworth, M. 1970. *An urban service for children based on analysis of Cambridgeport boys' conception and use of the city*. Doctoral dissertation, MIT, Cambridge, MA.

Southworth, M. 1990. City learning: Children, maps, and transit. *Children's Environments Quarterly* 7(2):35–48.

Tanner, T. 1998. Choosing the right subjects in significant life experiences research. *Environmental Education Research* 4(4):399–417.

Taylor, A.F., A. Wiley, F.E. Kuo, and W.C. Sullivan. 1998. Growing up in an inner city. *Environment and Behavior* 30(1):3–27.

Tolman, E.C. 1948. Cognitive maps in rats and men. United Nations. 1992. Agenda 21. New York: United Nations Publications. *Psychological Review* 55:189–208.

UNCHS (United Nations Center for Human Settlements). 1996. *An urbanizing world*. Oxford University Press, Oxford.

UNCHS. 1997. *The Istanbul declaration and the Habitat agenda*. UNCHS, Nairobi.

United Nations Population Division. 1998. World urbanization prospects: The 1996 revision. United Nations, New York.

Valentine, G. 1997. Gender, children and cultures of parenting. Pages 53–78 in R. Camstra, ed. *Growing up in a changing urban landscape*. Van Gorcum, Assen.

Van Andel, J. 1984/85. Effects on children's outdoor behavior of physical changes in a Leiden neighborhood. *Children's Environments Quarterly* 1(4):46–54.

Van Vliet, W. 1983. Exploring the fourth environment. *Environment and Behavior* 15(5):567–588.

Webley, P. 1981. Sex differences in home range and cognitive maps in eight-year-old children. *Journal of Environmental Psychology* 1:293–302.

Whiting, B.B., and C.P. Edwards. 1988. *Children of different worlds*. Harvard University Press, Cambridge, MA.
Wilhjelm, H. 2002. Large but not unlimited freedom in a Nordic city. Pages 161–182 in L. Chawla, ed. *Growing up in an urbanising world*. Earthscan Publications/ UNESCO, London, UK.
World Commission on Environment and Development. 1987. *Our common future*. Oxford University Press, Oxford.
Wurman, R. 1971. *Making the city observable*. MIT Press, Cambridge, MA.

19 "Ecological Thinking" as a Tool for Understanding Urban Ecosystems: A Model from Israel

SHOSHANA KEINY, MOSHE SHACHAK, and NOA AVRIEL-AVNI

To understand what we mean by "Ecological Thinking" (ET) we shall first follow briefly the evolution of the concept of "ecology." Historically, the term was coined by its founders *as a science that integrates the relationship among plants, animals, and their abiotic environment into a system of interactions* (Likens 1992). Over time, however, the concept was extended as a scientific discipline that studies the biotic–abiotic interactions at various levels of organization, developing such subdisciplines as population ecology, community ecology, ecosystem ecology, and landscape ecology. Though emphasizing a different level of organization, each subdiscipline uses a unified ecological approach (Allen and Hoekstra 1992). Once ecology was adopted by social scientists, such subdisciplines as human ecology, social ecology, and ecological economics developed (Hawley 1986; Machilis, et al. 1997). We contend that in spite of the diversity of use of the term ecology both in the biophysical and sociocultural sciences, scientists in both fields share a common "ecological perspective" for understanding biophysical and/or sociocultural phenomena or systems (Pickett, et al. 1994), which we term "Ecological Thinking" (ET).

Lately, a new ecological paradigm has emerged that emphasizes the openness of ecological systems to unexpected or episodic external influences; in other words, that systems can be subject not just to internal regulating factors, but to external factors as well. Accordingly, disturbances are regarded as important influences on ecological systems (Pickett and White 1985) rendering them rarely at equilibrium (Davis 1986). Moreover, *humans are common ecological agents of disturbance* (Turner, et al. 1990). The relatively recent recognition of humans as an important component of ecological systems leads us to regard ET *as a conceptual framework and a world view for understanding, management, and participation in complex natural and human-made systems embedded in the biosphere.*

In this chapter we suggest that Ecological Thinking should be a key goal of education. As compared with "critical thinking," "system thinking," and so on, all of which as we see them reside in the cognitive domain, ET involves the learners' value judgments, emphasizing their double role as both actors within the system and as reflectors of the system, aware of and responsible for their understanding and their actions. This mutual relationship between the individual or learner, and the context or learning environment, is further elaborated by Davis and Sumara (1997) who recast the cognizing agent as *part* of the context. "As the learner learns the context changes, simply because one of its components changes. Conversely, as the context changes, so does the very identity of the learner" (Davis and Sumara 1997). This notion of intertwined connectedness is congruent with our contemporary notion of ecology, where the environment evolves simultaneously with the species that inhabit it, and forms the basis of our pedagogical orientation. We suggest adopting ET as a keystone concept in ecological education and more specifically in urban ecosystem education, in order to provide students with tools for understanding, managing, and participating in the complex systems in which they live.

In summary, ET represents a challenging new direction for ecology education that stems from the following three assertions: (1) people's actions shape ecosystems, both because we are components within most ecosystems, and because we are important causers of their negative as well as their positive changes; (2) people's values shape their interactions with the world, through both our mental conceptualizations of reality as well as our physical actions; and (3) these parallel notions can be brought together by ET, an educational approach based on a cybernetic philosophy, recognizing and building on our dual roles in ecosystems.

Ecological Thinking in "Sayeret Shaked": A Case Study

To acquaint the reader with ET as basic to our new pedagogy, in particular with regards to urban ecosystem education, we choose to open with a case study that illustrates our pedagogy and more specifically, how we teach ET in a simple system with dominant plant, water, and soil components. We then present a case study of teaching about a small city.

Sayeret Shaked Park is an experimental site run by a team of scientists and environmental managers and supported by the JNF (Jewish National Fund). Their aim is to study the structure and function of the system in order to suggest techniques for ecological sustainability and proper system management (Shachak, et al. 1998). Situated in a semi-arid zone of 200 mm annual average rainfall, and as a result of thousands of years of grazing, it is predominately covered by a dwarf shrub plant community. Demonstration stations for the study of the area were established recently for teachers and university as well as high school students, where they can learn and experience ET.

The study begins with a presentation of a simple system diagram of the complex area, containing three interrelated components: organisms, resources, and landscape. Each component actually represents a subsystem, or a web of relationships. Thus, the aim of the study is to understand a web of webs. Starting with landscape, the first concept introduced is that of the "landscape patch," and the students identify the dispersed small shrubs on the slope area, with open spaces between them, as such patches. This triggers a discussion of the landscape patch as a system-unit, revealing important relationships between the system's components.

To study these relationships, the students' first activity is to simulate rain over "ecological-cells." These are different experimental units, each 100 cm × 50 cm, which are bounded off by plastic frames and consist of shrub and surrounding open areas, in different proportions. Measurements of input/output water indicate that about 40 percent of the "rain" does not penetrate into the open area soil but flows over as runoff, watering neighboring shrubs in cells where they occur. This small experiment is then translated into system language, and the water flow chain is drawn (Shachak, et al. 1995) (Figure 19.1, bottom portion).

The diagram leads the instructors to ask new questions, such as, *What makes the shrub a "sink" and the open space a "source" of water?* This leads the students to examine the patches more carefully, and they move from one station to another according to the nature of their specific questions. In station 1, students measure the accumulation rate of the airborne dust in different cells. They find accumulation in mounds mainly under shrubs, and their calculation yields the age of the mounds to be about a hundred years. They may then investigate the further consequences of these mounds. Because mounds act as sinks for runoff water and eroded soil particles from open areas, and shade from the shrub creates a special microclimate, these mounds are optimal areas for other organisms such as annual plants and small animals. Through their direct observations, students find that the open spaces are covered by microphytes (*Cyanobacteria*), which are responsible for the crust formation that prevents the infiltration of water into the soil. The dynamic relationship between crust patches and shrub patches is discussed, as well as the idea of organisms (in this case shrubs and microphytes) as "Ecosystem Engineers" (Jones, et al. 1994). They are now able to add the soil flow chain (Figure 19.1, top portion).

In station 2 students investigate snail populations, which are abundant upon and around the shrubs. They learn that snails actually feed on the crust *Cyanobacteria*, which are rich in nitrogen, and accordingly form a new hypothesis: Snail excretions act as fertilizer in the mound. In station 3, they examine the effects of fertilizers upon the abundance and variety of annual plants, and learn that nutrients are indeed a limiting factor in the Sayeret Shaked soil. In station 4 they study the shrub patch as shelter for seeds against heavy grazing, thereby acting as a "source" of seeds for the crusted patches. Students are encouraged to apply and demonstrate their understanding by adding to the diagram shown in Figure 19.1 and to incorporate a flow chain labeled "seeds." In this way, they illustrate the role and interaction of seeds in the ecosystem. This often triggers management questions such as *If we created pits, could runoff water be generated by the crust and accumulated in these pits to be used in order to grow trees?*

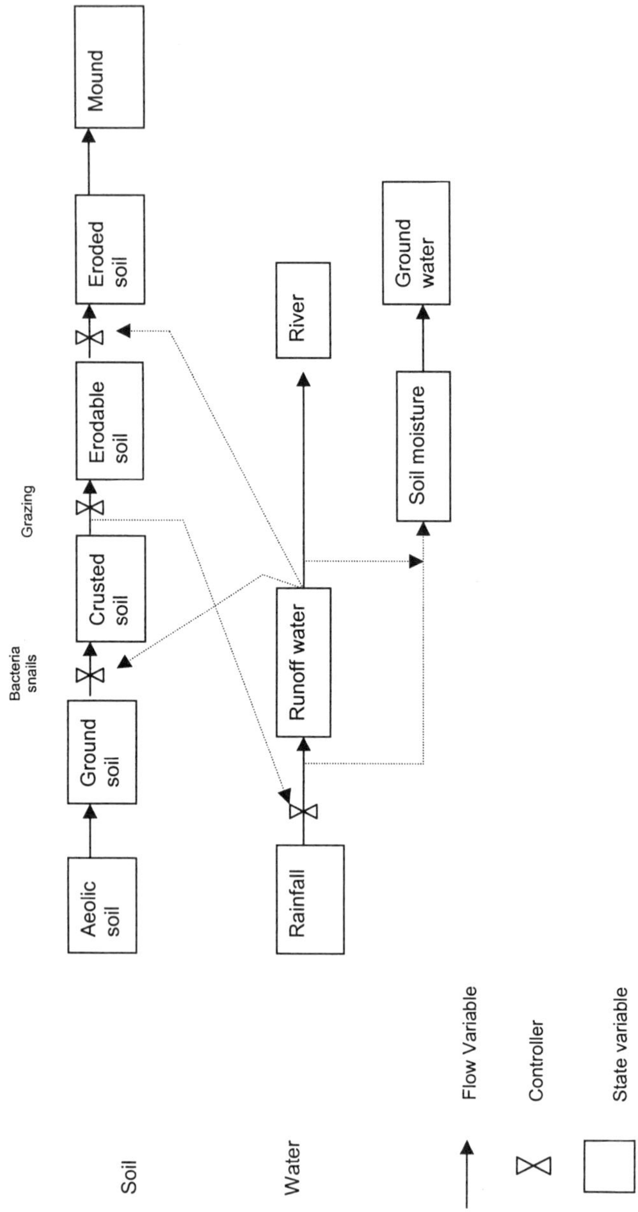

FIGURE 19.1. Water and soil flow chains.

The Sayeret Shaked case study was chosen to exemplify three aspects of ET. First, it is based on the real problem of desertification, an authentic phenomenon of the area that has been accelerated by people's negative interaction with the ecological system in the past (i.e., over-grazing). Second, it demonstrates how to introduce learners to the basic concepts and principles of ecosystem thinking. Third, as a value laden national project that aims to cope with the country's focal problem of how to inhabit the desert, or more specifically, how to reverse the process of desertification and move towards "savanization," it suggests a possibly sustainable solution. The planted trees could grow into a naturally watered park that would function as a green lung as well as a recreation area for the neighboring urban inhabitants.

Mizpeh-Ramon—A Developing City in the Desert: A Case Study

Having conceived ET as a new perspective or thinking tool, our next question is whether ET can be applied to the study of urban ecosystems and help us cope with complex human systems. To answer the question we choose again to revert to the concrete and use another case study to illustrate our pedagogy with respect to urban ecosystem education.

Mizpeh-Ramon is a small town built on the rim of the 40 km × 10 km Ramon Crater, in the Negev desert. Overlooking a most picturesque landscape, it is actually a part of the Nature Reserve Park Ramon. A curricular module was developed specifically for high school teachers, to prepare them to teach ET in their respective classrooms in Mizpeh-Ramon (Avriel and Keiny 1998).

Assuming the learners are already acquainted with basic ecological concepts due to their involvement in the Sayeret Shaked activities, the teacher introduces the idea of urban ecosystems and draws a general model of a city containing the three main subsystems: the ecological, the social, and the economical (see Figure 11.1 in Grove, et al. 2002, chapter 11 in this volume; Pickett, et al. 1997). The *intentionality* of man-made urban ecosystems as compared to natural ecosystems is emphasized, with their aim being to supply goods and services to their citizens. Thus, the focal question that stems from the urban ecosystem model is *How can the goods and services be increased to the citizens of Mizpeh-Ramon?*

To cope with the question, students are sent in teams to survey the small city and collect relevant data. Aided by maps, they carry out observations and interview some chosen key figures, with the aim of learning about the different subsystems. For example, they investigate the city's ecological elements (water, air, and energy); its economic assets (*How do the inhabitants of Mizpeh Ramon make a living?*); and the sociocultural component of Mizpeh-Ramon (municipal services, educational, and cultural institutions, etc.). Back in the classroom, each team debriefs to the plenary. A discussion guided by the teacher highlights the controversies that arise from looking at the city from these different perspectives, disclosing the interdependence between the different aspects they have investigated. This is the time to expose the participants to the history of the area (through archeological documents), which supplies evidence of the changes that have occurred over time. Translated into system language, these temporal changes construct new models, highlighting keystone concepts.

For example, the concept of landscape patches, introduced on a small scale in Sayeret Shaked, is generalized to the idea of a source-sink relationship where materials and/or organisms move from one area to another. New questions are generated, such as, *Can we identify a network of patches in every natural landscape mosaic that functions as a source–sink relationship?* Or, referring to urban ecosystems, *Could this same abstraction (i.e., source–sink relationship) be further applied to human-made ecosystems such as cities?* Deliberations lead the students to understand the connections between the town of Mitzpeh-Ramon and the neighboring crater as a source–sink relationship, which could be investigated by examining the flow of people, money, new enterprises, and so on.

Their next move was to study the *changes in* patchiness and source-sink relations that occurred between the city and the crater within the last 50 years. Starting in the 1940s, before the town was founded, the whole area was part of a single, more natural ecosystem with very different source-sink relationships. Between 1950 and 1980 Mizpeh-Ramon was a mining town, the crater serving as a source of minerals such as phosphates, silt and manganese. Today, Mizpeh-Ramon is mainly a tourist town, with the crater the main source of attraction and as such, something that should be rehabilitated and preserved. "Patch dynamics" is explored in the survey of nearby Avdat, the reconstructed archeological site of an ancient town of Nabbatians (nomads of Arab origin who inhabited desert areas during 700 B.C.–100 A.D.), throwing new light on the relations between man-made and natural ecosystems. A comparison between the ancient and the modern urban ecosystem shows that in both cases the urban ecosystem is dependent on outsiders: on farmers and merchants, who brought their products to the regional center Avdat, and on tourists attracted to the Mitzpeh-Ramon crater.

Having become familiar with the new ecological language, students can now use it as a thinking tool for coping with authentic questions they generate themselves. This final part of the module is characterized by a significant shift towards a higher-order learning level: from *conceptualization* or understanding, towards *contextualisation*, or taking responsibility for decision making. By inquiring into authentic, open-ended questions they themselves have generated, questions that relate to the relationship between the city and the surrounding desert, the students become genuinely involved in the process of decision making. Their value systems and ethical priorities become more explicit and the students become more aware of their decisions and begin to feel more responsible for the solutions they design. Two examples of students' inquiry studies illustrate how this final stage was practiced, by individuals or in small teams.

One team chose to study water as the most important resource in the desert. Their question was: *What are the sources of water available to a city in the desert?* Investigating the situation some 2000 years ago, they learned the Nabbatian's ingenious methods of collecting runoff water for all their needs. By developing runoff water agriculture they managed to grow their principal food, wheat, as well as fruit. The large and numerous winepresses in Avdat and its vicinity are clear evidence to the amount of grapes yielded. Thus, in a desert area where the average precipitation is 100 mm/year, the Nabbatians grew grapevines, which normally require 500 mm/year. In modern times with the population growth, Israel's urgent need was to attract people to settle in the Negev desert, which occupies two thirds of the country's total area. A national pipeline was laid from the country's main watershed, the lake of Galilee (a natural lake along the River Jordan in northern Israel), to the Negev that enhanced the cultivation of the desert. Today, with a population of 5–6 million, this water resource proves insufficient and new water sources are being exploited, such as recycling wastewater and desalinizing seawater. Figure 19.2 shows these three ways to boost the amount of available water, corresponding to these three eras in the Negev's history. Reflecting the students' understanding and knowledge

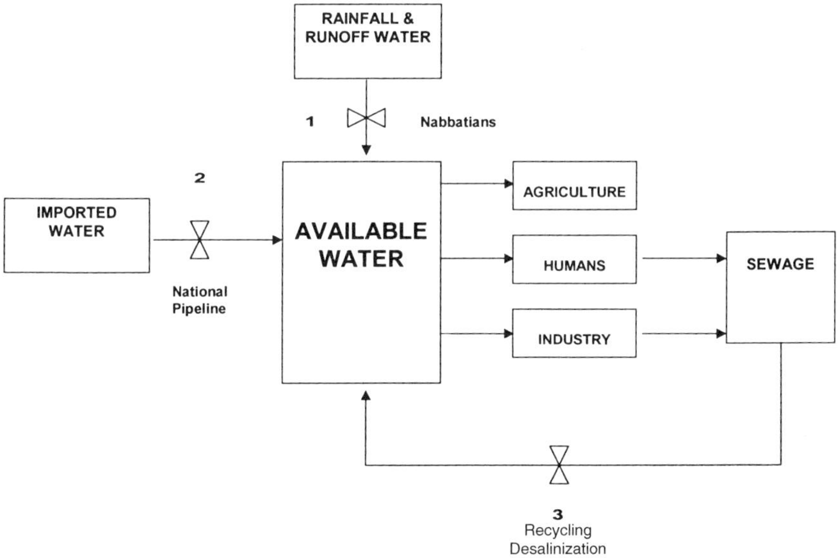

FIGURE 19.2. Changes in water supply to desert settlements.

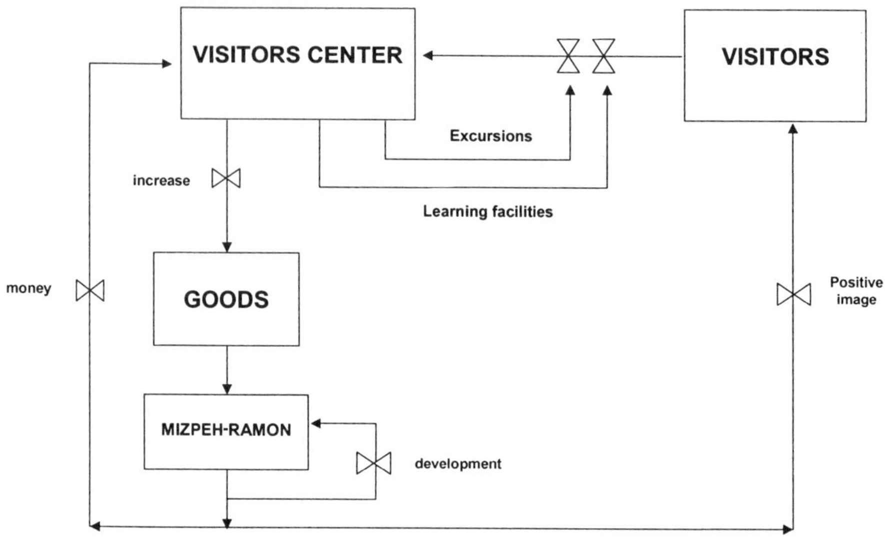

FIGURE 19.3. An urban ecosystem of Mizpeh-Ramon as a tourist town.

construction, the model triggers deliberations on Israel's developing conception of urban settlements in the desert, which go far beyond this specific case study.

The second example relates to the question of *How can the function of the "visitors'-center," situated as an observation spot overlooking the magnificent view of the crater, be improved?* A survey of the crater raises the students' awareness of its great potential as a learning resource. Its exposed geological morphology makes it suitable for the study of geology. A sculpture garden along its ridge, containing pieces of sculpture made by artists from different parts of the world, indicates the place's potential to stimulate artistic creativity. The team decided to plan a new brochure for the visitors' center that will inform potential visitors of an extensive learning program, including excursions in the crater.

Figure 19.3 represents the students' answer to the question in terms of a system diagram. It shows that as more visitors are attracted, more "goods" are gained. What are these goods: material or spiritual? How can they be used wisely in order to further develop the city? How can more visitors and maybe more settlers be attracted? These are some of the questions that require the team to make decisions. It is pertinent to emphasize that there is no one correct answer. The learners become involved in ethical discussions, and engage in deliberations that disclose their emotions and value priorities, which are usually tacit. The process thus heightens their awareness of their beliefs or conceptions as sources of their judgments. We believe that such unique learning experiences, that engage the learners both as insiders (inhabitants of Mitzpeh-Ramon) and as outside reflectors who understand the relationships within and without their urban ecosystem, will be significant in terms of students' personal growth and development. Our hope is that the impact of ET would shape their understanding, values, and action as responsible citizens who feel committed to the welfare of their environment, and actively involved in its sustainable management.

The New Pedagogy

Let us now try to characterize the learning activity in both case studies in order to formulate our pedagogical principles. Beginning with an authentic concrete phenomenon, the learners generate questions. With the teacher's assistance these are turned into researchable questions, which send the participants to collect relevant data. This is an example of *inductive learning*, where data from different sources (empirical and descriptive, theoretical and concrete, etc.) are being collected. Inductive learning leads to abstraction or conceptualization, which results in the construction of a theoretical model of a system. A higher-order level of understanding is achieved, namely, a higher-order interpretation of the particular situation. For example, consider the flow chains in Sayeret Shaked (Figure 19.1) or the dynamic relationship between Mitzpeh-Ramon city and the crater. As a result of abstraction, a wider scope of vision is gained which becomes a platform for generating new focal questions of a systemic nature (e.g., How can we grow trees in the semiarid environment? How can more people be attracted to live in the desert?). These are meaningful questions that stem from the learners' reference point. In other words, the study becomes student-oriented in the sense of being initiated by the learners' open-ended questions that emerged from their new interpretation of the situation. Each question defines the boundaries of the context. Oers termed this "activity as context" (Oers 1998). Accordingly, agent (subject, i.e., the learner) and object (i.e., the ecosystem) are not separate entities, but are dialectically related, mutually defining each other in the human activity. How a person acts on objects demonstrates how he or she stands in the world. Thus our pedagogy has not merely moved from being teacher-oriented to student-oriented, but from *conceptualization* to *contextualization*, which we assert is a higher level of learning. Initiated by the student, it reflects his or her reference point or field of interest. The learners' emotions, values, and deep beliefs, and not merely cognitive arguments, are recruited in the process of problem resolution. We believe this is what Foerster meant by coining the term "second order cybernetics" (Foerster 1992). This idea will be dealt with in the final part of the chapter, in which we expose the theoretical basis of our new pedagogy and then return to ET and the ecological framework.

Introducing Cybernetics as the Underlying Conception

The term "conception" will be used in this chapter interchangeably with *paradigm*, *worldview*, *ideology*, or *philosophy*. All of these terms convey a notion of a publicly shared context in which a body of knowledge develops. It consists of schemes of concepts, developed as a result of actions and interactions with the world. Anchored in the person's beliefs and basic assumptions, conceptions also influence our actions. The role of teachers' conceptions are part of this general structure, reflecting their beliefs and basic assumptions about learning and teaching, the goals of education, and what counts as knowledge (Keiny 1996).

Three perspectives, the ontological, the epistemological, and the ethical, characterize every conception. To introduce the cybernetic conception, we shall use the three perspectives as criteria to compare the cybernetic with the predominant positivistic conception.

The Ontological Perspective: What Is the Nature of the World?

According to the positivistic conception, the world, namely objects, events, and processes, exists independently of human perceptions and all thought or theory of them. The phenomena of nature are manifestations of the rules of nature. It is the role of the scientist to *reveal* them. In this sense, scientific theories are discovered.

According to the cybernetic conception, reality is human perception, what man has experienced only, since we have no other access to the world out there. In other words, *scientific theories are invented and not discovered*, and as such there is no one truth, or one correct representation of the world. Reality is the sum total of all the different conceptions of reality, carried by the participants in a particular context.

The Epistemological Perspective: What Accounts as Knowledge?

Knowledge, according to the positivists, is an external objective entity, consisting of structured bodies termed disciplines. Each has a distinct boundary, created by its specific concepts and methodology.

According to the cybernetic conception, knowledge is an individual construction and therefore subjective. We each construct our idiosyncratic concepts, which through social interaction, language, and discourse, "fit" but never completely "match" (Glasersfeld 1989). Knowledge, just like in Karosawa's Rashumon, is multifaceted. What is claimed to be scientific knowledge is that which the scientific community has accepted as the best interpretation of the world, leaving open edges for a possible new theory which will offer a wider explanatory power (Kuhn 1962).

The Ethical Perspective: What Is the Nature of Man and of Society?

The positivistic idea of society is highly hierarchical, distinguishing between different professional groups, between experts and nonexperts. Positivism also makes a distinct division between theory and praxis; in our case, between researchers as "knowers and thinkers," responsible for knowledge construction, and practitioners as "nonknowers," merely appliers of this knowledge (Cochran-Smith, et al. 1999).

In contrast, the cyberneticians' main principle is "interaction"—interaction in the sense of interdependence and interrelatedness between the different components of the world, in both the natural and the social contexts. Within the social setting, it means social interaction and collaboration. In this sense it is a foundation for a more democratic society, based on mutual respect toward the other and a recognition of their right to be different.

Yet, this kind of interaction is only *first order cybernetics. Second order cybernetics* relates to the relationship between humans as observers of their world and their world as the subject of their study (Maturana 1994). Humans are the only creatures aware of their and others' actions. In this respect they are both actors and reflectors on their action (Macmurray 1957). In Foerster's terms it implies taking responsibility for one's observations and interpretations of the world; in other words, for one's system of knowledge and personal conception (Foerster 1992; Maturana 1994). As compared to the positivistic values of objectivity and neutrality, the cybernetic aspired values are personal responsibility and involvement.

Ecological Thinking (ET) as the Core of Ecological Education

We hope that the reader who followed us along this journey has recognized the similarities between cybernetics and the contemporary ecological paradigm introduced above. Our main argument in this paper is that ET is a common denominator of the two, and that the cybernetic conception lends a sound philosophical and ethical basis to contemporary ecology. Embracing the natural world as including forested, rangeland, agricultural, and urban ecosystems, contemporary ecology recognizes humans as integral ecological agents. Moreover, humans' central role in the system is viewed in terms of second order cybernetics, as ones who are both actors within the system and reflectors, aware of and responsible for their knowledge and action (Keiny 1994). This double role or second order cybernetics is what we termed ET, and we suggest ET as an alternative tool for coping with environmental problems, in particular within complex urban settings.

This calls for a radical change in ecological education, one that entails a conceptual shift from positivism to cybernetics. Instead of isolating the natural environment, protecting its so-called equilibrium against humans' scientific and technological interventions, human interaction and involvement should be viewed as the fundamental principle of ecological education. Values such as personal responsibility and participation should become our primary goal of education. Ecological education, like social ecology or ecological economics, should be recognized as another human-ecological subsystem, ET being shared by them all. Stemming from the cybernetic conception, the new ecological pedagogy does not *apply* but *constructs* ecological knowledge out of the learners' direct interaction with reality.

When coping with authentic problems in complex settings such as urban ecosystems, ET as a tool implies that the learners are both insiders and outsiders. As insiders, for example as citizens of Mitzpeh-Ramon, they are actively interconnected with the economic, cultural, and natural subsystems of the city. As outsiders, the learners are self-reflective and responsible for the construction of a system model that reflects their system of knowing.

It is too early to evaluate our curricular modules to assess the congruence between our manifested goals and the learning outcomes. All we can say is that we have seen our learners become *aware* of their relationship within the system and better able to conceptualize it into a model, which could enable them to take responsibility for changing or preserving their urban ecosystem.

References

Allen, T.F.H., and T.W. Hoekstra. 1992. *Hierarchy: perspectives for ecological complexity.* University of Chicago Press, Chicago, IL.

Avriel-Avni, N., and S. Keiny. 1998. *Desert & desertification, a curriculum module.* "MATAS" publication, Ben-Gurion University press (in Hebrew).

Cochran-Smith, M., and S. Lytle. 1999. The teacher research movement: a decade later. *Educational Researcher* 28(7):15–25.

Davis, M.B. 1986. Climatic instability, time lag, and community disequilibrium. Pages 269–284 in J. Diamond and T.J. Case, eds. *Community ecology*. Harper & Row, New York.

Davis, B., and D.J. Sumara. 1997. Cognition, complexity and teacher education. *Harvard Educational Review* 67(1):105–125.

Haley, A.H. 1986. *Human ecology: A theoretical essay.* University of Chicago Press, Chicago, IL.

Jones, C.G., J.H. Lawton, and M. Shachak. 1994. Organisms as ecosystem engineers. *Oikos* 69:373–386.

Keiny, S. 1993. Teachers' professional development as a process of conceptual change. Pages 323–337 in I. Carlgren, G. Handalm, and S. Vaage, eds. *Research in Teacher Thinking and Practice*. Falmer Press.

Keiny, S. 1996. A community of learners: promoting teachers to become learners. *Teachers and teaching: theory and practice* 2(2):243–272.

Kuhn, T.S. 1962. *The structure of scientific revolutions*. University of Chicago Press, Chicago, IL.

Likens, G.E. 1992. *Excellence in ecology, 3: the ecosystem approach: its use and abuse.* Ecology Institute, Oldendorf/Luhe, Germany.

Macmurray, J. 1957. *The form of the personal.* Faber & Faber, London, UK.

Matron, H.R. 1992. Reality: The search for objectivity, or the quest for a compelling argument. In *The ASC workbook: language, emotion, the social and the ethical.* ASC Pub. Washington, DC.

Pickett, S.T.A., J. Kolasa, and C.G. Jones. 1994. *Ecological understanding: the nature of theory and the theory of nature.* Academic Press, San Diego, CA.

Pickett, S.T.A., and P.S. White. 1985. *The ecology of natural disturbance and patch dynamics.* Academic Press, Orlando, FL.

Pickett, S.T.A., and R.S. Ostfeld. 1995. The shifting paradigm in ecology. Pages 261–278 in R.L. Knight and S.F. Bates, eds. *A new century for natural resources management.* Island Press, Washington, DC.

Pickett, S.T.A., W.R. Burch, Jr., S.E. Dalton, T.W. Foresman, J.M. Grove, and R. Rowentree. 1997. A conceptual framework for the study of human ecosystems in urban areas. *Urban Ecosystems* 1:185–199.

Shachak, M., M. Sachs, and I. Moshe, 1998. Ecosystem management of desertified shrublands in Israel. *Ecosystem* 1:475–483.

Shachak, M., and C.G. Jones. 1995. Ecological flow chains and ecological systems: concepts for linking species and ecosystem perspectives. In C.G. Jones, and J.H. Lawton, eds. *Linking species and ecosystems.* Chapman and Hall, New York.

Turner, B.L., W.C. Clark, R.W. Kates, J.F. Richards, J.T. Matthews, and W.B. Meyer. 1990. *The earth as transformed by human action: global and regional changes in the biosphere over the past 300 years.* Cambridge University Press, New York.

van Oers, B. 1998. From context to contextualization. *Instruction & Learning* 8(6): 473–488.

von Foerster, H. 1992. Ethics and second order cybernetics. *Cybernetics & Human Knowing* 1(1):9–19.

von Glasersfeld, E. 1989. Cognition, construction of knowledge and teaching. *Syntheses* 80(1):121–140.

20 Systems Thinking and Urban Ecosystem Education

Gary C. Smith

Introduction

This chapter examines how systems-based holistic thinking can be used in urban ecosystem education (UEE). It is based on the author's experiences over the last six years working with teachers and students from two California schools, visiting school projects receiving Interdisciplinary Planning Grants for systems-based environmental education from the California Department of Education, and leading numerous workshops to teach educators about the use of systems thinking in environmental education. The changes in teachers and students the author and others observed appear to be so significant, and the degree of interest so high, that closely controlled research is indicated to help us understand the phenomenon. We need to more fully understand the answer to the question *Why does learning and using systems thinking have such a powerful positive impact on both teacher and student in terms of their self esteem and empowerment; their ability to learn and internalize content and processes and acquire skills; and their capacity to grasp complex issues and ideas?*

For the last few years, systems thinking has been widely discussed in the environmental education community. This may be due to the realization by educators that systems thinking can reconcile the apparent conflict in our ways of knowing: the thinking of the world into wholes (holistic) and the thinking of the world into parts (reductionistic/mechanistic). As an emerging way of "whole" knowing, the ideas and the activities that follow are used in very few American schools. Therefore, there are limited classroom activities to support this new way of thinking, and the writing of systems-focused curricula, as described here, has only just begun. This chapter will suggest that these shortages do not impede successful implementation of such instruction.

Systems-based, holistic thinking is a way of viewing the world. By using it, both teacher and student are able to bridge gaps between process and content, between abstract and real, as well as between the different subject disciplines. For students to be fully literate citizens in the future, educators will need to develop the mental dexterity in their students—and in themselves—to be able to consciously slip back and forth between reductive and systems thinking as the need warrants.

An important point to make here is that this writer has no interest in devaluing reductive thinking, only in adding systems-based holistic thinking to the bag of tools that students carry with them when they leave our schools. *Both* perspectives are essential. Both have strengths and weaknesses. So, the focus in this chapter is on helping teachers understand the value of systems thinking and how they can easily add it to their classroom instruction.

Systems thinking in this chapter does not refer to systems in a narrow scientific or mathematical modeling sense, nor would the use of this type of thinking be restricted to researchers of complexity theory and the like. Instead, it is described as an integral way of studying and interpreting the world in which complex systems are the context that frames the exploration. Experience with students and their teachers suggests that using systems thinking to study the environment has an extraordinary ability to make the chaotic and complex world we live in understandable. In learning and practicing this way of thinking and doing, students become hopeful, responsible, competent, and empowered citizens.

During the summer of 1999, Delaine Eastin, California's Superintendent of Public Instruction, gave a speech at the "Superintendent's Summit on Environmental Education" in which she asked a diverse assembly of educators to raise "connection" to a new level of visibility, application, and importance in every aspect of their work. As an example of its importance to students, she suggested that when students do not feel connected to their families, schools, communities, and peers it helps create the conditions for tragedies such as the recent killings of students, by students, in American schools. Her appeal to educate for connectivity supports a main thesis of this chapter: that connection, when brought to conscious practice through deliberate, repeated interjections into educational planning and instructional processes, is a powerful, even transformational, force in the lives of both educators and students (Eastin 1999). Eastin's request supports the argument of this chapter that some of the most fundamental literacies needed by citizens of the future are not presently in the American K–12 curriculum. Based in the power of connection, these literacies include the acquisition of new habits of mind, new sets of diverse skills, and alternate ways of perceiving and understanding.

The following quotes are from California students who have experienced instruction that is, *by design*, connection-forming:

> ISIS (Integrated Studies in Systems) has taught me how to take integration and diversity and systems thinking to the real world. Now I find myself trying to make connections in everything I do. It's a subconscious thing that happens. After you learn this way . . . it comes naturally.

Eleventh-grade student in the class of Pam Martin, the ISIS program at Lincoln High School, Stockton, California, in Lieberman, G., and L. Hoody 1998 (For more information on the ISIS Program see Martin 1999).

When Alex and I were making the graph we put down an idea and said, 'Oh! This connects with that.' So whenever you put down an idea you just think of everything we have done and it's easy to think of something. If you really think and try to remember it comes to your head. When I am in high school 5 years from now I think that it's really going to change my learning. Now I know how to say, 'We did this last week and this part is the same as that!' So now my work is much easier for me.

4th grade student in the class of Barbara Moreno and Judy Utvich, Open Charter Elementary School, Los Angeles, California

These quotations suggest that something very unusual and important is happening in the classrooms and in the lives of these two inner city students. What has transformed their experience is the use of systems thinking by their instructors—developing a strong sense of connectivity was a powerful outcome. Judy Utvich, a teacher of the fourth grader quoted earlier, substantiates the importance of systems thinking:

Since working with systems learning and systems teaching . . . things just pop out of the woodwork because your whole mind-set shifts and suddenly you see things in a different way . . . and kids are constantly bringing stuff in, "Oh, I have this article. Oh, I saw this on television," . . . because suddenly they are seeing the connections. . . . When you start looking, when you really think that everything is connected and then you start seeing evidence, there is a moment of awe and reverence.

Dr. Barbara Moreno, who team-teaches with Judy, has indicated that every aspect of her instruction has been changed by the use of systems thinking. This has led her to frequently ask herself and her students, "How is that connected?", "What is that connected to?", and "So what does that tell us—how does that information help?"

For these teachers, systems thinking has transformed the way they think, teach, and go about their everyday lives in very positive and generative ways.

Urban Ecosystem Education

Systems scientist John Holland has described how dynamic complexity characterizes cities. By describing the city in this way, he frames the challenge for those who seek to fully understand its function. To do so will require the very best of our intellect and our methods for understanding its functional processes, and it will require that we begin to study the city as a system both similar to and uniquely different from other systems.

. . . Cities have no central planning commissions that solve the problems of purchasing and distributing supplies. Nor do they maintain large reserves to buffer fluctuations. . . . How do these cities avoid devastating swings between shortage and glut, year after year . . . ? The mystery deepens when we observe the kaleidoscopic nature of large cities. Buyers, sellers, administrations, streets . . . are always changing, so that a city's coherence is somehow imposed on a perpetual flux of people and structures. Like the standing wave in front of a rock in a fast-moving stream, a city is a pattern in time. No single constituent remains in place, but the city persists (Holland 1995).

This chapter proposes that to learn to think in terms of connections and relationships is to use systems thinking and that systems thinking is a fundamental literacy of UEE, the educational study of the city as an ecosystem. From the systems perspective, to study the urban ecosystem students and teachers would expand their viewpoint beyond traditional ecological perspectives to include new perceptions, understandings, and methodologies from both systems thinking and systems science. Using these emerging ideas will help the students to more fully and accurately characterize the biological, cultural, social, and economic systems that function in urban settings. In addition, learners in UEE programs will study the commonalties and differences in the ways natural and built systems function and the relationships between them. Finally, these learners will confront the issues endemic in their communities, determine causality, and decide upon appropriate actions for restoring their communities so that they may enjoy a higher quality of life.

Using the ecosystem as a comparative conceptual framework to understand human-influenced areas will cast students and teacher into the rapidly changing, apparently chaotic, and integrated complexity that is the human community. It will stretch traditional discipline-based education and our understanding of ecosystems in new and fruitful ways. And it will require that teachers and students use a variety of disciplinary methods and "ways of knowing," which will be selected according to the needs of the situation.

UEE can become a core component of school curricula primarily by focusing on the relationships between the parts and processes of the systems of which humans are a part. The power of this context for learning can be used to establish a new, complementary root literacy and can enrich the ultimate goal of education: the development of a competency and disposition for creating sustainable communities. From this position at the core of education, UEE can take its place alongside such ideas as the pursuit of freedom, justice, and equality, which are embedded in instruction to perpetuate our way of life. No matter what our educational goals are in terms of community, systems thinking is essential to achieving them in the long term.

Systems Thinking in Education

There are two main ways that systems thinking can be used in UEE; indeed, in all of education: (1) as a mental model or metaphorical bridge between the known and the unknown—what is real and what appears to be real; and

(2) as a way of understanding and interpreting the relationships, structures, and processes that make up the complex systems in our natural and built environments.

To understand how systems thinking can serve as a metaphor, we must recognize the utility of metaphors in our lives. Most important to this discussion is their ability to help us interpret the world around us. Educators John Brown and Cerylle Moffett write that through metaphors, "meanings become unified and integrated into the underlying patterns that constitute our conceptual understanding of reality." (Brown and Moffett 1999).

As a metaphor, systems thinking is a way of interpreting the world in which the observer's point of attention is focused by the contextual lens of relationship and connection. The mental model for this metaphor is expansive thinking. The creative aspects of our mind learns to seek and follow relationships like strings across time and space: It is taught to pay close attention to relationships between subjects—and relationships between relationships. Buddhist leader Thich Nhat Hanh describes this way of knowing: "If you are a poet, you will see clearly that there is a cloud floating in this sheet of paper. Without a cloud, there will be no rain; without rain, the trees cannot grow; and without trees, we cannot make paper." (Hanh 1991) It is important to note here that just having a poetic sense of the world is not enough, but neither is having only a fact-based understanding. Hahn has connected the poetic with an understanding of how this system works. Systems thinking can provide an aesthetic perception and relationship-based understanding of urban communities and give context and meaning to a factual understanding of them. In this way, students can gain an integrated, and therefore more accurate and compelling, view of the communities in which they live.

Systems thinking also gives us conceptual tools for understanding and interpreting complexity. Through collaborative discussions at a meeting of about 40 professionals from a variety of organizations, interests, and subject disciplines, which was sponsored by the California Department of Education and facilitated by systems theorist Fritjof Capra, six primary descriptors for complex systems were identified. I have added another and also developed contextual definitions and examples (Table 20.1). These will help educators and students identify the presence of complex systems and to explore and interpret systems behavior. This modest amount of information will enable them to understand enough about systems to grasp complex phenomena such as multiple causality across time and space.

The basic foundation for systems instruction, as discussed in this chapter, was the functional criteria suggested by Capra (1994). He suggests that for students to become systems thinkers, they need to be able to shift from thinking in terms of parts to thinking in "wholes" and also to shift from thinking of structures or objects to thinking of processes (and to process thinking). These shifts address the core ideas and issues of UEE. To describe an urban ecosystem only reductively reduces its complexity to

TABLE 20.1. Systems descriptors identified by a California Department of Education discussion group, led by F. Capra, and the author.

- NETWORKS—Every component (part or relationship) is interconnected to every other component in some way. These interconnections can create interdependence and contribute to diversity and complexity.
  Examples from ecological systems—venation in leaves, plate tectonics.
  Examples from human communities—Internet, Neighborhood Watch, road systems.
- BOUNDARIES—The real or abstract separation between a system and its environment or between different levels of scale within systems—systems occur nested within other systems, each linked to the others across scale.
  Examples from ecological systems—cell membranes, city borders.
  Examples from human communities—amount of land available, workplace rules.
- CYCLES—Reoccurring events within or between systems. Cycles allow for the flow of materials and the revitalization and repair of individuals and systems.
  Examples from ecological systems—nutrient cycling, seasonal change.
  Examples from human communities—economic cycles, menstrual cycles, red/green traffic signaling.
- FEEDBACK LOOPS—They are chains of events in which an "output" of the event influences the first link at the beginning of the chain, either slowing down or speeding up initiation of the next event in the chain.
  Examples from ecological systems—resource depletion where growth depletes resources, which limits growth; schooling/flocking behavior of fish and birds.
  Examples from human communities—stop lights, democratic elections.
- FLOW—The flowing of such things as energy, matter, and information through systems of all sizes of scale. The flow creates an effect. The rates, amounts, and importance of flow-through varies greatly.
  Examples from ecological systems—the movement of water through an organism and energy through an ecosystem.
  Examples from human communities—water and energy through a building, money through a community, information, power in government.
- DEVELOPMENT—Processes at all levels of scale that create growth and generate new forms.
  Examples from ecological systems—succession, molecular interaction leading to new molecules, co-evolution.
  Examples from human communities—fund-raising, organizational/community infrastructure.
- DYNAMIC BALANCE—Changes that are continually occurring around an unfixed central point or temporary state of "well-being." These fluctuations may swing very widely or in small increments around this point. The focus of this phrase is on dynamic, rather than on balance.
  Examples from ecological systems—populations reacting to factors such as stress and the availability of food and water.
  Examples from human communities—traffic controls, financial market fluctuations.

isolated parts, which obscures the fact that it is made up of vast, interlinked complex systems.

While the term "systems" conjures up images of nonlinear equations, engineering diagrams, and complexity of the most abstract sort, teaching systems thinking takes very little knowledge of complexity to begin using it in instruction. That is because the power of systems thinking comes primarily from its way of changing the perception of those that use it, not from thoroughly understanding the phenomena of complex systems.

From a systems perspective, there are three primary types of relationships: those between "self and other"; those that occur between points in space; and those that occur between points in time (Table 20.2). The first relationship, self/other, refers to the relationship between two objects or beings. It is found at the very heart of all effective environmental education. The "self" referred to is often ourselves. When environmental educators refer to locus of control or expanding our sense of self to include other species, it is a part of this type of relationship. Other examples are the relationships between a human and a community, between the mind and a behavior, and between a species and a symbiotic partner. Expanding our sense of self to include other human beings and other species is a critical component of lifelong environmental education.

The space relationship refers to the dimension of distance, from the smallest space to infinity. The time relationship also stretches from the smallest fraction of a second to millenia or eons. For example, to find the source of groundwater pollution, we may need to cross a space of 20 miles and a time of 10 years. As a culture, we are myopic to events that cross time and space, so we are often surprised at the outcomes of our activities. With systems thinking, we search for causation of an event across both time and space, including looking across a variety of scales for time and space. This perspective helps us be much more open to understanding multiple causation and compounded—rippled—effects.

In Table 20.2, an overview of the scope and sequence of the content of complex systems has not been provided even though an initial list of descriptors was provided earlier. There are two primary reasons for not articulating this list further. First of all, some of the universal principles we might expect K–12 students to learn about complex systems have yet to be widely verified and agreed upon (e.g., is "self-organization" a universal principle?). Systems and social scientists and mathematicians are hard at work searching for these unifying principles and promise to have them in the next few years. Second, and more important, systems thinking is process thinking. It is a way of purposefully gathering and processing information and then turning it into meaning. To emphasize the details of complex systems would set this chapter up for misinterpretation by those educators who believe that content is superior to process and is an end in itself. This chapter supports the belief that process itself can be understood as content; that students should learn *from* the content, not learn *of* the content. Content, therefore, should be selected for its ability to provide rich experiences for thinking and generating new ways of understanding, which then places the processing of this content at the center of the learning experience (Acosta and Liebmann 1997).

TABLE 20.2. Systems thinking skills and attitudes in environmental education.

I General

A. Skills
- Identifying overarching ideas and principles
- Thinking and seeing relationally, systemically, and contextually
- Knowing when to use tools of systems thinking, holism, and reductionism
- Making decisions based on qualitative and temporal data
- Balancing the tensions of simultaneously holding opposites (e.g., interdependence and independence, flexibility and purpose, nature and culture, and leading and following)
- Integrating methodology and ways of knowing from multiple subject areas
- Intentionally nurturing imagination and intuition

B. Attitudes
- Appreciates metaphor
- Trusts both their intuitive and rational mind
- Tolerant of divergent thought
- Appreciates the complexity of action-reaction linkages
- Connects needs to actions
- Tolerant of ambiguity and uncertainty

II Self-Other

A. Skills
- Forming effective relational questions
- Linking self to other, inside to outside
- Forming a clear sense of boundaries between self and other
- Understanding and handling paradox successfully
- Using *both/and/also* logic
- Learning about learning processes and the cultural influences on them

B. Attitudes
- Able to be both self- and other-referential
- Cooperating for the common good
- Holds relational self-interest
- Tolerant negotiator of irreconcilable differences

III Space

A. Skills
- Seeing patterns and connections
- Linking cause and effect across space
- Crossing differences in scale

B. Attitudes
- Connects to global communities through local environments
- Has a strong sense of place
- Visualizing flowing-through and cycling in systems

IV Time

A. Skills
- Using creative and lateral thinking
- Linking cause and effect across time
- Designing backward from a vision of the future
- Predicting and forming educated guesses using scenario development and evaluation
- Making use of different forms and effects of change and influencing it
- Finding root causes

B. Attitudes
- Prepares to learn for a lifetime
- Plans the future for self and community
- Accepts and adapts to the unexpected and copes with uncertainty
- Respects impacts of change and is adaptable to it
- Respects cultures from other times
- Embraces and values error
- Hunts for and finds hope

Source: Inspired by a discussion paper for an online colloquium from Environment Canada, October 19–30, 1998 (Selby, 1998).

Teaching Systems Thinking

This section will describe two methods for using systems thinking, within the context of UEE, in the classroom curriculum:

1. Teaching about systems in general terms by relating them to students' lives.
2. Using systems descriptors to study the processes and parts of a system and to discern patterns of structure or behavior of systems encountered in real life.

(There are at least four other methods for studying systems in K–12 education that have been used or conceptualized.)

Method 1

The following activities, framed by questions to ask students, are the start of a compelling journey. It may not be possible to complete the full sequence of activities within a unit of study because of the rich opportunities for extension activities that emerge at every step of the way. This list of questions could become the focus of a year-long inquiry. They are answered most effectively through the collaboration of groups of students, who take part in sharing what they have learned with the class. Concept maps can be used by students to help them visualize relationships, understand connections, and discover previously unnoticed relationships.

A possible sequence of activity questions to pose to students:

- What is a system?
- What are some examples of systems?
- What do you know about systems?
- What do you know about natural systems? Human-built systems?
- How are they similar and different from nonliving systems?
- How are systems a part of your everyday life?
- How are these systems part of larger and smaller systems?
- When two systems are connected, what is the nature of the connection?
- What are the similarities and differences between natural and human-made systems?
- How have natural and human-made systems changed in your community in the last 10 years? Fifty years?

Advanced projects:

- Select an area in your community that is the least disturbed by humans. How are the systems in this area important to the larger natural community? To humans?
- In what ways is the natural area connected to larger natural communities?

- Select examples of relationships and discuss them (e.g., cause and effect, crossing levels of scale, and relationships across time and across space).

My experience with students and teachers that have followed this seemingly innocuous sequence of activities suggests that the experience has power far beyond what the reader might guess. The Moreno/Utvich classroom at Open Charter Elementary School offers some insights into how students think about complex systems. They keep a large student-generated systems concept map on the wall. Students are asked repeatedly throughout the year to see if there is any new information to add or modifications to be made on the map.

Experience to date shows that students are not daunted by the complexity of systems. The following is a list of two conceptual layers of system characteristics that were generated on a concept map by fourth- and fifth-grade students at Open Charter during the course of one year:

Systems are:

- Complicated—we don't always know what systems are
- Powerful—systems need energy to work
- Balanced—everything must be balanced for systems to work
- Changeable—systems can change and expand
- Interactive—systems have many parts that work together
- Natural and man-made—there are natural and man-made systems
- Everywhere—everything has a system
- Organized—systems have a cycle or method (can organize, make things work, have function)
- Connected—systems are connected

It is very interesting to note that over the last 5 years students at Open Charter have been consistent in what they have come up with at the beginning of each year and equally consistent in making few adjustments as the year ensued.

It is fascinating to compare their list, generated entirely from the students' personal observations, with those found in Chapter 11 (Systems/Models/Constancy and Change/Scale) of the *Benchmarks for Science Literacy* (AAAS 1993) for grades K–5. It would be of great value to discover the source of understanding that the students tap for their interpretations.

- Most things are made of parts.
- When parts are put together, they can do things that they couldn't do by themselves.
- In something that consists of many parts, the parts usually influence one another.
- Things change in some ways and stay the same in some ways.

- People can keep track of some things, seeing where they come from and where they go.
- Things can change in different ways, such as in size, weight, color, and movement.
- Some changes are so slow or so fast that they are hard to see.
- Some features of things may stay the same even when other features change.
- Some patterns look the same when they are shifted over, or turned, or reflected, or seen from different directions.
- Things change in steady, repetitive, or irregular ways—or sometimes in more than one way at the same time.

Method 2

The descriptors in Table 20.1 can function like themes in curriculum building, but are more powerful as organizing ideas. To start asking questions about any one of them can serve as an effective entry point into a systems study. But like a hologram in which every part of the system contains the whole of every other part, to study one inevitably leads to the study of them all.

What follows is the description of a method to use these systems descriptors to organize an investigation of a local community problem. This somewhat advanced series of activities is probably best used with upper elementary grades and above. Students are asked to discuss and choose one environmental issue from the school community or region (e.g., a leaking underground gas tank). In cooperative groups, students record their findings on large sheets of paper for reporting to the class. If all of the following were to be completed, it would probably take about six hours. Experience to date suggests that it is best to do this in one-hour blocks of time across many days of class.

1. Using concept mapping, identify the systems involved in the issue. (Check to make certain students are identifying the systems, not the causes. For example, groundwater is part of a system; a leaking gas tank is a cause.)
2. Answer the question: What relationships have been broken, modified, or added to the systems?
3. Identify where there is conflict in the systems—color red.
4. Select one of the systems and identify (using short phrases) the seven systems descriptors in the system—networks, boundaries, cycles, feedback loops, flow, development, and dynamic balance.
5. Identify other systems that are directly connected to this system.
6. Look further into the systems nesting within the system to identify another level of systems. There will be even more widespread connections.
7. By looking at the entire concept map, identify where the systems are "malfunctioning"—color blue.

8. Look at one of the systems involved in the "malfunction." Can you move across scale to larger and smaller systems that make up, or are made up of, the malfunctioning system? Any insights about the problem? From which of the seven systems descriptors does the malfunction arise?
9. Identify the leverage point where intervention to resolve the issue might be the most effective.

Group discussions and class reporting after each activity are very important. Students will come up with widely varying positions and interpretations, which can provide valuable new insights for the entire class.

A Caveat About Teaching Systems Thinking

Two important challenges to successfully teaching systems thinking have become apparent. First, it is both easy and our first inclination to require students to memorize the definitions of the systems descriptors, but that is not the goal. The teacher must trust the process approach and make sure students are given rich experiences with relationships, connections, and systems. Subsequently, this experience will lead students to the descriptors and enable them to develop understanding for how the descriptors connect to each other and to the systems that they describe.

The second challenge arises from the fact that all of the descriptors have both systems and reductionist meanings. When teaching about one of them, such as cycles, teachers must be mindful to help students see the functioning of the system as a whole and the relationships embodied in it. For example, having students describe the water cycle—the movement of water in a circular pattern from ocean to land to ocean, with labels for evaporation, rainfall, transpiration, and groundwater, is only weakly systemic. The teacher can help students by widening the context and adding more emphasis on relationships—by asking questions about how the water cycle interacts with other systems (e.g., living components of the ecosystem, human water diversion systems, human waste entering the cycle's flow, and changing landscapes and other cycling disrupters in their own community).

To know if the teacher is on track, there are some litmus test questions to consider: Will the question, lesson, or activity give students a deeper and broader context for understanding? Will it help them connect the subject to other knowledge they hold and situations they've experienced? Will it help them form new relationships in understanding? Will the students be able to apply what they have learned about relationships in new situations containing systems? If the answers are yes, the lesson probably has strong systems components.

Because of the transformational power of using systems-based questions in instruction, it is important to consider some further examples of their use in everyday classroom instruction. Most importantly, their use does not require new instructional materials, just an expansion of the anticipated outcomes of the preexisting lessons. To do so requires the use of targeted questions that get at relationships and connections within a systems context.

For example, a study of the history of slavery changes from a linear sequence of events into a group of relationship-based essential questions such as: What systems were in place that allowed slavery to flourish? What were the relationships at that time between the American federal government and its citizens? Are there similarities between the economic and political systems that accepted slavery then and today's environmental racism/justice issues where poor, minority neighborhoods have a much greater incidence of contamination from hazardous waste storage and Superfund sites?

What I have shown here is that the quality of the question posed, the searching for relationships, is the most basic, the most important, and the most powerful element of systems thinking for the K–12 educator. It is critical that the teacher consistently, repeatedly, and intentionally include the systems perspective in lessons. Relational questioning also holds the key to increasing the efficacy of curricular planning. For example, the theme of "sustainability" can be given power and meaning by writing essential questions that serve to guide the creation of lessons, such as: Where can we learn about sustainable communities? How do we live responsibly in our communities? How can we predict the impact of an action on our community? How is my future linked to the quality of my community?

Putting It All Together

There is strong anecdotal evidence that it takes only a small catalytic force to dramatically increase student success through systems thinking programs as described in this paper. In the case of Open Charter Elementary School, with a minimal amount of support on my part for them (less than 30 hours) in 1996, they made major changes in the way they think and teach, especially about the environment. It is noteworthy that the Moreno/Utvich classroom was selected by the State Education and Environment Roundtable in 1998 as the best example in the U.S. of a elementary classroom demonstrating the use of the environment as an integrating context for instruction (EIC) (Lieberman and Hoody 1998). Further studies were conducted on eleven California schools identified as using EIC. Lieberman and Hoody found that Open Charter's standardized test scores were higher than those of a control group for 20 of 33 results. Fourteen out of sixteen assessments specifically for the Moreno/Utvich fourth and fifth grade combination classroom were higher than those of a control group—and their mathematics score was only 1 percent below the control group's. The percentage *increases* indicated for students from this classroom compared to a control group were: Reading—up 8.3 percent (fourth grade) and 8.7 percent (fifth); Language—up 6.0 and 10.0; Using information—up 5.5 and 11.5; Listening skills—up 7.5 and 18.5; Science—up 5.3 and 10.7; Social studies—up 5.5 and 8.0; and Thinking skills—up 8.0 and 11.0 (Lieberman and Hoody 2000). It would appear especially significant that student scores were considerably higher during the second year with these same Open Charter teachers. While the Open Charter School did well on standardized testing overall, the data appears to show that something in the Moreno/Utvich classroom is making a difference. The two teachers eagerly report that the incorporation of systems thinking into their own thinking and into classroom instruction is primarily responsible for their success.

Good urban ecosystem education begins in the school and community where students live. It is rooted at home and seeks to create students eager and knowledgeable to make their communities safe, livable, and sustainable. Within the cores of our cities, questions that matter to students are related to issues of safety and survival, maintaining personal health, and developing an acceptable day-to-day quality of life. Focusing on local community issues and acquiring the associated basic urban skills are much more relevant to these students than saving a Brazilian rainforest or an endangered species, *yet the desired outcomes for students are much the same*:

- To care about and have empathy for other living things.
- To appreciate the interconnectedness of all life.
- To expand their sense of self to include their community.
- To feel connected to place.
- To understand and value complex systems.
- To find meaning, hope, and responsibility through establishing community relationships that can improve the community environment.
- To understand the relationship between the hidden infrastructures that sustain them and the natural systems that supply them and accept their waste.
- To find beauty and harmony within built environments.
- To feel pride in the human capacity for adaptation, invention, creation, and forethought.
- To value diversity, interdependence, and independence.

In urban ecosystem education, creating relationships serves as a unifying factor in creating viable programs; the environment serves as the context. The city serves as the setting and the creation of sustainable communities is the goal on the horizon. UEE uses the future to provide direction and systems thinking as the lens through which to see and interpret a changing world. Hope-building and learner-centered instruction serve as its core methodologies. And the pressures of rapid change and maintaining a high quality of life supply the motivational energy for it all.

References

AAAS 1993. *Benchmarks for Science Literacy*. Pages 262–279. Oxford University Press, New York.

Acosta, A., and R. Liebmann, eds. 1997. *Process as content* (Vols. I–III). Corwin Press, Thousand Oaks, CA.

Brown, J., and C. Moffett. 1999. *The hero's journey: how educators can transform schools and improve learning*. ASCD, Alexandria, VA.

Capra, F. 1994. *From the parts to the whole. Elmwood Quarterly, Elmwood Institute* [renamed the Center for Ecoliteracy], Berkeley, CA. Summer/Fall: 35–41.

Eastin, D. 1999. *Speech at "Superintendent's Summit on Environmental Education."* Sacramento, CA.

Hanh, T.N. 1991. *Peace is every step: the path of mindfulness in everyday life*. Bantam Books, New York.

Holland, J. 1995. *Hidden order*. Perseus Books, Reading, MA: 1.

Lieberman, G., and L. Hoody 1998. *Closing the achievement gap*. State Education and Environment Roundtable, San Diego, CA.

Lieberman, G., and L. Hoody. 2000. *Project Report: The effects of environment-based education on student achievement*. State Education and Environment Roundtable, San Diego, CA.

Martin, P. 1999. Integrated studies in systems. *Green Teacher* Fall: 20–24.

Selby, D. 1998. *Global education: towards a quantum model of environmental education*. Environment Canada. Internet website: *www2.ec.gc.ca/eco/education/Papers/selby.htm*.

21
Approaches to Urban Ecosystem Education in Chicago: Perspectives and Processes from an Environmental Educator

Carol Fialkowski

Any discussion of urban ecosystem education must include a consideration of its goal—what is educating about urban ecosystems meant to achieve? There already exists an internationally accepted goal statement for the field of environmental education: "The goal of environmental education is to develop a world population that is aware of and concerned about the environment and its associated problems, and which has the knowledge, skills, attitudes, motivations, and commitment to work individually and collectively toward solutions of current problems and the prevention of new ones" (UNESCO-UNEP 1976).

Building on this goal, the Tbilisi Declaration of 1978 established three broad, now commonly accepted objectives for environmental education: (1) to foster clear awareness of and concern about economic, social, political, and ecological interdependence in urban and rural areas; (2) to provide every person with opportunities to acquire the knowledge, values, attitudes, commitment, and skills needed to protect and improve the environment; and (3) to create new patterns of behavior of individuals, groups, and society as a whole toward the environment (UNESCO 1978).

Through the years of program development, research, and evaluation, five components have emerged as essential to quality environmental education: (1) Awareness—Helping students acquire awareness and sensitivity to the total environment and its problems; (2) Knowledge—Helping students acquire a basic understanding of how the environment functions; (3) Attitude—Helping students acquire a set of values and a feeling of concern for the environment and the motivation and commitment to participate in environmental maintenance and improvement; (4) Skill—Helping students acquire the skills needed to identify, investigate, and contribute to the resolution of environmental problems and issues; (5) Participation—Helping students gain experience in using their acquired skills to take thoughtful, positive action toward the resolution of environmental problems and issues (Hungerford, et al. 1980). These goals, objectives, and components give strong direction for program design and speak clearly to education in both the cognitive and affective domains.

Urban Ecosystem Education

Are these aforementioned goals, objectives, and components of environmental education applicable in an urban setting, or does urban ecosystem education have goals and objectives different from those that have been described?

In 1996, a white paper was developed for the North American Association of Environmental Education (NAAEE) for urban environmental education through a critique and consensus process involving hundreds of urban environmental educators over the course of years (Archie 1998). The field agreed that urban environmental education is rooted in and based upon the aforementioned goals and objectives. The guidelines for Urban Environmental Education (Archie 1998) state that, "quality urban environmental education is rooted in what experience and research indicate are the best practices of environmental education." In urban settings, these agreed-upon goals and processes are integrated with special considerations and emphases, and rooted in the community, drawing on the capacity and needs of the people as their driving force. While urban ecosystem education may require a new set of integrating approaches that deal with people, their beliefs, and institutions, along with the built and natural environments, the overarching goal, objectives, and components still apply.

Therefore, in answer to the opening question of the chapter, the goal and objectives have been agreed upon and described. What remains open to discussion and debate are the most effective techniques, strategies, and processes to achieve these goals. This chapter describes two approaches that use very different techniques and processes. Each approach has been designed to achieve both the cognitive and affective goals and objectives of environmental education.

Chicago Wilderness

Chicago Wilderness is a large-scale initiative, representing both an entity (i.e., an ecosystem) as well as an effort. The goal of the Chicago Wilderness initiative is to manage, maintain, and restore the biological diversity of the six-county Chicago metropolitan area, which includes parts of southeastern Wisconsin and northwestern Indiana.

Achievement of this goal is to be accomplished through management of, research on, and education about the metropolitan area as an ecosystem instead of a series of nonintegrated parcels of land separated by arbitrary political and county boundaries.

Having just celebrated its third year of existence, the Chicago Wilderness initiative currently comprises 150 agencies, organizations, and research institutions that have signed a Memorandum of Understanding to work together to achieve the aforementioned goal. These organizations range from federal agencies such as the U.S. Fish and Wildlife Service, the U.S. Forest Service, and the U.S. Environmental Protection Agency; to state agencies such as the Illinois Department of Natural Resources and the Illinois Natural History Survey; research institutions including The Field Museum, FermiLab, the Brookfield Zoo, and the Chicago Botanic Garden; and small nongovernmental organizations (Openlands, Friends of the Chicago River) and city agencies (Chicago Park District, Department of the Environment). Chicago Wilderness has no central mailing address, no paid director, and resists 501(c)(3) status. The essence of the coalition is to keep the "we" in our work. That work is achieved through teams—teams that each member organization participates in based on their respective area of interest and expertise.

The population of the area encompassed by Chicago Wilderness is conservatively estimated at eight million people. Research indicates that "only two in ten Americans report hearing about the loss of biological diversity, about the same proportion that registered awareness of the term in 1994" (Belden and Russonello 1996). That leaves approximately 6.5 million citizens for Chicago Wilderness to educate! Quite a job. It stands to reason that one program or one institution will never get the job done.

Therefore, the Education and Communication Team is attempting to organize the urban biodiversity programs of the 92 participating organizations into a pipeline to provide birth-to-death spiraled ecosystem educational opportunities. The mission of the team is to increase and diversify public participation in understanding and protecting the region's biodiversity by developing collaborative education programs, events, and professional development opportunities. The Chicago Wilderness Education group has approved a four-step process that identifies and prioritizes its directions and projects: Step 1—Identify existing resources that educate internal and external audiences about local biodiversity; Step 2—Identify gaps in understanding and prioritize projects to fill those gaps; Step 3—Disseminate existing and newly developed materials/programs/information through training and appropriate channels; and Step 4—Evaluate the effectiveness of the development, dissemination, and training.

Because the team has embraced the goals, objectives, and components of environmental education to guide its work, existing educational resources are charted, identifying the component addressed for each target audience. A matrix, as in the sample below, has been created that lists target audiences in the left column and component(s) across the top (Figure 21.1). (*Note*: Just one existing resource per audience is listed as an example of how the matrix is organized. For a complete matrix see the Chicago Wilderness Web Site at *http://www.chicagowilderness.org*).

Target Audiences	Existing Resources	Essential Components of Biodiversity Education				
		Awareness	*Knowledge*	*Attitude*	*Skills*	*Participation*
Adults	***Chicago Wilderness* Magazine**	X	X			
Teachers	**An Educators Guide to Chicago Wilderness: Biodiversity Kit**	X	X	X	X	X
Students (Pre-School–8th grade)	**Biodiversity Education Through Action (BETA) Project**	X	X	X	X	X
Teens	**UrbanWatch**	X	X	X	X	X

FIGURE 21.1. Matrix.

This type of analysis achieves a number of goals:

- The matrix reveals existing resource strengths and weaknesses.
- The full matrix has revealed that most existing programs emphasize awareness and knowledge. Few programs incorporate skill development, attitudinal exploration, and participation.
- Chicago Wilderness is rich in programs for school-aged students, weaker in programs for teens and teachers, and poor in outreach to adults.
- For some target audiences, such as policy makers, local planners, and politicians, there are currently no programs in place.
- The matrix firmly establishes that we are committed to achieving a common goal through an accepted set of teaching and learning practices (i.e., the components).

The goal is not to have all 92 organizations develop birth-to-death programs for all audiences, but rather, to think of Chicago Wilderness as an ecosystem itself in which all members have a niche, a role to play, and contributions to make toward the larger effort. Partnering and sharing of resources is a key strategy for ecosystem-wide education.

A few examples help illustrate how this model works.

The Chicago Botanic Garden has a fourth, fifth, and sixth grade program that is highly evaluated and very strong on the awareness and knowledge components. Evaluation results show that students participating in the 3-year program develop strong understandings of woodland, prairie, and wetland communities. A field trip per season per year to the community being studied, coupled with in-class instruction, provides knowledge of concepts such as community, diversity, adaptation, cycles, interdependence, and change (Figure 21.2).

The Nature Conservancy originated the Mighty Acorns program, which is strong in skill development and action through stewardship. Mighty Acorns has fourth, fifth, and sixth graders working with land managers and site stewards in the restoration of woodland, prairie, and wetland communities. Students acquire skills in fieldwork, study, and management, and apply these skills in positive action at restoration sites. Evaluation of Mighty Acorns indicates that the fieldwork and action components are very powerful experiences.

The Chicago Botanic Garden and The Nature Conservancy have formed a new partnership program, Biodiversity Education through Action (BETA), that integrates the individual program strengths. BETA combines the Mighty Acorns stewardship/action projects with the study of woodlands, prairies, and wetlands. Students participate in restoration activities each season at each grade level. Grade 5 students in the picture are carrying bags of prairie seed they have collected in the fall to sort, process, and prepare for planting in the spring (Figure 21.3).

The Junior Earth Team summer teen employment and training program of the Chicago Park District has now been integrated with career development and internships at the Field Museum and seven other Chicago Wilderness institutions that are looking to support urban youth through a sustained pipeline of environmental learning opportunities. Participating

FIGURE 21.2. Mighty acorns.

FIGURE 21.3. Biodiversity education through action.

teens may, for instance, be paid to work as a teaching assistant in outdoor and environmental programs in a Chicago Park in the summer, intern at the Shedd Aquarium in the education department during the school year, and move on to a mentorship and paid summer job at the Field Museum the next summer. Our hope is that by gaining experience with and exposure to a wide range of professional endeavors with strong opportunities for knowledge acquisition about local biodiversity, these young adults will become informed decision makers, leaders, and teachers in their communities.

For high schoolers, the state of Illinois Department of Natural Resources, a Chicago Wilderness partner, is supporting a coordinator to recruit and train high school teachers and students in EcoWatch programs. Students monitor forest communities in ForestWatch, collect and report data on the quality of rivers and streams in RiverWatch, and learn the appropriate procedures to collect baseline information for PrairieWatch.

Teacher training is now following the same collaborative process, with institutions partnering on trainings and sharing materials and resources. Research conducted in Illinois reveals that 92 percent of teachers have had no pre-service training in environmental education, and 65 percent do not infuse environmental education concepts in their class curriculum. Yet 90 percent of the responding teachers felt that it is important to integrate environmental education concepts into their teaching (Smith-Sabasto and Smith 1994).

Our vision is to provide the majority of teachers in the Chicago Wilderness region, most of whom grew up in cities and did not themselves have these kinds of learning experiences, with knowledge about and a comfort level with their natural surroundings; confidence in teaching about local biodiversity and attendant issues; and techniques to facilitate the development of skills, attitudes, and action strategies within their students. In many cases, this requires a very different way of teaching on the part of the teacher. However, with dozens of organizations working on the challenge in an integrated, focused way, realization of the vision becomes possible.

It is the aim of Chicago Wilderness to provide a series of learning and action opportunities at multiple levels to diverse audiences with varied learning styles. This pipeline is being developed with intentionality—not because we are so naive as to think all people will use it. If we don't provide it, however, there will never be the opportunity for sustained learning.

Is the Chicago Wilderness integrated education model displaying any signs of success? Although still in its infancy, some signs of achievement are visible. The formation and dissemination of the matrix with an articulation of the agreed-upon goals, objectives, and components of all local biodiversity programs has begun to "raise the bar" considerably on how institutional education departments do their work. Duplication and reinvention of programs is down, partnering is up, focusing beyond awareness building is increasing, and the inclusion of evaluation of all programs is on the rise. For example, the previously described Mighty Acorns program has expanded from one partner serving 25 schools in 1996, to 20 partners in seven counties serving 85 schools and 8000 students. A 3-year longitudinal evaluation study is underway to document program outcomes.

UrbanWatch

UrbanWatch, the second approach to be described, is an innovative "citizen-scientist" program that engages urban adults, teens, and educators in monitoring the quality of urban green spaces. Developed by The Field Museum of Natural History and the Illinois Department of Natural Resources (IDNR), UrbanWatch engages urban communities in ownership of, and empowerment about, decisions relating to their local environment, while collecting essential scientific data on the condition of urban green space (Figure 21.4).

The triangle depicts four levels of UrbanWatch engagement. At each level the target audience, components employed, and outcome are listed. The triangle also shows that the greatest number of participants are engaged in the information/awareness level, with participant numbers decreasing as the program complexity moves toward the apex.

The following diagram illustrates how the UrbanWatch program works—how a participant gets started and moves from one step to another (Figure 21.5).

The species, both indicator organisms for the BUS (step 3) and taxa in the step 4 blocks, have been selected by Field Museum scientists in part-

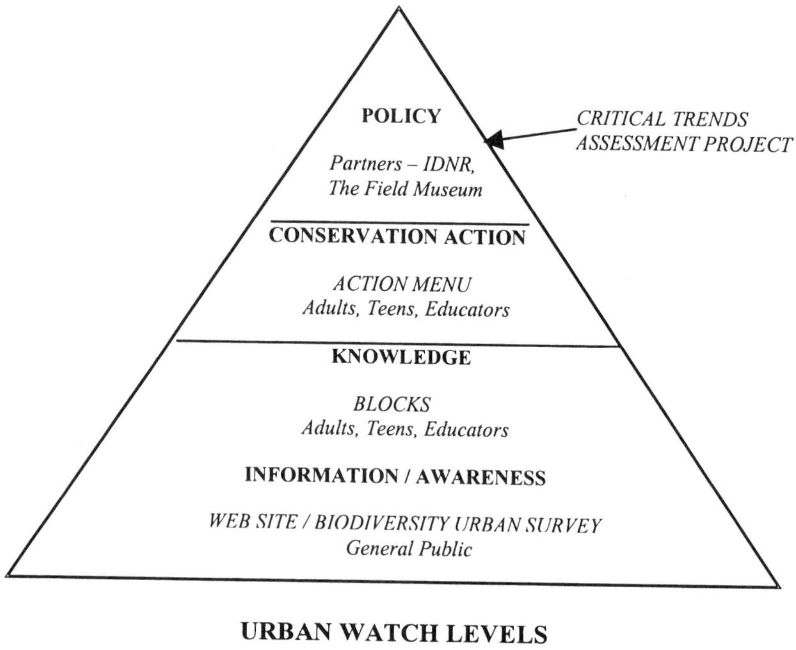

FIGURE 21.4. Overlay.

nership with the Illinois Natural History Survey. Both groups of scientists are conducting research on many of these species about which little is known in urban ecosystems—yet many are critical players in maintaining ecosystem health and function. In step 4, the blocks of specific organisms that are monitored are butterflies, two types of beetles, birds, trees and tree health, mushrooms and other fungi, and land snails and slugs.

UrbanWatch employs a problem-based approach to learning (PBL). This approach focuses on the investigation of real local issues of significance, on teamwork and cooperative learning, on self-selected problems that drive research and solutions, on skill development in issue investigation analysis and problem solving, and on solutions that ensure student involvement and action. Data is submitted online, so educationally there is a strong integration of the use of technology. Long-term plans call for students to share and overlay data between schools, generate their own questions for investigation, and use the program as part of the service learning and environmental science requirements for the Chicago Public Schools.

The data collected goes somewhere. Every two years the data are published by the state of Illinois in the Critical Trends Assessment Project, a state-of-the-state environmental report that summarizes all of the information known about Illinois's air and water quality and the status of prairie, forest, and wetland ecosystems. Data are also used by Field Museum scientists as part of their study and research on urban systems, and by local and state agencies to substantiate policy decisions.

Lessons Learned

UrbanWatch is in its final phase of development. To date, five pilots have been conducted with adults and seven Chicago public high schools. Three of the adult pilots occurred in the Lake Calumet area on the far south side of Chicago and neighboring suburbs. This is a heavily polluted industrial region formerly dominated by steel mills and ethnic communities hit hard by both environmental and economic hardships. The Field Museum scientists are conducting research studies in the area because, harbored between mills, factories, shipyards, and landfills are very unique and biologically rich natural areas.

The Urban Watch program involves a four-step process depicted in this model:

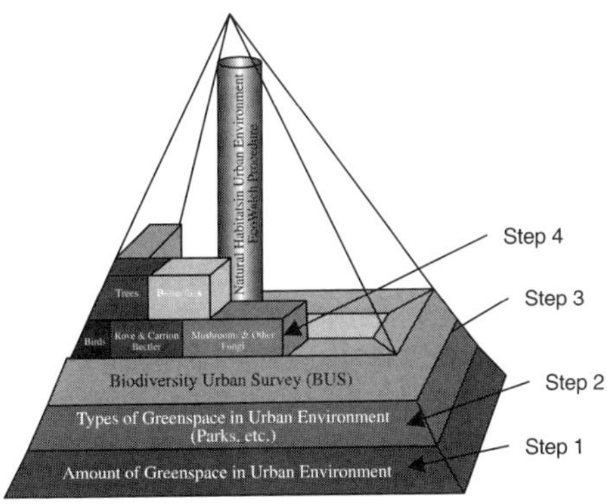

Step 1: Participation in Urban Watch begins by entering the website and clicking on "amount of green space." Participants then enter information about where they live or where they want to monitor. All documented urban green spaces in their community are shown on a map.

Step 2: Participants then choose which green space they want to monitor from the choices on the menu. Additional census and socioeconomic data about that area are provided.

Step 3: Participants conduct a Biodiversity Urban Survey (B.U.S.), the pre-survey of the chosen green space, recording the presence of selected quality indicator organisms. They calculate a biodiversity numerical index for the site based on observations. The survey can be conducted seasonally and the index averaged for a year.

Step 4: From the information gained in the B.U.S., citizen-scientists move on to learn more about their site and the organisms found there by conducting a Target Organism BLOCK protocol (focusing on tree health, mushrooms and other fungi, beetles, butterflies, birds, etc.). Participants submit collected data to scientists, on-line, for analysis.

FIGURE 21.5. Pyramid.

FIGURE 21.6. UrbanWatch.

As part of the museum's Lake Calumet Initiative, Dr. Alaka Wali, the staff urban anthropologist, has overseen ethnographic studies in the area. Many of the UrbanWatch pilot participants have been engaged in Dr. Wali's asset mapping project, which visually depicts the connections between existing community assets and outlines organizational perceptions, feelings, and understandings about the local environment. What the research has found is that residents "have a complex local knowledge system regarding environmental quality and processes. Regardless of the accuracy of his or her observations, every individual creates systematic explanations for changes they observe in the environment around them" (Babcock 1998).

Our pilots are revealing that residents are anxious to know about their local environment—what's there, whether it is "good," how it works, and what role the targeted organisms play. They are curious and eager, want a "healthy" environment, and crave information.

High school students, on the other hand, are astonished when they find any of the target organisms in their selected UrbanWatch monitoring site. Finding organisms and entering data online become automatic hooks for continuous learning (Figure 21.6).

Each school uses UrbanWatch in different ways. For example, Orr High School is employing a team of teachers, two in science, and one in technology, to implement the program as part of the field study program in the freshman environmental science course. Washington High School, in the Calumet region, has formed a math, science, and history team and is interested in using UrbanWatch to collect baseline data for a site restoration project. The monitoring data will allow students to analyze change over time and determine whether the restoration efforts have increased biodiversity as intended.

Surprisingly, teachers are not recommending the development of a separate curriculum package to accompany UrbanWatch. Instead they are recommending and supporting the hot links planned for the UrbanWatch website (*http://www.Fieldmuseum.org/urbanwatch*)—links to EPA's local air quality data, sociodemographic information available by census track, mapping capabilities through Geographical Information Systems (GIS), ARC Explorer, and more.

Simultaneously, the developers of UrbanWatch are interested in helping participants become informed decision makers and take positive action toward the resolution of environmental problems. Evaluation of RiverWatch and ForestWatch, also IDNR programs, has shown that those programs have achieved these results. UrbanWatch will be carefully evaluated to ensure it does the same.

Having worked as an urban environmental educator for more than 20 years, my experience indicates that educators spend a lot of time striving to design the one "right" educational approach or curriculum. With varied learning styles, multi-age needs and entry points, one technique will not suit all. The field would do well to consider the development of conceptual/collaborative matrices (e.g., the Chicago Wilderness model) or issue/action based programs that engage urban youth in learning by doing (e.g., UrbanWatch). In either case, programs built on strong foundations and research with longitudinal evaluation remain essential in the effort. By applying ecosystem "thinking" to our own work, and by realizing that all engaged in urban ecosystem education have a niche and an essential role to play, we have the potential to become a community of interdependent educators and learners striving to improve our practice for the benefit of the system as a whole.

References

Babcock, E.C. 1998. *Environmentalism and perceptions of nature in the Lake Calumet region*. Prepared for the Office of Environmental and Conservation Programs, The Field Museum, Chicago, IL:8.

Belden, Russonello, and Research/Strategy/Management. 1996. *Humans' values and nature's future: Americans' attitudes on biological diversity, an analysis of findings from a national survey*. Belden & Russonello Research and Communications: 6.

Hungerford, H.R., R. Peyton, and R.J. Wilke. 1980. Goals for curriculum development in environmental education. *Journal of Environmental Education* 2(3): 42–47.

Illinois Department of Energy and Natural Resources, and The Nature of Illinois Foundation. 1994. *The changing Illinois environment: critical trends, summary report of the Critical Trends Assessment Project (CTAP).*

Smith-Sabasto, N.J., and T.L. Smith. 1994. A study of the status of environmental education in Illinois public schools (K–12). The Illinois State Board of Education, Illinois Scientific Literacy Project, Springfield, IL:30.

UNESCO. 1978. *Final report: intergovernmental conference on environmental education.* Organized by UNESCO in cooperation with UNEP. Tbilisi, USSR:14–26. UNESCO ED/MD/49.

UNESCO-UNEP. 1976. The Belgrade Charter. *Connect: UNESCO-UNEP Environmental Education Newsletter* 1(1):1–2.

22
"Campus Ecology" Curriculum as a Means to Teach Urban Environmental Literacy

BRUCE W. GRANT

Knowledge of a place—where you are and where you come from—is intertwined with knowledge of who you are. Landscape, in other words, shapes mindscape.

—David Orr (1999)

We now live in a time of unprecedented ascendancy of the urban ecosystem type. Soon, more than half of the world's human population will live in urban environments. The ecological effects of these three billion urban dwellers extend way beyond the boundaries of the urban core and contribute greatly to ongoing degradation of planetary ecosystem services, geologically unprecedented rates of biological extinction, and global climatic perturbation (Wackernagel and Rees 1996). The global dominance of urban environments extends well beyond their ecosystem effects, however—the history of human culture is dominated by needs and technologies defined and devised in the urban core. The urban cultural/economic/educational footprint on the hinterland is arguably many times larger than its material/energy-based ecological footprint. The next 50 years will be of telling significance to humanity because the present generation of our citizenry must redesign key aspects of our basic cultural, economic, educational, and ecological framework to reduce our impact upon our environment and transform our society into one that is just and sustainable. This is a formidable task and because of the arguments above the success of these efforts depends greatly upon how well we can affect urban ecosystem sustainability.

An international movement is currently emerging to spur political, economic, and educational institutions toward ecological sustainability (Cortese 2000; Second Nature 2000; Filho 1999). The leaders of this movement have built upon the compelling and vivid lessons from the major social change movements of the twentieth century, such as the impact of local organizing, simple direct action, and creative use of the mass media. The take-home message is that large-scale reform begins with small steps at home. As it was succinctly put by David Orr (1995), "seemingly unsolvable global problems are often very solvable if approached at the right scale and with sufficient commitment." Thus, for me and for many others there is no better place to begin this process of cultural, institutional, and urban ecosystem redesign for sustainability than right here on the campus of higher education where we teach (Mansfield 1998).

Fundamentally the institutions of higher education are as much components of the urban ecosystems as are any other institutions in society. Therefore, the institutions of higher education contain all of the design flaws and unsustainable assumptions of the parental biosocial systems that they were designed to serve; however, higher education is also where visionary thinking and bold experimentation can be encouraged (Orr 1994; Cortese 2000), and is thereby in a unique position to wield tremendous leadership in the effort to educate our future citizenry about sustainability. According to Tom Kelly, former director of the Association of University Leaders for a Sustainable Future, "our campuses are overflowing with examples of ecologically irrational practices that are often economically and socially unsound as well. . . . This shadow curriculum is a constant, repetitive, and often unconscious educational force . . . in many cases working against the very principles of environmental literacy that we seek to engender in our students." (Kelly 1996).

This nexus of intentions and opportunities for innovation, leadership, self-reflection, and enlightened humanitarianism embedded within a structure of deeply unsustainable institutional design and practice creates a point for intervention in the lives of the students, faculty, staff, and community members that enter the sphere of higher education. Tom Kelly (1996) observed "by identifying and analyzing those examples [of unsustainability], formulating responses, and participating in their implementation, students are empowered and emboldened to take on issues of institutional change." Thus, the goal of these efforts is to look to one's home institution not only to affect institutional change locally toward sustainability, but more importantly to teach students how and why to engage in the process of change and thereby affect change toward sustainability at larger scales.

Campus Ecology Curriculum Overview

At Widener University, I teach a course to first-semester freshmen undergraduates entitled "Campus Ecology: Environmental Stewardship for the Twenty-first Century (FRS 101)" (*http://www.science.widener.edu/~grant/courses/campus.html*). This course meets 2 hours per week and is one of about a dozen "Freshman Seminars" at Widener in an innovative program created by Dean Andrew Bushko. My campus ecology curriculum consists of activities, such as demonstrations, discussions, and guided inquiries, which explore and attempt to improve the sustainability of Widener's campus ecosystem. My course goals are to enable students to better understand:

1. The physical, biological, economic, and social processes that determine the structure, function, and ecological design of the campus ecosystem and its interaction with the natural world within which it is embedded, which is "campus ecological literacy."
2. How to use methods of scientific inquiry (observation, literature manipulation, hypothesis formation, experimental design, data collection, modeling, analysis, and presentation) to construct the knowledge of the processes in (1), which is "scientific literacy."
3. Why and how to engage in the process of institutional change to move one's campus toward an ecologically just and sustainable ecosystem design, which is "human environmental literacy."

Ecological and human environmental literacy are sometimes used interchangeably; however, I prefer to restrict "ecological literacy" to understanding the system of multidisciplinary interactions described in (1). Developing ecological and scientific literacy equips one with the skills to read signposts of unsustainability (climate change, biodiversity loss, and environmental degradation), and identify ecological flaws in the social, economic, and political systems we have devised to interact with the natural world that have created these signposts. "Human environmental literacy" on the other hand, imposes the broader responsibility to use ecological and scientific knowledge wisely in personal decisions about how we interact with our environment, through both our consumer and disposer decisions, and requires our participation in the process that we all must undertake in our transition to a sustainable society. Environmentally literate people accept the realization that as authors of signposts of unsustainability, each of us has a profound moral obligation to act responsibly to correct what we have written. Human environmental literacy is thus synonymous with citizenship in its broadest sense.

Course Introduction

I begin the course by reading David Orr's foreword to Julian Keniry's *Ecodemia* (1995), including:

> creative and ecologically smart management can: reduce institutional operating costs; improve the quality of services ranging from food . . . to lighting; reduce waste and ecological impacts; rejuvenate local economies. . . . The fact that it is also the right thing to do is either an added bonus or the heart of the matter depending on your point of view . . . (Orr 1995).

I then lead a discussion of the ecological design of our classroom. Topics span a wide range of issues including lighting efficiency, motion detecting on/off switches, room heating and cooling issues, single pane glass windows, materials recycling versus "trash removal," toxics, and so on. My intention is to use the first class to expose students to the basics of making observations and asking questions on a very small scale. Then, we close with the message that once these skills are learned, they can be applied to systems of any scale.

Greening Our Lighting

We compare an 18-watt compact fluorescent bulb with 15- and 75-watt incandescent bulbs (see Figure 22.1a).

1. Students learn how each bulb makes light (incandescent lightbulbs produce light by heating a metal filament, whereas the compact fluorescent bulbs excite gasses to glow). I ask students to bring their hands near each lightbulb so they can feel the intense heat from the incandescent bulb, whereas the compact fluorescent bulb wastes very little energy as heat.
2. Students learn how to read an electric utility bill, and by making a few assumptions about lighting use, we calculate and compare the purchase and operations costs of each bulb (see Figure 22.1b). I show that the "payback time" to compensate for the higher purchase cost (about $15) of the compact fluorescent bulb is only a few months (Figure 22.1c). Thus students learn that one can reduce environmental costs and save money at the same time. Most students have never seen this type of demonstration before and are now less prone to the misinformation that environmental protection will always prove financially costly. In fact, many ask why everyone doesn't replace their incandescent bulbs with compact fluorescent bulbs. This leads to a discussion of how to design lighting fixtures and the recognition that most people are simply unaware of how to do these calculations and are wasting money and energy through ignorance. These are powerful lessons.
3. We then sketch out a system diagram of campuswide energy flow, and we discuss what one needs to know and how to pursue this analysis. Students also learn several case histories of energy-efficient lighting systems that have been adopted by campuses elsewhere (Keniry 1995; NWF 2000).
4. We then turn our attention to off-campus and discuss regional energy generation and carbon emissions from fossil fuels, the global carbon cycle, the greenhouse effect, acid rain, and other airshed pollution issues. Also, because about 40 percent of electricity in Pennsylvania is from nuclear power, we also discuss radioactive waste issues and the nuclear energy fuel cycle. We also discuss the science and economics of "alternative" renewable energy sources (hydroelectric, geothermal, solar, and wind power).
5. Last, we address the question of energy supply and demand. Shortages in energy stem from demand in excess of supply—but the question is, How do we respond to this? If one sees the shortage solely as a supply problem, then the solution is to increase supply. Because a sustainable society must rely on a finite base of renewable energy sources, however, our energy shortages must be ultimately solved by engineering finite energy

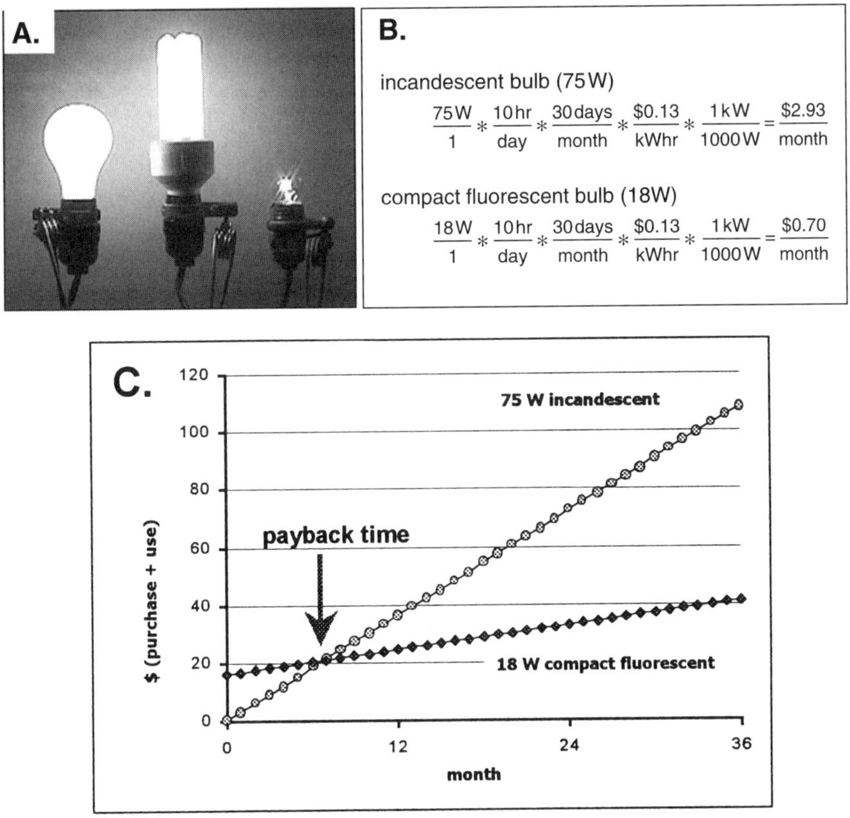

FIGURE 22.1. Greening our lighting. (A) Comparing a 15-watt compact fluorescent bulb to 15- and 75-watt incandescent bulbs. (B) Calculation of operations costs for ten hours/day, and (C) monthly sum of purchase plus use costs show that only a few months are required to “payback” the initially higher purchase cost of the compact fluorescent bulb.

demand, not by perpetual increases in supply. From this example, students glimpse the changes that are needed in world view to devote our efforts and technologies to a strategy of sustainability.

Greening the Paper Trail

We compare new paper from pine tree pulp with several grades of recycled paper and paper from alternative pulp sources.

1. Students compare the appearance, uses, and prices of various types of recycled paper, and we compare these to paper from several alternative paper sources (hemp and kenaf).

2. Students learn how much of what type of paper is used on campus per year for various purposes (of the 20 million sheets of white office paper Widener purchases every year, less than 1 percent contain any recycled fibers), where we obtain our paper, and where our paper goes once used. Students also learn several case histories of recycling and "closed-loop paper flow" that have been adopted by campuses elsewhere (Keniry 1995; NWF 2000).

3. We discuss the environmental costs of paper production processes. For new paper, we discuss forestry management issues of pulp production (biodiversity loss, habitat and watershed degradation, economic subsidies to the forest industry, etc.). Waste production for all paper also stems from the bleaching process (chlorine versus nonchlorine methods), and for recycled paper bleaching depends on the inks used and consumer demand for whiteness (e.g., soy-based inks, etc.). Students also learn about the economics of pricing of new versus recycled paper and learn that not all environmental costs have been adequately included in the consumer price of new paper. Most students have never heard of the difference between "internalized" costs and costs of market "externalities." This is a fundamental issue to the pursuit of sustainability, in which all prices of goods and services should reflect their true environmental costs. Students also learn that the external environmental costs of pulp production and paper bleaching are disproportionately borne by those living immediately downstream or downwind of these activities. This becomes an important opportunity to include issues of environmental justice. The lesson is that in a just and sustainable society, all people have equitable access to resources and no groups are disproportionately exposed to risks from waste.

4. We discuss the environmental costs of paper recycling versus disposal processes. Students generate a mass-flow system diagram to account for the 20 million sheets of white paper we purchase per year. Most of this paper ends up in a nearby incinerator in the Chester, Pennsylvania area, which is not known for its environmental sensitivity to its neighboring residents. This becomes another important opportunity to engage students in issues of environmental justice, especially since many of the recyclables that our campus does *not* recycle on a large scale (including paper, glass, metals, etc.) end up in the Chester incinerator and subsequently may contribute to reduced air quality and other toxicological concerns in our region.

5. We discuss the economics of paper recycling (and other materials) by comparing prices per ton of white, mixed, newspaper, and so on, and students learn some of the challenges to implementing recycling programs on our campus. In fact, this year several pilot programs are presently being developed at Widener to increase the flow of materials to local recycling centers. Students in my class are excited to learn of the immediacy and fledgling nature of these programs.

Brief Listing of Additional Green Topics

1. Greening the Grounds: I take the class on a campus grounds tour, and we examine issues in landscape design and management for sustainability.

Our campus lawns are dominated by a nonnative grass, have highly compacted soils, and are depauperate in invertebrates. We discuss the resulting effects on decomposition, soil nutrient cycling, runoff, and aquatic ecosystem stress and human health concerns downstream (another environmental justice issue). Students examine several model projects at other campuses.

2. Green Food: We examine campus food services in the main dining hall and in the numerous campus vending machines.

Students learn from where we get our food, and several key issues in agricultural practice (organic vs. nonorganic food, biotechnological foods, etc.). Students also learn key issues in food transport, food preparation and processing costs, and ecological footprint analysis (Wackernagel and Rees 1996). We also study the fates of waste food and the interconnections between food and nonfood solid and liquid waste streams.

3. Greening Transportation: We discuss how students get to school and from where they come.

We examine the economics of alternative modes of travel and the environmental issues surrounding transportation, including personal lifestyle choices about car size and patterns of use and larger-scale issues of urban sprawl.

4. Green Building Design: In this activity, we examine several web sites dedicated to sustainable buildings and the notion of "architecture as pedagogy" (Orr 1993).

According to David Orr, the intention of a new environmental studies building at Oberlin was to "create not just a place for classes but rather a building that would help to redefine the relationship between humankind and the environment—one that would expand our sense of ecological possibilities" (Orr 1998).

5. Hazardous Waste Issues:

The activity is for students to perform an "environmental audit" of the hazardous waste stream by documenting sources and sinks, exposures, and human health risks; and to analyze the efficacy of reduction initiatives. This is a problematic facet of this curriculum and should be approached with great care. The principal challenge is in gathering accurate data. Although I have had success with simple toxics issues (e.g., battery recycling), detailed toxicological studies require equipment and training beyond my reach. In addition, there is the possibility that students might uncover serious potential risks to toxic exposures, especially in older campus buildings. Without proper methods, however, the realities of these concerns are difficult to determine. I do not mean to argue that one is better off not knowing about these exposure risks; my point is that unless the data are really good, such a program is neither effective nor likely to engender administrative support. This last point is extremely important to devising a strategy of change, and I will return to this issue of operational ethics shortly.

Green Initiative Projects

These are original research projects in campus ecosystem design that students create and conduct.

1. Students learn the methods of science to pose questions, collect and analyze data (collaborating with campus operations officers), and draw conclusions. At the end of the semester, students present their results to their peers in an in-class symposium.
2. Students also learn how to devise and generate policy recommendations based on scientific data to submit to campus administrators, faculty, and other students with the aim of implementing their green initiatives. Through this process, they also learn that the viability of any particular recommendation, regardless of its environmental merit, is often deeply constrained by "economic and social reality" as defined by others. This provides powerful lessons about the process of institutional change and the diversity of individual stakeholder perspectives.

Related to this, I advise that great care must be taken to manage interactions between students and stakeholders (especially campus administrators). I advise setting clear ground rules to all contacts, and using role-playing activities in class. Students also should be made aware of the importance of following chains of command and of carefully noting what and when requests were made of the different stakeholders involved. Lastly, the instructor must follow up outside of the course to make sure that summary reports of students' projects are delivered to the appropriate people.

The projects in my freshmen course differ in two ways from projects in most other courses that I have examined (Second Nature, 2000, posted more than 500 course syllabi on campus ecology and sustainability. First, the goals of the projects in my course are (1) for students to understand the campus human ecosystem, (2) to understand the inquiry process of designing and conducting a project to improve campus sustainability, and (3) to understand why this is an important thing to do. The success or failure of any particular student in the class cannot be rigidly linked to their project's implementation because follow-up is a separate issue beyond the scope of this course. In fact, most of the other comparable courses are designed either for upper division undergraduates with declared majors (e.g., environmental science, studies, or environmental engineering) or for graduate students, whereas my students are freshmen from diverse programs and many are still undeclared. The fact that many of my students' projects are often continued by another student in a subsequent year or by students in Widener's student Environmental Club, or contribute to pilot programs by the university, is an added bonus (my students' projects are listed on the course web site at *www.science.widener.edu/~grant/courses/campus.html*). For my freshman course this cannot be a reasonable goal, however, and the students need to be made aware of that. I will return to this critical issue of "course outcomes" later.

The second way in which my course differs from many others is that I require students to select projects that serve the dual purpose of reducing campus environmental costs and campus operations costs. Admittedly, this can greatly constrain steps toward sustainability, but this becomes an important opportunity to show that green investment pays (Creighton 1998; Eagan and Keniry 1998) as well as to promote strategic thinking to build coalitions among diverse stakeholders to create change. I will return to this important ethical issue shortly.

Connection to Urban Environmental Literacy

Campus ecology curriculum teaches urban ecosystem ecology in microcosm and thereby enables students to attain urban environmental literacy. I have grouped the basic components of urban environmental literacy into four categories. Each of these components is either directly (**) or indirectly (*) educed in students through studies of campus ecology such as in my course.

Urban environmental literacy is:

1. Urban Ecosystem Science:
   People need to understand:
   ** how materials flow to, cycle throughout, and are exported from the urban ecosystem (such as water, nutrients, wastes, toxins, etc., moving through and transforming in air, water, soil, and living organisms), and how these flows change along the urban-rural gradient (McDonnell, et al. 1993);
   ** how energy flows to, throughout, and from an urban ecosystem;
   ** how different types of materials and energy interact as they move through the urban ecosystem, and how to engineer ecosystem mass and energy flow to serve human needs and prevent human distress;
   ** urban wildlife ecology (individual, population, evolutionary, and community), urban forestry science, urban biological conservation and management, and so on (Nilon and Pais 1997);
   ** agriculture in urban environments (Smit, et al. 1996);
   ** ecology of disturbance, fragmentation, and the roles of exotic species (Foresman, et al. 1997);
   * urban disease epidemiology (brownfields and public health connections, urban ecotoxicology, and the dynamics of human pathogenic epidemics);

** systems thinking necessary to understand how ecological systems are modeled; and
* the importance of temporal and spatial scales in modeling ecosystem processes (Grove and Burch 1997).

2. Urban Social Ecology:
People need to understand:
** how people form and prioritize their resource and service needs, and then act upon them through patterns of consumption, investment, resource use, waste production, and disposal decisions—which in turn affect ecosystem energy and mass flow to, through, and from urban ecosystems and the biophysical component of the urban ecosystem (Grove and Burch 1997; Pacala 1993);
** the diversity of stakeholder perspectives and abilities to control access and affect the decisions above (Grove and Burch 1997);
** the effects of local and global economics (especially pricing and marketing), public policy, and law upon the above, and the transformation of materials through industry into goods and services (complexly linking energy and labor with mass flows, as well as complexly linking local and distant markets) (Mander and Goldsmith 1996);
* human demography (including occupational and epidemiological effects) and the consequences of demography on resource use needs and decisions;
* the causes and consequences of the spatial distribution of domestic, service, and industrial development, and the resulting challenges to personal and commercial transportation, utility infrastructure, and residual "greenspace" management (Grove and Burch 1997);
** the causes and consequences of social inequity (e.g., inequitable investment in ecological quality and unjust occupational and domestic exposure to human health risks) (Dorman 1996; Grove and Burch 1997); and
** the operative social constraints on the process of inter- and cross-disciplinary problem solving in improving human ecosystem design. How to build consensus and devise a viable "pedagogy for change"?

3. Science Educational Literacy:
People need to understand:
** how to use the methods of scientific inquiry (observation, literature manipulation, hypothesis formation, experimental design, and data collection, analysis, and presentation) to construct an understanding of urban ecosystems (i.e., understand what scientists do) (Hogan 1994); and
** how to use the urban environment to teach about ecology (immersion experiences in city parks and preserves, wildlife observations [e.g., bird feeders], stream studies, urban gardening, etc.) and engen-

der in urban dwellers a sense of place in the natural world beyond the myopia of the built world (Orr 1994).

4. Urban "Land Ethic":
   People need to understand:
   ** that "a thing is right when it tends to preserve the integrity, stability, and beauty of the biotic [i.e., urban biosocial] community. It is wrong when it tends otherwise" (after Leopold, (1949, "biosocial" from Grove and Burch 1997, see also Callicott 1992);
   ** that at present we live in an ecologically unsustainable society based upon deeply flawed ecological assumptions, and we have a seriously degrading global environment as a result—especially in urban areas;
   ** that each of us has a profound moral and ethical obligation to act responsibly to rewrite our legacy, redesign the principal human ecosystem networks through which we interact with each other and with the natural world, and reduce our ecological "bootprint" and increase global social and environmental justice; and
   ** that transforming our urban environments into sustainable, safe, just, and desirable places to live is key to the flourishing of humanity in the new millennium.

Conclusion

My campus ecology course is one of hundreds that are intended to teach human environmental sustainability, which for me at Widener is synonymous with urban sustainability. My principal hope is that readers will be inspired either to adopt this course model, or find some other way to contribute to sustainability curriculum. In closing, I wish to discuss what I feel are the principal challenges to implementing any kind of "green campus" curriculum.

The Goal

If the overarching goal is to move us all toward human environmental sustainability (urban included), what steps do we take next? Campus ecology curriculum reveals a deep ethical conflict embedded in our society—the tension between actions based on rights/duties versus actions based on consequences/economic utility. Many "green initiatives" offer little help in navigating this maelstrom. For example, in our efforts to wean ourselves off of fossil fuels, certain economies of scale might suggest that it would be financially expedient to create centralized alternative energy sources (administered by the existing utility infrastructure); however, this may also maintain the socioeconomic sacrifice zone approach to resource access and risk management that have created widespread and often tragic examples of environmental injustice. Thus, utilitarian ethical constructs alone would provide no guarantee that switching from fossil fuels to alternative renewable sources is going to move us any closer to a just and sustainable society, depending critically upon how costs and benefits are calculated. The point is that little will have been accomplished if human suffering and inequity are hardwired into the new and more "efficient" biosocial ecosystem diagram. Clearly, this argues that issues of social justice, rights, and duties to others should be incorporated into our conception of any "green initiative" and form a basis of our intention with any "green action" (i.e., included in the calculation of utilitarian cost and benefit). Yet this is exactly the complaint of many who dismiss sustainability for being "a Trojan horse admitting radical environmental values" into economics and policy (Owens 1994; Burgess, et al. 1998) and marginalize issues of equity and environment while focusing on sustaining business growth (Eden 1994).

I raise this concern in my campus ecology class with the requirement that all of my students' projects must meet the dual goals of reducing campus environmental costs and reducing operations costs. I tell my students that we first need to build a track record of financial success to engender administrative support, and that by saving tuition we will attract student support. Many other campus ecology courses implicitly advocate the same ethic, and indeed a great deal of money has been saved. This bizarre ethical argument (based on a teleological suspension of belief), however, subtly teaches students that if justice is to be served, it must follow the saving of monies. Similar to Tom Kelly's shadow curriculum, this is a "shadow ethic"—we profess intentions of sustainability, but our actions speak to economic utilitarian efficiency without necessarily any real confrontation of the inequitable and unsustainable underlying assumptions. The shadow curriculum is the result of an unsustainable ethical model; are we responding to this challenge with an ethic that in fact casts a deeper shadow? As Einstein said, "We can't solve problems by using the same kind of thinking we used when we created them" (quote brought to my attention by Tony Cortese). In sum, the pedagogical challenge here is to engineer a curriculum (a path) that strategically builds a coalition to effect campus reform toward sustainability while illuminating the right reasons why. Intentions matter.

How Do I Know That My Students Are Attaining My Course Goals?

Unfortunately, the answer to the above question is that I do not know. My ignorance, however, is not unique among faculty whose campus ecology courses I have examined on the Second Nature web site. Too often, successes are measured in anecdotes or in terms of financial savings (see above). I think there are three categories of needs to address this question.

First, a clear set of learning outcomes and standards for campus ecology should be developed, which must map throughout the loop of goals → outcomes → activities → assessments → and back to goals. These outcomes and standards should also be aligned with urban environmental literacy components, such as the list I presented earlier. Second, many more research models specific to green campus initiatives must be published, with tools for assessment and course evaluation, that would enable faculty to pick up and use educational research methods to learn what worked and why in their classes. Last, whatever standards, assessments, and evaluations are developed must encompass courses designed to span a wider range of students. According to the Second Nature web site, my course is only one of five course syllabi (out of more than 500) that specifically target freshmen undergraduates, and mine is the only course designed for freshmen of any major. This discrepancy suggests that we need to think much more broadly, and strategically, about whom we want to attain human environmental literacy, and pool efforts accordingly.

Building Links to Urban Ecological Science, Urban Environmental Education, Social Justice Movements, and "Citizenship"

Despite the daunting challenges posed above, there are several promising directions for solving these challenges. First, there are many excellent service learning course models (e.g., at Brown University, the University of Pennsylvania, Middlebury College, Allegheny College, Antioch College, and others) that engage students in projects in the local community of their campuses. I look forward to research into student learning outcomes carried across these types of projects and in comparison to more campus-based ones.

Second, and perhaps more important, to be alive now is to participate in the period of the most rapid social, cultural, economic, and ecological change in all of human history. As we begin the third millennium of our calendar, we are witnessing monumental and long-awaited movements toward unity—such as the globalization of the economy (and the concerned dissention from the unified voice of major environmental and labor movements). A similarly monumental act of coalescence is occurring among the fields of ecological science, environmental education, and social justice—under the banner of sustainability. I enjoin that the urban environment shall be the proving ground of our understanding of sustainability—and, given the dominance of the urban ecosystem type in our world, this must be so.

Acknowledgments. I thank Alan Berkowitz, Karen Hollweg, Charles Nilon, and Ruth Cary for their inspirational words. I also thank Andrew Bushko,

Lawrence Panek, Itzick Vatnick, and Barbara Grove (Widener University), and all of my students in my Freshman Seminar (FRS 101). I also thank Widener University administrators James Rollins, Carl Pierce, Clayton Sheldon, and many others for their assistance in the campus ecology projects conducted by my students. A version of this chapter appears in e-poster form at (*www.ecostudies.org/cary8/grant/grant.html*).

References

Burgess, J., C.M. Harrison, and P. Filius. 1998. Environmental communication and the cultural politics of environmental citizenship. *Environment and Planning A* 30:1445–1460.

Callicott, J.B. 1992. Aldo Leopold's metaphor. Pages 42–56 in R. Costanza, B.G. Norton, and B.J. Haskell, eds. *Ecosystem health: new goals for environmental management*. Island Press, Washington, DC.

Cortese, A.D. 1999. *Education for sustainability: the university as a model of sustainability*. Second Nature, Cambridge, MA. (*http://www.secondnature.org/vision/vision.nsf*)

Creighton, S.H. 1998. *Greening the ivory tower: improving the environmental track record of universities, colleges and other institutions*. MIT Press, Cambridge, MA.

Dorman, P. 1996. *Markets and mortality: economics, dangerous work, and the value of human life*. Cambridge University Press, New York.

Eagan, D.J., and J. Keniry. 1998. *Green investment, green return: how practical conservation projects save millions on America's campuses*. National Wildlife Federation, Washington, DC.

Eagan, D.J., and D.W. Orr, eds. 1992. *The campus and environmental responsibility*. New Directions for Higher Education, No. 77. Jossey-Bass, San Francisco, CA.

Eden, S.E. 1994. Using sustainable development: the business case. *Global Environmental Change* 4:160–167.

Filho, W.L., ed. 1999. *Sustainability and university life*. Peter Lang Publ., Inc. Frankfurt, Germany.

Foresman, T.W., S.T.A. Pickett, and W.C. Zipperer. 1997. Methods for spatial and temporal land use and land cover assessment for urban ecosystems and application in the greater Baltimore-Chesapeake region. *Urban Ecosystems* 1:201–216.

Grove, M.J., and W.R. Burch. 1997. A social ecology approach and applications of urban ecosystem and landscape analyses: a case study of Baltimore, Maryland. *Urban Ecosystems* 1:259–275.

Hogan, K. 1994. *Eco-Inquiry: A guide to ecological learning experiences for the upper elementary/middle grades*. Kendall Hunt Publ., Dubuque, IA.

Kelly, T. 1996. *Learning from the "shadow curriculum"*—Message from the Director. The Declaration (May–August), Association of University Leaders for a Sustainable Future, Vol. 1 (2): 1–2. (*http://www.center1.com/ulsf/pubs/declare/othvol12.html*).

Keniry, J. 1995. *Ecodemia: campus environmental stewardship at the turn of the twenty-first century*. National Wildlife Federation, Washington, DC.

Mander, J., and E. Goldsmith, eds. 1996. *The case against the global economy: and for a turn toward the local*. Sierra Club Books, San Francisco, CA.

Mansfield, W.H. 1998. Taking the university to task. *World Watch* 11: 24–30.

McDonnell, M.J., S.T.A. Pickett, and R.V. Pouyat. 1993. The application of the ecological gradient paradigm to the study of urban effects. Pages 175–189 in M.J. McDonnell and S.T.A. Pickett, eds. *Humans as components of ecosystems*. Springer-Verlag, New York.

Nilon, C.H., and R.C. Pais. 1997. Terrestrial vertebrates in urban ecosystems: developing hypotheses for the Gwynns Falls watershed in Baltimore, Maryland. *Urban Ecosystems* 1:247–257.

NWF. 2000. *Campus ecology*. National Wildlife Federation. Washington, DC. (*http://www.nwf.org/nwf/campus/*).

Orr, D.W. 1993. Architecture as pedagogy. *Conservation Biology* 7:226–228.

Orr, D.W. 1994. *Earth in mind: On education, environment, and the human prospect*. Island Press. Washington, DC.

Orr, D.W. 1998. *Remarks at the groundbreaking of the Adam Joseph Lewis center for environmental studies*, September 25, 1998. Oberlin Online (*http://www.oberlin.edu/news-info/98sep/orr_remarks.html*).

Owens, S. 1994. *Land, limits, and sustainability: a conceptual framework and some dilemmas for the planning system*. Transactions of the Institute of British Geographers, New Series 19: 439–456.

Pacala, S.W. 1993. Part IV: A theoretical ecologist's perspective: Toward a unified paradigm for subtle human effects and an ecology of populated areas. Pages 307–309 in M.J. McDonnell and S.T.A. Pickett, eds. *Humans as components of ecosystems*. Springer-Verlag, New York.

Second Nature. 2000. *Education for sustainability web site*. Cambridge, MA. (*http://www.secondnature.org/*)

Smit, J., A. Ratta, and J. Nasr. 1996. *Urban agriculture: food, jobs and sustainable cities*. United Nations Development Programme, Publication Series for Habitat II (Vol. 1), New York.

Wackernagel, M., and W. Rees. 1996. *Our ecological footprint: reducing human impact on the earth*. New Society Publishers, Philadelphia, PA.

23
Ecosystem Management Education: Teaching and Learning Principles and Applications with Problem-Based Learning

HENRY CAMPA III, DELIA F. RAYMER, and CHRISTINE HANABURGH

A Challenge of Teaching—How Should We Present Material?

Williams and Young (1992) commented that "Science begins with an observation and a question and proceeds through a process of inquiry involving exploration and investigation, experimentation and analysis, and exposition and persuasion." If educators teach a science-based course, they must make sure students have an opportunity to experience the "process of inquiry" that Williams and Young (1992) describe. Problem-based learning (PBL) is one pedagogical approach that educators may use to expose students to the problem-solving process.

In the applied sciences, being able to apply ecological and management principles to problems, such as planning and managing urban ecosystems for their diversity of values, is perhaps as critical as understanding the principles themselves. As students move into careers they will be confronted with problems that need to be addressed, often without any guidance. So the experiences students have in classes applying concepts and principles to solve problems may be critical for them attaining their first jobs and becoming effective professionals. Educators, therefore, have the responsibility of presenting material to students in a fashion that simulates how they will use it as professionals to address future problems.

There are two general teaching-learning models educators use: subject-based learning (SBL) and PBL. In SBL, students first learn concepts and then see how concepts are applied by reading or discussing brief examples (Woods 1994). This approach to teaching has been the dominant model, especially in American education in the twentieth century, and allows educators to introduce a diversity of concepts in a relatively short period of time. One could argue that this is a *teaching model*; however, it may not be an appropriate *learning model* (Ryan and Campa 2000). In fact, this approach to teaching has been recently challenged as having limited applications for promoting learning in the classroom. Williams and Young (1992) describe acquiring knowledge in this context as a "spectator activity." Zundel and Needham (1998) commented that typically "foundation knowledge is taught independently of problem solving." These authors contend that the end result of not using PBL is that students often cannot integrate what they have been taught with other disciplines or use their knowledge to solve problems.

Problem-Based Learning

Wilkerson and Gijselears (1996) defined PBL as the process of initially presenting a problem to students to engage them as active participants in learning, then allowing them to collect data through their own research or by gleaning information from mini lectures provided in class. Lastly, the students share and defend their findings with peers. This model parallels how Williams and Young (1992) described the development of a scientific theory. Supporters of PBL contend that this approach to teaching can simulate the environment of working professionals of all types, making learning a participatory process that may lead to enhanced communication and critical thinking skills (Allen and Duch 1998; Harley, et al. 1998; Zundel and Needham 1998; Ryan and Campa 2000). In contrast to the benefits of using PBL, Gallagher et al. (1992) and Arambula-Greenfield (1996) commented that limitations of the pedagogical approach are teaching educators to be facilitators of learning instead of transmitters of information, and the potential unwillingness of students to accept responsibility for their own learning.

Our objectives are to: (1) discuss the PBL approach (with examples) and the theories of why it can enhance learning; (2) demonstrate how PBL can be used to teach ecosystem management principles and applications; (3) explain why PBL can enhance student learning about the ecological values associated with urban ecosystems; and (4) discuss outcomes of using PBL to enhance learning. We also discuss how PBL is used to teach ecological concepts and wildlife habitat management principles and their applications in an upland ecosystem management course that we have taught at Michigan State University (MSU) for the last seven years (Figure 23.1). In this discussion, we present how the "process of inquiry" used in our course parallels that described and recommended in the education literature for PBL. Based on our experiences and those of others at Missouri (M. Ryan) and Minnesota (K. Smith), we contend that undergraduates can learn "content" with PBL while simultaneously seeing how material can be used to solve problems.

An example of using PBL discussed in this chapter is having students work on an ecosystem analysis and management problem throughout a

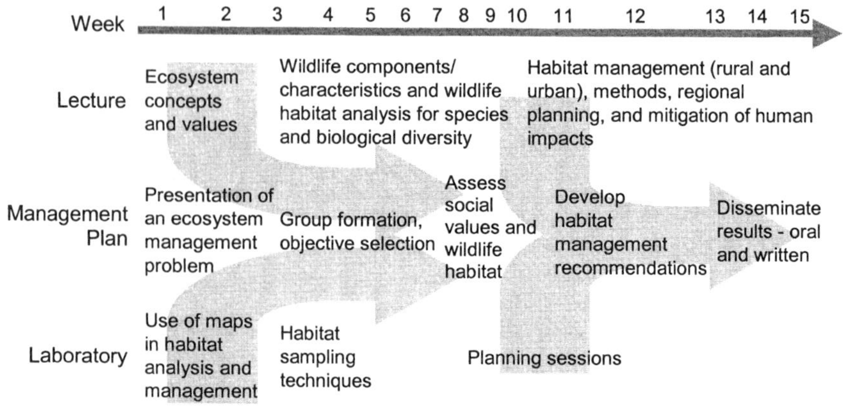

FIGURE 23.1. Course composition, including lecture and laboratory topics and how they are integrated with the problem of developing an ecosystem management plan.

semester. Using PBL on this assignment is essential to demonstrate to students the challenges of managing ecosystems to meet social and economic demands while simultaneously trying to maintain the ecological attributes of these areas.

The definition of ecosystem management we present to students is: a management approach that attempts to balance ecological, social, and economic objectives (Kaufmann, et al. 1994; Haufler, et al. 1996). The ecological objectives that we focus on during the course include maintaining or restoring elements of biodiversity within an ecosystem. The social and economic objectives focus on maintaining a human environment that is enhanced by the inclusion of a diversity of vegetation types while simultaneously providing socioeconomic opportunities for residents.

The ecosystem concept can be applied to a variety of biological systems that may vary in size, biological composition, and the dynamics that drive the system. Urban ecosystems, as discussed in this chapter, are those that have abiotic and biotic components that are intensely influenced by humans, compared to remote or wilderness ecosystems where humans play a relatively smaller direct role in ecosystem inputs, outputs, and processes. Within an urban setting, ecosystems can be viewed in many different spatial scales, from a single backyard within a subdivision to a network of green spaces throughout a metropolitan area. Urban ecosystems are simply at one end of a spectrum of ecosystems that experience varying degrees of human influences and use. This view of urban ecosystems is also discussed by Grimm, et al. (chapter 7 in this volume).

Phases of Problem-Based Learning and Their Applications for Teaching Ecosystem Management Principles

Giving Students a Problem

Perhaps the best way to force students to focus and learn material is to give them a problem. Arambula-Greenfield (1996) recommended that key ingredients to effectively using PBL are to use poorly developed problems and give students access to a library. Our experiences with the benefits of using poorly developed problems or very recent problems demonstrate that students cannot bring prior information or "baggage" with them to address the problem (i.e., they must research and learn new information) and as they collect information, they are able to better define the problem. In contrast to using poorly developed problems, well-defined problems that are typically used in SBL do not challenge students to research new information, think critically, and develop solutions—students are typically given the problem and solutions following some discussion of concepts (e.g., problem: decline of the black-footed ferret [*Mustela nigripes*], solutions: captive breeding, reintroductions, and social challenges).

Questions or problems can be initially presented to students in many formats—constructive controversies, case studies, and/or research questions. Johnson, et al. (1991) and Campa, et al. (1996) recommend constructive controversies to challenge students to initiate and use the research process, synthesize material, and then prepare and defend positions while working in cooperative informal, formal, or base groups (Johnson, et al. 1991). In essence, academic constructive controversies are controversial scenarios presented to groups of students that role-play as various stakeholders. Stakeholder groups must learn and synthesize material within their respective groups as well as gather information from other stakeholders so that information can be integrated and used to develop the *best* possible solution to the controversy. The best possible solution to a controversy must be agreed upon by all stakeholders, and students are encouraged to work cooperatively to develop their solution—it is not a debate. In the process of using controversies to promote PBL, instructors may ask students to switch stakeholder groups so they gain a better understanding of the concepts associated with the controversy (i.e., those concepts used and discussed by other stakeholders) (Johnson, et al. 1991). Johnson and Johnson (1992; 1997) mentioned that academic constructive controversies can be used to achieve academic objectives and enhance communication, critical thinking, and problem solving skills as well as students' abilities to work in teams. Examples of constructive controversies used in classrooms are the effects of acid rain (more research vs. we know enough) (Johnson, et al. 1991), managing beach ecosystems (develop a beach for recreation vs. develop a harbor of refuge for boats vs. preserve the endangered piping plover [*Charadrius melodus*] and threatened Pitcher's thistle [*Cirsium pitcheri*]) (Campa, et al. 1996), and management of old-growth forest (continue logging vs. alter logging to preserve the spotted owl [*Strix occidentalis*] vs. harvest the Pacific yew [*Taxus brevifolia*] for cancer research) (Campa unpublished data).

Case studies are an additional method that is used in business, veterinary science, human medicine, agricultural economics, and natural resources management to pose questions to students and facilitate PBL. The use of case studies typically involves a "case discussion," "discussing the case presentation," and student role-playing of stakeholders in a case (Swinton 1995). When using case studies, an instructor initially provides the case (i.e., all material) to the entire class for their review prior to any discussions. A benefit of using cases is that students can see how concepts are integrated and applied to address a real-world problem by having access to all of the material simultaneously (e.g., reading text and reviewing data). After reading and reviewing the material, the stakeholder interactions and solution(s) to the case are discussed by the instructor and class.

This process is in contrast to academic controversies, in which the processes of gathering and synthesizing material are dependent upon interactions among student stakeholder groups, because not all groups have access to the same information and groups must develop the best possible solution to the problem (Campa, et al. 1996).

Examples of how PBL (or "inquiry-based learning") has been utilized with cases such as ozone depletion and toxic waste disposal at the State University at New York are reviewed by Herreid and Schiller (1999). Examples of the applications of academic controversies versus case studies are discussed by Ryan and Campa (2000).

When instructors use academic controversies and case studies, the problems and stakeholder objectives are typically evident and some data are provided. This is in contrast to the approach of posing complex research questions to students to foster PBL. For example, perhaps the best way to teach students about the dynamic nature of ecosystems, their potential to provide wildlife habitat or recreational opportunities, and the consequences of altering ecosystems to meet human demands, is to give them an ecosystem to assess and manage. In our course, by giving students the responsibility to set objectives for an ecosystem, having them work through the process of investigating how to meet their objectives and present and defend their data and recommendations to peers, we achieve two teaching-learning objectives. First, students must learn the course *content*, presented in class discussions and lectures, to have a sufficient knowledge base to address their ecosystem management project. Second, as students work through the phases of PBL during the semester-long project, they experience how the course content is "used." These specific teaching-learning objectives parallel those presented by Harley, et al. (1999) for problem-oriented learning: (1) problems force students to find and collect information, and develop skills and (2) students must learn how to address realistic problems.

Working on a Team and Developing Objectives

PBL is similar to the working environment of professionals in many disciplines in that new problems develop daily, often when professionals lack the necessary information to address the problems. Additionally, to address complex and/or large-scale problems, professionals may have to work cooperatively in a team if they hope to develop effective solutions. Johnson, et al. (1993) stated ". . . groups outperform individuals, especially when performance requires multiple skills, judgments, and experiences."

In our course, the ecosystem analysis and management plan assignment is used to emulate a professional working environment where students work as formal cooperative learning groups, as described by Johnson, et al. (1991) and Smith and Waller (1997). In formal groups, students accept responsibility for their own learning as well as the learning of their team members—success requires that everyone master the material. Smith and Waller (1997) commented that formal groups can be used effectively to work on difficult problems throughout a course. Having the experience of working as an effective team member and "getting along with others" were noted by Kennedy (1996) as important attributes of individuals interested in working in natural resources management agencies.

Student teams in our course are formed when students "apply" to work with others to develop and address their management objectives—similar to how they would apply for a job. Students submit a letter of application to work with a group of their choice. Student letters address what they will contribute to the effectiveness of the team, how conducting the project will help fulfill professional and educational objectives, when are they available to work on the project and how they can be contacted, and their dedication to work on the project. All students in a group are required to review the letters of their group members and once they have decided they will work together, all of their letters must be stapled together and turned in. In essence, the packet of letters serves as a "contract" for students to work as a formal cooperative group. Students are told prior to sealing their contract (i.e., stapling their papers together) that any individual violating the contract (e.g., refusing to cooperate) will be "fired" from their group by the instructor. Fired students forfeit points on the assignment and are required to complete the management project alone—a task that is undesirable for one student due to the scale of the assignment.

In teams, students may select one of two urban "ecosystems" where they will conduct their ecosystem analysis and management project. One ecosystem is largely residential with some small businesses. The other ecosystem consists of multiple green spaces (e.g., an urban natural area, a cemetery, and small parks) interspersed with residential areas and businesses. Following the selection of their research area, teams must formulate their management objectives for the area. The freedom to select their own objectives is important for motivating students to become engaged in the PBL process and giving them experience developing management objectives.

In the process of developing objectives, student teams should be encouraged to talk to residents and nature area and industrial park managers to investigate what their ecological, social, and economic interests may be for the area students have selected. Possible objectives teams may choose for the urban green space ecosystem described earlier could be to: (1) enhance habitat for native songbirds, (2) enhance habitat for reptiles and amphibians, and (3) assess the effectiveness of habitat management recommendations over a 20-year period. These ecological objectives must be balanced against the social and economic conditions that are also part of the ecosystem. One example residents may be concerned with is dense bushes (i.e., planned as wildlife habitat) that may attract and provide cover for homeless people or others they perceive as undesirable or as a threat.

Information Exploration and Investigation—What Information Is Needed to Evaluate Objectives?

Harley et al. (1998) mentioned that students may learn most effectively when they are engaged in activities that "motivate learning." What better way to motivate someone to master material than by confronting them with a controversy or a complex research topic? Robles and Nakamura (1999) commented that in a PBL environment students can master their use of class material (e.g., concepts, principles) in a professional frame work if they are required to collect their own empirical data or to exchange data.

The decisions involved in deciding upon what data needs to be collected are central to the problem-solving process in natural resources management. Although specific data needs differ depending on the objective (i.e., whether natural resources managers seek to work with urban residents to maintain biodiversity or evaluate the effects of an urban development project on songbird habitat) the "process of inquiry" is the same. The information exploration and investigation phase of PBL, as it applies to developing an urban ecosystem management plan, requires that individuals gain an understanding of the fundamental concepts of wildlife habitat components and characteristics, limiting factors, and habitat assessment methodology.

We give students a writing assignment designed to give them experience gathering, organizing, and synthesizing information—an important component of PBL (Harley, et al. 1998; Robles and Nakamura 1999). Each student writes a one-page paper summarizing the necessary habitat components (e.g., food, cover, space) and characteristics (e.g., how components are distributed—juxtaposition) for a selected wildlife species. For the assignment, students must learn how to search the Internet and library for scientific literature, concisely summarize their information, and submit their paper in journal style. This assignment is independent of their ecosystem analysis and management project so that individual students develop research skills that will help make them a more effective and accountable team member.

After student teams decide upon their ecosystem and objectives for their project, they need to research what the specific habitat requirements may be for the taxa they will manage. These habitat requirements must be *quantified* using information in the scientific literature. Team objectives and a summarized list of habitat requirements are then presented to the instructors in a 20-minute planning session as potential habitat-based limiting factors for their objectives.

Each team has two additional planning sessions with the instructors later in the semester to address the other main parts of their management plan: (1) field sampling methods and data analysis, and (2) results and management recommendations. These sessions are an integral component of the course for students and instructors. They are deadlines to facilitate completing the plan by the due date and allow each team to have individual guidance. By having these sessions, we can direct the progress of individual teams, identify and solve potential team problems early, and gain knowledge of each team's progress.

The support and feedback students receive from the instructors during the planning sessions are components of PBL that enable students to meet objectives that may be initially beyond their intellectual range. Harley et al. (1998) referred to this type of instructor support in a PBL environment as "scaffolding." Scaffolding may involve an instructor "coaching" students through the learning process by breaking a complex problem into smaller, more easily researchable topics or problems (i.e., objectives, methods, results).

Experimentation and Application of Procedures and Data Analysis

Robles and Nakamura (1999) commented that with PBL, students have an opportunity to practice their knowledge of material and apply it in a professional format by being required to use their knowledge to collect data. One motivation for academics to provide this type of learning experience for students is to meet employers' needs for graduates with greater technical skills.

One example of how to facilitate students mastering technical material is by requiring them to use habitat assessment methods to collect their own data. In contrast, having students learn habitat assessment methods and data analysis solely by reading a text would be equivalent to teaching a child to ride a bike by giving them a book on bicycling. Comments from our students on course evaluation forms indicate that they like this learning approach since it gives them "hands-on" experiences with research techniques and the summarization of data.

During laboratory periods we present the types of field sampling designs and habitat sampling and data analysis methods that students will need to use to conduct their project (Figure 23.1). Habitat sampling methods are first described in a classroom using visual aids, then demonstrated to students outside. Students are then given a chance to use the field sampling methods while the instructors are available to answer questions. After students learn the sampling methods, we show them how data sets are summarized and follow up by giving them worksheets with sample problems and applied questions that can be answered using their data.

Students gain further familiarity with sampling designs and habitat sampling methods while they conduct their own ecosystem management projects. Developing their own sampling design and collecting data are critical for our students to: (1) develop critical observational skills, (2) gain experience with sampling designs and methods, (3) understand how data are interpreted to guide habitat management decisions, and (4) understand what characteristics in an ecosystem can provide wildlife habitat and how habitat can be impacted by changing land use practices. In essence by working through this phase of PBL students have an opportunity to learn about many of the ecosystem characteristics that Grimm, et al. (chapter 7 in this volume) commented are common among all types of ecosystems (i.e., pristine and urban) and important for individuals to understand. Learning about ecosystem characteristics occurs because students are required to recognize and quantify compositional (e.g., species present, species diversity) and structural (e.g., heights, growth forms) attributes, availability of abiotic resources, and the boundaries or spatial distribution of biotic (e.g., types of vegetation) and abiotic resources in order to evaluate their objectives.

Exposition of Results

Egan (1998) conducted a survey of forestry alumni at Mississippi State University to quantify their views on curriculum improvements needed to develop the skills necessary for successful employment. Results indicated a greater need for better communication skills. Murphy, et al. (1998) made a similar conclusion concerning new college graduates, stating ". . . employers still look for graduates with extensive technical backgrounds, today they also stress the need for graduates with excellent written and oral communication skills, strong critical thinking skills, and teamwork and leadership skills." The need for better communication skills among college graduates is not restricted to the natural resources fields. For example, Karl Smith (University of Minnesota, personal communication) commented that a historical concern with engineering education was the lack of attention paid to developing effective communication skills.

One advantage of using PBL is that educators can easily develop and integrate a variety of communication assignments by having students summarize or synthesize their responses to a problem. For example, in our course student performance on their management plan is assessed with written materials (i.e., progress reports) provided during their three planning sessions, their final written management plan, and an oral presentation. These products are similar to the ones used to assess the productivity of natural resource professionals.

The Outcome—Lessons Learned?

A critical aspect of using a pedagogy such as PBL to enhance learning, or of implementing a habitat management practice to improve habitat conditions, is *assessing effectiveness* of meeting the desired objectives. In ecosystem management education, teachers attempt to enhance student knowledge of the ecological principles of ecosystem management while integrating them with what is known about socioeconomic values. An evaluation of the effectiveness of this educational objective would provide information on how well students can use what was presented in a course. Our course objectives can be grouped into three categories—objectives that focus on assessing current conditions of an ecosystem, evaluating how human activities and natural events influence different types of ecosystems, and planning management activities for an ecosystem to conserve wildlife habitat. We believe that students who successfully complete the course fulfill these objectives and are able to judge student products based on evidence of their learning.

Our ecosystem management class has evolved over the last seven years. Many of the class changes reflect shifts in the natural resources management fields and the results of new research. We have made other changes in an attempt to refine the process of educating students and to help them better meet course objectives. Improving the PBL format of the course and identifying which elements of the course are especially valuable come from our use of multiple assessment methods.

One area where we have found that students could improve the quality of their management plans is in providing stronger support for their proposed habitat recommendations, based on their objectives, data, and the scientific literature. Groups often propose habitat management recommendations that are too general. For example, a management plan may include maintaining a grassland in a small urban natural area for songbirds by periodically burning it to set back succession. However, students may fail to elaborate on *how* this will benefit specific grassland birds, what the implications may be to nearby residents, or how to gain support from local urban residents for the proposed manipulation. One way to help students develop better support for their recommendations, which we implemented in 1999, is to present information on specific habitat management techniques using case studies with supportive references. We think this approach helps students learn about how habitat management techniques are used, what the expected results may be, and where to find additional information on wildlife habitat management. For example, our discussion of forest management includes a case study of using municipal sewage sludge as a method to enhance wildlife habitat quality and serve as a sludge disposal alternative. Included with this presentation was a list of references that describe the circumstances and opportunities of using this management practice.

An additional limitation of student management plans is that they often lack a clear and logical justification for their objectives, even though it is presented as one of the criteria on which plans will be graded. Objectives presented by students often lack a clear description of their scientific basis, ecological values, or human benefits. One reason for this may be that setting objectives is the first step groups must address in their management plans and at that point, students still may be struggling with how to work effectively as a team. Another possibility is that while the habitat management plan assignment is developed to be as realistic as practical, the scenario lacks the real-life budgetary constraints and varied public demands that make management agencies or wildlife consultant firms accountable for their decisions. Without these two constraints, students choose to manage for the species and ecosystem conditions they are most interested in. To overcome this problem, we have started spending more time emphasizing the importance of having meaningful justifications for objectives and ask students to revisit their objectives when developing habitat management recommendations for their ecosystem.

Overall, conclusions from class interactions, student self-assessments, and standardized course evaluations indicate that the course was "a lot of work" compared to other courses of equal credit, but that they also "learned a lot" from working on their management plans. For example, toward the end of the semester when students give oral presentations of their management plans to the class, some of the toughest questions for groups come from students in the class, as opposed to the instructors. Having just written a management plan themselves, students are able to think critically about, and question, the strengths and weaknesses of how data were collected, what data represent, and the recommendations that are made. Kunkel (1992) stated that the needs of education should include confidence building, perseverance, teamwork, and problem solving, as well as critical thinking. We think that the level of student engagement while conducting the project and during the oral presentations demonstrates the development of many of the attributes Kunkel (1992) mentions.

Further support for the success of using PBL to effectively teach ecosystem management principles is implicated by student self-assessments throughout the semester. For example, we asked our students to respond to the following question in a one-minute, in-class writing assignment during the second and the last week of the course: "How comfortable do you feel making wildlife habitat management decisions?" At the beginning of the course, only 19 percent (n = 11 out of 59) of the students felt they were fairly confident making wildlife management decisions but would need input and information from others, and 81% (n = 48 out of 59) felt they were not confident at all. During the last week of class, 67 percent (n = 33 out of 49) felt they were "much more" or "more" confident making wildlife habitat management decisions than when class started, 29 percent (n = 14 out of 49) were fairly confident and only 4 percent (n = 2) were not confident at all. These results indicate that students gained confidence in conducting an ecosystem management plan throughout the semester—a task that many students considered difficult, but worthwhile given their professional aspirations. These results are also supported by student responses on standardized course evaluation forms used at MSU.

On MSU's evaluation form, students are asked to respond to a series of questions and statements using ranks—5 being "superior" and 1 being "inferior". The following results are descriptive statistics compiled for a few of the questions on the evaluation forms from 1993 to 1999. For example, students felt: (a) "The course was demanding of my time." (mean rank = 4.1; range = 4.4 [1999] to 3.8 [1993]), (b) "The course increased my knowledge of the subject." (mean rank = 4.4; range = 4.6 [1999] to 4.0 [1993]), and (c) "Course deserves an overall rating of:" (mean rank = 4.5; range 4.7 [1998] to 4.2 [1993]). These results indicate that students find the course demanding, however, they also appear to learn a great deal and think highly of the course. The trend of improving evaluations over time may be attributed to improvements in our abilities to use PBL throughout the class and an increase in the use of case studies (without role-playing) in the lecture component of the course.

Conclusions

We feel that educators can implement PBL through the use of academic controversies, case studies, or complex research questions to teach students the necessary natural resources management principles for managing ecosystems as well as to demonstrate how principles can be applied in urban ecosystems to address ecological, economic, and/or social questions. When using PBL to teach students about the components of urban ecosystems, educators may not be able to cover the breadth of material they can in traditional SBL using a lecture format; however, we contend that PBL will help students uncover their own information and help them build many lifelong skills (i.e., critical thinking, problem solving) that will be valuable to them as professionals.

References

Allen, D., and B. Duch. 1998. *Thinking towards solutions: problem-based learning activities for general biology*. Saunders College Publishing.

Arambula-Greenfield, T. 1996. Implementing problem-based learning in a college science classroom. *Journal of College Science Teaching* (Sept./Oct.):29–30.

Campa, H., III, K.F. Millenbah, and C.P. Ferreri. 1996. Lessons learned from fisheries and wildlife management: using constructive controversies in the classroom. Pages 235–244 in J.C. Finley and K.C. Steiner, eds. *Proceedings of the first biennial conference on university education in natural resources*. The Pennsylvania State University, University Park, PA.

Egan, A.F. 1998. Forestry education and employment: views from alumni of a southern forestry school. Page 216 in C. Heister, compiler. *Proceedings of the second biennial conference on university education in natural resources*. Utah State University. Logan, UT.

Gallagher, S.A., W.J. Stepien, and H. Rosenthal. 1992. The effects of problem-based learning on problem solving. *Gifted Child Quarterly* 36:195–200.

Grimm, N.B., L.J. Baker, and D. Hope. 2002. An ecosystem approach to understanding cities: familiar foundations and uncharted frontiers. Chapter 7 in A. Berkowitz, C.H. Nilon, and K.S. Hollweg, eds. *Understanding urban ecosystems: a new frontier for science and education*. Springer-Verlag, New York.

Harley, H.D., C.D. Seals, and M.B. Rosson. 1998. *A formative evaluation of scenario-based tools for learning object-oriented design*. *http://www.acm.org/crossroads/xrds5-1/eval.html*.

Haufler, J.B., C.A. Mehl, and G.J. Roloff. 1996. Using a coarse-filter approach with species assessment for ecosystem management. *Wildlife Society Bulletin* 24: 200–208.

Herreid, C.F., and N.A. Schiller. 1999. *http://ublib.buffalo.edu/libraries/projects/cases/case.html*.

Johnson, D.W., and R.T. Johnson. 1992. *Creative controversy: intellectual challenge in the classroom*. Interaction Book Company. Edina, MN.

Johnson, D.W., and R.T. Johnson. 1997. Academic controversy: increase intellectual conflict and increase the quality of learning. Pages 211–242 in W.E. Campbell and K.A. Smith, eds. *New paradigms for college teaching*. Interaction Book Company. Edina, MN.

Johnson, D.W., R.T. Johnson, and E.J. Johnson. 1993. *Cooperation in the classroom*. Interaction Book Company, Edina, MN.

Johnson, D.W., R.T. Johnson, and K.A. Smith. 1991. *Active learning: Cooperation in the college classroom*. Interaction Book Company, Edina, MN.

Kaufmann, M.R., R.T. Graham, D.A. Boyce Jr., W.H. Moir, L. Perry, R.T. Reynolds, R. Bassett, P. Mehlhop, C.B. Edminster, W.M. Block, and P.S. Corn. 1994. *An ecological basis for ecosystem management*. U.S. Department of Agriculture. Forest Service General Technical Report. RM–246.

Kennedy, J.J. 1996. Early career development of Forest Service fisheries managers. *Fisheries* 11:8–13.

Kunkel, H.O. 1992. Overview. Pages 1–15 in *Board on agriculture, national research council*. National Academy Press, Washington, DC.

Murphy, B.R., D.W. Smith, W.E. Ensign, T.M. Wildman, C.A. Bailey, and E.J. Pert. 1998. Page 235 in C. Heister, compiler. *Proceedings of the second biennial con-*

ference on university education in natural resources. Utah State University, Logan, UT.

Robles, C.D., and R.R. Nakamura. 1999. *http://www.nsf.gov/cgi-bin/showaward?Award=9653261.*

Ryan, M.R., and H. Campa, III. 2000. Application of learner-based teaching innovations to enhance education in wildlife conservation. *Wildlife Society Bulletin*. 28:168–179.

Smith, K.A., and A.A. Waller. 1997. Cooperative learning for new college teachers. Pages 185–210 in W.E. Campbell, and K.A. Smith, eds. *New paradigms for college teaching*. Interaction Book Company, Edina, MN.

Swinton, S.M., ed. 1995. *Teaching and learning with cases*. Michigan State University, East Lansing, MI.

Wilkerson, L., and W.H. Gijselears, eds. 1996. *Bringing problem-based learning to higher education: theory and practice*. Jossey-Bass, San Francisco, CA.

Williams, P.H., and A.L. Young. 1992. Teaching science as inquiry. Pages 204–207 in Board on Agriculture, National Research Council. *Agriculture and the Undergraduate*. National Academy Press, Washington, DC.

Woods, D.R. 1994. *Problem-based learning: how to gain the most from PBL*. McMaster University, Hamilton, Ontario.

Zundel, P., and T. Needham. 1998. The problem-based learning module: computer aided professional education in forestry. Page 251 in C. Heister, compiler. *Proceedings of the Second Biennial Conference on University Education in Natural Resources.* Utah State University, Logan, UT.

24
Using the Development of an Environmental Management System to Develop and Promote a More Holistic Understanding of Urban Ecosystems in Durban, South Africa

Debra C. Roberts

Setting the Scene

Africa—the word itself conjures up dramatic images of savannas and sunsets; of dusty gorges cradling the secrets of humankind's earliest beginnings. The reality is, however, very different and infinitely more complex. On one hand Africa is a continent ravaged by civil war, political upheaval, poverty, and human suffering—on the other it is a member of the world community, buffeted by a host of global pressures and trends. Key among these (for the purposes of this chapter) is the rapid urbanization of its population.

Globally it is anticipated that within the next decade more than half the world's population, an estimated 3.3 billion, will be living in urban areas (World Resources Institute, et al. 1996). Between 1990 and 2025, this number is expected to double to more than 5 billion people. Almost all of this growth—an estimated 90 percent—will occur in countries of the developing world, with Africa and Asia currently exhibiting the most explosive growth (World Resources Institute, et al. 1996).

Not surprisingly, this rural-to-urban metamorphosis has impacted on the fabric of Africa's cities. The combination of often uncontrolled growth and a steady deterioration in the quality and distribution of public services and infrastructure has affected the majority of Africa's cities to varying degrees. In South Africa this situation has been further exacerbated by the policy of apartheid which produced an urban form "fragmented, racially structured and in which the vast majority of the poor are located on the urban periphery and the affluent in the core." (Hindson, et al. 1996).

It is against this backdrop of escalating human need and declining built and natural resources that urban ecosystems must be contextualized and understood in Africa. Just as Evernden (1992) in his treatise *The Social Creation of Nature* highlights that the concept of ecology is an ambiguous one, strongly influenced by social perceptions: "What matters is not what ecology is, but how it functions, how it is perceived and used", so too the understanding of the role, function, and importance of urban ecosystems differs between cultural and socioeconomic contexts. The purpose of this paper is to review the manner in which the concept of the urban ecosystem has been interpreted and employed in the development of an environmental management system for an African city.

The South African Perspective: The Emerging Sustainable Development Paradigm

In South Africa, the understanding of urban ecosystems is constantly being broadened as the country reenters and begins to partake in international debates. Pivotal to this process is the current global policy focus on sustainable development. Although difficult to define, the sustainable development concept is useful in that it highlights the interdependence of the social, economic, and ecological factors that make up the urban ecosystem (see Figure 24.1)—as well as the need to achieve a balance between these (often) competing factors if humanity is not to endanger its long-term future on the planet.

Building on this concept, a potentially important tool in managing and planning urban ecosystems is the Local Agenda 21 mandate contained within chapter 28 of Agenda 21. (Agenda 21 is the global action plan for achieving environmentally, socially, and economically sustainable development endorsed at the United Nation's Earth Summit held in 1992). The Local Agenda 21 mandate requires local authorities—as managers of the world's urban ecosystems—to undertake "a consultative process with their populations" and achieve "a consensus on a 'local Agenda 21' (i.e., a local sustainable development action plan) for the community" (United Nations Conference on Environment and Development 1993). The *raison d'etre* of Local Agenda 21 is that a new form of urbanism is required, which recognizes the critical role that urban ecosystems play in achieving global sustainability, and which focuses on local level action aimed at ensuring that a sustainable balance is achieved between environmental, social, and economic needs.

Durban—At the Frontier of the "New Urbanism"

Durban is the largest port and city on the east coast of Africa, with a population of some 2.5 million people, and covers an area of 1,336 km^2 (Spatial Development Framework Steering Committee, 1998). In 1994, Durban became the first city in South Africa to accept Local Agenda 21 as a local government responsibility. The overall aim of the Durban Metropolitan

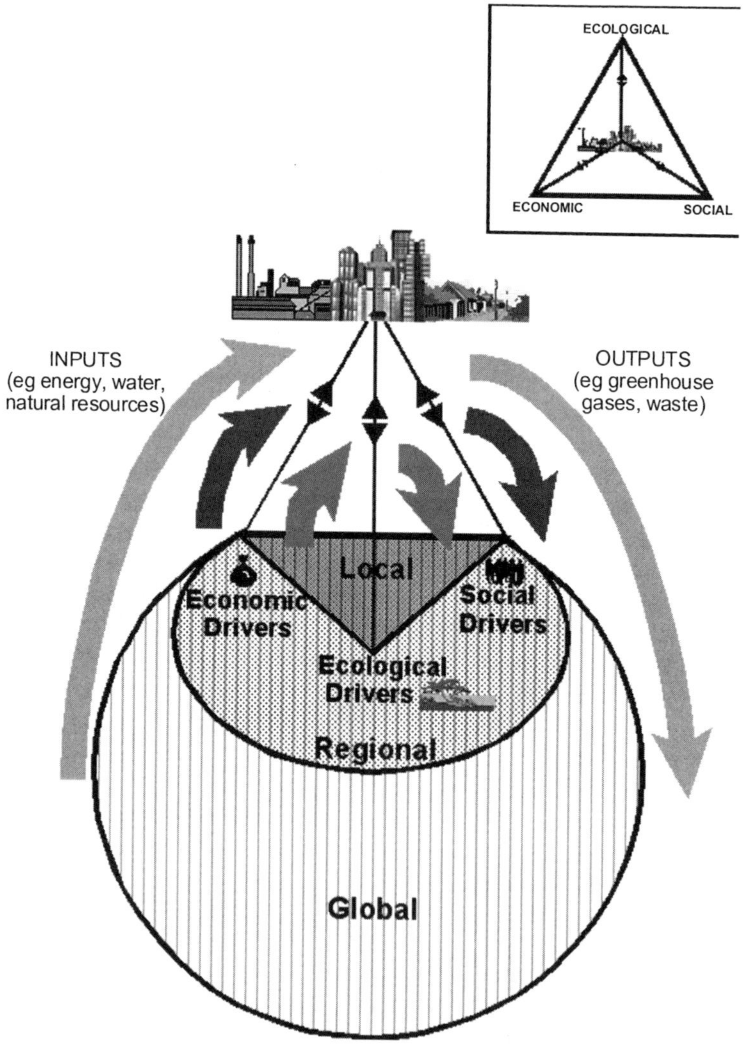

FIGURE 24.1. The urban ecosystem.

Council's (DMC) Local Agenda 21 program is the development and implementation of an environmental management system to ensure the long-term sustainable development of the city. The following analysis of the first two phases of Durban's Local Agenda 21 program (an updated description of Durban's Local Agenda 21 program can be found at *http://www.durban.gov.za/environment*) highlights the manner in which this broader urban planning initiative has provided a direct and indirect vehicle for promoting and developing a more holistic understanding of the urban ecosystem concept.

Phase 1: Assessment and Prioritization

The objective of Phase 1 of the Local Agenda program was to develop a better understanding of the state of the economic, ecological, and social components that make up the urban ecosystem in the Durban Metropolitan Area (DMA). To this end, the first State of the Environment and Development Report (SOEDR) for the DMA was commissioned in 1994 and completed in 1996.

Research Partnership

The SOEDR was undertaken as a partnership project between the then Durban City Council and two research organizations: the CSIR (Council for Scientific and Industrial Research—a quasi-governmental entity) and ISER (the Institute for Social and Economic Research—University of Durban-Westville). This partnership approach provided the opportunity for interactive planning, but also highlighted the sectoral bias of the three institutions and the critical need to develop a more holistic understanding of urban ecosystems among practitioners and decision makers.

Encouraging Interaction and Broader Horizons

In order to address this challenge, 17 key environment and development sectors were reviewed during the preparation of the SOEDR. These sectors were chosen to represent all aspects of the urban ecosystem in Durban (i.e., terrestrial resources, atmospheric resources, fresh water resources, marine resources, urban form, housing, transport, water supply and sanitation, waste, energy, economy, education, health, safety and security, governance, city finances, and the legal framework).

In order to encourage broad-scale discussion and interaction between the specialists and practitioners, and to help them identify links and interrelationships between the sectors, three techniques were used:

- Firstly, specialist panels were convened for each sector (except for the municipal financing, metropolitan economy, and legal sectors). These panels provided an opportunity for people with different perspectives and experiences of the urban ecosystem to share their views in open debate. Each panel consisted of a facilitator, specialists from the sector, service providers and other participants who were selected because of their experience and ability to think strategically vis-a-vis development and environment issues in the Durban area. As far as possible attempts were also made to ensure the presence of community experts on these panels. For example, community representatives who depended on polluted water sources were included in the water resource panel.
- Secondly, expert authors were appointed to prepare sector reports for each of the 17 sectors reviewed in the SOEDR. These authors were required to prepare a report, which included a concise statement of the

state of the environment and development with respect to the sector, and identified linkages between the sectors, as well as opportunities for a move toward more sustainable development and practices in the metropolitan area. Again, a number of the experts, because of their specific sectoral bias, found it very difficult to contextualize the contribution of their sector within an understanding of the urban ecosystem as an interactive platform for ecological, social, and economic forces, and within the sustainable development debate generally.
- Recognizing that the parochial and sectoral views held by the various stakeholder groups were exacerbated by a previous lack of dialogue, a 2-day away Cross-Sectoral Workshop was convened in June 1995. This was attended by the project team, the authors of the sector reports, key panelists (from the specialist panels), and community members (representing the three case study areas) drawn from the advisory committee (see Getting Everyone Involved and Using Community Experiences). For many, this provided the first exposure to a multisectoral mix of viewpoints on the urban ecosystem and underlined the need for cross-sectoral dialogue as a fundamental building block in promoting a more holistic understanding of the urban ecosystem.

Getting Everyone Involved

To further facilitate interaction and partnership-building between stakeholders around the urban ecosystem concept, three key stakeholder groups were convened during Phase 1:

- *The Interim Steering Committee*: This consisted of a small group of 10 key power brokers (representing business and industry, environmental Non-Governmental Organizations (NGOs), civic organizations, labor, the research sector, development agencies, and local government) convened in September 1994—prior to formal initiation of the project—to review the initial research proposal for the SOEDR. The proposal was substantially reworked on the basis of their review to emphasise the need for community participation in defining the issues and problems to be addressed in the study.
- *The Advisory Committee*: Once the study was formally initiated a much larger consultative forum was established. Representatives of the Advisory Committee were drawn from a broad range of stakeholder groups (NGOs and environmental interest groups, Community Based Organizations (CBOs), local authorities, development agencies, business and industry, academics, and professionals). The Advisory Committee was first convened in March 1995, and its primary function was to provide a forum for reporting back and reviewing the research process and findings. The committee originally consisted of 20 nominated representatives, but expanded to a final total of 95 due to changing representation within sectors and the emergence of new stakeholder groups during the political transition.

An important observation made during this period was that discussion of substantive issues relating to problems and opportunities within the urban ecosystem were often displaced by discussions regarding the legitimacy of the overall study process and representation on the committee (Roberts and Patel 1997). It was clear that apartheid's lingering legacy of distrust among government, civil society, and business and industry posed a significant obstacle to constructive discussions regarding the urban ecosystem and its planning and management. This situation was exacerbated by the limited human and financial resources available to the project team to intervene to build greater trust, capacity and understanding among the stakeholders groups.

- *The Inter-Service Unit Network*: A committee was also formed to facilitate discussion amongst local government officials. The committee consisted of 40–50 local government officials drawn only from the nine service units of the central municipal structure and was first convened in September 1995. Following the first democratic local government elections (held in June 1996) this group was expanded to include representatives from the six new local council areas within the DMA and the new Durban Metropolitan Council. The aim of this forum was to provide an opportunity for local government officials to discuss the concept of sustainable development and the manner in which it should influence planning and management of the urban ecosystem. It also provided an opportunity for report-back and outside review of the research process for the project team. The network was, however, beset by the same difficulties as the other stakeholder groupings (e.g., sectoral bias, lack of capacity, etc.).

Using Community Experiences

In order to test the depth of general society's understanding of urban ecosystems, three case studies (in the areas of Inanda, Cato Manor, and the Durban South Basin) were initiated in order to develop a better understanding of peoples' perceptions and attitudes regarding environment and development issues. Several different approaches were used in developing this understanding.

- Firstly, leadership interviews were held with 104 interviewees within the case study areas, which investigated the concept of environment and development, and examined priorities, resources, and problems in different areas.
- Secondly, focus groups were convened. This was an important intervention, as the views of community leadership do not necessarily always reflect the diversity of views within a community. Twenty-two focus groups comprised of ordinary residents of 16 different communities within the three case study areas were conducted. The focus groups were convened to reflect differing levels of socioeconomic development and different gender perspectives, and also to examine the views of the youth in the three different areas.

During the focus group sessions, it became clear that different socioeconomic and cultural backgrounds would affect the way in which the more holistic concept of urban ecosystems was understood. For example, it was found that the very poor, whose survival depends closely on direct access to natural and other resources in their immediate environment, have a clear understanding of the interdependence of environmental health and quality of life (Hindson, et al. 1996). When asked "What are the most and least important environmental and development problems in your area?" a township leader answered as follows:

> They are all important because they all have an important impact on our lives. For example, if waste is not removed, it becomes a health hazard. Narrow roads lead to high accident rates and lack of housing leads to unhealthy conditions—squatter camps. They are all important. There are none that are the least important. (Hindson, et al. 1996).

There were also contrary views. The following comment, taken from a discussion which took place in a men's focus group in a township, highlighted conventional views about the urban ecosystem (Hindson, et al. 1996):

> To be honest I personally wouldn't care less about the environment. That is a white man's thing. For instance I won't conserve water, electricity, etc. Those things are of no material benefit to me or the next black person because we have been deprived for many years. Conservation is for the elite, the ones who have been enjoying life for the last couple of years.

This latter view is very much a product of apartheid policy, which denied people a role in managing and planning the resource base of the country, creating a high level of disregard and ignorance. This mind-set remains a significant hurdle in trying to promote a more holistic understanding of the urban ecosystem concept in South Africa.

Prioritization Process

Following the completion of the SOEDR, a prioritization exercise using workshops was undertaken with local communities in the six local council areas, the Advisory Committee, and the Inter-Service Unit Network to identify action areas for sustainable development in the DMA. The following needs emerged as priorities:

- to promote peace, safety and security in the metropolitan region;
- to improve water and sanitation management;
- to develop an integrated housing policy;
- to establish a structure to coordinate land use, transportation, and environmental planning; and
- to institutionalize the Integrated Environmental Management procedure.

Phase 2: Policy Formulation and Planning

Step 1: Actions Initiated by the Environmental Branch

In Phase 2 of Durban's Local Agenda 21 program, a package of projects was initiated by the Environmental Branch to address the strategic priorities highlighted by the Phase 1 process; that is, the need for:

- institutional change (to ensure that local government has the skills and resources necessary to plan and manage the urban ecosystem effectively);
- environmental information to guide strategic planning and development (to develop a clearer understanding of how the urban ecosystem functions);
- tangible products and poverty reduction through environmental improvement (to demonstrate how sustainable urban ecosystems can improve quality of life); and
- capacity building (to ensure that all stakeholders understand sustainable development and urban ecosystem concepts).

Key Pressure: Need for Institutional Change

In Durban the driving forces for institutional change are:

- a changing statutory and policy environment at the national and provincial levels, which is increasing the environmental management responsibilities of local government; and
- stakeholders are increasingly concerned about, and want to be involved in, the management of the urban ecosystem.

Strategic Response

Durban Metropolitan Environmental Policy Initiative (DMEPI). In order to manage the urban ecosystem in a sustainable manner, the DMC must restructure and resource the institutions that currently deal with the metropolitan environmental management function. To initiate this process, the DMC approved the development of the city's first environmental policy in 1997 and a review of the human and financial resources that would be required to implement the policy.

In line with the participatory nature of Durban's Local Agenda 21 program, the Durban Metropolitan Environmental Policy Initiative (DMEPI) involved a broad range of stakeholders in the preparation of the new policy and related institutional structures. Learning from the shortfalls of Phase 1, the stakeholder group convened for DMEPI was a smaller, more representative grouping (known as the Review Panel) with clearer terms of reference.

The fact that the Review Panel had a clear function and was able to influence the policy development process resulted in the creation of a committed and energetic stakeholder forum, whose debates broadened the capacity of participating members to understand the plethora of issues affecting the planning and management of urban ecosystems. Capacity building also occurred outside of the Review Panel through a series of community workshops held in response to the NGO and CBO sectors' view that the lack of awareness of broader environmental issues in local communities limited their meaningful engagement in environmental policy development.

This broader capacity-building around basic environmental issues proved to be an essential prerequisite for meaningful participation by previously disadvantaged stakeholders in the DMEPI process. Following the workshops, the quality of community representatives' participation in public events was noticeably improved, a fact which appeared to be directly related to the increase in understanding and confidence developed through the capacity-building workshops (Common Ground Consulting 1998).

Key Pressure: Need For Environmental Information to Guide Strategic Planning and Development

In the DMA there is a growing need for improved environmental information because:

- A range of existing environmental and development opportunities and challenges exist within the city and are potentially in conflict; and
- New statutory obligations require local governments to produce strategic plans that will ensure *inter alia*, that the principles of environmental sustainability and sustainable development are applied to local development and planning.

Strategic Response

Durban South Basin Strategic Environmental Assessment (SEA). The Durban South Basin is the economic heartland of the DMA, contributing more than 60 percent of the DMA's gross geographic product and 30 percent of the industrial land in the DMA (Economic Development Department 1998). It is also environmentally degraded, experiencing a range of problems, from air pollution and waste disposal issues to the depletion of important natural resources. This not only undermines the quality of life of residential communities in the area, but negatively impacts on the competitiveness of the business environment. Furthermore, the close interface between residential and industrial activities that exists in the area has created tensions between residential communities, local government, and industry regarding the desirability of future industrial development.

The aim of the SEA was to assess the environmental, social, and economic problems and opportunities in the area, and to propose sustainable development guidelines and management options to address both current problems and to guide future development. The study addressed the current situation and long-term development scenarios for the area (e.g., petrochemical expansion, dig-out port development, and a mixed residential-light industry option) and assessed their impact on: living environments, air quality, waste generation, the natural environment, and disposal and institutional and infrastructural needs.

Learning from experiences with the Advisory Committee in Phase 1 of the program, all reasonable efforts were made to ensure that stakeholders (particularly local communities) played an informed and capacitated role in the study. As a result, various new elements were incorporated into the education and capacity-building program for the study based on suggestions made by the South Durban Community Environmental Alliance (SDCEA), an alliance of community and environmental organizations in the South Industrial Basin. These included, *inter alia*:

- **a schools education program** involving information sharing in a workshop environment and an environmental competition aimed at familiarizing young people (at the secondary school level) in the area with the key issues affecting the urban ecosystem in their area; and
- **a field-worker development program** (including a training course) for five individuals from residential zones within the study area. This was aimed at building capacity among local communities in terms of environmental issues and facilitating informed participation in the SEA process. (Participative Solutions Africa 1998)

A difference of opinion, however, arose over the role of the field-workers and the strategic direction of the education and capacity-building program between the project team and SDCEA. This culminated in the withdrawal of SDCEA from the management of the education and capacity-building program of the SEA. SDCEA claimed that the program had been unsuccessful and that they had been excluded from its design and execution. This was despite the fact that SDCEA had been involved in every aspect of the program from its inception and had the opportunity to monitor and redirect the program during the monthly project meetings with the project team. This clearly indicated that while a great deal of effort and resources might be allocated to developing a broader understanding of the urban ecosystem, these efforts are easily marginalized and undermined through conflict, lack of trust, and power struggles between stakeholder groups.

Durban Metropolitan Open Space System (D'MOSS) Framework Plan. In addition to being a manufacturing centre, the DMA is a tourist city located within a region of high biodiversity on the East African coast. The need for an open space plan that ensures the conservation and appropriate management of the important natural resource base of the city has therefore been identified as an urgent and strategic priority by the city's planners (Spatial Development Framework Steering Committee 1998).

Preparation of the plan involved identifying, mapping, and quantifying all potential metropolitan open space in order to create an inventory of the open space assets. The roles of the different open space assets in delivering goods (e.g., water for consumption), services (e.g., waste treatment), or benefits to the residents of the DMA were then evaluated (Markewicz English 1998). In the DMA the services delivered by open spaces (i.e., ecosystem services) can be categorized into 17 different service types: gas regulation, climate regulation, disturbance regulation, water regulation, water supply, erosion control, soil formation, nutrient cycling, waste treatment, pollination, biological control, refugia, food production, raw materials, genetic resources, recreation, and cultural. The replacement value of these services is conservatively estimated to be about half the value of the current operating budget of the Durban Metropolitan Council.

The use of resource economics to communicate the value of the natural components of the urban ecosystem in a way that is more easily understood by most stakeholders is particularly important in a city such as Durban where concern for "green" issues and the natural resource base is still seen as the prerogative of an elite white minority, and of little import to the larger, poorer black majority. Tools such as resource economics therefore must be investigated as means of turning abstract concepts about the urban ecosystem into concrete realities and consequences for as broad a range of stakeholders as possible.

Key Pressure: Need for Tangible Products and Poverty Reduction Through Environmental Management

Because of the high levels of poverty and unemployment within the DMA, key driving forces in environmental management include:

- people needing to see the benefits of improved environmental management; and
- poverty as a significant cause of environmental degradation and resource depletion.

Strategic Response

Community Open Space Development. This project focused on local-level project opportunities (linked to the bigger policy and planning initiatives of the overall program) that had the potential to benefit local communities directly and create opportunities for advancement through job creation, transfer of skills, and son on. In terms of the development of the D'MOSS Framework Plan, for example, it was realized that previously disadvantaged communities would need to be able to identify and experience the value of open spaces before they were likely to support the preparation and implementation of such a plan. In working toward this goal, community open spaces were developed in five previously disadvantaged communities as a pilot project.

Initially meetings were held with the local councillor for each area and with the local Parks Department representative in order to identify projects which would likely enjoy community support. This approach resulted in the development of open space areas that provide a valuable recreational and environmental education resource to communities that previously did not have access to such resources. There was also extensive community involvement in the development of the open spaces, to ensure community support for the projects. Community groups and schools were involved in the design of the open space and local labor and contractors were used wherever possible to ensure employment creation. The project capitalized on the fact that people learn "from the soles of their feet up" and that making urban ecosystem issues (i.e., the protection of urban open space) relevant and meaningful to peoples' day-to-day lives provides a positive platform for ongoing learning. The ultimate failure of the project lay in the fact that funds were subsequently not provided by local government for the maintenance and further development of these areas. This illustrates the potential power of political decisions to undermine sustainable development initiatives.

Key Pressure: Need for Capacity Building

The major driving force is:

- A lack of understanding of environmental and sustainable development issues among stakeholders and decision makers.

Strategic Response

Local Agenda 21 Outreach Program. The aim of this project is to complement the project-based work of the Local Agenda 21 program and facilitate ongoing learning about the urban ecosystem by making available to stakeholders a range of media products (e.g., booklets, posters, pamphlets, and glossaries). The focus of these publications is a review of key environmental and sustainable development issues in the DMA and the role of the Local Agenda 21 program and projects in addressing them. Capacity-building opportunities (e.g., workshops, schools programs, field-worker training, and training programs for city officials) are also being provided where resources permit and partnerships can be developed with other institutions. Innovative approaches such as community theater are also currently being explored as a means to further develop the understanding of sustainable development and its link to human well-being and to address the limitations of printed media in areas where illiteracy rates are high. An initial pilot theater project has, however, revealed the limitations of theater's value as a mass communication tool, in that it is generally more effective with smaller, contained audiences and because it is difficult to produce a dramatization that is effective with diverse audiences.

Conclusion

In any management system, there is a requirement for continuous monitoring and evaluation to determine progress in meeting objectives and to highlight where changes in objectives and strategies may be necessary. This is where the Durban experience is most limited. A combination of scarce resources (human and financial) and crisis management (born out of the process of political transition and a legacy of unsustainable planning and development) have meant that there have been no structured or quantitative attempts to monitor or evaluate the success of the Local Agenda 21 program to date. A subjective analysis, however, suggests that, at the very least, the program has helped promote a more holistic understanding of the urban ecosystem at a metropolitan level, for example:

- Local councils' acceptance and endorsement of the new Durban Metropolitan Environmental Policy (structured around the following six themes: environmental management systems; development and planning; human health and safety; pollution and waste management; environmental resources management; and education, training and awareness) suggests an awareness among decision makers of the fact that social, ecological, and economic concerns are of equal significance in ensuring the effective planning and management of the urban ecosystem;
- Impending local government restructuring in Durban will see this more holistic understanding of the urban ecosystem help to create more effective networks among officials working within the social, ecological, and economic sectors. This in turn will result in the city's human and financial resources being more effectively harnessed and directed towards achieving sustainable development; and
- Initiatives and projects within the Local Agenda 21 programs have contributed to the emergence of environmental pressure groups within civil society which utilize local councils' commitment to sustainable development and Local Agenda 21 to challenge policies, strategies, or developments they believe are unsustainable from a social, economic, and/or ecological perspective.

These initial successes confirm the potential value of Local Agenda 21 as a tool for ensuring a more holistic understanding of the urban ecosystem and facilitating the achievement of global sustainable development through local action. In advancing a more holistic understanding of the urban ecosystem, the following rules of thumb were highlighted by the Durban experience:

- Put It On Center Stage: Use mainstream processes that people are already involved in (e.g., such as the urban planning and management process) to generate opportunities to discuss the urban ecosystem concept with stakeholders;

- Getting People Together: It is important for people to be able to share different understandings and experiences of the urban ecosystem;
- Different Strokes for Different Folks: There is a need to use different tools to communicate the concept to different stakeholder groups. These tools can be as varied as street theater and resource economics;
- Recognize the Capacity Mismatch: There is a need to empower all stakeholders. This applies as much to politicians and the heads of industry as it does to individuals and community-based organizations;
- What's In It for Me?: Acknowledge that social context and day-to-day needs will impact significantly on the way the concept of the urban ecosystem is understood. Approaches to developing a broader understanding of the term must be relevant and sympathetic to need and social circumstance;
- Need to Break the Mold: Sectoral bias among "experts" involved in the development of environmental management systems at the local level often limits the development of a broader, more integrated understanding of urban ecosystems;
- Put Your Money Where Your Mouth Is: Building capacity and understanding of complex concepts such as urban ecosystems takes time and money and must be appropriately resourced;
- Getting Down and Dirty: The concept of the urban ecosystem as a product of synergistic social, economic, and ecological forces is often difficult to grasp. Getting stakeholders involved in processes and practical projects that are linked to the planning and management of the urban ecosystem can, however, help to demystify issues;
- Build Partnerships: Social and political tensions and/or distrust can be significant limiting factors keeping us from a broader understanding of urban ecosystems. Developing an understanding of urban ecosystems must therefore be preceded by the building of viable partnerships between stakeholders; and
- Need to Know What Is Out There: Accurate, up-to-date data are required on all aspects of the urban ecosystem.

Acting on these guidelines is vital, for as Elizabeth Dowdeswell, the executive director of UNEP noted at the City Summit, held in Istanbul in 1996: "One thing is clear: The fate of cities will determine more and more not only the fate of nations but also our planet. We can afford to ignore the issue of the sustainable management of our cities only at our own peril" (Dowdeswell 1996).

References

Common Ground Consulting. 1998. *Review of capacity building workshops for the Durban metropolitan environmental policy initiative*. Unpublished Report.

Dowdeswell, E. 1996. Editorial. *Our Planet* 8(1):3.

Evernden, N. 1992. *The social creation of nature.* John Hopkins University Press, Baltimore, MD.

Hindson, D., N. King, and R. Peart. 1996. *Sustainable development in the Durban metropolitan area.* Indicator Press, Durban, South Africa.

Markewicz English. 1998. *Durban metropolitan open space system framework plan.* Unpublished report.

Participative Solutions Africa. 1998. *South Durban strategic environmental assessment.* Public Participation, Education and Capacity Building Report. Unpublished Report.

Roberts, D.C., and Z. Patel. 1997. *Durban's local agenda program—planning for a sustainable future.* Paper presented at the Pathways to Sustainability—Local Initiatives for Towns and Cities International Conference. Newcastle, Australia.

Spatial Development Framework Steering Committee. 1998. *Durban metropolitan area spatial development plan.* Volume 1: Spatial Development Framework. Urban Strategy Department, Durban, South Africa.

United Nations Conference of Environment and Development. 1993. *The Earth Summit.* Graham and Trotman/Martinus Nijhoff, London, UK.

World Resources Institute/The United Nations Environment Program/The United Nations Development Program/The World Bank. 1996. World Resources. *A guide to the global environment.* Oxford University Press, Oxford, UK.

Section IV
Visions for the Future of Urban Ecosystem Education: Themes

ALAN R. BERKOWITZ, KAREN S. HOLLWEG, and CHARLES H. NILON

How can the promise of urban ecosystem education can be achieved? The collection of chapters in Section IV, in addressing this question, represent a similar diversity of viewpoints that has characterized the perspectives and substance of the entire book. The authors give both pragmatic and visionary glimpses of the future, synthesizing what we have learned from science and education, from theory and practice. They excite us with the possible, inspire us with compelling examples, and challenge us with the obstacles, gaps, and needs they identify.

The chapters paint vivid pictures of the future, but each does more than that, offering us a synthesis of all of the themes addressed in the book. In considering how the promise of urban ecosystem education can be achieved and how we might get there, the authors also had to touch on the topics of the earlier sections, at least in some way. Each author sets out why understanding urban ecosystems is important and what they mean by understanding cities as ecosystems, and each lays out what might be the goals of urban ecosystem education as they look to the future.

In painting their visions for the future, the authors also identify the gaps that exist in our intellectual foundations and practical application, and concrete needs for action. These, in aggregate, can comprise the agenda for the nascent field of urban ecosystem education. *What new research should be undertaken by urban social and natural scientists to discover and clarify the basic concepts and perspectives we all need to allow us to more clearly grasp the principles on which understanding can be built? Do we need more research on how people learn and develop understandings about complex systems like cities? Do we know how to integrate a theme like urban ecosystems into the K–16 system so that it is not an added-on, peripheral topic? How can the next generation of teachers be better prepared to teach ecology, social studies, geography, history, and the like in ways that truly embrace urban ecosystems and effectively integrate the relevant disciplines? What must happen for mainstream curricula and textbooks to incorporate urban ecosystems? How can we develop and sustain partnerships for urban ecosystem education?*

As in the other sections of the book, the authors in this section take different perspectives and speak to somewhat different audiences, while also sharing much common ground. We start with a reminder of the global dimension of urban ecosystems and a compelling argument by Golley (Chapter 25) for the importance of teaching for the common good as a core component in any education work focused on cities. His scope encompasses the entire globe, and he places the frontier in the context of past work in cities by scientists and designers. Golley's writing goes beyond his ecological roots to inspire all educators and scientists to look for ways to use urban ecosystem education to foster environmental citizenship for the betterment of the human condition.

Bybee (Chapter 27) speaks most directly to educators involved in formal education, to researchers, teachers, and administrators. He gives a blueprint for integrating urban ecosystem education into education reform, developing the rationale and laying a framework for using urban ecosystems as a theme and vehicle for achieving recognized academic goals. Placed in the context of the evolution of our scientific understanding of human ecology, he provides a detailed and practical curriculum that might be used and adapted in many educational contexts, and concrete recommendations for infusing urban ecosystems into the education agenda.

Burch and Carrera (Chapter 26) and Agyeman (Chapter 28) reflect on opportunities for environmental educators in and out of schools, and for scientists, educators, activists, and citizens who work in and with community organizations and everyday people in cities. Both chapters show how the creation of dynamic, community-centered partnerships for research and education can achieve broad educational, community revitalization, and sustainability goals. Burch and Carerra's wonderful examples from Baltimore and beyond, are complemented by Agyeman's examples from Detroit and Great Britain. The vital role of genuine participation in research, education, policy making, and management in working toward a more sustainable and equitable society is championed in both of these chapters.

The final two chapters in the book represent two attempts at synthesis. The chapter by Peter Cullen (Chapter 29) is based on the final remarks he presented at the end of the Cary Conference where he was asked to summarize what he heard and learned. As an ecologist and science administrator seasoned at creating innovative programs at the interface between research, policy, and the public, Cullen's insights are illuminating and challenging. In our concluding chapter (Berkowitz, et al., Chapter 30) we have attempted to bring together the rich and diverse ideas from the formal and informal discussions during and after the Cary Conference in addressing the question, quite literally, "Where might we be 10 years from now?"

25
Urban Ecosystems and the Twenty-First Century—A Global Imperative

Frank B. Golley

Introduction

The city, which is so dominant in our thinking today and the focus of so much attention, is a relatively new phenomenon, if one has the several-hundred-thousand-year history of the human species in mind. The evolution of the town or village into this new form of living space occurred less than 10,000 years ago, approximately coterminous with the invention of agriculture. Cities appeared in several different places on the Earth at about the same time.

Ancient cities were not all dense, crowded, dirty places. The hanging gardens of Babylon were one of the seven wonders of the ancient world. Athens and Rome were cities of great architecture that awe us even today. Cities were more than magnificent gardens and architecture, however. New challenges of life induced new thought; experimentation and progressive administration were features of cities. When they appeared they signified something new in the world which was a fundamental contrast with the natural landscape and the agricultural countryside. This may be one reason why the Mongol hordes razed cities to the ground when they captured them. Could they sense the competition of the urb with their nomadic society of the Asian grassland and desert and so destroyed it where possible?

The contrast between city and country creates a dualism which shapes our views of human life and nature. The city contains the centers of government, religion, commerce, and education. Cities engender an urban point of view which is different from the perspectives of the nature-oriented hunter-gatherers or nomads, with their sacred groves and caves, or the farmer caught in the annual calendar rituals of plowing, sowing, and harvesting. The city dweller is not just free of the direct controls of environment but lives in a world of human design and construction. The city is an invention of humans to satisfy human needs. Of course, it is shaped by its larger environment, but the city also *creates* an environment which shapes its citizens through the experience of living in the midst of the city and through education. One source of the estrangement of Western societies from the natural environment is their urban perspective. Even the agriculturalist is urbanized to the extent that the city is the source and the market for the machinery, products, and information used by and produced on the farm.

Because of this centrality of the city, it has been the focus of Western scholarly thought throughout history. Cities are a challenge to us. Understanding these old and large human-built systems is not only challenging because of the problems we find in cities or the numbers of people living there. The challenge is also intellectual and pedagogic. How do we understand something so complex and so shaped by human desire, design, and dreams? How do we express our understanding of pattern and centrality and convey these concepts to our fellow citizens, including small children, so that they can become successful urban dwellers? How do we ecologists reach across the disciplinary gulf that separates the natural and the social sciences to create something new and useful? Addressing these questions is the object of this introductory chapter and the chapters that follow. Examining the cities as ecosystems provides a different and maybe even unconventional perspective of a familiar topic and provides a new habitat for ecological exploration.

Who Studies Cities?

Challenged by problems and intrigued by opportunities, the urbanist has created a field characterized by diversity of many disciplines addressing aspects of city life. The natural sciences seldom play an important role in these studies because the city is largely a social construction. Instead, the ecologist studies organisms in their natural habitat and constructs ecosystem models that predict how forests, rivers, or even oceans may act under changing environmental conditions. The invention of urban ecology and the application of the models derived from the study of nature to the city is relatively new.

Beyond an occasional paper, urban ecology really becomes visible as a subdiscipline in the post–Second World War period in Europe. Ecologists were challenged by the destruction wrought by massive bombing of cities. They described the processes of recovery of urban plant and animal life from military insult. Herbert Sukopp talks about the history of these studies in his paper, Stadtokologische Forschung und deren Anwendung in Europa (1987) and in the volume he edited with Slavomil Hejny (1990), *Urban Ecology: Plants and Plant Communities in Urban Environments.* These studies mainly described relationships of species with environmental conditions or with change in community structure over time and space, but they raised other questions which are under study today. For example, Kowarik (1990) discussing the changes in the plant communities in the former West Berlin notes that the flora of the city contains 839 native species and 593 aliens brought to Berlin directly or indirectly by human activity. The ecologist observing the failure of alien species to invade intact, undisturbed communities asks, *Does the urban environment provide new niches for species? Does the disturbed nature of the urban environment after the war provide opportunities for alien species to compete successfully with native species?*

Most urban ecological studies have addressed the ecology and biology of species and populations. There have been fewer attempts to apply the concepts of ecosystem ecology in cities. Several of these urban ecosystem projects were organized by the International Biological Program, or IBP, which, among a variety of projects, tried to develop an understanding of the biological productivity in ecosystems representing large regions, called biomes. Almost all of these studies focused on sites in natural landscapes. For example, Maxime Lamotte (1978) led a team of scientists to study the savanna ecosystems in Ivory Coast; George Van Dyne (1980) led a similar group studying the short grass prairie in Colorado. An exception to this focus on natural systems was the work led by Paul Duvigneaud in Belgium. Duvigneaud and his colleague S. Denayer-De Smet (1977) developed a description of the biological productivity of Belgium, which contained a section focusing on the city of Brussels. Duvigneaud (1974) published his conception of urban ecology in the Memoirs of the Royal Botanical Society of Belgium. Clearly the IBP paid inadequate attention to cities, but at least the first internationally organized urban ecological studies were carried out and became visible to the public.

The IBP was a fixed-length program and it ended officially in 1974. As IBP came to an end, UNESCO initiated a new program called Man and the Biosphere. This program was led by Francesco di Castri and it used the experience of the IBP to address areas representing research gaps in our understanding of ecosystem ecology. MAB began a program on urban studies, which started as one of the smallest sections of UNESCO's Environmental Division but gradually became one of the largest (Spooner 1986). Sukopp (1990) comments, "At the start investigations of energy and water played an important role. Great hopes were set on the potential of systems science." These hopes were never fully realized, in my opinion. Descriptions of energy flow and water usage in urban areas, such as in Hong Kong (Boyden, et al. 1981), were interesting in themselves but were not connected to practical issues or environmental problems. Others apparently shared my opinion because the MAB urban program later shifted its attention to issues such as migration to cities, administration of cities to enhance quality of life and reduce negative environmental impacts, and management of urban green areas. This shift was in part stimulated by the UNESCO prediction that half of the humans in the world would live in great cities by the year 2000.

In the USA an organization called The Institute of Ecology or TIE was formed in the late 1960s to carry out applied studies in ecology. Among these was a consideration of the urban ecosystem (Stearns and Montag 1974), which was organized into eight task groups: (1) Goals for the urban ecosystem; (2) Physical structure and function; (3) Resources: water, energy and materials; (4) Population processes; (5) Institutional structures; (6) Case studies of cities and urbanized regions; (7) Indicators of urban ecosystem function and health; and (8) Systems capacities, limits, and intersystem linkages. Each task group was asked to consider four general questions: *What is the urban ecosystem? What do we think we know about it? What do we need to know to improve our understanding of it?* and *How can we apply what we already know?*

The study yielded a book, *The Urban Ecosystem: A Holistic Approach*, which presented 26 recommendations from the task groups. The first of these states: "We recommend that high priority be given the development of specific social means for implementing the goals of urban society. Changes in attitude, behavior, and lifestyle should be encouraged through education." Here is a direct linkage between the research of ecologists in the 1970s and the interest in urban ecosystem education which underlies this volume. These ecologists proposed that education in urban ecosystem principles and concepts will help citizens make decisions that will enhance their quality of life. Their proposal is a challenge because questions about how society will decide upon directions for change and how the various elements of society will work together to accomplish change need to be addressed and answered. It is not enough to pass the responsibility for creating citizens capable of solving urban problems onto professional educators alone. The task is much larger, although education will play a key role in what is achieved.

The Urban Ecosystem

My intention in this chapter is to apply the general concept of the ecosystem (Hagen 1992; Golley 1993) to the city in an international context. Since there is little hard data from which to make comparisons, my discussion will be general. Lack of knowledge of urban ecosystem ecology is a conspicuous gap and deserves our attention. Presumably the LTER studies (see Pickett, chapter 5; Grimm, et al., chapter 7; and Grove, et al., chapter 11 in this volume, Grimm, et al. 2000; Pickett, et al. 2001) will stimulate other work and will lead eventually to comparative studies which could be used by administrators to judge performance under a variety of conditions and could be used by citizens to inform their decisions. That is not possible currently.

Urban ecology began at a time when studies of system energetics were popular and the energy flow in cities was an obvious goal of these early investigations. Later, studies of matter cycling became popular and these were also conducted in cities. Probably the prime example of urban ecosys-

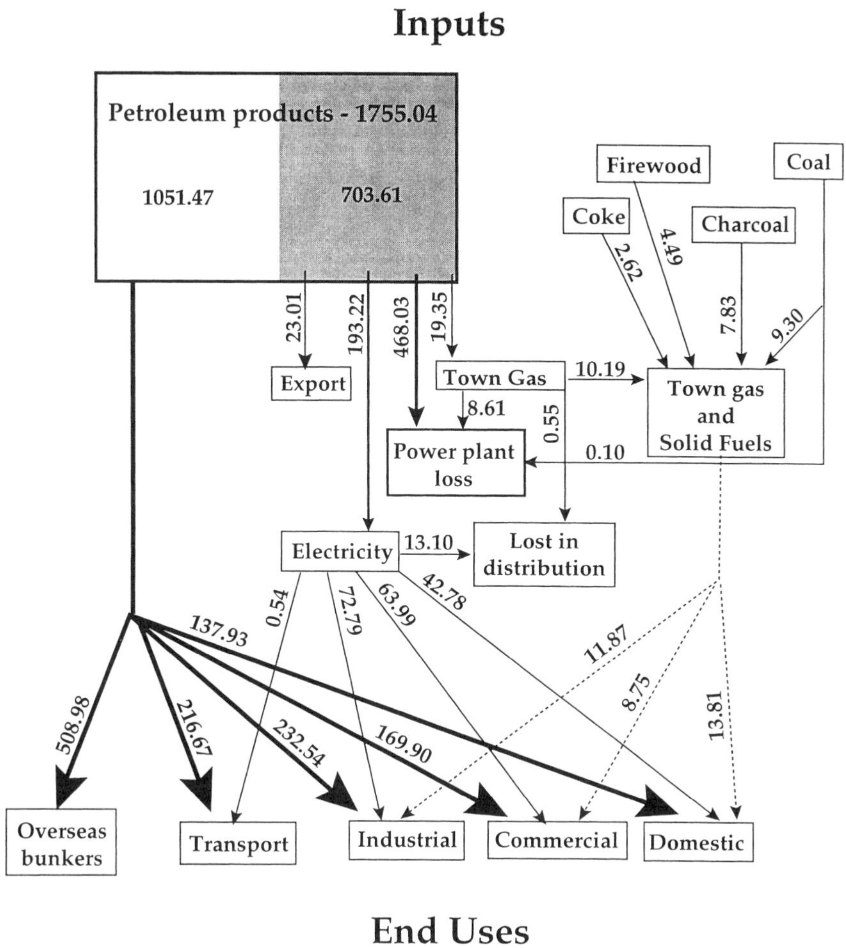

FIGURE 25.1. Extrasomatic energy flow chart: Hong Kong, 1971 (MJ $\times 10^8$). Redrawn from Boyden et al. (1981), permission of Australian National University Press.

tem ecology is Steven Boyden and colleague's (1981) study of the metropolitan region of Hong Kong. Even though it stands almost alone, a taste of the type of data gathered in the study conveys a sense of what was and can be accomplished.

In this project, the city was considered as a processor of energy and phosphorus which were related to the health and well-being of the inhabitants. The majority of energy input to Hong Kong was from petroleum products (Figure 25.1). The annual input of energy was 1755×10^8 MJ. However, this rate of input is deceptive. Hong Kong acts as a transfer point for distribution of petroleum and other products in the Asian region. Thus, the fuel that enters the system is partly transferred to overseas bunkers or storages

(Figure 25.1) and is not used in the metropolitan region itself. Domestic energy use (in percentage of the total) is 7.8 percent, commercial 9.5 percent, industrial 13.1 percent, transportation 12.2 percent, and the power loss in distribution is 26 percent. These statistics emphasize the fact that Hong Kong is a commercial city that lives by transferring products across its boundaries as it collects and sells products worldwide. This pattern also is shown in the cycling of phosphorous. The inputs and outputs are many times the soil and water stocks internal to the city.

These data suggest that the city is analogically like a parasite. It depends upon other regions to provide its inputs and accept its outputs. It functions as a node in the global economic system. Even though Hong Kong's energy use is numerically large, the per capita use of energy (total energy divided by the population) is only about one tenth that of the United States and only about double that of China as a whole.

The population of Hong Kong is quite dense, as any visitor can attest. In 1971 the population density of the built-up area was 380 persons per hectare; in the most dense district, Wanchai, it reached 2288. The area of maximum density is about ten times as crowded as areas of maximum density in London or Paris. The life expectancy for Hong Kong residents is approximately the same as for people in the United States, but the rate of death in the United States is higher. The rate of death in the United States in 1973 was similar to Hong Kong for malignancies but was about four times greater for heart disease.

These comparisons indicate some of the contextual differences between Hong Kong and other cities and give us dimensions with which to evaluate conditions across a gradient of difference. In this sense they may be useful for broad political or economic decisions on the part of government; however, they seem to be much less useful at the level of the citizen or the school. The significance of the statements depends upon their quantitative value, which becomes significant only in comparison with other values. For example, we are impressed that the most dense district of Hong Kong is more dense than the most dense district of New York City, but the difference is not meaningful in itself. It becomes meaningful when we translate this density into the design criteria for some new housing projects in Hong Kong that provided two square meters per person. Two square meters is a statistic that is personal. We can draw two square meters on the ground and imagine living within that space. Of course, rooms were larger than two square meters because families made up of several people shared a single room and maybe a balcony for cooking. Further, we are surprised when we learn that families live reasonably well within such limited spaces, and upon being questioned say that life is pleasant and satisfying. Numbers of such families commit little social crime nor do they act as vandals. In this cultural situation the structure of space, the role of the individual and the family, the nature of work, class and status create a very different attitude toward the structure of the city than do those same properties in a city in

America, for example. Sociocultural characteristics are crucial in understanding these differences. If there are rules about human need for and use of space, they do not appear to be general across cultures.

This kind of comparative, energy-focused analysis is useful because it makes clear the connection of the city to its hinterland, from which it derives resources and into which it deposits products and wastes. The hinterland is the environment in which the city is embedded. It may be large or small. In the case of Hong Kong, it is probably the entire planet. Eugene Odum has often said in my presence, "The city is a parasite on its region." A parasite lives in or on a host and usually obtains its requirements from the host without killing it. Odum is suggesting that cities feed on their hinterlands but that the relationship is seldom fatal.

The approach taken by Steven Boyden is grounded in his study of the history of Western civilization from a biological point of view (Boyden 1987). In this book Boyden traces the development of the city primarily in England, and especially the tremendous overcrowding due to changes in agricultural practices, population growth, and industrialization. The conditions in late eighteenth and early nineteenth century London, for example, were terrible from our modern point of view. Overcrowding, inadequate housing, lack of waste disposal, poor food, and so on were the normal conditions of life for the lower classes and the poor (Mayhew 1861). Boyden points out that these conditions caused a few reformers to create what he calls secular humanitarianism, which involved a sense that the citizens were responsible for the condition of life of the people as a whole. This sense led to the government replacing the church as the organization concerned about the poor and the institution of many measures including sanitary sewers, garbage collection, street lighting, a police force, and many elements of urban life considered routine today. The public health services have their origin in this period. Boyden treats these changes as a form of human adaptation to new conditions created by the crowded city.

Boyden also terms the modern urban period (the twentieth century) as the high energy phase of human development, in which humans depend upon fossil and nuclear fuel to supplement the organic fuels of earlier times. High energy inputs cause changes in the technometabolism of the city (Boyden's phrase), its social organization, and the personal well-being of the cities' inhabitants. As a consequence, the success of urban management using the information generated by science reversed a situation characterized by high death rates, poverty, and intense pollution and changed it to an urban environment where health conditions, education, and opportunity far exceeded that of rural hinterlands surrounding the city. Our present inability to manage the technometabolism of urban areas, however, threatens these gains with a return to large numbers of urban poor incapable of escaping poverty and experiencing increased ill health and much higher levels of pollution.

Urban Biology

It is essential that we apply our modern concepts of scale and ecological hierarchy to the city. The city is embedded in a region from which it obtains many resources, such as water, inhabitants, and often a large part of its food supply. The hinterlands also receive the wastes of the city. The metropolitan region itself has significant properties, including administration, and a psychological sense of place, made apparent by sports teams or traditions, which distinguishes a city from another place. And then the city can be subdivided into various hierarchical levels which emphasize one aspect or another of concern to the researcher or manager. Many urban ecological studies have focused on a fine scale of urban ecology. I am referring to studies of the natural history and biology of urban green spaces, like parks, roadsides, middle- and upper-class lawns and yards, river margins, and wetlands.

There are many examples of this level of study. For example, Joan Nassauer (1997) has shown how reconstruction of urban waterways in the Minneapolis–St. Paul complex in Minnesota has converted a deteriorating area into a valuable green space that has raised the environmental and economic values of communities associated with the waterway. Even a shopping mall that had been built on a filled-in lake was removed and the lake recreated, complete with wetlands—a reversal that is hopefully prophetic. In another example in the Chicago hinterland, reconstruction of the flood plain of the Desplains River, which formerly entered the city as a channeled stream and often flooded, has resulted in valuable green parkway and a capacity to hold floodwater, reducing the impact on the urban area. Akira Miyawaki described in a recent lecture how the impact of the recent earthquake in the city of Kobe, Japan, which caused widespread damage and human death, was much less severe where trees were associated with houses and other buildings. Trees reduced the spread of quake-caused fires by separating patches of flammable structures, held up collapsed structures allowing people to escape, and held the soil after the fires, thereby reducing soil loss.

Thus, ecologists' studies of the vegetation of urban areas, the aquatic life and flow characteristics of waterways, or urban wildlife all provide a basic foundation which can be used by managers and the public to make decisions when this information is needed. Clearly, the city as a whole needs GIS-based information systems, which organizes all this ecological information so that it is readily available. As an example of the value of such information, Roy Welch, of the Center for Remote Sensing and Imaging Processing at the University of Georgia, has developed GIS systems for urban fire departments so that firefighters can locate exactly where fires occur and have displayed before them in the station or in their fire trucks all the connections to water, gas, and sewer lines, types of structures nearby, presence of explosives, and so on, so that they can be informed about what they will encounter as they drive to the scene of the fire. Ecological problems, while often not as serious in the short term as are fires, should be treated in a similar fashion. This is the reason why it is necessary to monitor the species occurring in the city and measure species abundance. In some cases these measurements can be the duty of schools, who report their finding to the managers and local newspapers. For example, schools in Portland, Oregon, have undertaken the responsibility of collecting and mapping data on tree species and have presented their findings to government authorities and the public to influence decisions regarding street trees. All of these data can be collected, summarized and displayed on hierarchical GIS systems to be used in the newspaper in the same way that weather patterns are presented, and can be used by the administration and citizens for decision making and management of the city.

Urban Design

In the past, cities appeared on suitable sites over a long time span so that the inhabitants had the impression that the city had existed forever. Indeed, in the Middle East we can see how one city was built on the remains of a former city that had been destroyed in earlier conflicts. City upon city formed strata of hills or "tells" illustrating the history of the place. Yet cities must have had an origin in time. Most probably, they evolved from villages and towns that were located in places that had special advantages, such as a river crossing, a sheltered bay along the ocean, or a break between mountain and piedmont. Nevertheless, cities also were designed and constructed anew in ancient times. The construction of these cities required human intention, thought, choice, and aesthetic principles. What were the principles upon which city design took place?

Boyle Huang (1960), one of the respected traditional Chinese architects on Taiwan, describes the form of the Chinese imperial city. This city is based on the body of Lord Buddha. In the center, at the top of the city where Buddha's head would be located, the king's quarters were placed, symbolizing the king as the mind of the city. To the left and right, symbolizing the arms of Buddha, were the quarters of the administrators and the soldiers. At the center of the city were located the religious complexes, the great temple and the monasteries and minor temples. This location symbolized the navel of the Buddha where the *chi* or spirit is located in Chinese culture and medicine. Finally, symbolizing the legs of Buddha, the quarters of the merchants, artisans, and workers were located in the bottom half of the city. All of these quarters were surrounded by a wall patrolled by the soldiers, pierced by only a few gates.

In our time new cities have been constructed in a number of places to respond to increased demands for housing, growing populations, and new administrations based on aesthetic and architectural principles appropriate to our modern culture. I have studied several of these cities with the goal of understanding the designers' ecological perception to fit structure and space to resource and population. Most designers are architects and their training focuses them on buildings, roads, and other aspects of the built environment. If landscape architects are involved in the design, the green environment is included, but many landscape architects focus on design at the fine scale of the garden or the immediate surroundings of buildings, and do not use the ecological concepts of scale and hierarchy to extend the design into larger scales and eventually into the region or hinterland into which the city is embedded. This is why the projects of Joan Nassauer (1997) are important, and those of Ann Spirn (1984), who worked with water and wetlands in the Boston area. These landscape architects have taken the concepts of landscape ecology and applied them at multiple scales.

My experience has led me toward a different point of view about urban design. I have asked how has the designer thought about the needs, resources, and constraints in a new city. This is an opportunity to organize space in an ideal way and to overcome all of the problems of the past that are carried into and restrict what can be done in the present. Design of a city in a new landscape, such as the plains of the Punjab where Le Corbusier designed the city of Chandrigar in India, allows the designer to think afresh. What is in their mind, what is the consequence of their design? I have been surprised at the results of my studies. To make the point, I will contrast two examples of design: Chandrigar, India, and Vallingby near Stockholm, Sweden.

Chandrigar

When India and Pakistan were created, the state of Punjab in northern India was divided. The Indian portion did not have the capital and it was necessary to create a new state capital; the French architect Le Corbusier was chosen to design the new city. I studied Chandrigar in the 1980s, long after it was constructed. Le Corbusier had designed a modern city on the dusty plains of Punjab. The city contained great avenues, raw concrete surfaces decorated only with primary colors, with the grand government buildings on the city edge in a parklike neighborhood. Bicycles, bullock carts, herds of cattle and sheep, and pedestrians filled the avenues, with only an occasional car or truck or bus. I was concerned about the clerks of Chandrigar, who were required by the design to cycle across the city to their offices in the governmental complexes and home again each evening. The distances seemed enormous.

I visited a machine shop in the city and found each lathe or shear covered by a huge black umbrella. A torrent of water came through the roof and was being directed to the gutter of the road by sandbags. When I expressed amazement at this unexpected hydrological situation, the manager responded "All roofs leak in Chandrigar." Apparently the designers did not fully understand how the extremes of the Punjab climate would affect the rates of expansion and contraction of the great steel-reinforced concrete beams which support the roofs of the city.

My reaction was negative. It was, I thought, an example of the imposition of the architect's private and cultural perspective on a people, who had to cope with his vision to their inconvenience. The costs seemed large. I viewed the government clerk who had to bicycle across the city to his desk in the complex at the margin of the city; the effect of the lack of a city center on the perception of the inhabitants; the inconvenience of leaking roofs; and fungus- and algae-covered concrete walls as failures in urban design. Yet when I surveyed opinions of citizens, I found that they were delighted to live in Chandrigar. They thought that their city was modern and global. After all, it had been created by a famous European architect. Living in Chandrigar differentiated them from other Indians and made them modern. They liked this new city.

Vallingsby

The second lesson comes from Stockholm, which is built on a series of islands and peninsulas along the Baltic Sea. As Sweden modernized its agriculture, like other European countries, it experienced migration of inefficient farmers and the marginal rural poor to the city or to America. In Sweden the process of urban expansion challenged the designers to deal with urban growth in a complex landscape and achieve social objectives of equality in housing, employment, and education. Stockholm's solution was to build a collection of satellite cities around Stockholm, connected to the city center by high-speed trains. I studied several of these satellite cities, and I will comment on one of them called Vallingsby.

In the case of Stockholm, the new towns were built with a considerable variety of housing, a city center with a train station, restaurants, and shops, parking of automobiles placed on the periphery so that the center was restricted to pedestrians, and so on. The arrangement of houses protected and preserved the native landscape. Citizens could return to their apartment in the evening, remove their city clothes, dress to ski, and then cross-country ski in the forest before supper. Or in the summer, they could swim in the Baltic Sea. The design seemed sensitive to human needs and natural values. I considered Vallingby to be an attractive and successful design.

Again, I was surprised. Interviews with the inhabitants told me that while they welcomed the advantages of the new town, it was boring. Everybody was the same age, and there was little social variety. This was especially serious for teens, who were inclined to go into the central city on the train, where they might get into trouble but at least have an adventure. This led to conflict with their parents.

What lessons did I draw from these studies? Cities clearly serve a variety of functions to meet human needs, and our judgments about the positive and negative features of a city depend upon our weighting of these functions and needs. Jane Jacobs (1961) made this point in her eloquent book, *Death and Life of Great American Cities*. She pointed out that the in-town neighborhood, where people sat on their stoops in the summer evening while the children played in the street and an all-night restaurant on the corner attracted customers at all hours appeared cluttered, dangerous, and crowded to a middle-class observer from the suburbs. From the urban point of view, however, the neighborhood was safe, interesting, healthy, and satisfied many of the needs of an active community. The diverse, old-fashioned buildings, the street life, and the mix of urban people represented, according to Jacobs, viable communities. In contrast, some observers thought that removal and replacement of the messy streets and buildings by rectilinear, cleaned, and mechanized high-rise apartments would meet human needs more effectively. Not so! The high-rise represented sterile spaces that became locations where crime and dissatisfaction were common. These experiences suggest that in making judgments about the value of place, we need to begin from the point of view of the inhabitants and not impose on the evaluation our judgment based on theory or inappropriate generalization from limited experience.

The Ecosystem Model of the City

The advantage of the ecosystem model is that it permits us to organize the variety of functions of the city in a coherent fashion, recognizing the deep structure of place, controls of flows of energy, materials, information, and the maintenance of conditions for human health. What the model does not do is focus on the social implications of humans living together in close proximity. The urban social perspective needs to be combined with the ecosystem point of view to wholly address the urban ecosystem. Sociality is the cement that connects the parts of the urb together into a whole. Sociality can be positive or negative and our objective is to reduce the negative effects of commerce, crime, and politics and accent the positive mutualistic actions that build relations among people.

In a sense our goal is to recognize and increase the common good of the city. The common good is a phrase which is positive and references the American Constitution, and therefore is more than mere words. I was questioned by students about the meaning of this term at the Cary Conference. They asked if I meant a symbiotic or mutualistic relationship among urbanites, in the sense that ecologists use these terms. Actually, I was thinking about a habit of mind that might lead to mutualistic actions. I was thinking about an active sense of belonging to a group that would generate the voluntary effort to be positive and contribute to the whole. I think that if we can focus education on the common good we will have contributed something important to sustainability of our environment.

So the logic proceeds in this way. The city is a human construct; it exists for reasons of human need. An understanding of the city as an ecosystem is necessary to manage the city to be a healthy, productive, and satisfying place to live. The huge concentration of humans in cities is a new phenomenon that is growing worldwide. Today approximately half the people alive live in cities and this trend will probably continue. The poor and disadvantaged on every continent migrate to the city seeking improvement in their lives. While this increase in population is usually treated as a problem, let us accept it for what it is—a social phenomenon. In the city people move with or against other people. The connections that make up the urban ecosystem are social more than anything else. These connections can build and maintain the common good, but it does not just happen. Education is needed to build a comprehensive understanding of the urban ecosystem using both a mechanical flow perspective and a social connection perspective. I am using the term education in a broad sense. Education will be informal and formal in that it might shape signage, aesthetic design, greenness, and also be part of school curricula. It should be an education in values, because values underlay all decisions we make.

Urban Environmental Education

If you accept my argument, what would urban environmental education look like? If we approach this question from an ecosystem perspective we will consider three major divisions, plus, of course, retaining some of the current curriculum. I will not address the current curriculum; that is a matter for later discussion. Rather, I am suggesting that in cities we change our emphasis on the liberal arts and science to another form of education that focuses on relating humankind and the environment. Justification for such a radical idea is that our survival may depend upon it. The three divisions involve the *biota* (that is, humans and human society as well as other species), the *environment* (including the built environment), and the *regional context* in which the city is embedded, which supplies it air, water, food, and materials and accepts its waste products.

The biotic element in the city has biological, psychological, and social needs that must be fulfilled for a healthy and satisfactory life. Everything in the city is supplied to the individual. The individual cannot supply herself or himself with the requirements for life as she or he might be able to do in the countryside. The citizen needs to understand how each flow is accomplished and be able to visualize the patterns and the costs of supplying these necessities to the citizens. Humans also have social and political needs. They must be given tools to convert prejudice and intuitive understanding into concrete ideas and language. They must understand the legal, political basis of urban life and how courts, police, and social services work. In short, they must become "street-smart" or "city-smart" in the vernacular.

The environmental aspect which interacts with and shapes the human mind is the next general topic for education. *What is the history of place?*

What is beauty? What is change? How are flows constrained? How can physical and chemical problems be solved? Think of the ancient cities of dry landscapes, such as are found in Yemen. The construction of the buildings created and took advantage of winds, moving air in tunnels beneath the ground and providing cool air to the structures. These technologies even produced ice for human consumption. Their adaptiveness compared to Phoenix, for example, is profound. People need to learn about such places, so that they can ask questions about how to improve their own cities. Physical solutions to problems are everywhere.

Finally, thinking in this cartoon fashion, we must place the ecosystem into a context (using context in the sense used by Claudia Pahl-Wostel [1995]). This means that we must learn about the sources of the materials we require from the region outside of the city. By pulling materials from the hinterlands, cities impact and shape their context. The urban sprawl that is occurring in many places is extending the urban system into the countryside, with all kinds of extensive disturbances and costs. This is becoming a global phenomenon. In a sense, the context of the modern megacity is the ecosphere itself.

It seems to me that these three parts form the basis for urban ecosystem education. Of course, students of different ages can cope with different parts of the analysis, but if the work focuses on real matters that are age-appropriate, it should hold the attention of all students.

Summary

Summarizing, the urban ecosystem fits into a larger system which supplies it with inputs and receives outputs. For great commercial cities the entire planet may be its source; for smaller cities the hinterlands may consist of a biome or landscape. At the finest scale the ecologist will study urban ecotopes or patches of homogeneity within the urban landscape. At this scale the presence and abundance of species is of high significance. At all scales, the ecosystem ecologist is concerned with flows of energy, matter, and information between ecosystem components. While the theoretical basis of urban ecosystem studies is not very different from studies of forested watershed or other systems, urban systems are different because they are created by humans to serve human purposes. That means that they are accessible to logical analysis in a way fundamentally different from natural systems that have evolved within a logic that the ecologist seeks to understand and explicate.

The core of urban ecology is the interaction between human beings, so ecology, as a natural science, must reach out across the barrier between the natural and social sciences and create a union with those subjects that make humans their main concern. This will probably be the most troubling step in creating an urban ecosystemology, because ecologists tend to be aggressively partisan in regard to method and theory. In the city, ecologists must learn to understand and utilize qualitative data from social studies of organization, human well-being, culture, economics, and history. This is a large challenge but the ultimate payoff is the creation of a science which is really dedicated to solving human problems in an holistic fashion. This has been the long-term goal of ecologists. It is ironic that the goal may be easiest to obtain in the humanized urban ecosystem.

References

Boyden, S., S. Millar, K. Newcombe, and B. O'Neill. 1981. *The ecology of a city and its people: the case of Hong Kong*. Australian National University Press, Canberra, Australia.

Boyden, S. 1987. *Western civilization in biological perspective: patterns in biokistory*. Oxford Science Publications, Clarendon Press, Oxford, UK.

Duvigneaud, P. 1974. L'ecosysteme "Urbs." *Mem. Soc. Roy. Belg.* 6:5–35.

Duvigneaud, P., and S. Denayer-De Smet. 1977. L'ecosystme urbs: La ecosysteme urbain Bruxellois. Pages 581–599 in P. Duvigneaud and P. Kestermont, eds. *Productive biologique en Belgique*. Scope, Gembloux.

Golley, F.B. 1993. *A history of the ecosystem concept in ecology*. Rutgers University Press, New Brunswick, NJ.

Grimm, N.B., J.M. Grove, C.L. Redman, and S.T.A. Pickett. 2000. Integrated approaches to long-term studies of urban ecological systems. BioScience 70: 571–584.

Huang, B. 1960. *A history of Chinese architecture*. 114 pp. In Chinese.

Jacobs, J. 1961. *The death and life of great American cities*. Random House, New York.

Kowarik, I. 1990. Some responses of flora and vegetation to urbanization in Central Europe. Pages 45–74 in H. Sukopp and S. Hejny, eds. *Urban ecology*. SPB Academic Publishing, Den Hague.

Lamotte, M. 1978. La Savane Preforestiere de Lamto, Cote d'Ivoire, in M. Lamotte and F. Bourliere, eds. *Problems de ecologies*. Masson, Paris.

Mayhew, H. 1861. *London labor and London poor: a cyclopedia of the condition and earnings of those that will work, those that can not work and those that will not work*. C. Griffin Co. 4 volumes, London, UK.

Nassauer, J.I. 1997. *Placing nature: cultural and landscape ecology*. Island Press, Washington, DC.

Pahl-Wostel, C. 1995. *The dynamic nature of ecosystems: chaos and order entwined*. John Wiley, Chichester, England.

Pickett, S.T.A., M.L. Cadenasso, J.M. Grove, C.H. Nilon, R.V. Pouyat, W.C. Zipperer, and R. Costanza. 2001. Urban ecological systems: linking terrestrial ecological, physical, and socio-economic components of metropolitan areas. *Ann. Rev. Ecol. System* 32:127–157.

Spirn, A. 1984. *The granite garden*. Basic Books, New York.

Spooner, B. 1986. *MAB urban and human ecology digest*. UNESCO, Paris.

Stearns, F., and T. Montag. 1974. *The urban exosystem: a holistic approach*. Dowden, Hutchinson and Ross, Stroudsburg, PA.

Sukopp, H. 1987. Stadtokologische Forschung und deren Anwendung in Europa. Dussel. Geobot. Kolloq. *Dusseldorf* 4:3–28.

Sukopp, H., and S. Hejny, eds. 1990. *Urban ecology: plants and plant communities in urban environments*. SPB Academic Publ., The Hague.

Van Dyne, G. 1980. Reflections and projections, in A.I. Breymeyer and G.M. van Dyne, eds. *Grasslands, system analysis and man*. Cambridge University Press, Cambridge.

26 Out the Door and Down the Street–Enhancing Play, Community, and Work Environments as If Adulthood Mattered

William R. Burch, Jr., and Jacqueline M. Carrera

Introduction

This paper will first explore some lessons learned on the frontlines of urban ecology; most of the data come from our experiences in the City of Baltimore. For these core data, however, we provide a context of snapshots about other projects in other places and with other activities that have a similar goal: using the place where people live as the proper means for learning about ecology, self, and community.

The Parks and People Story

The Parks and People Foundation (P&P) is a private, nonprofit organization that serves open space needs and services in Baltimore and its region. It has worked on four projects over the past decade that attempt to help people redefine themselves as part of their ecosystem. The programs begin where people live and provide the means for their understanding of their community's nature, its pattern and processes, its ecology, and their role as restorers and guardians of the community's quality of life. A brief outline of the four projects may provide something of the scope of our experience:

1. *Kids Grow* is an after-school and summer camp program for Baltimore City elementary and middle school children that introduces concepts—such as tree identification and function, gardening approaches, stream restoration, wildlife identification and habitats, and woodland function—in a way that facilitates a long-term appreciation for and stewardship of the urban natural environment and suggests some possible occupational futures working within that environment.
2. *Community Forestry* is a program through which the P&P staff works with Baltimore City neighborhood groups to promote and assist them in planting and caring for street trees and restoring vacant lots. This program

is designed to give people the tools that they need to create a healthier neighborhood environment and possibly to generate some income as well.

3. *Revitalizing Baltimore* is a cooperative project coordinated by P&P that involves the U.S. Forest Service; Maryland Department of Natural Resources; Baltimore County; Baltimore City Department of Recreation and Parks, Department of Public Works, and Department of Planning; neighborhood groups; and three watershed groups. The watershed is the unit of action and the aim is to connect biophysical restoration of these stream valleys with the sustained revitalization of the adjacent neighborhoods and communities. It is developing a model of interorganizational, interdisciplinary, public and private cooperation to improve the quality of human and natural environments in urban areas. It is environmental education for the policy makers.

4. The *Baltimore Ecosystem Study* (BES) is a NSF-funded Long-Term Ecological Research Project that is part of a network of 22 such study sites in the United States, with Baltimore and Phoenix being the only two urban sites. Work by P&P prepared the ground for connecting the BES scientists and the region's many local interests, agencies, and neighborhoods. As the organization that will remain long after BES has completed its work, P&P will continue the school science education program being implemented by the BES. Further, it will have a critical role in sustaining good connections between the local communities and the field research scientists, in seeing that the scientific findings of the BES are interpreted so they may be of use to local policy makers, and in helping to ensure that the research serves neighborhood needs and those of the three volunteer watershed associations. Also, P&P and BES have developed a science-mentoring program for local community youths and seniors. These unique projects are environmental education at both the knowledge production and distribution ends.

Parks and People has learned many lessons from its environmental education programs. Four of them seem most relevant for the readers of this book. These are:

1. We need to go beyond the classroom and create networks of communication among a larger audience that includes all decision makers—individuals, community groups, government agencies, and businesses.

What will happen if we pick up all the trash with our school kids at a cleanup? More trash will likely be there tomorrow. We cannot just put a Band-Aid on the problems that we see in the cities. These problems are symptoms of a larger, systemic problem and it will take more than just us and our class/program participants to address it. Neither can we count on traditional educational venues to find the people. We need to break out of our institutional glass walls and go where the people are. We need to have a much better understanding of our decision makers. *Who are they? How will our data impact their bottom line or goals? How do these priorities stack up? Does everyone have all the information that they need to make informed decisions?*

In Baltimore the demand for park and recreation services keeps expanding at the same time the recreation and parks budgets keep getting cut. The mayor and city council recently cut $3 million out of a very thin $19 million budget for the city's recreation and park system. How can we sustain years of neglect of the 6,500 acres of parkland and 42 recreation centers? We need more than just children and decision makers to become aware of the value of nature and open space in urban environments.

2. Involving a larger audience means collecting meaningful data and translating the data into information that is useful (relevant) to people, whether they be students, mayors, or CEOs of large corporations.

The difference in urban ecosystem education is that the data being collected can be used to change the way people behave and the way cities are managed. It is ecosystem education with a major purpose beyond simply identifying a plant or insect species. In our research we need to ask: *Are we just teaching students for the sake of giving them information or do we want to see better management of our resources? Are we going to take on issues such as safety and security, poverty and drugs? Are we going to rewrite textbooks? Are we going to provide cost analyses for the actions suggested by our research data? How much is this information worth? For whom? And for when?*

3. Urban Ecosystem Education goes beyond the basics.

While it is important to teach students of all ages about all of the scientific "ologies," it is equally important to teach them to play out cause-and-effect scenarios, and where possible have the research directed to recurring issues confronting people in real life. Linkages between these many "ologies" are the critical test of science involved in real life.

4. Partnerships are key in achieving urban ecosystem education.

Scientists cannot do it alone. Educators cannot do it alone. Community organizers cannot do it alone. Mayors and public servants can't do it alone. To have effective, efficient, and sustainable ecosystem education we must bring all of these people—scientists, funders, educators, politicians, officials, and businesses—into discussion with residents as to how we ALL own these environmental challenges.

We know that values are a difficult subject for many scientists and educators. However, attention to them is critical. The need to confront this reality has best been articulated by Charles Jordan, the recreation and parks director in Portland, Oregon, "What we do not value we will not care for. What we do not care for we will not own. What we do not own we will lose." In short, values are the essential linchpin of any hope of developing a long-term education program in ecology.

The Structure of Our Learning

The Baltimore Parks and People experience suggests four key elements for effective ecosystem education—political connection, reality connection, knowledge integration, and collaboration with all the relevant players. We do not have absolute answers as to how to design and implement programs that serve such a wide variety of constituents, clients, and information needs. What we do have are some stories about what we have learned thus far.

It is important to note some of the broad goals we began with, some of the mechanisms we sought to employ, and our good fortune in beginning with an interest in the art of science. Our specific goals or output for the development of individuals in our education in environment programs—Project RAISE (Raising Ambition Instilling Self-Esteem), Outward Bound Urban Resources Initiative (OBURI), Kids Grow, the Baltimore Ecosystem Study (BES) Neighborhood Science—might be summed by the seven Cs shown in Table 26.1. As you will note, the training outputs for the individual move from serving the needs of the person to serving the needs of the community. The community or neighborhood becomes a critical learning environment. Community development action research then becomes an open effort to systematically understand and enhance six community goals or COAPES (Table 26.2).

The seven Cs and COAPES might be seen as our working hypotheses. Our approach tries to combine necessary services for individual growth that connect to community development needs. Critical to trying to make this connection is the role of the artist as one who points out the possibilities. As Paul Sears noted years ago, the ecological scientist climbs and climbs

Table 26.1. The seven Cs—specific goals individual development in our education in environment programs.

1. Career—One must become literate in understanding the rules, roles, and relationships necessary to succeed in the environment of adult work.
2. Confidence—Self-awareness and self-esteem are critical for the individual to sustain resilience in adapting to the changes in one's life environments.
3. Cooperation—Rugged individualist myths are the luxury of the rich whose social welfare is brokered by the luck of kinship and situation. Those of lesser opportunity need an environment of cooperation with friends and neighbors who help us up when we are down and keep us humble once we are up there.
4. Citizenship—How to assert access to the common environment of civility that confers rights and responsibilities in regard to public resources and private opportunity.
5. Community—Service to self begins with service to the cultural and social environment of one's neighborhood and/or larger ethnic traditions.
6. Connections—Ecological science helps us to perceive those connections in our networks that are most critical for given opportunities.
7. Complexity—There are no magic bullets, no simple answers for finally getting one's environment right—there are only some processes that work better than some others. The job is never done.

TABLE 26.2. Goals for community development from education in environment (COAPES).

1. **C**ontrol—Over necessary resources and accountability for their maintenance (e.g., to be information producers about the structure and function of the community ecology rather than consumers of information provided by others).
2. **O**ptions—Awareness that there are a wide array of alternative solutions and consequences of these solutions and that these can be systematically examined. Greening can have side or unanticipated consequences (e.g., in Nepal the shift from central to community management of forests has enhanced a form of green democracy where unscheduled castes participate in resource decisions.
3. **A**ccess—The means of access to common resources are identified and sustained by community institutions; e.g., regular service from police and garbage pickup or community management of city green spaces.
4. **P**articipation, as a learning process—The community system is dynamic and ongoing, learning is continuous and forever if a community is to adapt and survive. Hence the ecology of the community is the learning environment.
5. **E**quity—The distribution of the burdens and gains for improved or sustained community ecosystem productivity requires a negotiated balance (e.g., the improved condition of a neighborhood can lead to the displacement of the residents who made it happen). The issue is which institutions are there to support continued occupancy by the restorers?
6. **S**ustainability—The means by which desired benefit opportunities do not exceed the threshold of cultural and ecosystem resiliency (e.g., natural, cultural, and other capital regimes are continually invested and only the "interest" is the base of consumption).

the mountain of understanding only to arrive at the top and find that the footprints of the poet have gone before. In our program in Baltimore we needed that guidance, and we needed strong reminders of what was working and what was not in the full dimension of feeling as well as "rational" planning. In Baltimore we are fortunate to have the insightful help of Stephanie Graham, a professional photographer whose nature photography captures the nature of human nature in wildland and urban ecological settings. Like the poet described by Sears she showed us the summit of possibility and helped us to evaluate our progress toward that summit. She helped us all—pupils and teachers—to understand ourselves and our shared mosaic of opportunity.

Origins of Our Learning—Other Stories, Other Places

Our story begins in New Haven in the early 1970s when cities were put at risk of destruction through "slum clearance" projects connected with the construction of the Eisenhower Interstate Highway Defense system, which saw the parks and playgrounds, the homes and neighborhoods of the poor as the ideal venues for highway right-of-way takings. Residents of these neighborhoods experienced the interstate highway systems as one more means in the marginalization and removal of black, poor, and elderly residents. It was in this era that Burch first heard cities referred to as a frontier. The term was coined by an African American scholar who was discussing the great postwar migration from the rural south to the urban north and he was worried that this frontier was going to be stolen from those needy pioneers.

In the forestry and natural resource business we wondered if we had a role on this "new frontier." At that time Burch had graduate students who studied whether experiencing nature through an outdoor camping program might have a role in building self-esteem and leadership skills among inner city children (Allen 1977). And in 1975 the U.S. Forest Service and some other agencies sponsored a major conference in Washington, DC, called "Children, Nature, and the Urban Environment" (USDA Forest Service 1977). It was a symposium fair that was child and city centered, with nature relegated to a more decorative aspect—the elevators were festooned with plants, one of which was poison ivy. I think that mistake must have been symbolic of how many of the perplexed foresters attending the gathering viewed the odd mixture of architects, radicals, psychologists, sociologists, child welfare folk, geographers, poets, and gobs of children running all about the hallways and researchers running on and on about their findings in the meetings.

This was our first exposure to the theories of Roger Hart on how children can be their own researchers and guides to understanding their environments. It clicked with our own experience in the empirical reality of our own children and those we had worked with. His insight became our guide from then on. Hart has taken his notion to many corners of the world, stressing how important it is to the global environment of the future to have children doing research on their own environments. He notes, (Hart 1997) ". . . educators usually assign children to work on environmental projects rather than involving them in identifying problems themselves and collaborating with them in finding solutions. . . . [continuing his argument for local environmental research by children] genuine ecological understanding involves an understanding of environmental phenomena "in place"—that is, in their complex spatial relatedness to one another. If one accepts the theories of developmental psychologists that young children (at least those under 10 years of age) require direct interaction with phenomena to understand them, then it follows that children must first investigate small-scale, local ecosystems."

Play and the Creation of Social Capital

From Hart we draw the notion that children's play—their exploration of self and its connection to a wider context of society and nature—is their work, their initiative in preparation for a full and useful adulthood. And as we seek to ensure that the means for conducting science be entrusted to ordinary citizens, so must we trust children to have capacities for learning beyond our expectations. Our connections between these ecologies of

TABLE 26.3. Six kinds of capital with examples.

1. Natural Capital (basic elements of survival)—Water, air, forests, plants, soils, landscape form, aesthetics.
2. Human Capital (individual development/empowerment)—Education, health, nurturing (socialization), skills training, responsibilities.
3. Economic Capital (material productivity)—Money, trade centers, means of production, distribution of goods and services.
4. Social Capital (reciprocity)—Organizations such as churches, synagogues, temples, schools, etc.; social centers and nodes of exchange, gossip, and debate; public spaces that link elements of the social, e.g., across generations, household structures, and kinship affiliations.
5. Cultural Capital (identity)—Language, tradition, myths, pride, laws, regulations, and informal norms of social conduct—shared knowledge and ways of knowing.
6. Infrastructural Capital (built social environment)—Sewer systems, water systems, streets, bridges, and other collective structures that engineer and shape public life.

development have been enhanced by an emergent theory of capital regimes stimulated by Robert Putnam's work on social capital (1993). For our community "development" purposes we consider six kinds of capital that interconnect in ways that determine the trajectory, shape, and sustainability of a given social order (Table 26.3).

The reader will recognize that these notions are but expansions and extensions of various theories of wealth production put forth by Mills, Marx, Schumpeter, and others. The difference is that in a tertiary, consumerist global economy the issues of welfare and wealth, equity and accountability become as confused as they were at the start of the industrial order. By expanding the notions of capital we include critical functions formerly assumed away by traditional economics and thereby left out of social accounts and causal interpretations. By expanding the rules of fiscal capital to include these other forms of capital, we may consider the consequences and trade-offs when one spends a certain capital or invests it or draws upon its "interest" and so forth. Putnam's point was that the usual development action of simply providing fiscal capital often failed unless there were available strong forms of social capital to make the financial investment funds actually work.

What Have We Learned—Final Examples for a Different Future in an Indifferent Society

In 1989 our Baltimore Urban Resources Initiative (URI) began the Raising Ambition Instilling Self-Esteem (RAISE) program in partnership with Outward Bound and Parks and People to connect the various capital regimes (assets) available in our poorer communities. Our teachers were Yale graduate students and they worked with children at risk of dropping out of school. The Yale students trained, worked with, and helped to encourage youths to carry out neighborhood activities such as community forestry: vacant lot clean up, restoration, and conversion to gardens. Along the way they learned silvics, botany, hydrology, geometry, and social ecology and they learned it all without knowing they were being "environmentally" educated. Their self-esteem improved. The graduate students' sense of worth improved. And fewer kids were at risk of dropping out of school. In 1992 this program became Outward Bound Urban Resources Institute (OBURI).

In 1991 the New Haven URI began a more formal association with a public school program to provide some of the same educational efforts. We were fortunate to have a dedicated alumnus and teacher, Susan Swensen, who has worked tirelessly with the teachers and administrators at Edgewood School, located near Edgewood Park. As Leigh Shemitz (1998), then director of URI notes, "Over 500 young people gained valuable exposure to the natural world and learned to integrate this natural environment into their own sense of neighborhood and community.... The program was especially notable for reaching the kids who were harder to reach. The children who did not thrive in a traditional classroom often came alive and developed a sharper focus when let outside to team.... The dozen Masters in Environmental Science graduate students who participated in the project gained experience in teaching, outreach, and curriculum development that has been a valuable part of their graduate experience. Several of these students went on to careers in environmental education, based directly upon their exposure with the Edgewood Park/School Project." A variety of reports detail the learning curve of this specific project (Esser 1996; Hayes 1994; Filardi and Hayes 1994; Honigfeld 1994; Jacoby 1995; Roy 1993; Yelin and Swensen 1997).

In 1993 the USDA Forest Service funded a program in Baltimore called Revitalizing Baltimore. This program had a variety of dimensions, one of which was Kids Grow. This was a more systematic and directed program than RAISE and was based in three neighborhood recreation centers. Outward Bound continued its program of developing self-esteem and community service called CORE (Communities Organizing to Revitalize the Environment). In 1994 Kids Grow served 22 students at three recreation centers. These participants were "... gaining confidence in themselves and their abilities as they learned about the ecosystems of Baltimore and the Chesapeake Bay region. Then they created growth in West Baltimore, both physically through planting gardens and trees and politically through community organizing and bringing local concerns to elected officials and the media" (Parker 1998). In 1995 there were 50 students, in 1996 there were 100 and in 1997 there were 105 summer students from seven communities between the ages of 11 and 13, and 60 day-school pupils between the ages of 8 and 13. These pupils were mostly from neighborhoods with high overall risk indicator rankings—significant minority populations, large youth populations, high proportion residing in subsidized public housing, and high proportions of low-income residents. This program is continuing to learn how to match development of human capital with necessary community development (Parker 1998).

Kids Grow by 1997 had the following kinds of impacts: 1,165 students from six schools were trained in the program, 11 educators were provided 40 hours of training, seven specific communities were impacted, including 1,000 persons of color, eight natural resource professionals were involved, and the students learned about natural resource professions in hydrology, horticulture, forestry, park management, fisheries, and wildlife ecology. Five students from previous trainings were teachers in the 1997 program, two assistant instructors took positions in youth education and one instructor is working as an educator with the Chesapeake Bay Foundation. Students created and maintained habitat gardens at six sites, did stream surveys and cleanups and earned student service credits for their work in their communities. There has been a continuing waiting list, as more and more parents want their children involved in the program. And these requests come from some of the most at-risk communities in the city.

The lessons from Kids Grow suggested that we needed more intensive, long-term action research and monitoring work in a specific neighborhood to identify the most critical connections in community ecology. From the Kids Grow program we knew that one could combine a variety of "education in the environment activities"—preparation for careers, science training in the service of one's community, youth and economic development through microenterprises, restoration of environmental quality, and community revitalization. In 1998, BES\USDAFS scientist Dr. Morgan Grove initiated discussions with Mr. Clayton and Mr. Elroy, the founders of the Rose Street Community Center (RSCC) in the Madison East End neighborhood of Baltimore. This was a neighborhood plagued with poverty, crime, drugs, and social disorder, yet it had a strong base of families trying to survive, to seek a better life for their children. In September 2000, the RSCC program became a separate operation and the BES efforts shifted to the Washington/Pigtown neighborhoods but continued with similar approaches.

In the RSCC program, a group of young people were involved in designing research in the community and helped to carry out mapping and inventory studies in their neighborhood. Youth were trained in the use of computer technology to take the field maps and convert them to GIS, rendering them more legitimate in the eyes of decision makers. They explored ideas for developing microenterprises that would remain in the neighborhood and give real careers to some of the residents. Individual and community pride have grown as the vision of the founders has been expanded by the youths into their "vision of reality" program. Much trial and error learning was going on in these initial phases but so far the Hart theory is being explored in daily increments. Of course, the efforts in Baltimore and New Haven are not alone but operate within a context of similar programs directed to "education within the learner's necessary environment."

A Larger Context for Our Learning

Our Baltimore experiences draw upon lessons from a wide range of venues and programs. Therefore, it is important to note some of the great diversity of learning and techniques they offer. They all are community based. They trust children and local people to conduct scientific (or quasi-scientific) efforts to design or monitor or understand the nature of their neighborhood ecology.

A good place to start is a cluster of four community gardens in the South Bronx that draw inspiration from their volunteer leaders and technical help from a Parks Department program—Green Thumb—and a similar program provided by The New York Botanical Garden.

Abu Talib is the leader of a 90-member community farm established on an abandoned lot in the shadow of Yankee Stadium. This program is botany in the service of family development. A short distance away but in another culture is Batey Borincano whose leaders are Mr. Dimas and Ms. Olga Cepeda. This is a Puerto Rican–style casita in the shadow of the old Bronx Courthouse, and is a resource for sustaining ethnic pride and traditions. It was recently under threat of being auctioned for sale by the Giuliani administration and then saved by help from Bette Midler and the Trust for Public Land. The next garden is attached to a park/playground of the city and is an orchard, flower, grape, and berry refuge gracing an empty space between two apartment buildings. Its dynamic leader is Mr. Al Qinones, who is dedicated to providing a safe place and role models for the children and the parents of the neighborhood. This incredible oasis park was also scheduled for auction. The fourth example from the many stories of Green Thumb is a program affiliated with the Lorraine Hansberry School. Here one enters the usual prisonlike architecture of public schools in the city and then with the guidance of Mr. Sebert Harper the visitor is escorted to the rear of the building and enters a dazzling green space along the Bronx River that was recently filled with trash and less than desirable occupants. The children helped to clean up the site and design it, and planted the flowers and trees. Now, they use it as an outdoor example of landscape restoration having true community educational value. The program is part of the regular school curriculum.

We want to highlight four other examples of programs that could help guide us on this new frontier of science and education. The first gives most attention to what is called "process" learning (Grant and Grove 1997) that challenges the traditional "content as curriculum" paradigm with a model of "research as curriculum." An example was a research project, designed by eleventh and twelfth graders, called 'White Tailed Deer—Its Impact on Forest and Field Vegetation." The students' learning process examined a common problem and at the same time they learned the methods and approaches of science. This program expects students to be guided by trained scientists while learning techniques of research in a way that makes them creators rather than consumers of science information.

Another example is a youth-based environmental justice program called The Natural Guard that was founded in New Haven in 1990 by Ritchie Havens. Its goal is to "empower urban youth to see themselves as key forces for creating change—fostering an appreciation for themselves, each other, the local environment, and the world around them" (Sciulli 1999). The program considers key environmental issues to be unemployment, lack of opportunity for poor families, and the disproportionate impact of environmental degradation on communities of color. Its environment is the place where children live, which means it addresses threats from lead poisoning, contaminated needles, and violence on the street. It has an overall goal of helping urban youth become stewards who are connected to and concerned about local and global environments and have the knowledge and skills and motivation to change things. It has summer and after-school components that help youth learn about the biosocial ecology of their city and neighborhoods, acquire the knowledge of what to do, and teach what they have learned to others in the community.

A third example is a research project/intervention carried out in six villages in Northern Thailand from November 1995 to February 1996. The purpose was to change teaching, learning, and school/community relations. They did this by involving students in studies of local village problems related to forest management. "Fifth and sixth grade students were taken out of school and into their communities to study real-world problems. Communities became laboratories for information gathering, and their human and physical resources were used to enhance students' understanding of concepts taught in class" (McDonough and Wheeler 1998). This change also transformed the schools in teaching orientation, made them better integrated into the community, and provided necessary research information to better manage the community forests.

The fourth example comes from the last Buddhist kingdom in the Himalayas—Bhutan. The unity of the nation and its ability to weather the challenges of the external world has been based upon three interlocking elements—the kingdom, the religion, and the environment. This ecology of social sustainability has been advanced in the primary school curriculum where the arts, sciences and humanities are based within the environment of the village school. The environment is the education. This is not a surprising factor, as any visitor to the country can tell you—in Bhutan the people are the environment and the environment is the people. The notion of separation between society and nature simply does not hold. The future, however, may be less promising, but that is a story for another time.

Conclusions

We have tried to sketch some contours of possibility for the new frontier of education and science suggested by the book title, providing some lessons that might help us to better map where we might go forth humbly yet boldly. We have traveled to many exotic venues where seldom has an environmental educator ventured. Unlike Lewis and Clark, however, we think that the learning will *not* be found only in the wisdom that is brought in by outside experts, but rather will be mutual discoveries best found within the rich cultural and biophysical environments occupied by these communities. We scientists and educators have much to learn from these overlooked frontiers and their resident experts—seniors, children, and youth. It is best if we get on with it, that we have the "nerve of failure," that we learn from our failures and never doubt our optimism. Most of all we must trust the people—adults and children, rich as well as poor—to have the ability to apply the methods of scientific learning to their own problems, even if such learning may lead them to challenge the established order of government and science.

Acknowledgments. We would like to thank the many people who inspired us and the many who advised us as well on the development and preparation of this paper: Ruth Allen, Alan Berkowitz, Alma Bell, David Bowes-Lyon, Colleen Murphy-Dunning, Shawn Dalton, Vicki Fabiyi, Barbara Grove, S. Graham, Morgan Grove, Roger Hart, James Jiler, Cheryl Jordan, Joaquin F. Lequia, Janet Parker, Rich Pouyat, Joanne Sciulli, Leigh Shemitz, Erika Svensden, the citizens in the neighborhoods of Baltimore and New Haven, and the community gardeners of the Bronx.

References

Allen, R. 1977. *Social organization, group cohesion and persistence of children's outdoor programs: field study in sociobiology*. Ph.D. Thesis, New Haven, CT.

Esser, L. 1996. *A study of the environmental values of current and former students in the Edgewood Park/School Project*. Yale URI Working Paper 34, New Haven, CT.

Grant, B.W., and B.F. Grove. 1997. *A teacher's manual for use in pre-college environmental education: an ecological inquiry framework*. Widener University Department of Biology and Science Teaching Center, Chester, PA.

Golley, F. 1993. *The history of the ecosystem concept in ecology*. Yale University Press, New Haven, CT.

Hart, R.A. 1997. *Children's participation—the theory and practice of involving young citizens in community development and environmental care*. Earthscan Publications, Ltd, London, UK.

Hayes, E. 1994. *Learning about the natural world in urban classrooms: the education of decision makers*. Yale URI Working Paper no number, New Haven, CT.

Hayes, E., and C. Filardi. 1994. *The education of decision makers—community forestry and environmental education activity guide*. Yale URI Working Paper no number, New Haven, CT.
Honigfeld, H. 1994. *The Edgewood Park/School Project: innovation in urban environmental education*. Yale URI Working Paper 13, New Haven, CT.
Jacoby, J. 1995. *The Edgewood Park/School Project: Intern Report*. Yale URI Working Paper 18, New Haven, CT.
Kotkin, J. 1999. For retailers in some city centers gentrification is a four-letter word. *The New York Times* 27 June 1999:BU–7.
McDonough, M.H., and C.W. Wheeler. 1998. *Toward school and community collaboration in social forestry—lessons from the Thai experience*. Academy for Educational Development, Washington, DC.
Parker, J. *1997/1998 year-end reports of kids grow*. Parks and People, Baltimore, MD.
Putnam, R.D. 1993a. *Making democracy work: civic traditions in modern Italy*. Princeton University Press, Princeton, NJ.
Putnam, R.D. 1993b. The prosperous community: social capital and public life. *The American Prospect* 13:35–42.
Roy, K. 1993. *New Haven Conservation Corps: a curriculum and resource guide*. Yale URI Working Paper 14, New Haven, CT.
Sciulli, J. 1999. *New programs for the natural guard*. The Natural Guard, New Haven, CT.
Shemitz, L. 1998. *Brief synopsis of work in education outreach in the Edgewood Park/School Project*. Yale URI Report, New Haven, CT.
USDA Forest Service. 1997. *Children, nature and the urban environment—proceedings of a symposium fair 19–23 May 1995*. Upper Darby, PA: General Technical Report NE–30.
Yelin, J., and S. Swensen. 1997. *A pond ecosystem curricula guide for interactive, interdisciplinary, experiential environmental education—expanding the Edgewood model*. Yale URI Working Paper 36, New Haven, CT.

27
Integrating Urban Ecosystem Education into Educational Reform

RODGER W. BYBEE

Developing an understanding of the urban ecosystem requires perspectives from the natural, technological, and social sciences as they specifically apply to the dominant influences of humans in cities. I approach the theme of integrating urban ecosystem education into educational reform by first presenting a brief history of human ecology, an approach that succinctly states my view for urban ecosystem education. I then answer the question: *What should citizens know, value, and do as a result of urban ecosystem education?* In order to help establish perspectives on the problem of integrating urban ecosystem education into educational reform, I describe a course of study on the topic. A separate section includes this description. In the final section, I address the difficult issues of integrating urban ecosystem education into the educational system.

Human Ecology: Origins of a Perspective for Education

A perspective such as the urban ecosystem complements the theme of human ecology. A review of human ecology provides a rationale for the emerging need to integrate this perspective into the educational system. Further, the historical perspective of human ecology identifies some of the ideas upon which to design policies, programs, and practices for urban ecosystem education. The following discussion is based on earlier discussions of human ecology (Bybee 1984; 1993).

Charles Darwin (1809–1882) provided the intellectual origins of human ecology when he published *On the Origin of Species* (1859). Darwin's work established the relationship between organisms and environments in the natural selection of traits that enhance the organisms' ability to survive and reproduce. German biologist Ernst Haeckel (1834–1919) first identified ecology as a separate field of study in his 1869 text, *General Morphology of Organisms*. Darwin's theory and the emergence of ecology as a field of study in the late nineteenth and early twentieth centuries established the initial perspectives for human ecology. While the biologists were developing the basic concepts of ecology, they usually implicitly included humans but without explicitly describing human ecology. Historically, biologists have been rather late in recognizing humans in their study of environments. Biology contributed to the intellectual birth of ecology and the early development and synthesis of a conceptual foundation for aspects of human ecology (Young 1974).

Primarily work by sociologists contributed to the early development of human ecology as a discipline. Robert E. Park, Ernest Burgess, and Roderick McKenzie are usually recognized as the founding fathers of human ecology. Park and Burgess first used the term "human ecology" in their 1921 book, *Introduction to the Science of Sociology*. These authors drew heavily on the work of biologists such as C.R. Darwin, F.E. Clements, and C.S. Elton. For a reader interested in comprehensive collections of writings in human ecology, I refer to George Theodorson's *Studies in Human Ecology* (1961), Gerald Young's *Origins of Human Ecology* (1983) and a long essay by John A. Moore in his "Science as a Way of Knowing" series (1984).

In a 1936 article entitled "Human Ecology," Robert Park described human ecology as having a biological base with sociological applications. I quote Park's summary of human ecology:

> Human ecology is an attempt to apply to the interrelations of human beings a type of analysis previously applied to the interrelationships of plants and animals. The term "symbiosis" described a type of social relationship that is biotic rather than cultural. This biotic social order comes into existence and is maintained by competition. In plant and animal societies competition is unrestricted by an institutional or moral order. Human society is a consequence and effect of this limitation of the symbiotic social order by the cultural. Different social sciences are concerned with the forms which this limitation of the natural or ecological social order assumes on (1) the economic, (2) the political, and (3) the moral level. (Park 1936)

The summary expresses Park's biological and sociological orientation. The organizing concept is competition, a restatement of Darwin's idea of "the struggle for existence," and an idea that dominated social thought during this period. For Park, human society was organized on two levels: the biotic, based on competition and cooperation; and the cultural, based on communication and consensus. The biotic level influenced human adjustments and spatial distributions of populations. Human ecologists of this period studied the processes that maintained and changed the biotic balance and social equilibrium.

R.D. McKenzie (1926) described human ecology as even more distinctly sociological. In McKenzie's view, human ecology was concerned with the spatial grouping and sustenance relations of humans occupying a geographical area. Interactions among humans centering on space and sustenance result in sociological changes in the population, such as mobility, employment, segregation, and dispersion. I quote a summary from McKenzie:

> Human ecology concerns the process of spatial grouping of interacting human beings of interrelated human institutions. Ecological distribution is the resultant of competing forces and changes in distribution are measurable by the rate of mobility, of change of residence, of employment or of any utility. Many factors of general or local significance affect ecological organization and may be classed as geographical, economic, cultural, technical, and political. (p. 141)

In reading this work, I note the distinctly sociological tone of the definition and the lack of any suggested problems for study.

Other social science disciplines also adopted an ecological perspective. For example, early papers with a human ecological perspective were published in geography (Barrows 1923; Thornthewaite 1940), anthropology (Sayce 1938), and psychology (Lewin 1944). At mid-twentieth century, the social science literature included the human ecological themes in economics (Boulding 1950; 1966), political science (Pfaltzgraff 1968; Rodman 1980), cultural anthropology (Orlove 1980), and environmental psychology (Craik 1973).

Three points emerge from this historical review. One, early development of human ecology was primarily in the social sciences. Two, there was a conspicuous absence of biological ecologists contributing to human ecology. And third, although the social scientists were using the term *human ecology*, there was a great deal of conflict over academic territory and little study of any significant problems.

Critics of human ecology expressed their views in the late 1930s and early 1940s. Milla Alihan's 1938 book, *Social Ecology: A Critical Approach*, was central to the reassessment. She criticized the sociologists' use of the term environment as too elastic, the overextension of biological concepts to social situations, too much accretion of social concepts with ecological concepts, and the inappropriate use of competition as the central process in human relationships. She also lamented the apparently incomplete understanding held by the sociologists of uniquely human qualities, such as purpose and desire, and the role these may play on interactions with the environment.

I was particularly impressed with a 1944 criticism by Amos H. Hawley entitled "Ecology and Human Ecology." Hawley lists the "aberrant intellectual tendencies," which had dominated human ecology:

> The most dominant of these may be described as: (1) the failure to maintain a close working relationship between human ecology and general or bioecology; (2) an undue preoccupation with the concept [of] competition; and (3) the persistence in definitions of the subject of a misplaced emphasis on "spatial relations" (Hawley 1944)

Hawley perhaps missed the point that spatial relations are a key element in the overlap between the social and natural sciences. Cities are spatial entities, and spatial relations are critical both within them and in explaining their interactions with surrounding parts of the landscape.

At this time in history, according to Hawley, the challenge of human ecology was to differentiate itself from other disciplines by defining and tackling its own set of problems (Hawley 1944). With the important (and noted) exception of spatial relations, Hawley's 1944 assessment was accurate. Indeed, it was not until the 1960s when problems of populations, resources, and environments became dominant and were no longer preempted by other disciplines, that human ecology became an important area of study. Hawley later describes his view of human ecology: "In simplest terms, human ecology is the descriptive study of the adjustment of human populations to the conditions of their respective physical environments" (Hawley 1944). In terms of this essay, I note that one of the key features of human adjustment or nonbiological adaptation is through the development of technological devices, tools, and systems. Hawley did not emphasize the important role of technology in human ecology. He did, however, recognize the role of adjustment as a part of the dynamic system of interactions between humans and their environments, and described human ecological problems:

> This [population adjustment to resources and environment] resolves itself into a number of related problems, such as: (1) the succession of changes by which an aggregate passes from a mere polyp-like formation into a community of interdependecies; (2) the ways in which the developing community is affected by the size, composition, and rate of growth or decline of the population; (3) the significance of migration for both the development of the community and the maintenance of communal structure, together with the factors which make for change in the existing equilibrium and the ways in which such change occurs. (Hawley 1944)

Amos Hawley's discussion of population, resource, and environment interactions anticipated very well the 1960s, when human ecology became an important, indeed essential, field of study.

In sum, earlier thinkers placed the idea of human ecology into scholarly discourse. Limitations of their thinking stemmed, at least in part, from simplistic ecological models; for example, placing too much emphasis on competition and not incorporating a rich and deep scientific understanding. They did, however, bring ecology to the study of humans. In the next decades, especially the 1960s, all of society developed a greater understanding of ecology primarily through the writings of scientists.

Human Ecology Gains Public Attention

Some individuals have always understood the relevance of ecology and its relations to the human condition. It was not until the 1960s, however, that ecology became a science with personal meaning for the general population. Why? Because population, environment, and resource problems became important enough to demand public attention. These were human ecological problems, and they were thrust into social prominence by individuals describing significant problems for the general public.

If one person and publication were used to identify the significance of human ecology for the public, they would have to be Rachel Carson and her book *Silent Spring*, published in 1962. *Silent Spring* immediately drew national attention. Simply, clearly, and forcefully written, it carried the message that the use of chemicals, particularly chlorinated hydrocarbons, had grave consequences. Chapters in *Silent Spring* provide the scientific background and examples of chemical effects on soil, on plants, on birds and fish, and on humans.

Scientists and the public both questioned the message of *Silent Spring*; however, time has provided support for most of Rachel Carson's arguments. More importantly, environmental problems gained national attention, and a human ecological perspective was introduced to the public.

The Population Bomb, written in 1968 by Paul Ehrlich, is another significant book in the recent development of human ecology. Ehrlich's book, also powerfully written, demonstrated how population growth was related to numerous other problems, such as food shortages, environmental degradation, and weak economic growth. In the book, Ehrlich makes the point that we must find ways to implement the "birth rate solution," or the traditional "death rate solutions" (e.g., pestilence, famine, and war) will prevail.

Garrett Hardin's 1968 essay "The Tragedy of the Commons" is a classic statement in human ecology. This article continues to be insightful and informative about the relationship among populations, the physical environment, and cultural issues related to carrying capacity. He argues using an analogy to a New England town commons. The tragedy occurs because herdsmen ask—*What is the utility to me of adding one more animal?*—a question of personal gain, not social responsibility or environmental accountability. Of course, the herdsman adds an animal. He will receive all the profit from the extra animal, but the effect of overgrazing will be shared by all other herdsmen. Hardin also confronts the difficult problem of avoiding the tragedy. He concludes that "mutual coercion mutually agreed upon by the majority of the people affected" is the only reasonable solution. This position clearly is not strictly biological. Hardin is suggesting changes in things such as laws or taxes to avoid exceeding the carrying capacity of air, land, and water.

The list of significant contributors to human ecology has to include Barry Commoner. As early as 1963 he published *Science and Survival* as a warning of environmental problems. His most notable contribution was probably *The Closing Circle*, published in 1971. Where Ehrlich focused on population and Hardin on carrying capacity, Commoner directed his attention to the environment. Commoner consistently presented an holistic understanding required of human ecology.

Ten years after *Silent Spring*, a Club of Rome report entitled *The Limits to Growth* (1972) was published. *The Limits to Growth* was innovative in its approach, using computer modeling, and far-reaching in its conclusions. A great debate centering on the study immediately emerged. Although some individuals and groups criticized aspects of the study, the symbolic value of the title and the fundamental findings of the study remain significant in the development of human ecology.

The Global 2000 Report to the President: Entering the Twenty-First Century (Barney 1980) was published eight years after *The Limits to Growth*. *The Global 2000 Report* was a massive undertaking by several United States government agencies to assess the probable changes in the world's population, natural resources, and environment through the end of the century. President Carter's intention was to use the result in long-term planning and policy development. Findings from this study included:

- Rapid growth in world population will hardly have altered by 2000. The world's population will grow from 4 billion in 1975 to 6.35 billion in 2000, an increase of more than 50 percent;
- The large existing gap between the rich and poor nations will continue to widen;
- World food production is projected to increase 90 percent between 1970 and 2000. Most of the increase will go to countries with an already high per capita food consumption;
- There will be significant losses of world forests due to use for forest products and fuel wood;
- Agricultural soils will deteriorate due to erosion, loss of organic matter, desertification, salinization, alkalinization, and waterlogging; and
- Atmospheric changes due to carbon dioxide and ozone-depleting chemicals could alter the world's climate and upper atmosphere by the year 2050.

The Global 2000 Report generally confirmed what many had been saying for years. There is an urgent need to recognize the earth's carrying capacity while maintaining a decent quality of life for humanity. One concluding statement summarizes well the human ecological view.

> Vigorous, determined new initiatives are needed if worsening poverty and human suffering, environmental degradation, and international tensions and conflicts are to be prevented. There are no quick fixes. The only solutions to the problems of population, resources, and environment are complex and long-term. These problems are inextricably linked to some of the most perplexing and persistent problems in the world—poverty, injustice, and social conflict (p. 4).

I have referred to *The Global 2000 Report* at some length because of the timelines of projections for the Year 2000 and because the conclusions generally support the positions of others (e.g., the authors and publications reviewed here).

This review would be incomplete without mentioning the work of Lester Brown and the Worldwatch Institute. One of Lester Brown's first important works was *The Twenty-Ninth Day*, published in 1978. The book begins with the dimensions and consequences of ecological stress. Brown then expertly presents the biological and social means of accommodating human needs and numbers to the earth's resources. In 1981, Lester Brown published *Building A Sustainable Society*. The concept of sustainable society is an idea and a vision that other authors had, but did not articulate. Brown discussed sustainability in *State of the World* (1984), the first in a series of yearly reviews of progress toward a sustainable society:

> Sustainability is an ecological concept with economic implications. It recognizes that economic growth and human well-being depend on the natural resources base that supports all living systems. Technology has greatly expanded the earth's human carrying capacity, most obviously with advances in agriculture. But while the human ingenuity embodied in advancing technology can raise the natural limits on human economic activity, it cannot entirely remove them (p. 1).

Brown continues with a definition of sustainable society. "A sustainable society is one that shapes its economic and social systems so that natural resources and life support systems are maintained" (p. 2).

Personal Reflections on Human Ecology

I developed this portion of the essay as a personal exploration of several questions. *What is human ecology? Does human ecology have any application to education?* These questions are related to a larger and more important question concerning education—*What is the place of human ecology as a perspective in education, in particular urban ecosystem education?*

What is human ecology? The discussion includes various definitions offered by different authors to develop a general conception of human ecology. Very broadly, human ecology considers the interrelationship of human beings and their environments. Authors do, however, tend to slant their definition of human ecology toward their discipline. Concerning an educational perspective, I recommend that human ecology not be set apart from the general discipline of ecology. This point also was made by Edward Kormandy (1984).

Indeed, Eugen Odum (1977) has suggested that ecology may emerge as a new integrative discipline. He summarizes:

> The new ecology, then, is not an interdiscipline, but a new integrative discipline that deals with the supraindividual levels of organization, an arena that is little touched by other disciplines as currently bounded—that is, by disciplines with boundaries established and strongly reinforced by professional societies and departments or curriculums of universities. Among academic subjects, ecology stands out as being one of the few dedicated to holism. (Odum 1977)

Odum stresses the ecosystem level of study and uses a holistic methodolgy. Humans are as much a part of the integrative approach as other organisms.

In fact, Odum (1975) points out that the new ecology links the natural and social sciences. This position, the integrative link between natural and social sciences, seems to me an excellent way to think of urban ecosystem education. It clearly includes human aspects of ecology without positing human ecology as a new or unique discipline. The term *human ecology* suggests that both the human sciences and the natural sciences are important in the study of urban ecosystems.

A Framework for Urban Ecosystem Education

Turning to the question of human ecology as a perspective in education, the first, and perhaps most important, conclusion is that human ecology does indeed have a place in education. All of the reasons reviewed for human ecology's importance justify including human ecology as a part of education, and the urban ecosystem provides an excellent educational context for a majority of students.

Throughout the essay, several points have arisen that begin forming the *educational framework*. The human ecological perspective in urban ecosystem education would include

- studying *significant problems* such as population growth and distribution, environmental quality, and resource use;
- introducing the *role of technologies* in ecosystems;
- using an *ecosystem level of study*;
- *balancing holistic and reductionistic methods of study*; and
- using an *integrative approach* that presents human ecological study as the linkage between natural and social sciences.

Human ecology is too important to be left out of education. The courage demonstrated by individuals who developed the field of human ecology now has to be shown by educators.

What Should Students Know and Be Able to Do as a Result of Urban Ecosystem Education?

What content should educational programs about urban ecosystems include? An answer to this question can be found in the *National Science Education Standards* (NSES) (NRC 1996) and *Benchmarks for Science Literacy* (AAAS 1993), which provide the essential content and contexts for the design of instructional materials and professional development experiences. Use of these standards as the core for the design of curricula should be supplemented by the North American Association for Environmental Education Guidelines, Excellence in Environmental Education (NAAEE 1999), the National Council of Teachers of Mathematics (NCTM

2000), and the International Society for Technology Education Standards (ISTE 2000). Linking the content of these standards in school programs would provide a solid, integrative approach to urban ecosystem education.

Several educational ideas clarify this discussion of content, curriculum, and the issue of implementing urban ecosystem education into educational reform. The following discussion presents content from the NSES. Content is what students should learn; it is not the curriculum. Curriculum is the way content is organized and emphasized (and in my view, assessed). Curriculum includes the structure, organization, balance, and presentation of the content; although my review of content will, of necessity, have an organization that does not imply any particular curricular structure. I discuss the latter in another section.

Answering one fundamental question helps integrate urban ecosystem education into educational reform—*What is the content of urban ecosystem education?* The aforementioned national standard documents do not present a category of content identified as "urban ecosystem." I propose that the *National Science Education Standards* do have content that could form the basis for a curriculum on urban ecosystems (see Table 27.1). In presenting this content from the NSES, I will begin with what is perhaps the most familiar and, in my view, important.

Rather than approaching urban ecosystems from a discipline-based view, I recommend beginning with category 2—Science as Inquiry (Table 27.1). This approach establishes the legitimacy of urban ecosystems within science and provides the means to approach other content through the processes of investigations in the urban setting. Table 27.2 presents the abilities from the Science as Inquiry standards for grades 9–12. I recommend that problem-based inquiry be used as a defining attribute of urban ecosystem education. The historical discussion of human ecology identified the lack of such an orientation as a weakness (McKenzie 1926) and my own resolution of this criticism with the recognition of pressing problems associated with population growth, environmental degradation, and resource use (Bybee 1993). In the context of urban ecosystem educational programs, inquiry will

TABLE 27.1. Eight categories of national science education content standards.

1. Unifying Concepts and Processes in Science
2. Science as Inquiry
3. Physical Science
4. Life Science
5. Earth and Space Science
6. Science and Technology
7. Science in Personal and Social Perspectives
8. History and Nature of Science

Source: (NRC 1996).

TABLE 27.2. National science education standards for science as inquiry: Grades 9–12.

As a result of their activities in grades 9–12, all students should develop the abilities of scientific inquiry and understandings about scientific inquiry.

Fundamental Abilities and Concepts That Underlie This Standard

Identify questions and concepts that guide scientific investigations. Students should formulate a testable hypothesis and demonstrate the logical connections between the scientific concepts guiding a hypothesis and the design of an experiment. They should demonstrate procedures, a knowledge base, and conceptual understanding of scientific investigations.

Design and conduct scientific investigations. Designing and conducting a scientific investigation requires introduction to conceptual areas of investigation, proper equipment, safety precautions, assistance with methodological problems, recommendations for use of technologies, clarification of ideas that guide the inquiry, and scientific knowledge obtained from sources other than the actual investigation. The investigation may also include such abilities as identification and clarification of the question, method, controls, and variables, the organization and display of date, the revision of methods and explanations, and the public presentation of the results and the critical response from peers. Regardless of the scientific investigations and procedures, they must use evidence, apply logic, and construct an argument for their proposed explanation.

Use technology to improve investigations and communications. Students' ability to use a variety of technologies, such as hand tools, measuring instruments, and calculators, should be an integral component of scientific investigations. The use of computers for the collection, analysis, and display of data is also a part of this standard.

Formulate and revise scientific explanations and models using logic and evidence. Student inquiries should culminate in formulating an explanation or model. In the process of answering the questions, the students should engage in discussions and arguments that result in the revision of their explanations. These discussions should be based on scientific knowledge, the use of logic, and evidence from their investigation.

Recognize and analyze alternative explanations and models. This standard emphasizes the critical ability to analyze an argument by reviewing current scientific understanding, weighing the evidence, and examining the logic thus revealing which explanations and models are better and showing that although there may be several plausible explanations, they do not all have equal weight. Students should appeal to criteria for scientific explanations in order to determine which explanations are the best.

Communicate and defend a scientific argument. Students in school science programs should develop the abilities associated with accurate and effective communication including writing and following procedures, expressing concepts, reviewing information, summarizing data, using language appropriately, developing diagrams and charts, explaining statistical analysis, speaking clearly and logically, constructing a reasoned argument, and responding to critical comments through the use of current data, past scientific knowledge, and present reasoning.

Source: (NRC 1996).

be facilitated by the availability, familiarity, and meaningfulness of the urban ecosystem.

Next, I would identify the content most directly associated with the life sciences, especially ecology. In the science education standards that would be the content in life science standards (see Category 4, Table 27.1); for example, organisms and environments at grades K–4; populations and ecosystems at grades 5–8; and interdependence of organisms at grades 9–12 (NRC 1996).

With science as inquiry and ecological content from the NSES, I recommend going to the standards on Science in Personal and Social Perspectives (see category 7, Table 27.1). Table 27.3 presents content for the grades 9–12 standard on Science in Personal and Social Perspectives. These standards could establish a foundation for curricula on urban ecosystems. Of necessity, one would have to complement science content with environmental education, mathematics, technology, English language arts, and social sciences. For example, use of the English language arts for organizing information and developing proposed explanations; mathematics for accurately and appropriately representating data and using algebraic and geometric models; and social studies for addressing citizenship skills and citizens' rights and responsibilities.

In addition, knowing about and understanding the urban ecosystem requires incorporating technology as a domain of study (Table 27.3). One unique feature of cities is the degree to which humans have relied on technology as a means of adapting to and constructing an environment to accommodate their needs and aspirations. In addition to the technology standard from the NSES and the technology sections from *Benchmarks for Science Literacy* (AAAS 1993), the content standards being developed by the "Technology for All Americans" project (ITEA 2000) should be an integral aspect of urban ecosystem education. Technology is an intrinsic part of urban ecosystems, and while it cannot countervene natural laws, it does shape and reflect human values. In a basic sense, technology is the nonbiological means by which humans adapt. Humans use technology to move from one place to another, exchange information, construct buildings, to extend the senses, to grow food, and in general to shape environments to accommodate our goals. It is clear that the results of changing the environment are complex and unpredictable, and they have costs, benefits, and risks that attend different individuals and groups.

For urban ecosystem education, students should develop some understanding of ideas such as: systems and subsystems, materials, constraints, optimization, risk assessment, and probability. Many of these ideas have been discussed in national documents (NRC 1996; AAAS 1993). Some of the topics that might be considered in an urban ecosystems curriculum include: materials, manufacturing, energy sources, energy use, design processes, communication, information processing, and health technology.

What Might an Urban Ecosystem Education Program Include?

In this section, I address the question—*How would one design a program for urban ecosystem education?* To address this challenge, I use the content standards—Unifying Concepts and Processes in Science (see category 1,

TABLE 27.3. National science education standards for science in personal and social perspectives: Grades 9–12.

Fundamental Abilities and Concepts That Underlie This Standard

Personal and Community Health

Hazards and the potential for accidents exist. Regardless of the environment, the possibility of injury, illness, disability, or death may be present. Humans have a variety of mechanisms—sensory, motor, emotional, social, and technological—that can reduce and modify hazards.

The severity of disease symptoms is dependent on many factors, such as human resistance and the virulence of the disease-producing organism. Many diseases can be prevented, controlled, or cured. Some diseases, such as cancer, result from specific body dysfunctions and cannot be transmitted.

Personal choice concerning fitness and health involves multiple factors. Personal goals, peer and social pressures, ethnic and religious beliefs, and understanding of biological consequences can all influence decisions about health practices.

An individual's mood and behavior may be modified by substances. The modification may be beneficial or detrimental depending on the motives, type of substance, duration of use, pattern of use, level of influence, and short- and long-term effects. Students should understand that drugs can result in physical dependence and can increase the risk of injury, accidents, and death.

Selection of foods and eating patterns determine nutritional balance. Nutritional balance has a direct effect on growth and development and personal well-being. Personal and social factors, such as habits, family income, ethnic heritage, body size, advertising, and peer pressure, influence nutritional choices.

Families serve basic health needs, especially for young children. Regardless of the family structure, individuals have families that involve a variety of physical, mental, and social relationships that influence the maintenance and improvement of health.

Sexuality is basic to the physical, mental, and social development of humans. Students should understand that human sexuality involves biological functions, psychological motives, and cultural, ethnic, religious, and technological influences. Sex is a basic and powerful force that has consequences to individuals' health and to society. Students should understand various methods of controlling the reproduction process and that each method has a different type of effectiveness and different health and social sequences.

Population Growth

Populations grow or decline through the combined effects of births and deaths, and through emigration and immigration. Populations can increase through linear or exponential growth, with effects on resource use and environmental pollution.

Various factors influence birth rates and fertility rates, such as average levels of affluence and education, importance of children in the labor force, education and employment of women, infant mortality rates, costs of raising children, availability and reliability of birth control methods, and religious beliefs and cultural norms that influence personal decisions about family size.

Populations can reach limits to growth. Carrying capacity is the maximum number of individuals that can be supported in a given environment. The limitation is not the availability of space, but the number of people in relation to resources and the capacity of earth systems to support human beings. Changes in technology can cause significant change, either positive or negative, in carrying capacity.

Natural Resources

Human populations use resources in the environment in order to maintain and improve their existence. Natural resources have been and will continue to be used to maintain human populations.

The earth does not have infinite resources; increasing human consumption places severe stress on the natural processes that renew some resources, and it depletes those resources that cannot be renewed.

(Continued)

TABLE 27.3. *Continued*

Humans use natural systems as resources. Natural systems have the capacity to reuse waste, but the capacity is limited. Natural systems can change to an extent that exceeds the limits of organisms to adapt naturally or humans to adapt technologically.

Environmental Quality
Natural ecosystems provide an array of basic processes that affect humans. Those processes include maintenance of the quality of the atmosphere, generation of soils, control of the hydrologic cycle, disposal of wastes, and recycling of nutrients. Humans are changing many of these basic processes, and the changes may be detrimental to humans.

Materials from human societies affect both physical and chemical cycles of the earth.

Many factors influence environmental quality. Factors that students might investigate include population growth, resource use, population distribution, overconsumption, the capacity of technology to solve problems, poverty, the role of economic, political, and religious views, and different ways humans view the earth.

Natural and Human-Induced Hazards
Normal adjustments of earth may be hazardous for humans. Humans live at the interface between the atmosphere, driven by solar energy, and the upper mantle, where convection creates changes in the earth's solid crust. As societies have grown, become stable, and come to value aspects of the environment, vulnerability to natural processes of change has increased.

Human activities can enhance potential for hazards. Acquisition of resources, urban growth, and waste disposal can accelerate rates of natural change.

Some hazards, such as earthquakes, volcanic eruptions, and severe weather, are rapid and spectacular. But there are slow and progressive changes that also result in problems for individuals and societies. For example, change in stream channel position, erosion of bridge foundations, sedimentation in lakes and harbors, coastal erosions, and continuing erosion and wasting of soil and landscapes can all negatively affect society.

Natural and human-induced hazards present the need for humans to assess potential danger and risk. Many changes in the environment designed by humans bring benefits to society, as well as cause risks. Students should understand the costs and trade-offs of various hazards—ranging from those posing minor risk to a few people to major catastrophes with major risk to many people. The scale of events and the accuracy with which scientists and engineers can (and cannot) predict events are important considerations.

Science and technology are essential social enterprises, but alone they can only indicate what *can* happen, not what *should* happen. The latter involves human decisions about the use of knowledge.

Understanding basic concepts and principles of science and technology should precede active debate about the economics, policies, politics, and ethics of various science- and technology-related challenges. However, understanding science alone will not resolve local, national, or global challenges.

Progress in science and technology can be affected by social issues and challenges. Funding priorities for specific health problems serve as examples of ways that social issues influence science and technology.

Individuals and society must decide on proposals involving new research and the introduction of new technologies into society. Decisions involve assessment of alternatives, risks, costs, and benefits and consideration of who benefits and who suffers, who pays and gains, and what the risks are and who bears them. Students should understand the appropriateness and value of basic questions—"What can happen?"—"What are the odds?"—and "How do scientists and engineers know what will happen?"

Science and Technology in Local, National, and Global Challenges
Humans have a major effect on other species. For example, the influence of humans on other organisms occurs through land use—which decreases space available to other species—and pollution, which changes the chemical composition of air, soil, and water.

Source: (NRC 1996).

Table 27.1). I intentionally did not include these standards in my earlier discussion of standards and benchmarks. I want to provide two different examples of approaches. My purpose in describing an approach to standards-based programs is to suggest that urban ecosystem education programs may have different content and especially different curriculum emphasis (Roberts 1982). The prior discussion centered on content from the standards on inquiry, life science, science in personal and social perspectives, and science and technology.

The standards for unifying concepts and processes of science complement the more discipline-based subject matter perspective. The conceptual ideas and procedural abilities in these standards provide students with productive and insightful ways of thinking and integrating ideas that will help them investigate and explain natural and designed systems in the urban environment. The proposed course of study (Table 27.4) is only an example based on one set of unifying concepts and processes. My purpose is not to suggest a complete curriculum; rather, it is to provide a concrete example of how one might design a program for urban ecosystem education. If a curriculum such as I describe were actually developed, it would have sequences of activities, suggested readings, educational technology, use of the web, and other approaches to learning. Here, I provide a sense of the conceptual flow of a course.

Implementing Urban Ecosystem Education

Using the urban ecosystem as a central focus for introducing students to important concepts associated with, for example, populations, resources, and environments does, on the face of it, make sense. The largest numbers of students are in urban centers and their cities are the environments they experience, know best, and that have the most meaning for them. Alone, this clear, and to some, compelling justification does not have the power to compete for space in an already crowded science curriculum. Integrating urban ecosystem education into educational reform will involve addressing a complex set of factors that include the district science standards (i.e., a syllabus or curriculum framework that does *not* include urban ecosystem education), the current program (i.e., textbooks and instructional materials that are outdated and infrequently replaced), and the teaching practices (i.e., use of lecture, a non–activity-based approach).

An essential first step is to articulate what is meant by urban ecosystem education and what would be included in such a program. To date, most dialogue has been among those who understand the idea of urban ecosystems and that dialogue has centered on why there is a need and who might change. It is now time to address the question of what is the actual content of urban ecosystem education. I have tried to point to human ecology as a possible orientation for an answer to this issue. Earlier discussions of a

TABLE 27.4. Investigating the urban ecosystem: A proposed course of study for science and technology (Grades 9–12).

Course Objectives

As a result of activities in grades 9 and 10, all students should develop understandings and abilities aligned with the following concepts and processes:

- Systems, order, and organization
- Evidence, models, and explanation
- Constancy, change, and measurement
- Evolution and equilibrium
- Form and function

Unit I: Introducing the Urban Ecosystem

The natural and designed world of the urban ecosystem is complex; it is too large and complicated for students to investigate and comprehend all at once. Students learn to define small portions of the urban environment for the convenience of investigation. The units of investigation can be referred to as "systems." A system is an organized group of related objects or components that form a whole. In the urban setting, systems can consist, for example, of organisms, machines, information, health, transportation, and education. In urban ecosystems education, particular emphasis could be placed on a class of forms that are particularly important for cities and identified in earlier discussions (e.g., spatially defined systems such as the single home, the neighborhood, the school and its grounds, and city parks).

The goal of this unit is to think and analyze in terms of systems. This will help students keep track of energy, objects, organisms, and events as they study the urban environment. They will learn that even simple systems encompass subsystems and learn to identify the structure and function of systems, the ideas of feedback and equilibrium, and the distinction between open and closed systems.

Science assumes that the behavior of the universe is not capricious, that nature is the same everywhere, and that it is understandable and predictable. Through investigations of selected aspects of the urban environment, students can develop an understanding of regularities in systems, and by extension, the universe; they then can develop understanding of basic laws, theories, and models that explain the world. An assumption of order establishes the basis for cause-effect relationships and predictability.

Prediction is the use of knowledge to identify and explain observations, or changes, in advance. The use of mathematics, especially probability, allows us to quantify our degree of certainty.

Order—the behavior of units of matter, objects, designed systems, humans, or other organisms in the environment—can be described statistically. Probability is the relative certainty (or uncertainty) that we can assign to selected events happening (or not happening) in a specified space or time. In science, reduction of uncertainty occurs through such processes as the development of knowledge about factors influencing objects, organisms, systems, or events; better and more observations; and better explanatory models.

Types and levels of organization provide useful ways of thinking about the urban ecosystem. Types of organization include the classification of functional systems such as transportation, communication, and information. Spatial systems, as identified above, could be described as type systems and located in a hierarchy of organization from simple and local to complex and regional or even global.

Living systems also have different levels of organization, for example, cells, tissues, organs, organisms, populations, and communities. The complexity and number of fundamental units change in extended hierarchies of organization. Within these systems, interactions between components occur. Further, systems at different levels of organization can manifest different properties and functions.

TABLE 27.4. *Continued*

Unit II: Investigating the Urban Ecosystem

Students learn that evidence consists of observations and data on which to base scientific explanations. Using evidence to understand interactions in urban ecosystems allows individuals to predict changes in natural, social, and designed systems.

Models are tentative schemes or structures that correspond to real objects, events, or classes of events, and that have explanatory power. Models help scientists and engineers understand how things work. Models take many forms, including physical objects, plans, mental constructs, mathematical equations, and computer simulations.

Scientific explanations incorporate existing scientific knowledge and new evidence from observations, experiments, or models into internally consistent, logical statements. Different terms, such as "hypothesis," "model," "law," "principle," "theory," and "paradigm" are used to describe various types of scientific explanations.

As students develop and as they understand more science concepts and processes, their explanations should become more sophisticated. That is, their scientific explanations should more frequently include a rich scientific knowledge base, evidence of logic, higher levels of analysis, greater tolerance of criticism and uncertainty, and a clearer demonstration of the relationship between logic, evidence, and current knowledge.

Unit III: Understanding Change and Constancy

Although most things are in the process of becoming different—changing—some properties of objects and processes are characterized by constancy, including the speed of light and the total mass and energy in the universe. In urban environments, changes might occur in spatial systems, transportation systems, energy systems, and educational systems. Interactions within and among systems result in change. Changes vary in rate, scale, and pattern, including trends and cycles.

As in any study of science, students should develop an understanding of the essential forms and functions of energy in urban ecosystems. Energy can be transferred and matter can be changed in systems, including the urban ecosystem. When measured, however, the sum of energy and matter in all closed systems, and by extension the universe, remains the same.

Changes in systems can be quantified. Evidence for interactions and subsequent changes and the formulation of scientific explanations are usually clarified through quantitative distinctions; that is, measurement.

Scale should be introduced to help students understand the different characteristics, properties, or relationships within the urban ecosystem and what might happen as they increase or decrease.

Rate involves comparing one measured quantity with another measured quantity, for example, the speed of automobiles on city streets as 30 miles per hour. Rate is also a measure of change for a part relative to the whole, for example, change in the city's birthrate as a part of the city's population growth.

Unit IV: Form and Function in Urban Ecosystems

Form and function are complementary aspects of objects, organisms, and systems in the natural and designed world. The form or shape of an object or system is frequently related to use, operation, or function. Function frequently relies on form. Understanding of form and function applies to different levels of organization. Students should be able to explain function by referring to form and explain form by referring to function.

The urban environment provides numerous excellent examples to investigate the relationship between form and function. This unit should give special emphasis to spatial systems and public works systems (e.g., water supply and waste processing) and unique qualities of those systems in the urban environment.

(Continued)

TABLE 27.4. *Continued*

Unit V: Adaptation, Evolution, and Equilibrium in Urban Ecosystems
Adaptation in human systems occurs in a variety of ways. Some of the ways and means of human adaptation include, of course, biological changes. In the study of urban ecosystems, biological adaptation should be one of a variety of adaptations in human systems including economical, political, social, and technological. One important idea that should be developed in this concluding unit is that of sustainable systems and societies. That is, the adaptation of personal, economical, political, social, spatial, and technological systems so that natural resources and ecosystems are maintained. Study of adaptations leads into the larger concepts of evolution and equilibrium.

Evolution is a series of changes, some gradual and some sporadic, that accounts for the present form and function of objects, organisms, and natural and designed systems. The general idea of evolution is that the present arises from materials and forms of the past. Although evolution is most commonly associated with the biological theory explaining the process of descent with modification of organisms from common ancestors, evolution also describes changes in the urban ecosystem.

Regardless of technology, social forces, and other issues, students studying urban ecosystems should understand that some fundamental scientific concepts will prevail. Equilibrium is one unifying concept that students should understand. Equilibrium is a physical state in which forces and changes occur in opposite and offsetting directions; for example, opposite forces are of the same magnitude, or offsetting changes occur at equal rates. Steady state, balance, and homeostasis also describe equilibrium states. Interacting units of matter tend toward equilibrium states in which the energy is distributed as randomly and uniformly as possible.

proposed course of study (Table 27.4) also provide a concrete example of what is meant by, and how one might design a program for, urban ecosystem education.

To effectively implement an urban ecosystem perspective requires applying critically important ideas. My first recommendation is to support the extant standards (national, state, or local) that promote an urban ecosystem perspective. National organizations such as the National Research Council (NRC), the American Association for the Advancement of Science (AAAS), the North American Association for Environmental Education (NAAEE), and the International Technology Education Association (ITEA) have developed standards that apply to an initiative such as urban ecosystem education. The power of standards lies in their capacity to change all the essential components of the educational system, namely curriculum, instruction, assessment, and the professional education of teachers. Beginning with current standards and adapting them to the themes and topics of urban system education brings this innovative idea into line with contemporary educational reform. This enhances the probability of acceptance and eventual implementation.

My second recommendation is to elaborate what would be involved in an urban ecosystem education program and to describe where and how this idea fits in the educational system. Should ideas and processes of urban ecosystem study permeate all grades, or is it explicitly for certain grades such as elementary, middle, or high schools?

Although my examples focused on high school, there is no reason that the urban ecosystem could not be introduced in elementary or middle school. Is this a science course? An integrated science, mathematics and technology course? Or an integration of science with other school disciplines, such as social studies? For those outside the group proposing urban ecosystem education, there is a clear need for a concrete view of what such a program looks like, where it is located in the educational system, and what objectives it will achieve.

Although I have discussed school programs, my third recommendation is to include the entire educational system, not just schools. For example, museums, science and technology centers, and environmental education programs all contribute to urban ecosystem education. In brief, recognize the fact that citizens are educated by a diverse array of institutions and groups, referred to as the informal education community.

The following is my fourth recommendation. Although we must initially think broadly and abstractly about urban ecosystem education, implementing this perspective requires us to eventually act narrowly and concretely. For example, what does this topic mean for teacher education? What does this theme mean for the professional development of current elementary, middle, and high school teachers? What recommendations could we make for programs in urban museums, science, and environmental education centers? Implementing the urban ecosystem view will be advanced by presenting policies, programs, and practices for those who have responsibility for implementing the theme in the educational system.

Finally, a fifth recommendation centers on the issue of actually bringing about broad-scale educational change based on the urban ecosystem theme. This is, perhaps, the most difficult challenge of all. Getting innovations to any significant scale in educational systems is, at best, difficult and almost never occurs (Elmore 1996). Some pitfalls we face when addressing the problem of getting to scale include (1) we convince ourselves of an important idea but seldom work to establish the idea (e.g., the content of the urban ecosystem) in key policy documents such as state frameworks, and critical leverage points such as state and local assessments; (2) we often talk to colleagues and other educators about innovations but seldom explain to the public why urban ecosystem education is important; (3) we often implement a new program without developing educational support for sustained professional development and maintenance of the innovative program; (4) we develop new supplies of instructional materials without changing the *demand* for instructional materials to include the general topic of urban ecosystems; and (5) we seek short-term fixes instead of taking a long-term steady work approach to the implementation of urban ecosystem education that is evolutionary.

Conclusions

Integrating urban ecosystem education into the educational system presents a major challenge for the twenty-first century. An ecological understanding and approach to implementation, combined with the essential nature of the goals, hold promise of integrating the urban ecosystem perspective into educational reform.

References

Alihan, M. 1938. *Social ecology: a critical approach*. Columbia University Press, New York.

American Association for the Advancement of Science (AAAS). 1993. *Benchmarks for science literacy*. Washington, DC.

Barney, G. 1980. *The global 2000 report to the President*. U.S. Government Printing Office, Washington, DC.

Barrows, H.H. 1923. *Geography as human ecology*. Association of American Geographers 13:1–14.

Boulding, K. 1950. *An ecological introduction: a reconstruction of economics*. Wiley, New York.

Boulding, K. 1966. Economics and ecology. In F. Dorling, and J. Milton, eds. *Future environments of North America*. Doubleday, New York.

Brown, L. 1981. *Building a sustainable society*. W.W. Norton, New York.

Brown, L. 1984. *State of the world: 1984*. W.W. Norton, New York.

Brown, L. 1978. *The twenty-ninth day*. W.W. Norton, New York.

Bybee, R.W. 1984. *Human ecology: A perspective for biology education*. National Association of Biology Teachers, Reston, VA.

Bybee, R.W. 1993. *Reforming science education*. Teachers College Press, New York.

Carson, R. 1962. *Silent spring*. Fawcett Crest Books, Greenwich, CT.

Commoner, B. 1971. *The closing circle*. Bantam Books, New York.

Commoner, B. 1979. *The politics of energy*. Alfred A. Knopf, New York.

Commoner, B. 1976. *The poverty of power*. Alfred A. Knopf, New York.

Commoner, B. 1963. *Science and survival*. Viking Press, New York.

Craik, K. 1973. Environmental psychology. *Annual Review of Psychology* 24: 403–422.

Darwin, C. 1859. *On the origin of species*. Appleton, New York.

Ehrlich, P. 1968. *The population bomb*. (Revised, 1978). Ballantine Books, New York.

Ehrlich, P., and A. Ehrlich. 1974. *The end of affluence*. Ballantine Books, New York.

Ehrlich, P., and A. Ehrlich. 1981. *Extinction*. Ballantine Books, New York.

Elmore, R. 1996. Getting to scale with good educational practice. *Harvard Educational Review* 6(1):1–26.

Haeckel, E. 1869. *General morphology of organisms*. C. van der Post, Utrecht, Germany.

Hardin, G. 1968. The tragedy of the commons. *Science* 162:1243–1248.

Hawley, A. 1944. Ecology and human ecology. *Social Forces* 122:398–405.

Independent Commission on Environmental Education. 1997. *Are we building environmental literacy?* George C. Marshall Institute, Washington, DC.

International Technology Education Association (ITEA). 2000. *Standards for technological literacy: content for the study of technology*. Reston, VA.
Kormandy, E.J. 1984. Human ecology: an introduction for biology teachers. *The American Biology Teacher* 46(6):325–329.
Lewin, K. 1944. Constructs in psychology and psychological ecology. In Lewin, et al. *Authority and frustration*. University of Iowa Press, Iowa City, IA.
McKenzie, R.D. 1926. *The scope of human ecology*. Papers and Proceedings American Sociological Society, Twentieth Annual Meeting. Washington, DC.
Meadows, D., et al. 1972. *The limits to growth*. Universe Books, New York.
Moore, J.A. 1984. *Science as a way of knowing: human ecology, volume II*. American Society of Zoologists, Baltimore, MD.
National Research Council (NRC). 1996. *National science education standards*. National Academy Press, Washington, DC.
Odum, E.P. 1975. *Ecology: the link between the natural and the social sciences*. Holt, Rinehart, and Winston, New York.
Odum, E.P. 1977. *The emergence of ecology as a new integrative discipline*. Science 195:4284.
Orlove, B. 1980. Ecological anthropology. *Annual Review of Anthropology* 9:235–273.
Orr, D. 1991. *What is education for?* In Context. 27.
Orr, D. 1991. *Ecological literacy: education and the transition to a postmodern world*. State University of New York Press, Albany, NY.
Orr, D. 1994. *Earth in mind: on education, environment, and the human prospect*. Island Press, Washington, DC.
Park, R. 1936. Human ecology. *American Journal of Sociology* 42:1–15.
Park, R., E. Burgess, and R. McKenzie. 1921. *Introduction to the science of sociology*. University of Chicago Press, Chicago, IL.
Pfaltzgraff, R. 1968. Ecology and the political system. *American Behavioral Science* 11:3–6.
Roberts, D. 1982. Developing the concept of curriculum emphasis in science education. *Science Education* 66(2):243–260.
Rodman, J. 1980. Paradigm change in political science: an ecological perspective. *American Behavioral Science* 24:49–51, 64–74.
Sayce, R.V. 1938. The ecological study of culture. *Scientia* (LXIII):279–285.
Theodorson, G., ed. 1961. *Studies in human ecology*. Harper and Row, New York.
Thorthwaite, C.W. 1940. The relation of geography to human ecology. *Ecology Monographs* 10:343, 347–348.
Vayda, A., and R. Rappaport. 1968. Ecology, cultural and non cultural. In J. Clifton, ed. *Introduction to cultural anthropology*. Houghton Mifflin, Boston, MA.
Young, G. 1974. Human ecology as an interdisciplinary concept: a critical inquiry. *Advances in Ecological Research* 8:1–40.
Young, G. 1983. *Origins of human ecology*. Hutchinson Ross, Stroudsburg, PA.

28
The Contribution of Urban Ecosystem Education to the Development of Sustainable Communities and Cities

JULIAN AGYEMAN

In the late 1980s, I was working in the environmental health department of a large inner London borough, as an environmental education adviser. It was my job to work within a broad educational constituency of schools, colleges, and adult education and community groups in an attempt to involve them in a wide range of local sustainability issues related to the functioning of the urban ecosystem. The theory was that if people became involved in a particular issue, they would develop transferable skills, understanding, knowledge, values, and confidence, such that they might be able to apply them in different situations and become active citizens, or in contemporary parlance, "stakeholders" in their community. Two particular instances outline some of the possibilities as well as the dilemmas, conflicting agendas, and ill-defined and often contradictory goals of urban ecosystem education in relation to the development of sustainable communities and ultimately cities.

On one occasion, a group from the London Wildlife Trust (LWT) contacted me regarding working with communities living on housing estates (housing projects) which were becoming a greater focus of their work. I briefed them about a particular estate in south London that was known locally as a rough place in need of building maintenance, and I set up a meeting between LWT and the tenant association. Virtually the first thing the eager conservationists from LWT said to the tenants as they rolled out a variety of colorful and ambitious plans was, "OK. Where shall we put this nature garden? Do you want a pond here, or there? Where do you want your native hedgerow and wildflower meadow?"

The tenants looked on, confused. Was LWT wrong to suggest that funding be devoted to a nature garden when it quickly transpired that Maslow's (1943) Hierarchy of Needs was being invoked and that leaking roofs, broken windows, and damp conditions were clearly the tenants' environmental priorities over a nature garden? The LWT staff were doing their jobs as ecologists with funding earmarked for "conservation." It was, however, not the most appropriate environmental intervention that could have been made in the tenants' lives when such lower-order needs as safety and secure housing were not being fully met.

On another occasion around the same time, I was at a conference and heard Friends of the Earth UK's then-director, Jonathan Porritt, describe his concept of "*the privilege of concern*": the notion that it is a privilege afforded to the relatively wealthy to be able to release oneself from the daily grind to get active in environmental protection and conservation. This came in stark contrast to the next speaker who quoted the bishop of Chicago, who had said that his people need to "consume for at least 100 years," and when they'd finished consuming to the point that the U.S. middle classes are at now, then perhaps they too could share in "the privilege of concern." On reflection, it is perhaps disingenuous to think of concern as a privilege. After all, as Dunlap (1997) notes there is wide public concern and support for environmental protection. In other words the *concern* is shared, not privileged to the few. But is concern enough? One can sit in a room and be *concerned*. I would like to propose a more action-oriented concept: the "*privilege of choice*." This is the privilege of being able to release oneself from the day-to-day in order to *choose* to behave in an environmentally friendly or sustainable manner. This is a fundamental aspect of the achievement of sustainable communities and cities and has two prerequisites. First, people need to be assisted in meeting their lower-order needs (Maslow 1943) such that they can release themselves to *choose* sustainability. Second, we need to provide them with clearer *understandings, signposts* and *choices*. Both should be *concerns* of urban ecosystem education, although ultimately in the case of the former, concern must be tempered by the reality of what can be achieved.

These two experiences in London illustrate that in order to contribute fully to the development of sustainable communities and cities, urban ecosystem education (whether or not it is seen as a subset of environmental education which may, in itself, be a subset of the new contender, "education for sustainability" [Agyeman 1994]), must be broad and robust enough to help tenants develop the advocacy and political literacy skills required to improve the quality of their lives and help ecologists to put scientific ecology into a wider, social needs perspective. The first stage should be to develop a clear understanding of the parameters, structure and functioning of the urban ecosystem in terms of its integrated environmental, economic, and social components. If eighth grade students at Pistons Middle School in Detroit identify safety as a parameter of their urban ecosystem, do we tell them that it's not, and that they should restrict their learning to just the biological dimensions of ecosystems? The pedagogical approach proposed here is through participatory research and learning where the teacher is co-learner and the learner co-researcher. However, as Wals (1996) and others have argued, developing an understanding alone is not enough. The second stage must be the integration of understanding with agency, which can lead to action. From creating community gardens to getting people out of their cars, from reclaiming the streets so that they are safe for people to walk them to reducing waste, ultimately, lifelong urban ecosystem education should equip people to act. Although it cannot solve all the world's ills, urban ecosystem education should help people identify strategies to curb both the spiraling consumption of the wealthy, and the deepening poverty of the poor. In a world of education reform, it must offer school-aged children "real-world experiences (that) are the foundations of effective learning" (Williams and Agyeman 1999, p. 27). It must do all of these things, and more. It must be democratic, it should be solutions-oriented, it should be values-based, it should develop an environmental and political literacy, and it should inculcate a sense of agency (and urgency).

In short, urban ecosystem education must aim to become part of a wider lifelong learning process which helps people to envision and define what sustainable communities and cities are in both the short and longer term. Through participatory research it must also help people identify strategies to achieve their visions. Wals (1996) suggests that this can happen if individuals and communities *construct, transform, critique*, and *emancipate* their worlds in an existential way. He goes on to explain these terms, noting that people should be able to "*construct* in the sense of building upon the prior knowledge, experiences and ideas of the learner . . . *transform* in the sense of changing, shaping, influencing the world around them regardless of scope or scale . . . *critique* in the sense of investigating underlying values, assumptions, world views, morals, etc . . . *emancipate* in the sense of detecting, exposing and, where possible, altering power distortions that impede communication and change" (Wals 1996, p. 301).

The chapter continues by building on this foundation. It outlines two more key concepts in urban ecosystem education: sustainable communities and cities, and participation. Participation is examined through Arnstein's (1969) Ladder of Participation and an assessment of its benefits is made. Following on, key process issues for urban ecosystem education (and educators) are elucidated, namely that the main interests agree on the appropriate level of participation, that there is a common language to discuss issues and develop ideas, and that appropriate methods (including techniques, structures and programs) are used to get as much agreement as possible on desired outcomes. I then provide two examples that illustrate some of these issues, one from Pistons Middle School in Detroit, and the other, "Gardening for Health" from Bradford, UK. I finish with some concluding remarks.

While there is much disagreement on the meanings and interpretations of sustainability and sustainable development (see, for instance, Satterthwaite [1999]), there is widespread agreement on the need to move toward sustainable communities and cities, despite some imprecision in defining these entities (Hempel 2000). The implication is that sustainable communities, and ultimately sustainable cities, are predicated on raising the quality of human life in a just and equitable manner, while living within

TABLE 28.1. Characteristics of a sustainable community.

A sustainable community seeks to:	
Protect and Enhance the Environment	• Use energy, water, and other natural resources efficiently and with care • Minimize waste, then reuse or recover it through recycling, composting, or energy recovery, and finally sustainably dispose of what is left • Limit pollution to levels that do not damage natural systems • Value and protect the diversity of nature
Meet Social Needs	• Create or enhance places, spaces and buildings that work well, wear well, and look well • Make settlements "human" in scale or form • Value and protect diversity and local distinctiveness and strengthen local community and cultural identity • Protect human health and amenity through safe, clean, pleasant environments • Emphasize health service prevention action as well as cure • Ensure access to good food, water, housing and fuel at reasonable cost • Meet local needs locally wherever possible • Maximize everyone's access to the skills and knowledge needed to play a full part in society • Empower all sections of the community to participate in decision making and consider the social and community impacts of decisions
Promote Economic Success	• Create a vibrant local economy that gives access to satisfying and rewarding work without damaging the local, national, or global environment • Value unpaid work • Encourage necessary access to facilities, services, goods, and other people in ways which make less use of the car and minimize impacts on the environment • Make opportunities for culture, leisure and recreation readily available to all

Source: DETR 1998.

the limits of ecosystems (see Table 28.1). Roseland (1998) utilizes a set of descriptors, which develop these aims. He argues that sustainable communities are based on "the efficient use of urban space, on minimizing the consumption of essential natural capital, on multiplying social capital, and on mobilizing citizens and their governments towards these ends" (p. 24). There is not the space here to unpack these descriptors fully; however, Table 28.1 presents the characteristics of an ideal sustainable community that espouses environmental, social, and economic goals. The Local Government Management Board in Britain developed these characteristics in the early 1990s as a result of community consultations. What is interesting is that they are virtually identical to those developed by the Institute for Sustainable Communities in Vermont, which were subsequently used by the former

President's Council on Sustainable Development. They have clear implications for urban ecosystem education that will be touched upon in the discussion that follows.

Perhaps the most fundamental concept to the development of sustainable communities and cities worldwide is that of community and citizen participation. But what is participation? Participation is both a means (*a process*) and an end (*a product*). It is neither a sufficient means in and of itself, nor is its attainment (whatever than means) an indicator of sustainability. At its best, however, the notion of community and citizen participation represents the full embodiment of the link between democracy and civil society by enabling citizens and their communities to both envision, and then to develop the understandings, skills and confidence required to actually develop sustainable communities. Discussing citizen initiatives in a global context, Korten (1996) notes that "there are signs throughout the world of a political and spiritual awakening of civil society to the reality that national and global institutions are pursuing agendas at odds with the needs of people and living things. Countless citizen initiatives prompted by this awakening are coalescing into a global political movement for transformational change" (p. 46). At its worst, according to Blake (1999, p. 272), the notion merely "persuades people of their responsibility to follow prescribed patterns of environmental behavior without giving them any effective tools to enable change."

U.S. planner Sherry Arnstein (1969) defined a "Ladder of Participation," with eight levels of citizen participation (Figure 28.1), from the lowest, *manipulation* (a case of nonparticipation, where public relations is the primary goal), through *consultation* (attitude surveys, neighborhood meetings, and public hearings), to the highest, *citizen control* (where citizens set policy, govern, and manage). At the level of citizen control, one would expect those involved to have reached an advanced stage in terms of Wals's (1996) four dimensions of change: that is, they would have the ability to

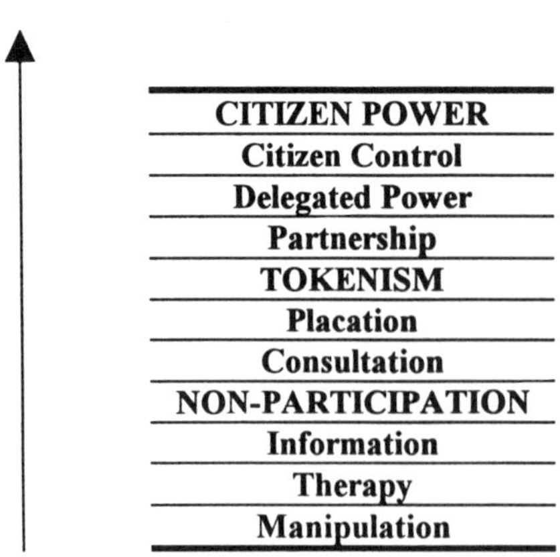

FIGURE 28.1. The ladder of citizen participation. (Arnstein 1969; redrawn by permission of the *Journal of the American Planning Association*, Vol. 35, No. 4, 1969).

construct, *transform*, *critique*, and *emancipate* their worlds. The achievement of citizen control would be impossible without these abilities.

Clearly, if it is to be effective, and to realize its full potential as envisaged at Cary Conference VIII and in the foregoing discussions, lifelong, formal and informal urban ecosystem education must be both *rooted in*, and *informed by* local experiences and understandings. This in turn will assist communities, researchers, classroom teachers, and others in the development of locally defined participatory research agendas and interest bases as the examples later in this chapter show. As the former President's Council on Sustainable Development (1996) has stated, however: "Time and time again, community leaders told us that a fundamental component of implementing sustainable development locally is having people come together to identify a community's needs and then work toward collaborative solutions. Accomplishing this requires both political leadership and citizen involvement" (p. 87). While urban ecosystem education can contribute to citizen involvement and participation, we must be aware that sympathetic and brave political leadership is fundamental to achieving our ends, because as Arnstein (1969, p. 216) notes "participation without redistribution of power is an empty and frustrating process for the powerless."

The benefits (and problems) of inviting the participation of local people and communities will be interpreted differently by the various interests involved (e.g., researcher, classroom teacher, funder, citizen, local politician). However, the wider benefits in terms of those involved in urban ecosystem education include:

- People who are involved in a process with which they identify are more likely to feel empowered. This will develop their understanding, knowledge, trust, will, and confidence. In short, it will build their capacity (i.e., help them to apply their understanding such that they can participate in other projects or program;
- Fresh ideas, innovations, approaches, and/or perspectives may emerge from a diverse range of inputs to a program;
- A program or project may get help in kind or other resources if its ownership is shared more widely;
- Ownership or being a stakeholder in a process or program helps people feel that they are part of a long-term solution, not just an add-on or accessory; and
- Participation establishes multistakeholder partnerships and builds "social capital" ("features of social life, networks, norms, and trust that enable participants to pursue shared objectives" Putnam [1996] *http://www.prospect.org/print/V7/24/putnam-r.html*).

Implications for Urban Ecosystem Education Practice

The key process issues for urban ecosystem education (and educators) are therefore:

- That the main interests agree on and achieve the appropriate level of participation;
- That there is a common language to discuss issues and develop ideas; and
- That appropriate methods are used to get as much agreement as possible on desired outcomes.

The Appropriate Level of Participation

Using Arnstein's ladder (Figure 28.1), it is clear that in different cases, different levels of participation will be more appropriate. The key is to get all parties to agree on the appropriate level. Wilcox (1994) suggests five different possible levels of participation: *information* (informing people), *consultation* (asking certain, carefully chosen questions), *deciding together* (discussing and giving choices, but not necessarily the responsibility for action), *acting together* (deciding and acting together), and *supporting local initiatives* (helping others develop and carry out their own plans). Not every project or program in urban ecosystem education could, or even should lead to citizen control. Some highly specialized, scientific or technical ones may be best applied at the *information* level, where clearly others could *support local initiatives*. However, if urban ecosystem education is to make its greatest contribution to sustainable communities and cities, the level of participation must be explicit, and more importantly, it must be achievable and ultimately achieved.

A Common Language to Discuss Issues and Develop Ideas

A prerequisite to developing a common language is developing the ability among urban ecosystem educators to listen to local experiences and understandings. As was argued earlier, this will assist researchers, classroom teachers, and others in the development of locally defined participatory research agendas and interest bases. Listening is quite a skill and is certainly crucial in situations where "expert" and "lay" publics interact. Irwin (1995) takes this a stage further when he notes that "it is clearly important that we should consider the possibilities for an approach to science and expertise which offers at least the potential for a dialogue between scientific and citizen groups. Is it possible for a 'citizen-oriented science' (or 'citizen science') . . . to emerge?" He continues that "any kind of citizenship which neglects the knowledges held by citizen groups will be restricted in its practical possibilities" and that "there will be no 'sustainability' without a greater potential for citizens to take control of their own lives, health and environment." (p. 7) Clearly, "citizen science" is a component of the move toward sustainable communities and as such is a longer-term goal. In the shorter term, however, urban ecosystem education must be informed by the thinking behind "citizen science" and also avoid, as much as possible, the use of overtly technical language.

Appropriate Methods Are Used

Wilcox (1994) suggests that methods appropriate for participation come in three main categories: (1) techniques, (2) structures, and (3) programs.

Techniques

At the technique level, there are short-term interventions often used by consultants and trainers. At their most basic these may be communication materials and simple workshop sessions, but they also can be more complex methods of decision making. There are many possible interventions that could be used in *both* formal and informal urban ecosystem education. Such interventions can be very useful ways of focusing efforts on involving people, but the temptation is to see them as "quick fixes": they are not. Developing a participatory framework takes time, and the following techniques will usually need to be part of a long-term program, or related to a structure, or both. FAO (1994) and the International Institute for Environment and Development (IIED 1995) offer a list of participatory learning and participatory action research techniques that place people in the research process. Participatory learning and action approaches are sometimes called collaborative approaches (see for example chapter 3 by Bryant and Callewaert in this volume; Allen, et al. 1998).

The simplest techniques in participatory research include: brainstorming, case studies, community surveys and profiles, describing visual images, experimentation, field visits and excursions, information collection, interviews, local histories, linkage diagrams, mapping and modeling, memory games, participatory discussion, practical demonstrations, preference ranking, presentation by a resource person, problem solving, skits or plays, songs, systematic or transect walks, timelines, trend and change analysis, and Venn diagrams.

At a more complex level of participatory research, techniques include "Action Research and Community Problem Solving" (AR&CPS). This builds on the work of Lewin (1946), the "founder" of Action Research, which is usually thought of as "a form of self-reflective inquiry undertaken by participants in social (including educational) situations in order to improve the rationality and justice of their own social or educational practices, their understanding of these practices and the situations in which these practices are carried out" (Kemmis 1985, p. 42). Community Problem Solving "describes the realm in which action research can be employed in the context of environmental education" (Wals 1996, p.302). Taken together, Wals (1996) explains that "AR&CPS represents an inquiry process that enables teachers and students to participate more fully in the planning, implementing, and evaluating of educational activities, aimed at resolving an environmental issue that the learners themselves have identified." Crucially for proponents of urban ecosystem education, Wals (1996, p. 302) continues that "the definition of an environmental issue largely depends on the perceptions and experiences of the learner and on the context in which education takes place." What this means is that what the facilitator considers to be an "environmental" issue, and what the students consider to be an issue may be very different. This was certainly the case in the example of Pistons Middle School in Detroit below.

Structures

In terms of *structures* for participation, there are many, as Wilcox (1994) notes, ranging from the more temporary working parties and advisory committees that are usually set up to achieve a goal, or specific outcome or output, to partnership organizations like development trusts and community-based coops. These are more permanent organizations that may or may not develop into other organizations and structures. However, the development and management of the structure for participation must be clear at the outset, or it must be regularly reviewed.

Programs

Programs (as opposed to *projects*, which are usually for shorter terms) are longer-term processes for developing participation. They are usually planned consensually, over a period of time, and may involve (locally recruited) staff devoted partly or wholly to the program, as well as the use of techniques and structures that are described earlier.

Examples

How do the concepts and key process issues for urban ecosystem education and educators pan out in terms of examples? Each of the following examples illustrates different approaches and facets of good urban ecosystem education practice, which are appropriate to the different contexts in which they occur. The one from the U.S. is based in the formal education sector and uses the AR&CPS approach that is, in Wilcox's (1994) classification of *methods of participation*, a *technique*. In terms of *level of participation*, it falls somewhere between Wilcox's (1994) *deciding together* (discussing and giving choices, but not necessarily the responsibility for action) and *acting together* (deciding and acting together). The other case study is from the United Kingdom. It focuses on informal or community education and illustrates collaborative community greening involving different cultural groups. In Wilcox's (1994) classification of *methods of participation*, it is about developing the correct *structures* and *programs*. In terms of *level of participation*, it falls somewhere between Wilcox's (1994) *acting together* (you decide and act together) and *supporting local initiatives* (helping others develop and carry out their own plans).

Example 1—Pistons Middle School, Detroit

Wals's (1996) study in inner-city Detroit provides an excellent example of the workings of AR&CPS. He worked with eighth grade students who were very aware of the problems in their area: crime, drugs, pollution, poverty, poor housing, homelessness, and more, and were frustrated at their own feelings of impotence. The facilitators (Wals and his team) initially met with a science teacher and a social studies teacher. Teachers were surprised that their role was, along with students, to *develop* the program, not just *implement* it as was often the case when outsiders came in. A one-day workshop was held in which past AR&CPS projects were shared and decisions were made (prior to going into the classroom) on: a timeline (two-hour sessions twice a week for two to three months); parental notification/involvement; a community walk; topic generation and selection; the number of topics (one was decided upon); action taking; and evaluation. The steps taken in the program were as follows:

Step 1—Identifying Issues of Mutual Concern

A community walk was selected as the best process for generating ideas. Both the students and teachers were wary of the walk because of security and safety issues, but agreed to do it. They talked to people in the neighborhood, took photographs, and wrote in their journals. The experience was very positive and was a good starting point. A long list of issues/problems came up which students reflected upon, and eventually they unanimously selected school safety, which, as was mentioned previously, may challenge some peoples' notions of urban ecosystem education. When they decided to interview the principal, he claimed it wasn't an issue, which disappointed but reinvigorated the students. They widened their survey to encompass staff and students and the principal had to accept their findings.

Step 2—Analyzing a Particular Issue

Initially, safety was defined and understood by the students in terms of short-term solutions. They came up with statements about the school, for instance, it has "no metal detectors" or "no ID" (i.e., they were focusing on symptoms of the problem, not causes). In their desire to come up with solutions quickly, they thought that the mere identification of these safety shortfalls was sufficient. The students clearly thought that the project was over and that they needed to move on. The teachers and facilitators tried to slow the students down, to get them to research at the library, conduct local interviews, and scan the local press. This made the students think that the teachers and facilitators were preventing them from moving from developing an understanding to applying that understanding (i.e., taking action). This led to a dilemma for Wals and his team. Should they let the students go on, knowing that they needed more information and possibly risking losing some of them (the "experience first" argument) or should they continue to slow them down (the "information first" argument)? They decided to take the former approach.

Step 3—Generating Potential Solutions

The students pushed on investigating the pros and cons of issuing everyone ID passes. They looked at other schools in Detroit and put together a plan that they presented to the principal only to hear that they had not looked at costs and that there was no money to implement the plan. They looked into this, including how to raise the money and undertook a similar process regarding having more and better security guards and student monitors. However, as they got further into their investigations, they began to realize that the issue of school safety is intimately related to community safety and that a safe neighborhood would probably mean a safe school. A paradigm shift was occurring: they changed from looking at ad hoc solutions to looking at problems. *Why do people become violent? Are there other ways to resolve conflict?* The class realized they needed to research these issues. They invited specialists, spoke with mothers who had had children killed, and visited criminal courts. No longer was the goal to install metal detectors or implement an ID system; it was to educate the school about the destructiveness of violence in the community and to find creative ways of resolving conflict.

Step 4—Implementing a Solution

The paradigm shift from ad hoc solutions to education and communication was to be carried out by a mix of rap and skit. The process culminated in a "Stop the Violence" rap/skit in front of the entire school.

Step 5—Evaluation of Results

Wals (1996, p. 311) is careful to note that "it is important to emphasize that in most AR&CPS projects, the point is not that students actually completely resolve a problem (although this has been done) but that they take action to alleviate it." In this case, awareness raising was the solution chosen by the students; however, evaluation was built into the process at Pistons: The teachers and facilitators met frequently and the students reflected on their experiences via their journals. A spin-off of the project was a change in teaching style and utilization of community resources in teaching.

The value of the AR&CPS approach is that it helps people develop transferable skills, understanding, knowledge, and confidence. These are the building blocks required to participate in the envisioning, defining, and development of sustainable communities and cities. Note also that the students made links between school and wider, community safety. This recognition of embeddedness, of interlinkage between the characteristics of sustainable communities, is critical (Table 28.1).

Example 2—Gardening for Health

In 1992, the United Nations Commission on Environment and Development (UNCED) agreed to Agenda 21, a program for global sustainable development (UN 1992). Signatories committed themselves to depositing a national plan for sustainable development by 1994. Even though Agenda 21 is the global agenda, the UNCED organizers were persuaded that, according to the principle of subsidiarity, as the level of governance closest to people *local governments have a vital role to play in educating, mobilizing, and responding to the public to promote sustainable development.* The International Council for Local Environmental Initiatives (ICLEI) duly wrote Chapter 28 of Agenda 21, which recognizes and relates to this pivotal role. Local Agenda 21 (LA21) was born.

Bradford is a city of 350,000 in West Yorkshire, UK, with a large Asian population and an active community LA21 program. The fact that Asian women are at a particularly high risk of coronary heart disease was causing concern among community health workers. This led them to join with environmentalists and members of the Bangladeshi community in developing a comprehensive package of ideas to turn gardening into a means of urban greening, improving vitality, and keeping fit.

In 1994, a community health worker for the charity HeartSmart, assisted by the Bradford Community Environment Project (BCEP), developed the idea for a project that would respond to the needs of the local Bangladeshi community, in which approximately half the adults do not read, write or speak English. This means that women in particular have little contact with the outside world and are acutely isolated from many statutory and other agencies. Gardening for Health's aim was to raise awareness of health and environmental issues and provide an opportunity for community participation and empowerment.

The community health worker researched the idea and found that no one else was addressing these linked environmental and health issues, so she enlisted the support of a small group of women at the Bangladesh Parishad community center which provided a base for the project. Overcoming start-up difficulties with support and horticultural expertise from a BCEP community environment worker, together with help from other volunteers from Shipley College, Shipley and Baildon Volunteer Bureau, and the Allotments Action Group, the group began to restore a derelict allotment (an abandoned, municipally-owned area formerly used for growing vegetables). In the process, the women increased in confidence and physical fitness (most could hardly manage 20 minutes of physical work at first), with the additional benefit of an abundant harvest in the first season. The project grew from a hesitant beginning to a firm footing with regular commitment from about eight women in a short time.

The community health worker coordinated and supervised the program, supported by the HeartSmart steering committee; the Bradford Environmental Action Trust (BEAT) is also represented on the program management committee. The program illustrates some of the concepts and key process issues for urban ecosystem education. It is a participatory program that is firmly rooted in and informed by locally defined agendas and interest bases. Expertise, where needed, is *on tap*, not *on top*; i.e., expert horticultural input is provided by BCEP's community environment worker in a "citizen science" model. The women participating in the project decide on what action will be taken, how, when, and where. These women have moved from being (literally) silent, to being stakeholders in their local environment.

In terms of achievements, the women have contributed to the efficient use of urban space. There also have been changes toward healthier lifestyles, with the women taking regular exercise and building up their stamina, growing and eating healthier food, and providing social support for each other. All are features of the move towards sustainable communities, whose characteristics were described earlier. The women have successfully grown Bangladeshi vegetables and herbs, as well as British vegetables, and have also begun to grow fruit and vegetables in their own back gardens, illustrating the transferability of skills which urban ecosystem education can inculcate. The flow of gardening expertise has also been two-way. BCEP's community environment worker, although initially relied upon by the women, has learned about the formation of irregular shaped beds and an absence of monoculture, because the Bangladeshi women mix the seeds together, knowing that they can recognize the plants once they grow. They are now multipliers: the source of gardening advice in the wider community. In October 1998, the community health worker from HeartSmart won a British Broadcasting Corporation (BBC) Good Food Award for the best community educational program.

Conclusions

The clear message for urban ecosystem education (and educators), from both the theory elucidated above and the examples presented, is that if it is to help people *construct, transform, critique* and *emancipate* their worlds (Wals 1996) towards sustainable communities and cities then *it must start where they are*. The ability to listen to local experiences and understandings, challenge them, and develop participatory research agendas and interest bases where the teacher is co-learner and the learner is co-researcher, will assist in developing a robust learning pedagogy for classroom teachers, researchers, and others involved in urban ecosystem education. Only in this way will urban ecosystem education fully contribute to the development of sustainable communities and cities.

Williams and Agyeman (1999, p. 463) argue that "effective urban environmental education involves listening closely to the perceptions and priorities of urban residents, and creating programs that are relevant and sustainable over the longer term . . . it must be rooted in local communities in order to take up the challenge of making the urban environment a more livable and sustainable system." There will be dilemmas, conflicting agendas, and ill-defined and often contradictory goals, but dealing with *resonant* issues first—such as security in the example of Pistons Middle School in Detroit, or health issues, as in the Bradford example—will more likely develop people's transferable skills, understanding, knowledge, values, and confidence, such that they might be able to apply them in different situations. This will, ultimately, lead to a broader partnership toward achieving the kinds of environmental, social, and economic security that are characteristic of sustainable communities.

References

Agyeman, J. 1994. Next step: education for participatory democracy? Page 52 in *Annual Review of Environmental Education 1993*. 25 year Special Edition Council for Environmental Education, Reading UK.

Allen, W.J., K. Brown, T. Gloag, J. Morris, K. Simpson, J. Thomas, and R. Young. 1998. *Building partnerships for conservation in the Waitaki/Mackenzie Basins*. Landcare Research Contract Report LC9899/033, Lincoln, New Zealand.

Arnstein, S.R. 1969. A ladder of citizen participation. *Journal of the American Planning Association* 35(4):216–224.

Blake, J. 1999. Overcoming the "value-action gap" in environmental policy: tensions between national policy and local experience. *Local Environment* 4(3):257–278.

Department of Transport, Environment, and the Regions. 1998. *Sustainable local communities for the 21st century*. DETR, London, UK.

Dunlap, R. 1997. Public opinion and environmental policy. Pages 63–114 in J.P. Lester, ed. *Environmental politics and policy (2nd edition)*. Duke University Press, Durham, NC.

FAO. 1994. *The group promoter's resource book*. Food and Agriculture Organization, Rome.

Hempel, L.C. 1999. Conceptual and analytical challenges in building sustainable communities. In D.A. Mazmanian, and M.E. Kraft, eds. *Towards sustainable communities: transition and transformations in environmental policy*. MIT Press, Cambridge.

IIED. 1995. *RRA Notes 21: Special issue on participatory tools and methods in urban areas in the series: participatory learning and action notes PLA 21*. IIED, London.

Irwin, A. 1995. *Citizen science*. Routledge, London.

Kemmis, S. 1985. Action research. Pages 35–42 in T. Husen, and T. Postlethwaite, eds. *International encyclopedia of education: research and studies* Vol I A–B. Pergamon, Oxford.

Korten, D. 1996. Civic engagement in creating future cities. *Environment and Urbanisation* 8(1):35–49.

Lewin, K. 1946. Action research and minority problems. *Journal of Social Issues* 26: 3–23.

Maslow, A. 1943. A theory of human motivation. *Psychological Review* 50:370–396.

President's Council on Sustainable Development. 1996. *Sustainable America: a new consensus for prosperity, opportunity, and a healthy environment for the future.* U.S. Government Printing Office, Washington DC.

Putnam, R.D. 1996. The strange disappearance of civic America. *The American Prospect* 24: (Winter).

Roseland, M. 1988. *Toward sustainable communities*. New Society Publishers, Gabriola Island.

Satterthwaite, D., ed. 1999. *The earthscan reader in sustainable cities*. Earthscan, London, UK.

UNCED. 1992. *The global partnership for environment and development: A guide to Agenda 21 Geneva*. UNCED. April 1992.

Wals, A.E.J. 1996. Back alley sustainability and the role of environmental education. *Local Environment* 1(3):299–316.

Wilcox, D. 1994. *The guide to effective participation*. Partnership, Brighton.

Williams, E.J., and J. Agyeman. 1999. Educating for a more livable urban environment. Pages 26–30 in *EEducator*. North American Association for Environmental Education, Washington DC.

29 Perspectives on the Future of Urban Ecosystem Education: A Summary of Cary Conference VIII

PETER CULLEN

The Eighth Cary Conference has broken new ground in focusing on urban ecosystems and how we can educate society about them. In the last 30 years there has been a growing concern about the need for ecologically sustainable development, and the realization has dawned that much of what we are now doing is not sustainable. While there have been many international conferences and agreements pledging support to the goal of sustainability, progress has been remarkably slow. The Eighth Cary Conference was based on four major propositions:

- Most of the world's population now lives in urban areas, and in many of these areas we are both failing to provide the basic services needed by humans and failing to sustain the biological communities;
- To move towards sustainability will require significant changes of behavior by large numbers of people;
- Changing behavior requires knowledge of what should be done, but also requires attitudes to change. Humans need to reevaluate short-term, personal values and look at longer-term consequences to our communities; and
- Education is the key to providing knowledge and appropriate skills and for helping individuals reexamine and redevelop their value systems.

The challenges are to establish what are the critical ecological concepts, and to decide how and when they should be communicated through the processes of formal and informal education.

The combination of greed and ignorance has been a potent force in causing environmental degradation. It is widely accepted that the science of ecology can help us to understand many of our impacted ecosystems, and that this understanding can guide our attempts at wiser management or even restoration. The influence of ecology is apparent in many areas, including agriculture, forestry, and watershed management. The challenge now for ecology is to use the tools and understandings already developed and apply them to the vastly more complex situation of urban communities, where most of the world's citizens live and where we may be furthest from sustainability. These tools will in many cases be found wanting, but they will develop and improve. Hopefully we will develop a better predictive capacity to guide planners and managers, and to repair some of the damage we have done.

The Cary Conference provided challenges to educators as well as to ecologists. The challenge is to work with ecologists to decide on the key ecological processes that control and shape urban ecosystems, and then to develop new and better ways to help urban communities understand these processes, their underlying principles, and their consequences.

One overarching issue was the central role of our human value systems, and how differences in values made dialogue more difficult. Some participants were concerned that the urban systems under consideration were so complex that "objective" science could not cope. Others felt it was essential for ecologists to engage in understanding complex human-dominated ecosystems if we were to have any hope of moving towards sustainability. This tension permeated the conference. While some asserted the objectivity of science, others appreciated that different disciplines with different assumptions, constructs and scales often arrived at different understandings, indicating that our objectivity was bounded by our cultures and assumptions. This issue was not resolved because participants chose to explore the insights and perceptions of different traditions and be tolerant of the differences. Indeed, the differences provided the intellectual stimulation of the conference and in the end were widely valued.

The Importance of Understanding Urban Ecosystems

One of the key questions for the conference was "Why is it important to understand urban ecosystems compared to other things competing for space in the curriculum?" As the world's population grows rapidly beyond 6 billion inhabitants, we continue to see migration from rural areas to urban communities. More than half of the world's population now lives in urban areas. The pressure this growth has put on our ability to provide food, water, energy, drainage, sanitation, and clean air has been well documented. Even in well-developed countries we are often not able to provide all of our citizens with their basic needs. In the undeveloped world the problem is frightening. Many of the world's urban inhabitants do not have these services at even a basic level. Misery and reduced life expectancy are the consequences of our failures to provide these services. The waste of human capital should alarm even the economic rationalists amongst us; the damage to the human spirit should concern us all, as it demeans us all.

It is clear that if we are serious about sustainability we need to address the issues of urban society head-on. Urban communities draw food, fiber, and water resources from a vast hinterland. They concentrate waste products and release them in a limited area, causing problems like eutrophication through nutrient excess at the same time as they cause nutrient depletion in our farmlands. Nutrients, toxicants, energy, and waste products all are concentrated in some ways by urban communities and cause problems for the environmental services on which we all depend.

Urbanization not only has impacts on ecosystem function through changes in the pathways and the scale of cycling of materials but it also makes major structural changes as we replace producers with consumers and decomposers, and reduce and simplify habitat and hence biodiversity. These changes to the structure and function of ecosystems as we urbanize have certainly had major impacts on the ecosystem services on which we depend. We have often overwhelmed the capacity of the urban ecosystem to provide us with clean air and water. Our man-made systems themselves often collapse, causing widespread illness, misery, and death. Outbreaks of cryptosporidia and giardia in major urban communities now are common. In many cities there are substantial periods when we are unable to supply food, water, and energy sufficient to meet human needs. As urban communities get bigger, the impacts of these system failures become more devastating.

What Does an Ecosystem Understanding of Cities Mean?

Few ecologists have focused on the ecology of cities since the pioneering work of Boyden, et al. (1981). An issue of the journal *Science* in 1997 (Vol 277, Number 5325) focused on human dominated ecosystems, and had papers addressing the global impacts of land use change and the contributions of ecology to agriculture, fisheries, and forests. It is notable that no paper addressed the issues of urban ecology, and the issue of urbanization was mentioned merely as one of the land-use transformations that were having impacts. It is apparent that we have not yet developed any serious schools of "urban ecology," although the major U.S. studies conducted in Baltimore and Phoenix provide a promising start.

The challenge of working on systems that are not only dominated by humans but where we might need to predict human behavior also has been an issue. There are differences of opinion as to whether we will ever have robust models that let us predict human behavior. We are probably not yet ready to try and build in social drivers to predictive models. I believe we need to develop our understanding of the biophysical elements and of the social drivers themselves before we try to integrate them. We need to be working in parallel with social psychologists, however, to ensure that we can combine insights from both areas at an appropriate stage.

Much ecology to date has been ambiguous as to whether humans are part of the natural ecosystem or whether they transcend it (see McIntosh 1985). This arrogance needs to be laid to rest; humans are not only a major component of ecosystems but are an almost totally dominant organism, causing impacts on almost all other organisms as well as the structure and functioning of systems.

One issue that needs to be addressed is whether urban ecosystems are fundamentally different from nonbuilt ecosystems. It seems important for us to explore the similarities and differences between built and nonbuilt ecosystems. Clearly there are structural differences. Many of the ecosystem functions however are similar, although they may operate with different control mechanisms. As with most ecosystem studies, scale is a critical element. In urban areas we have structural hierarchies from the individual house to the district to the city. We can expect that different processes will dominate at different scales. These different size units give us a basis for comparative studies of mass and energy movement in various urban communities. Knowledge of such fluxes in itself might provide useful feedback and cause individuals to reflect on their behavior. Information on average and peak water or energy use comparing current use with that of previous years, or from other areas (Grimm, et al., Chapter 7 in this volume) could be gathered on a real-time basis, providing almost continuous feedback that might influence behavior. It can readily be done using local media if the information can be provided in an accessible form.

Defining boundaries for urban ecosystem studies will be a challenge, because the footprint of impact of the urban area commonly greatly exceeds the urban area itself. Food, fiber, water, and energy are commonly drawn from a large area to support an urban area. Some of these resources may come from other continents and this may increase with current trends in globalization. The zone of impact of waste materials and waste energy from urban areas can also be substantial, and again globalization is leading some communities to export intractable or dangerous wastes large distances.

One interesting ecological concept of use is that of a home range for an organism (Chawla and Salvadori, Chapter 18 in this volume). It was reported that the home range of city children determines the range of experiences of the children. Limited experiences may well lead to limited opportunities. Chawla and Salvadori reported that disadvantaged children in urban areas have a much smaller home range than do other children.

The importance and conservation of biodiversity is a major issue for environmentalists, and the role of diversity in regulating the stability of ecosystems is a major issue for ecologists. If humans understood this issue better it might lead to a greater tolerance of differences within human society.

Ecology has always been an integrating discipline despite its roots in plant or animal biology. As we attempt to tackle issues in the urban area, it is important that we appreciate that we will not make progress when locked into our disciplinary silos. Collaboration between disciplines is fundamental to progress in ecosystem ecology, as opposed to the simpler plant and animal ecology fields (Likens 1998). Many of the larger problems now facing society are not amenable to solution through disciplinary research, and require the intellectual contributions of several disciplines if progress is to be made. We are learning and developing new models of generating knowledge that are focused on understanding a complex practical problem through the effective interaction and collaboration of a variety of scientific disciplines (Gibbons, et al. 1994). This new mode of knowledge production builds on the traditional discipline-oriented model of research and requires all of the normal quality control systems of peer review but also judges the usefulness of the research findings.

The emerging model might be thought of as transdisciplinary, rather than just multidisciplinary. It does not depend on various disciplinary specialists working in isolation with the occasional interconnection. It relies, rather, on frequent interaction and stimulation across disciplinary boundaries. Partnership and collaboration across boundaries are key features (Cullen, et al. 1999). The interaction of different observations, interpretations, and insights is a critical part of the creative process. The tolerance of multiple interpretations and the views of others is important. Trust is one critical outcome of an effective dialogue between disciplinary experts. It must also be appreciated that effective cross-disciplinary dialogue and stimulation takes time. It is a classic example of where it is necessary to "go slow to go fast."

The sorts of multidisciplinary teams needed to advance our understanding of urban ecosystems do not lend themselves to the traditional discipline-based university department. Consequently they will tend to be transitory, assembled for a period of perhaps five years to address a particular problem, and may be disbanded when the problem that brought them together is resolved. This multidisciplinary, interactive model of learning can be applied at the community level as well. Urban ecosystems can only be understood using the tools and perspectives of a number of disciplines, and bringing these viewpoints together can be achieved by people who are part of the system and see it from various perspectives. Community members may in fact be better integrators than the disciplinary experts.

The power of a simple, diagrammatic conceptual model of the system being investigated must not be underestimated. All researchers have conceptual models, but often do not make them explicit. If these models can be shared and negotiated, issues like the meaning of terms, scale, critical drivers, and so on are argued and resolved rather than remaining as hidden barriers to communication. This process allows people to challenge their own assumptions and the baggage they bring to any new problems. Robust argument lets people test alternative views, clarify the appropriate scale with which to view the problem, and be comfortable in changing their position or assumptions.

The process of developing a conceptual model puts the model building at the center of the learning process. The model becomes a simple statement of the collective understanding of how the system works and is in fact the first stage in articulating hypotheses for later testing. The model is not about putting all possible pathways and interactions into a comprehensive overview; it is a synthesis of views about subsystems brought to the process by disciplinary specialists. It demands that these views be brought together at a common scale. It is a working model that can be further developed as insights grow through the process. The process makes the players' various assumptions obvious, and allows them to be debated.

There are other broader elements of science that can contribute to a better and more sustainable society. It is important for us to increase community literacy of science. We need to emphasize the importance of careful observation, of measurement and of speculation to establish hypotheses, and the subsequent testing of these hypotheses as we seek the truth. We need a community that comes to judgments using such processes. They need to appreciate that multiple explanations and disagreements are part of the process and are an essential part of our search for truth. They need to appreciate that different assumptions and different paradigms may lead to differing solutions, and that teasing apart and understanding such differences is the way to improve understanding. When they hear scientists disagree, they should be encouraged to ask why rather than assume one of the protagonists is ill-informed or corrupt in some way (Cullen 1998).

Providing Learning Opportunities About Urban Ecology

An underlying assumption of the conference was that we need a much broader societal understanding of ecosystem principles as they apply to urban areas if we are to change the behavior of sufficient individuals to make a difference. We need to develop a political constituency with a greater environmental perspective; we need to assist power elites to understand not only the importance of these issues, but to give them confidence that certain actions will lead to improved outcomes. If education is to lead to changes in behavior it must go well beyond understanding. The learning opportunities we provide must impart skills so that people know how to do things in other ways. They must also lead to individuals and communities reexamining and reformulating values and attitudes. This wider concept of education is fundamental to progress. Our experiences in trying to change behavior in the area of human health has clearly shown that understanding is not enough by itself. Providing knowledge is necessary, but is not sufficient. As Bloom and Krathwohl (1984) appreciated, educational objectives must encompass knowledge, skills, and attitudes, and it is better that each be made explicit.

The emphasis of our educational activity must be on learning rather than teaching if we are to produce useful outcomes. This provides a massive challenge to all of us to identify, communicate, and foster an understanding of the insights of ecology as they impact on urban communities.

Our experiences to date with both formal and informal education have made it clear there are major differences between teaching and learning. The outcome we seek is learning; the opportunities we provide are designed and may be delivered by teachers. Transmitting a message is not enough; it must be received, understood, and acted upon. It is likely that there are different learning processes involved in team-based experiential learning than in the more traditional master–pupil situation. The traditional master–pupil relationship may inhibit free-ranging speculation of ideas and connections in a way that does not happen with a group of peers.

There are real problems for the development of learning materials. There is as yet no professional body that determines the critical content of environmental education programs, especially those looking at urban ecosystems. Our ecological understanding is growing rapidly, so there is as yet no agreement among ecologists as to what should be taught. If there was agreement, it is not likely it would be stable as new work is completed, and new insights emerge. This poses a particular challenge in educational systems dominated by standards that may provide a real straitjacket to change and the introduction of new ideas. There is a major challenge for those in the formal education system where there is pressure to include all sorts of topics in an already-crowded curriculum. Conference participants recognized, however, that influencing the formal curriculum was necessary although not sufficient to induce the changes we must make. Informal education of many types was identified as another way to help people see and hopefully understand the ecosystems of which they are the dominant part.

Changing human behavior gives us a clear goal for education. This is a critical issue. We can spend a lot of time arguing about what should be taught, by whom and when, but unless we are very clear as to why we are teaching about urban ecosystems we will probably make inappropriate choices. For over a century we have been told about the iron laws of economics that drive our society. As the world's population reaches past 6 billion, however, the rigid laws of economics look positively flexible compared to the iron laws of ecology. Our society must learn to understand this.

There was agreement at the conference that education must go beyond facts to help the learner seek cause-and-effect relationships and to understand the dynamic nature of the world we live in. The content and "facts" were seen as important, but were also a vehicle to help people appreciate interconnectedness and relationships, as well as developing skills of observation and questioning. The learning opportunities we seek to provide go well beyond the immediate content.

It must be recognized that many of our communities have very low levels of scientific and ecological literacy. This provides a challenge, but urban systems provide a good vehicle to teach ecological literacy because of their immediacy and relevance. Examples abound if only we can help people see them and question them. We know that curiosity to solve or understand a problem is a key motivator for learning. We need to teach basic skills of observation and the making of simple connections. It is often a surprise to urban residents living in flood-prone areas when they get flooded. Observing the topography and vegetation patterns are likely to give clues to such hazards, yet many urban dwellers are not sensitized to look for such clues. If they were sensitive to the risks they could easily check historical records, but few do.

The perception of relevance is critical to achieving educational outcomes, and inquiry-based learning is one of the educational fashions of the times. In science, however, we appreciate the difficulty of framing the right question—one that is amenable to testing through observation or experiment, and one that yields useful insights into the issue of concern. Given that framing an appropriate question is a key scientific skill, the challenge of providing high quality education through inquiry-based learning is obvious.

While it is obvious that school children and their teachers are necessary targets for education, it is clear that we need to have a much wider focus on the community and perhaps even decision makers in the community. There is little doubt that community recognition and ownership of both problems and their possible solutions is a prerequisite to moving forward on many issues. This is evident from case studies presented in this volume for Chicago by Fialkowski (chapter 21 in this volume) and for Durban by Roberts (chapter 24 in this volume), which moved beyond school education to the community and to groups of decision makers.

The Landcare movement in Australia is a good example of communities taking ownership and responsibility for an issue, and doing something about it. Landcare involves local rural communities banding together to solve problems with the support and assistance of the various agencies. A Landcare group is a voluntary organization of local community members dedicated to combating land degradation, protecting wildlife, and managing the land in a responsible way. It involves neighbors working together to identify, analyze, and then tackle a problem of common concern to all of them. Landcare groups are significant in that they draw upon common and pre-existing social values and networks, and yet empower the individual to action. They allow the development and articulation of a new land management ethic, but also provide a framework through which government support and expertise can be made available in a cost-effective manner compared to previous extension models. There are now more than 3,000 local voluntary Landcare groups in Australia, and the movement has now been developed in urban and urban-rural fringe communities. It has been slower to develop in urban areas because of the lack of agreement as to what is the most critical problem in the urban areas, and the slowness of agencies to provide facilitation and support services in urban environments.

The environmental movement has long espoused the adage "think globally, act locally." This Cary Conference challenged this notion, and the idea of "think locally, act locally" may well be more useful (Shu, Chapter 4 in this volume). The global immensity of the problems we face are intimidating. If we can refocus them to a local level that everyone can see and appreciate, we may have more success with changing people's behavior.

There needs to be further dialogue at a technical level between experienced ecologists and educators to identify the central ecological ideas that are important to educating people about urban ecosystems. These need to be central ideas that impact on the structure and function of urban areas. It is unlikely they will focus on the odd bit of remnant nature that has survived in a city, or the impressive capacity of weeds to colonize bare ground, although these ideas may be part of the curriculum. We need to conceptualize the urban area as a functioning complex system rather than as a collection of artifacts and remnants. This process will identify those bodies of knowledge for which there is reasonable agreement, but it should also identify the areas and topics that are not agreed on, and may even identify topics about which we have very little idea. These will usefully inform the research agenda.

Looking Back

The challenge for the participants at the conference was to explore the contributions that might be made by ecologists and educators to address the monumental issue of global sustainability. There was widespread agreement about several issues.

- The disciplines of ecology and education are fundamental to moving towards sustainability;
- To achieve sustainability we are going to have to change human behavior with regard to resource use, waste disposal, and the maintenance of ecosystem services at appropriate scales;
- Understanding is a prerequisite for changing human behavior, but in itself is not sufficient;
- Values need to change. Understanding of ecosystems and social processes is critical to the sorts of dialogue that lead to changes in values;
- The required value shifts will lead us towards greater awareness of environmental citizenship and the need for environmental justice. Poor ecological literacy in the community makes environmental citizenship difficult;
- Urban systems are familiar to many people and are a useful vehicle for teaching many important things beyond ecological principles, such as reasoning, scientific literacy, and the need for collaboration; and
- Not only are urban systems a useful teaching model because of their immediacy to so many people, but because they are the location of our greatest problems, they are an appropriate place to use our intellectual capacity if we want to make a difference.

Looking Forward

The conference provided the start for an important discourse between ecologists and educators about education in urban ecosystems. It was an innovative and exciting idea to bring these groups together to start the dialogue; however, there is much still to be achieved.

- There is not yet agreement as to the key ecological concepts that need to be communicated and understood to serve as a basis for our actions and decisions. One interesting suggestion was that the Ecological Society of America might undertake a study of urban ecosystems for its "Issues in Ecology" series to draw together our present understandings;
- There is considerable work required by ecologists to start developing and using tools to address this most complex of systems. Ecological studies of urban areas need to be undertaken to test the idea that such studies can produce useful insights. Social psychologists need to be involved if we are to develop any understanding of the drivers of human behavior;
- Educators need to develop more effective ways of providing opportunities to learn through experiential learning. This provides challenges for teachers in an area where the existing knowledge base is developing rapidly. The educational opportunities needed will be found in both formal and informal education;
- Better collaboration is needed between educators and ecologists, and both groups need to recognize and appreciate the contributions made by the other group. Trust and improved understanding will result from this collaboration;
- Feedback to local communities is one obvious way forward. Information on water use, energy use, and waste generation can now be provided in almost real time, and communities can compare themselves to previous years or to other benchmarking communities.

Our ongoing failures to meet the material needs of so many of our global citizens puts at risk the very fabric of society. We are now an urbanized society and must learn to address these issues in the places that really matter, our cities. The increasing divide between rich and poor has traditionally been expressed in monetary terms, but now may well also be expressed in terms of global information. Education, perhaps more than ever before, will divide the "haves" from the "have-nots." History has shown us that the disadvantaged will rebel and act in ways that threaten the sustainability of our societies. We can all observe the impacts on the sustainability of our society of drugs, crime, and ethnic cleansing. Social and community sustainability is as critical to our future as is sustainability of our biophysical systems.

References

Bloom, B.S., and D.R. Krathwohl. 1984. *Taxonomy of educational objectives*. Handbooks 1 and 2. McKay, New York.

Boyden, S., S. Millar, K. Newcombe, and B. O'Neill. 1981. *The ecology of a city and its people: the case of Hong Kong*. Australian National University Press, Canberra, Australia.

Cullen, P.W., R.H. Norris, V.H. Resh, T.B. Reynoldson, D. Rosenberg, and M.T. Barbour. 1999. Collaboration in scientific research: a critical need for freshwater ecology. *Freshwater Biology* 42:131–142.

Cullen, P. 1998. The role of science and scientists in environmental conflicts. *Australian Journal Environmental Management* (Nov.): 55–59.

Cullen, P. 1994. A bottom-up approach to integrating land and water management. Pages 525–534 in *Integrated land & water management proceedings Stockholm Water Symposium*. Stockholm, Sweden.

Gibbons, M., C. Limoges, H. Nowotny, S. Schwartzman, P. Scott, and M. Trow. 1994. *The new production of knowledge*. Sage Publications. London, UK.

Likens, G.E. 1998. Limitations to the intellectual progress in ecosystem science. Pages 247–271 in M.L. Pace, and P.M. Groffman, eds. *Successes, limitations, and frontiers in ecosystem science*. Springer-Verlag, New York.

McIntosh, R.P. 1985. *The background of ecology*. Cambridge University Press. Cambridge, UK.

30
Urban Ecosystem Education in the Coming Decade: What Is Possible and How Can We Get There?

ALAN R. BERKOWITZ, KAREN S. HOLLWEG, and CHARLES H. NILON

The overarching goal of urban ecosystem education is to help people to understand cities as ecosystems. With this book, we have tried to contribute to the development of this field of education, giving substance to its ideas, motivations, approaches, and potentials. In this final chapter, we ask the questions, "Where might we be 10 years from now?" and "What challenges need to be addressed, problems solved, and pathways explored in order to get there?" In choosing to paint the pictures we do, we are not really attempting to *predict* the future, but rather to use descriptions of the "possible" as a way of synthesizing what we've learned and what we strive for, and then to frame critical thinking and provoke plans regarding how we might get there. We hope that by taking the positive voice, the creative optimism of the Cary Conference held at the Institute of Ecosystem Studies in April 1999 might help inspire in educators and scientists the great leaps that clearly are wanted and needed in the days and years ahead.

The chapter poses, and then elaborates on the following predictions of what might be 10 years from now. These predictions relate directly to a set of concrete recommendations generated in discussion groups at the Cary Conference (Table 30.1):

Prediction 1. The idea that *cities are ecosystems* will be pervasive in society.
Prediction 2. Educators and scientists will have a richer, deeper, and more pragmatic understanding of what is meant by the concept of an "urban ecosystem."
Prediction 3. The importance of a broad understanding of urban ecosystems will be recognized, appreciated, and well articulated.
Prediction 4. People everywhere will have many and diverse opportunities to learn about and apply their understanding of urban ecosystems.
Prediction 5. Urban ecosystem understanding will be developed in many places via participatory models linking scientists of many sorts, urban residents, managers, and decision makers.
Prediction 6. People will understand the vital role well-designed and -managed cities play in sustaining life—human and nonhuman—on earth.

TABLE 30.1. Key recommendations from Cary Conference discussion groups.

Recommendation 1. Advance urban ecosystem education by fostering broad public understanding that *cities are ecosystems.*

This understanding is essential in order to achieve key societal goals: improve urban communities, reduce cities' negative impacts on other ecosystems, and help achieve a sustainable future for humans and other organisms on earth. It will involve work with school-based and non–school-based programs, media outreach and public participation to reach adults along with young people. A diverse and inclusive network of individuals, programs, and organizations interested in promoting an understanding of cities as ecosystems will be needed.

Recommendation 2. Advance the practice of urban ecosystem education by:

a. Incorporating urban ecosystems into the formal and nonformal education agenda for a broad diversity of audiences.

School administrations should adopt learning outcomes and curricula that emphasize urban ecosystems. Likewise, we must provide teachers with new curricula and support to focus their instruction on important ideas in understanding urban ecosystems (e.g., mass balance of a school, a neighborhood, a metropolis), incorporate urban ecosystem examples in diverse textbooks, and promulgate research that shows the advantages and limitations of urban ecosystem and project- and case-based teaching.

b. Developing an educational model that utilizes a collaborative process for participatory research, community discovery, and action in urban ecosystems.

A participatory, community-based approach to learning about and taking actions to improve urban ecosystems is a promising educational tool in need of development, analysis and refinement, and implementation. Such an approach will teach environmental civics and democracy through practice, empower students to use science to understand, organize and improve their community, and contribute to the development of systems thinking. We need good examples of Participatory Action Research in cities that are documented, assessed, and show learning on the part of students/participants, benefits to the community, *and* benefits to the scientists involved.

c. Promoting continued work between educators and natural and social scientists on urban ecosystem education.

To understand urban ecosystems, we need concepts and ways of knowing from natural, physical and social sciences, and the humanities. Curriculum integration and innovative instruction are required, accepting a commitment to depth over breadth and the need to provide interdisciplinary and integrated training to future teachers, citizens, professionals, and students. We must champion new partnerships between scientists, educators, students, and citizens in developing collaborations where research, education and stewardship go hand-in-hand. To do this, all involved need new kinds of rewards in order to sustain their work. Likewise all involved need to learn how to use systems thinking as an integrator.

Recommendation 3. Advance the conceptual foundations of urban ecosystem education by:

a. Considering age appropriateness, concept development, and pedagogy in advancing the practice of urban ecosystem education.

We need to provide learners with a variety of ways to be successful in building their understanding of urban ecosystems, where all learn as they learn best and are challenged to stretch and grow in new ways, too. Our work must be grounded in the insights into pedagogy, learners, and educators from the fields of educational and cognitive psychology, sociology, and anthropology. We must motivate people to create a sense of place within the urban environment and to support that sense of place with connections through learning, activity, and participation in the local community. Part of this effort will require new and expanded education research.

(Continued)

TABLE 30.1. *Continued*

b. Clearly defining the concepts and ways of knowing involved in understanding urban ecosystems.

Social science insights must be thoroughly integrated with the traditional ecosystem perspective to build an understanding that embraces the biota, the physical environment, ecological interactions, controls, and boundaries, along with humans and their institutions, the built environment, information and behaviors, social structure, money and power hierarchies.

c. Basing the practice of urban ecosystem education on a critical analysis of the broader social framework within which we operate.

While cities contribute to the wealth and well-being of the countryside, in material ecological terms, cities are largely parasitic on nature. The consumer habits of (particularly wealthy-country) urbanites both drain the increasingly global hinterlands of their resources and pollute the global commons, including the immediate urban environment, with the poor and cultural minorities both within and outside of cities suffering the worst consequences. Urban ecosystem education must elaborate a new vision of cities as positive physical and social environments for people that contribute to the overall integrity of the ecosystems that sustain them and of the ecosphere as a whole.

Prediction 1. The Idea That Cities Are Ecosystems Will Be Pervasive in Society

The Vision

Ecological scientists have long recognized the power of big ideas, and it is not surprising that the ecosystem concept falls near the top of many ecologists' (Cherret 1989) and educators' (AAAS 1993; NRC 1996) lists of the most important big ideas in ecology. The concept embodies much of what is most central to an ecological world view—it promises a comprehensive detailing of *all* components of a system, a thorough identification of interactions and feedbacks, a careful balancing of the books, and an honest appraisal of the system's position in time and space. The term "ecosystem" already appears in many places outside of strictly academic science—ecosystem services, ecosystem management, ecosystem health—and the concept may be less intimidating and bear fewer negative stigmata than do similarly large concepts such as biodiversity (Elder, et al. 1998).

We envision a future where the insinuation of the ecosystem concept into the common lexicon and the communal consciousness continues, and where more and more different kinds of people use the concept and apply it in diverse ways to cities. Interestingly, this can happen even without clear or consistent or deep definitions of what "urban ecosystem" really means, either in the hands of those using the concept, or in reference to an objective or current, scientifically-based definition. Consider, for example, the ways that the notions of "survival of the fittest" and "watershed" ramify through public discourse and thinking. Consider, also, how more and more people are talking about something being "bad for the ecosystem" where they used to talk about something being "bad for the environment."

The "cities are ecosystems" idea will provide a unifying and inspiring context for those who choose to strive to inspire people to think beyond single factors and simple solutions. While the concept, by itself, doesn't presume a valuation of the other parts of a system that an ecosystem perspective forces us to confront and deal with, nor a moral code for their treatment (be they other organisms, poor people not usually considered, distant ecosystems receiving the city's wastes, etc.), it does provide a more honest and intentional arena for these to be considered and evaluated.

Challenges for Urban Ecosystem Researchers and Educators in Achieving the Vision

Thoughts about what we really mean, in detail, about this concept of "city as ecosystem" are explored more fully in the next section. Here, let's consider challenges to the broad process or phenomenon of insinuating a concept into the common lexicon.

Essentially, this is *not* going to be one of those fields where scientists figure it all out, then craft it into curriculum and public relations messages, and then disseminate it to the masses. The development of the idea in various public and nonscientist sectors will, in some cases, outstrip scientific progress in urban ecosystem science. This is because we live in a very environmentally aware world, and ecology concepts, partial understandings, etc., are legion.

The challenges, therefore, will be whether we, as scientists and educators, can have a positive influence in insinuating the idea in the right places, and help prevent the wrong ideas from being adopted. There are many pitfalls in thinking about cities as ecosystems. For example, some using the idea include in their thinking implicit modifiers to the term *system* that have important implications for understanding. If, for instance, one presumes that systems are "self-sustaining" or "internally balanced" or "healthy for the organisms in them that we care about," then they might be unwilling to even consider a city an ecosystem at all, or might have incorrect expectations of urban ecosystems when they do. We take up the issues of conceptual pitfalls more in discussing prediction 2.

The challenge and role for urban ecosystem educators is to work toward an evolution of thinking, rather than aspiring to achieve a stable endpoint of comprehensive, systemic understanding. Thus, what we need to foster is the development of and appreciation for an evolving concept of the urban ecosystem that becomes more and more detailed, with more components considered (other organisms besides people, other costs or benefits besides monetary or material), more connections and feedbacks understood, more elements distant in space, and more parts distant in time. The term's two parts contain the essence of its meaning—"eco" (cities *have* nature; i.e., components including but in addition to humans) and "system" (there are many connections within and between components).

Another key frontier is within our science itself. Although scientists do not need to all agree on what is meant by "cities are ecosystems," there should at least be agreement that cities *are* ecosystems to send a clear message to educators and the public at large. The debate about what this means that takes place *within* the academy should not be construed by the rest of society to imply that scientists do not know what they are talking about, nor that the idea of the urban ecosystem is just a subjective notion. Establishing the validity of the idea that cities are ecosystems within science will be determined by, for example, whether urban ecosystems can be the subject of regular grant proposals within funding agencies rather than just special "urban" competitions, whether urban ecosystems are used in ecology and sociology textbooks, or just "urban" textbooks, whether there are questions about cities as ecosystems in the standardized tests given to students in biology, earth science, and history.

Prediction 2. Educators and Scientists Will Have a Richer, Deeper, and More Pragmatic Understanding of What Is Meant by the Concept of an Urban Ecosystem

The Vision

Over the next ten years we envision tremendous growth in our scientific understanding of cities as ecosystems. This growth will take place with unheralded levels of lay participation (discussed more in prediction 5). Progress will take place both within the major disciplines that already are addressing cities—geography, sociology, history, ecology—but also at the overlap between fields (Figure 30.1). This overlap will be one both of ideas and of people working together to fuse their disparate perspectives. We do not envision that this evolution of understanding will yield a single definition of what an urban ecosystem is. Rather, understanding cities as ecosystems will represent a wonderful celebration and demonstration of the positive richness of a diverse and sometimes dissonant set of approaches. This diversity of perspectives was represented at the Cary Conference and pervades this book, and hopefully it will persist and continue to evolve. For example, a number of alternatives to the framework shown in Figure 30.1 were discussed [e.g., Venn diagrams with biology, technology, and humans (from Poland); with environment, economics, and society (from England); or with wildlife populations, habitat, and people (from resource management)]. Some prefer to consider an ecosystem approach to cities to be more of a perspective or way of thinking, whereas others look at concrete, physical entities they define as cities or urban areas.

Running through virtually all of these new perspectives on the ecology of cities will be the disposition, skills, backgrounds, and abilities to look at cities, or parts thereof, using more complex, rich, and diverse frameworks

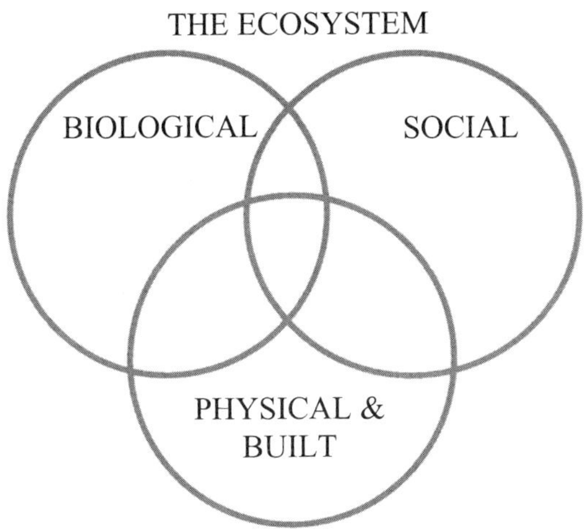

FIGURE 30.1. One conceptual model for the urban ecosystem, involving the overlap between the three dimensions or disciplines: biological (including ecology), physical (including the abiotic environment and the human-built environment), and social (including human institutions, social order and cycles, and socioeconomic and cultural resources).

and dimensions than were previously used. Thus, people accustomed to thinking about crime in the city will also think about trash and air pollution; people thinking about urban birds will consider issues of poverty and how history has shaped the urban landscape; people thinking about childhood asthma will consider the functioning of the information dissemination system and the distribution of environmental burdens; people thinking about whether to spray mosquitoes for West Nile Virus control will automatically think of potential impacts on the poor and infirm, on adjacent aquatic ecosystems, and on nontarget organisms.

As people become more adept at applying this new ecosystem perspective to cities, they will become more facile at the following:

- Identifying the boundaries and the time scale of the ecosystem that are suited to the particular needs they are addressing. This might be a neighborhood, the city proper, an entire metropolitan area, or a watershed within an urban/suburban region;
- Comprehensively detailing basic ecological components within the system, and then looking at the important interactions among components and with the world outside the system they've defined. The emphasis on "ecology" in its broadest sense insures that we will include organisms—including people—in our thinking and teaching. The emphasis on being comprehensive means that when adding new components to our conceptual models from unfamiliar disciplines, we are thorough. Thus, ecologists incorporating people in their thinking will include human

institutions and units of study important to people (e.g., ideas, social structure, money and power hierarchies, cultural structures, beauty, etc.), and sociologists adding an environmental dimension will look at more than just a few charismatic species to consider all important ecological components and processes (e.g., nutrient cycling and not just pollutants, microbes and not just birds, human population and community ecology, etc., in the ecological realm). This broader, ecological perspective will force us to consider those sometimes overlooked in thinking about cities—the poor and dispossessed; the less visible plants, animals, and microbes; or things distant in space and time;

- Arriving at some kind of meaningful balancing of the books using inputs, storage, and outputs for key things people care about in cities. The tools will be at hand for understanding the tremendous exchange of materials, energy, ideas, wealth, and services that link cities with the local and global hinterlands;
- Placing the system in a broader, historical context in order to understand its long-term dynamics, including processes such as urbanization and urban sprawl;
- Placing the urban systems into broader spatial contexts. Urban ecosystem understanding will provide us—scientists, educators, and society —with the tools and frameworks we need in order to think rationally about efficiencies and trade-offs in whole new ways. Urban ecosystem understanding will weigh in on key confusions about the relationships between population, wealth, consumption, and impacts. Do cities cause high levels of consumption and thereby, environmental impacts? Some believe that cities devour parts of many different hinterlands as judged, for example, by calculating their ecological footprints (see Grimm, et al., Chapter 7; Rees, Chapter 8 in this volume). Even though some of cities' resource consumption is for their own "use," however, other consumption enables cities to produce goods and services to export to other cities and to rural areas. What is the relative importance of these two types of uses of hinterland-based resources? Perhaps even more importantly, how efficient are cities, versus other modes of human dispersal, at producing these goods and services? Would spreading people out into less dense settlement patterns have less impact?; and
- Considering cities in aggregate, applying ideas of metapopulation and landscape mosaic analysis to whole regions, building on models from geography.

Understanding of cities as ecosystems will be advanced by applying each of the four major approaches of ecosystem science—comparison, long-term study and monitoring, experiment, and theory/modeling (Carpenter 1998)—especially as the biological and social sciences merge and help bring to light all of the components of the system at hand. For example, we envision the development of data sets for comparing neighborhoods, watersheds, green spaces, and so on, along important gradients from the urban core to the suburban and rural fringes of many cities over the coming decade. Also, data sets that include a fuller suite of ecological, sociological and physical parameters will enable us to compare whole urban watersheds or whole cities. These data sets will be available to scientists, educators, and citizens, allowing people to ask questions about variation from place to place, from city to city. Synthesis will come not just from ecosystem modelers and theorists, but from historians and landscape ecologists, and from the crucible of praxis. Indeed, in cities, the application of the experimental approach to ecosystem science is synonymous with practice, with planning and/or following up on changes people make in real neighborhoods and communities. Thus, urban planners, landscape designers, environmentally oriented community activists and urban decision makers will be, even unbeknownst to them, carrying out for scientists the very sorts of ecosystem experiments needed to yield important synthesis of our understanding of cities.

Challenges for Urban Ecosystem Researchers and Educators in Achieving the Vision

Successfully crossing the frontier of interdisciplinary thought and practice is crucial to the development of urban ecosystem science and education. How can scientists from different disciplines work together, combine their perspectives and ideas, and craft a truly interdisciplinary world view? This question plagues many within science here at the cusp of a new century. Scientists need a common language, but also must preserve the richness and nuance of their own tongues. Each discipline needs its own strong foundations in the solid earth of its domain while also needing to know how to build bridges into and across the rarely navigated spaces between them. The consequences of failure are clear. As one conference participant put it,

> [Ecologists should] provide ecological metaphors for many functions, structures, processes, and organizations of the city. These metaphors will give people multiple frameworks for understanding the complexity of the world. Caveat—If ecologists offer a sliver—a narrow understanding rooted only in classical ecosystem study—you will be ignored. Teachers will go to geography and sociology to teach/understand cities. The benefit of the sliver will be out shadowed by the larger pieces of the pie.

In achieving the rich understanding of cities as ecosystems that we envision, a number of important pitfalls must be avoided or otherwise dealt with.

1. People think that the idea of urban ecosystems is synonymous with human, or even impacted, ecosystems. However, all ecosystems on the globe are impacted by people, even the most distant arctic or desert hinterland. Likewise, humans are sustained by resources and ecosystem services

provided by virtually all systems on the globe. Indeed, even in ecosystems like the deep sea where no humans dwell, humans are the top and dominant predator. The levels of our dependency and impacts have much less to do with our urban habitation patterns and much more to do with human consumption patterns, especially those of the wealthy few. Thus, even though people may blame cities for the environmental costs of high population and high resource consumption—costs to poor people (and others) who live in cities and to the ecosystems that receive the wastes of cities—urban ecosystem science can give us the conceptual tools we need to disentangle which of these costs are unique or endemic to cities, versus those things having to do with high consumption or high concentrations of poverty separate from the urban condition. Understanding urban ecosystems provides the intellectual key to sorting out the roles of cities per se, versus the more general consumption of resources by people, as the cause of human impact on the globe. It points us to an appreciation of the ecological advantages of cities, the social and ecological costs and benefits of cities to city residents, and to other places and people.

2. "Ecology" is not always easy to see in cities. As places where, by definition, there are high densities of humans and their structures, and enormous throughput of materials and energy orchestrated by people, cities can seem devoid of other organisms and ecological processes. Likewise, many human-defined problems in cities seem impervious to ecological thought. There are at least four rejoinders, all of which are worth considering: (1) Not all problems will have a rich ecological dimension; (2) Ecology concepts and ecosystem thinking have surprising utility when applied as metaphors or models even to strictly human systems. The chapters by Wolford (Chapter 10), Grove, et al. (Chapter 11), and Melosi (Chapter 12) give good examples of this cross-fertilization of ideas; (3) An understanding of human ecology in a larger-scale context is essential even when addressing issues at a local level that might seem devoid of ecology. Thus, the ecological perspective forces us to think about our ecological impacts and about ecosystem services provided by distant ecosystems. These connections are more than just physical or biological, but also spiritual (e.g., urban people's connections to the rural landscape or the distant wilds); and (4) Other organisms and ecological processes, in fact, pervade every place on earth, even the most seemingly abiotic. From the rich suite of organisms harbored by our own bodies, to the pests that infest all human habitations, and the incessant press of living things in the soil, sidewalk cracks, wasteplaces, and the like, every place has at least some "ecology."

3. It often is difficult to define a system. Faced with the complexity of urban areas, it can be difficult to come up with a system that isn't too simple to be realistic or too complex to be understandable. A more subtle stumbling block to thinking about cities as ecosystems is the unspoken expectations some people bring to what they call systems: that each is self-organized, self-sustained, or self-contained, and that *all* elements of the

system interact with each other. This can lead to misconceptions about the actual functioning of the system, which in reality might be loose, more connected to elements external than within, unstable, or unsustainable. On the other hand, people's awareness of the unsustainable or imbalanced nature of cities can make them reluctant to call cities ecosystems in the first place, due, again, to this implicit expectation that systems must be balanced or sustainable. Interestingly, both human inventiveness and human institutions can help here. Science gives us a way of thinking about systems where our assumptions can be made explicit. Thus, some of us consider a city an ecosystem, but recognize that it is neither sustainable nor closed; while others prefer to define the city explicitly as part of a larger, more functionally defined ecosystem that includes the hinterlands it is connected to (see Rees, Chapter 8 in this volume). Likewise, human institutions can give "reality" to systems that otherwise seem ill-defined. For example, the city itself might seem like a strange ecological system to some, but it is a very real and viable political system; or the collection of urban wilds might seem a very weak "system" of green spaces, but the agencies and organizations that manage and advocate for them provide this system tangible reality.

4. The interdisciplinary work required in order to understand urban ecosystems has both conceptual and pragmatic challenges. Scientists and educators need reward systems, institutional support, training, and other forms of help and encouragement to nurture collaboration and cross-disciplinary thought. Likewise, we need to wrestle with questions of how to and under what circumstances we can achieve a *completely* holistic view—whether it is good enough simply to take a *more* holistic view—whether there are common currencies that allow us to bridge the disciplines, or if we lose important understanding by so doing.

5. There are limits to the utility of the ecosystem perspective in understanding cities. While some view the ecosystem perspective as exceptionally broad and inclusive, others point out that there may be important features of cities that are obfuscated or missed by taking such a view. In promulgating a broad understanding of cities as ecosystems, purveyors of this view must be cognizant and honest about its limitations.

Prediction 3. The Importance of a Broad Understanding of Urban Ecosystems Will Be Recognized, Appreciated, and Well Articulated

The Vision

We envision a future where the importance for lots of different kinds of people to understand urban ecosystems will be recognized, appreciated, and well articulated. This will spur on and provide a clear rationale for

TABLE 30.2. Goals for urban ecosystem education that emerged at the Cary Conference.

1) Urban ecosystem education will help build environmental citizenship (David Orr 1991). It . . .
   a) fosters a sense of place and connects urban people with an expanded home range.
   b) enables all people to read the newspaper and understand the ecosystem ramifications of articles about cities, urbanization, sprawl, etc.
   c) helps people understand their impact on the environment.
   d) teaches systems thinking, enabling people to understand their world and its internal and external connections and dynamics.
   e) builds mutual understanding of each others' homes among urban and rural peoples.
   f) cultivates a citizenry that thinks locally *and* globally, and acts locally.
2) Urban ecosystem education helps society work toward creating sustainable human systems. It . . .
   a) provides a means for helping people strike an intentional balance between individual wants and needs and the common good.
   b) gives an ecological and systematic basis for a critique of current economic frameworks as they constrain the achievement of sustainability in general and sustainable human habitation patterns in particular.
   c) shows us the value of the other living things, besides people, in cities.
   d) helps us champion and maximize the ecological advantages and minimize the ecological costs and impacts of cities.
3) Urban ecosystem education facilitates community cohesion, provides people with access to power, and stimulates socioeconomic revitalization of urban communities. It . . .
   a) improves urban kids' ability to interface with society.
   b) stimulates community awareness.
   c) fosters participation and engagement in society.
   d) cultivates social change.
   e) opposes the forces of domestication with education for liberation.
4) Urban ecosystem education enriches our lives with understanding of the unique and important ecosystems cities represent. It . . .
   a) helps people see that cities, like all places, can be viewed and appreciated as sacred ecosystems.
   b) celebrates the beauty of nature and the human spirit in all their guises and combinations.
5) Urban ecosystem education is for everyone, and not just for teaching urban kids about nature in the city. It . . .
   a) teaches us about the ecology *of* the city as an ecological system.
   b) provides useful understandings for everyone—urban, suburban, and rural—recognizing that we all are linked to urban ecosystems and urbanization in many ways.
   c) helps us understand other people in other places.

urban ecosystem research and education. From this vision comes a set of goals for urban ecosystem education (Table 30.2) that might guide us into this future.

Knowledge is power, and those who have it wield this power over the environment, over other people, over regions or nations, and over their own decisions and futures. We have to recognize that striving for more knowledge among more people doesn't mean we're unrealistic and expect all people to have all knowledge. There always will be variation in the distribution of knowledge. No one person will have a complete understanding. The idea, however, is to strive for a world in which these asymmetries are reduced wherever they limit important things—oppressed people's rights for a quality life, and future generations' (human and nonhuman) rights. This is what it comes down to, correct? Thus, we posit that understanding urban ecosystems will be embraced as a key right for all people, and that it will be perceived not just as a conceptual "nice to know" kind of thing, but that the understanding will be expected to be useful, to empower the people wielding it to make a better world. The desired kinds of understandings will be well articulated and will demand of educators, managers, policy makers, and scientists an urban ecosystem research and education enterprise that fosters these understandings.

In our envisioned future, there will be a broad appreciation of the pivotal role urban ecosystems play in the globe. Over half of the world's people already live in cities, yet they take up less than 2 percent of the globe's surface (O'Meara 1999). These 3 billion plus urban people and their industries and activities consume more than ¾ of the total human use of wood and building materials, and around ⅔ of the total human use of water (O'Meara 1999). In 10 years, no one will doubt that cities are important ecologically, nor that the global expansion of human population in cities represents an unprecedented change in the structure of the global environment.

Cities produce goods we really cherish and quite possibly couldn't do without—culture, ideas, commerce, products—and create many forms of wealth, capital, and conditions crucial to humanity—social capital, safety and security, the village it takes to raise a child, etc. In our imagined future, the call for urban ecosystem understanding will be a call to understand more fully the benefit/cost ledger sheet for producing these goods, and a call to help us apply our knowledge to the ultimate question: How can we maximize the good and minimize the costs?

The rationale promulgated for urban ecosystem education will address, directly, a redirection of knowledge to help us reverse the incredible inequalities that exist within cities in access to the amenities they contain and exposure to the ills they generate, whether caused by spatial, racial, or economic differentiation. It also will be a call to appreciate more fully the beauty of urban ecosystems—as a whole, and the sub-ecosystems they harbor. These are unique ecosystems, with amazing things in surprising places.

The fact that there are social institutions already set up to manage the same entities that urban ecosystem scientists and educators might identify is tremendously important. Cities have the potential for rational, ecosystem-based decision making and management, both because they exist as political entities and because in some ways people already understand that there is a need for complex thinking in and about cities.

Challenges for Urban Ecosystem Researchers and Educators in Achieving the Vision

In order to achieve the envisioned level of recognition of the importance of understanding cities as ecosystems, a number of challenges must be addressed head-on.

1. Urban ecosystem education is *not* just for city folks. Cities are important for everyone to understand because there are so many connections—material and otherwise—among *all* people and among cities and surrounding ecosystems. Understanding these connections is a crucial part in understanding the ecology of human existence.
2. Knowledge about cities as ecosystems is important for *all* city folks, and not just the rich, the privileged, or the powerful. Due to many factors, including historical and cultural differences in exposure to environmental experiences or amenities, urban ethnic minorities are significantly underrepresented in environmental science professions (Leatherberry and Wellman 1988; Adams and Moreno 1998; National Science Foundation 2000). Likewise, in many cities access to knowledge is not equitably distributed. These problems can result in a perpetuation of inequities and disempowerment in cities if not addressed directly. Cities are the natural classrooms for urban dwellers to learn ecology. For them, it makes sense to first learn about urban ecosystems and then expand learning to apply the concepts to less familiar ecosystems. Because they bear the brunt of the almost inevitable environmental burdens that the crush of humanity and their resource use creates in cities, city dwellers need all the ecological knowledge-as-power they can get. As much as possible, *all* people living in cities should be the beneficiaries of the wonderful positives that cities create, including knowledge and appreciation of the urban ecosystem.
3. Recognize that more than just understanding goes into the decisions people and institutions make and the actions they take. Knowledge isn't all there is to power. Further, while advocating for increased understanding and application of the urban ecosystem concept, educators and scientists must also allow for the fact that incomplete knowledge can be dangerous if applied unknowingly.
4. Will the purveyors of urban ecosystem understanding have the ear of decision makers and managers? Consider, for example, those making the decision here in 2002 about what to do with the 16-acre World Trade Center site in New York City. Will people who see the city as an ecosystem have their say? Imagine an ecosystem theme to the plan—redeveloping the site in concert with the ecosystems surrounding it, recognizing history and

context, and designing buildings that celebrate New York's position as one of the most energy-efficient states in the United States (Geller and Kubo 2000) by taking energy efficiency to a new level in every form (the buildings' materials and processes, operations and maintenance). What a statement of the country's potential for achieving energy self-sufficiency and excellence this could be.

Or consider the decisions being made in every urban area in the United States right now by individuals, families, small suburban and rural communities, cities, and states that, in aggregate, shape the complex process of urban sprawl. Will knowledge of the ecological and sociological ramifications of these decisions, to those nearby and beyond, to those of tomorrow and of the next half century, be available and utilized?

Prediction 4. People Everywhere Will Have Many and Diverse Opportunities to Learn About and Apply Their Understanding of Urban Ecosystems

The Vision

Urban ecosystems will be infused into the education agenda of both formal and nonformal institutions, from schools and their curricula, school systems, their mandates and high-stakes assessments, universities and their degree requirements, teacher training programs, museums, community-based organizations, and educational TV, to the Internet. Through these avenues, learners will have the opportunity to ask their own questions about cities, to compare their findings with those of others, to cut across the traditional disciplines of science and even across the science/humanities gulf. They will learn things they can actually discover and apply in their homes and neighborhoods.

In this imagined future 10 years from now, we envision an education system . . . where urban ecosystem examples are used to teach basic ecology concepts in textbooks and curricula, . . . where ecology is taught as part of social science and history, and not just in the natural sciences, . . . where kids are tested on their understanding of how their own city, and nearby or major cities in their region, actually function and interact with surrounding ecosystems and human institutions. A diversity of pathways will spread urban ecosystem understandings through the schools, using teacher-to-teacher, school-to-school, top-down *and* bottom-up approaches.

Urban ecosystems will pervade not just the formal education system, but the full suite of informal education and communication programs and institutions—urban nature, community and environmental education center programs, adult/continuing education programs, media, public information pathways, and the like. Weather reporters will report on their city's "ecosystem health" parameters much like they now report on ozone or rainfall patterns. Imagine hearing "Today's carbon signature was down significantly from yesterday's level, as voluntary switching to bicycles public transportation, and electric cars took a big jump."

These opportunities will be guided by the very best knowledge available about how people learn, how this changes as people grow, and how it varies among different kinds of learners. It must build on insights from research about teaching and learning in systems thinking, the natural and social sciences, and at the interface among disciplines. Teaching and learning will start early with accessible emphasis on initial, prerequisite concepts (see Roseman and Stern, Chapter 16 in this volume) and on systems and critical thinking, and will develop from there. Through these age-appropriate experiences, children will develop a sense of place, engaging all of their senses. In order to do this, the programs must break children out of the shackles that limit their mobility in the environment, expanding the spatial scale and scope of their world as they grow, affecting their exposure to local goods or ills and their access to learning opportunities (see Chawla and Salvadori, Chapter 18 in this volume). Likewise, different strategies will be implemented in primary vs. secondary school, with their different time and space scales (see Keiny, et al., Chapter 19, and Fialkowski, Chapter 21 in this volume). Opportunities for people to learn about urban ecosystems will be provided by teams of educators, scientists and community members, reflecting the diverse disciplines and perspectives needed to grasp the subject at hand.

We envision a future where the collection of people and institutions that need to understand urban ecosystems for practical applications and decision making will be involved in setting the research agenda, and will have the requisite data, information, and concepts they need for sound decision making. This will be widely available to society through a "knowledge system" that, in its own way, mirrors the structure and function of an ecosystem or nested set of systems, and will poise cities to play a more pivotal role in working toward sustainability (see prediction 6). This knowledge system might be the neighborhood (i.e., a geographically defined entity), or the planning system (i.e., a functionally defined system), or something even more complex than this (i.e., the political/decision-making system). This frees us, to a certain extent, from the obvious limitations on what any one individual can learn and guides us to think about and bring into being a system of diverse knowers and discoverers. Thus, in our envisioned future, *everyone* will *not* have a complete ecosystem understanding of cities, but collectively we will.

Challenges for Urban Ecosystem Researchers and Educators in Achieving the Vision

There are many challenges to infusing into the education agenda an understanding of urban ecosystems—in general and of people's own particular place in specific. The vision requires a form of education that is geared to fostering citizenship, empowerment, and relevancy, and one that blurs the distinction between school learning and real-world, scientific learning. It almost certainly will require at least some learning-by-doing, through working, experimenting, and interacting with others in the real world. This can tip the apple cart of the status quo and many forces are lined up to thwart such a radical direction in education. Imagine . . . a community group wants to plant street trees to see what role vegetation can play in crime and/or drug reduction, but finds itself hemmed in by an underfunded city agency and a school system that doesn't see how this project fits into its curriculum.

Considerable research in teaching and learning is needed to guide teaching practices in classrooms and in the field, curriculum development, teacher training, and learner assessment in urban ecosystem education. Emerging from discussions at the Cary Conference were a number of questions about the ways that individuals learn about things as complex as urban ecosystems and how teachers might best foster such understandings (Table 30.3). For example, even though lip service is paid to the idea of thinking globally but acting locally, we have scant evidence for the strengths and limitations of teaching about global issues on the one hand, and about the educational values and limitations of studying the local environment on the other.

With something as multidisciplinary as urban ecosystem education, we must confront what are likely to be vexing differences in the ways different disciplines are or should be taught. How do vocabulary and pedagogy differ for each of the disciplines involved? What do these differences mean when it comes to teaching tools for decision making, building a world view, fostering ways of knowing and questioning, or safeguarding objectivity and rigor?

The current focus on reading and math in formal education may constrain adoption of an urban ecosystem emphasis. One of the consequences of the incredible press on teachers is that, unless they find ways to integrate instruction in the basic skill areas with teaching about the environment (e.g., through project-based, interdisciplinary, or integrated instruction), topics like urban ecosystems can get pushed aside entirely or relegated to only a small part of their curriculum and their schedule. One of the biggest challenges for our formal education system is to find ways to teach about subjects that require long-term, integrated instruction spanning many years. Fortunately, visionaries like those at the American Association for the Advancement of Science, with their multidisciplinary and multiyear maps for conceptual learning, lend optimism that we can meet this challenge (AAAS 2001, and see Roseman and Stern, Chapter 16 in this volume). Nonformal education may be in a unique position to champion interdisciplinary approaches as well.

TABLE 30.3. Frontiers for urban ecosystem education: research questions about teaching and learning about cities as ecosystems.

Questions about learning:

- What is the link between engagement and motivation (e.g., in a local environmental problem) and students' achievement of measurable goals (e.g., performance on standardized tests)?
- What is the link between cognition and affect in teaching and learning about urban ecosystems? When is understanding a necessary (if not sufficient) component of caring and action? How does affect (caring) influence one's understanding and one's disposition to learn more about it?
- How can students be meaningfully involved in data collection about urban ecosystems, and be guided to go beyond that to do analysis and build real meaning from their observations?
- Can we provide a scholarly basis to address the debate about thinking globally and acting locally? Is there a trade-off here? For example, do urban children perceive urban nature differently than a tropical rainforest? If we must at least start with the local, how and at what point in education can we connect people to the global, either transferring local knowledge or developing new understandings?
- What is the interplay between meaning making and value formation in approaching and understanding urban ecosystems? If we want our decisions to be based on our values and we want knowledge to help us be clear about whether our values will actually be achieved by a decision, then what are effective ways to deal with values in our teaching about urban ecosystems?

Questions about assessment:

- How can we assess the sorts of complex learning outcomes we're most interested in achieving, such as attainment of systems thinking, understanding of complex dynamics, or integration of disparate disciplines as they apply to local neighborhoods and communities?
- How can we assess learning about the local and the variable, versus the "standardized/generic" and the "single right answer"?
- Can we document long-term outcomes in such a way that we can positively influence education policy.

Questions about learners:

- What do people currently know about urban ecosystems, what can they do at different ages along the lines of their key skills and abilities, and what are the important sources of variation, both among learners (different learning styles and intelligences, etc.) and over time through development and growth?
- When and for whom do distant and "exotic" ecosystems represent foreign and inaccessible subjects of study rather than enticing and intriguing subjects for learners?
- What, if any, are the connections between conditions in the urban environment and people's ability to learn? This has both a health dimension (e.g., does environmental lead contamination or environmentally caused or exacerbated asthma limit learning?) and a socioeconomic dimension (e.g., does access to wealth relate to access to information?).

Questions about teaching:

- Does it work to use the Internet or other technologies to help scale up, understand global, etc.?
- How can we see this at a system level (i.e., whose understanding is being acted upon if you coerce people to act without understanding)?
- How should teaching of a complex subject like urban ecosystems, requiring integration across disparate disciplines, best take place? Via integrative, capstone courses or after disciplinary courses have been taken? Infused (perhaps just superficially, or as examples) in each disciplinary course? As an integrated theme cutting across curricula? Through a stand-alone, immersion-type learning experience taught by an interdisciplinary team?

Prediction 5. Urban Ecosystem Understanding Will Be Developed in Many Places Via Participatory Models Where Scientists, Students, Teachers, and Other Urban Residents, Managers, and Decision Makers All Work Together

The Vision

We stand at a unique time at the crossroads of science, education, and social discourse that augers well for the development of new participatory models of research and learning. Ecosystem scientists are forging new partnerships with managers and policy makers (e.g., Lubchenco, et al. 1991) and together are exploring new modes of public participation (e.g., the West Philadelphia Landscape Project described in Chapters 2 and 13, or initiatives in Arizona described in Chapter 7, in Chicago in Chapter 21, in colleges and universities in Chapter 22, and in South Africa in Chapter 24). Rural and urban development specialists are championing various models of participatory research and assessment for research and "extension" (Brown 1985; Chambers 1994; Warner, et al. 1998; Freudenberger 1999; Ison and Russell 1999). Education is emphasizing learning through inquiry and service (Hart 1997; Shumer and Cook 1999). New technologies (the Internet, etc.) are fundamentally altering and potentially democratizing the way knowledge and information are distributed and accessed. Segments of the academic community that had previously been more singularly focused on "basic" science now are broadening their roles in society, joining other parts of the academy with long histories of application, such as extension and the land grant system in the United States. Definitions of scholarship and institutional reward systems will be broadened for scientists and educators.

As part of this movement, urban ecosystem education will find itself at a singular historical position, able to achieve unique things, to be a leader in the exploration and proliferation of participatory models. Faced with a pressure to democratize science and its fruits, to break down barriers between science, education, and community, to recognize the value of diverse perspectives in building rich understanding, urban ecosystem science will champion new models of science-education-community interaction (Figure 30.2, parts B and C), transcending the traditional "dissemination" model from science *to* education and society (Figure 30.2, part A). We envision a future where a participatory, community-based approach to learning about and working in urban ecosystems is a research *and* educational tool that will see tremendous development in the coming decade. Such an approach will teach environmental civics and democracy through practice, empower students to understand and use science, help those who strive to organize and improve their communities, and contribute to the

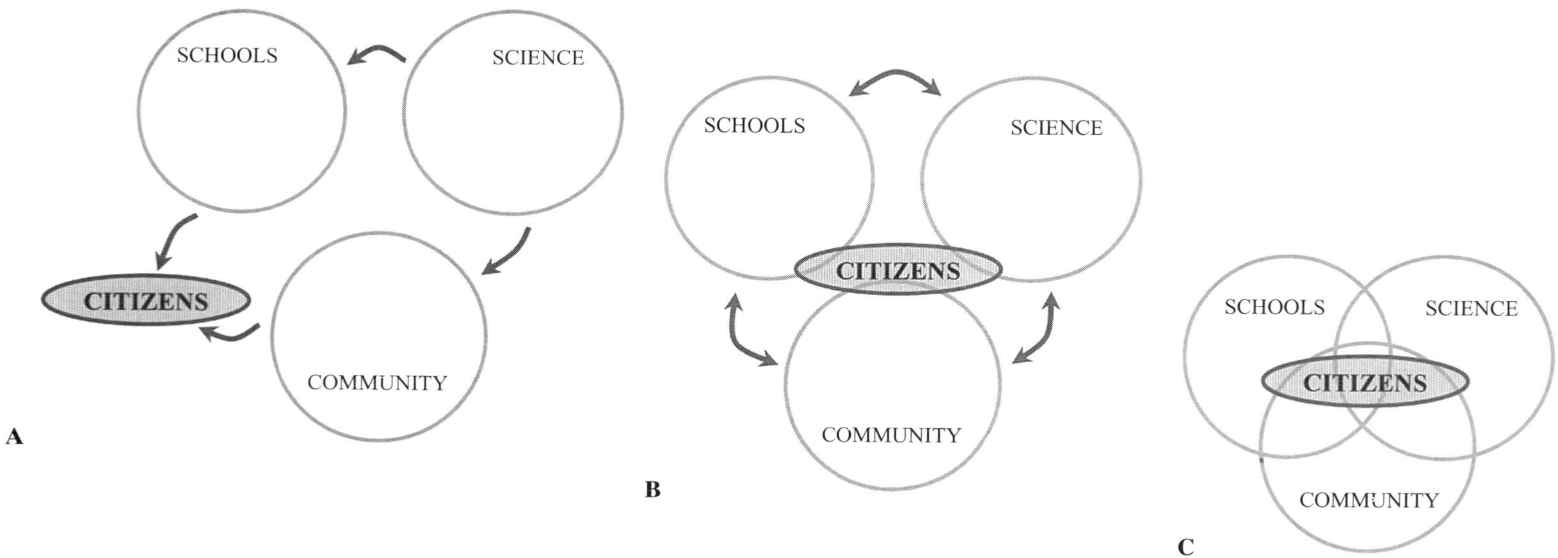

FIGURE 30.2. Three models of collaboration among science (research scientists and institutions), schools (educators and administrators at formal and nonformal institutions), and the community (including resource managers and policy makers, community organizations, and institutions), and their links to each other and to citizens (students, adults, voters). Arrows show the flow of knowledge and information. (A). The traditional dissemination model from scientists to educational and community institutions, and thence to citizens. (B). A mutualistic, interactive model where each sector gives and takes with the other, and citizens receive benefits from all three (arrows not shown). (C). An integrated or overlap model where sectors share common goals, personnel, questions, and activities with each other and with citizens.

development of systems thinking, while also providing scientists with key benefits—access to the systems they wish to study, information about the systems from those who live in them, and improved confidence that the questions they are investigating are of pressing importance to citizens or managers.

Why are we interested in the overlap zones and connecting arrows between science, education, and community (Figure 30.2)? The complexity of the urban ecosystem and questions about it, and the daunting nature of the multidisciplinary frontiers being explored by scientists, make participation by urban residents and decision makers essential. Simply stated, the scientific "experts" cannot fully know enough about what's important, and the fact that people and knowledge are thoroughly embedded parts of the system demands genuine inclusion in the scientific enterprise. Just as farms and issues of rural development now are a fertile ground for proliferation of new models of participatory discovery and learning, so urban ecosystem studies will represent a very positive and very rich reinvention of traditional town-gown relationships between the academy and cities.

What are some examples of the kinds of things that participation and collaboration will produce in our imagined future 10 years from now?

- We envision that collaborations between the schools, scientists, and the community will identify new learning outcomes about understanding urban ecosystems for formal education systems. As we work together to figure out what we mean by "understanding urban ecosystems," the requisite concepts, skills, and habits of mind will be infused into curriculum and standards.
- We envision continuation and expansion of the trend for scientists who work in cities to involve students and teachers, youth and elders, managers and decision makers in their research.
- We envision an expansion of the current trend that bridges the interface between the school and the community, where the community becomes even more of the curriculum and the classroom, and where service learning in the schools serves more and more of the community. Imagine . . . schools being hired to monitor environmental conditions, to craft solutions to community problems, and to help compare difficult choices. Likewise, the community will provide more career ladders into environmental jobs in their broadest sense, and schools and informal educational institutions will provide expanded services to the community beyond the narrow curriculum and school day, beyond the static exhibit and the short museum visit.
- We envision participatory approaches infused into formal and community-based education, perhaps following the steps identified in discussions at the Cary Conference (Table 30.4). These participatory programs will be based on well-articulated learning goals and their outcomes will be assessed and documented for everyone's benefit.

TABLE 30.4. The stages of a participatory research/education program, from discussions at the Cary Conference.

1. Multi-way information sharing among students, educators, community members, scientists, policy makers, etc. All available expertise is used, including local and "outside" experts, resulting in the identification of what we do and don't know.
2. Identification of issues. Discover what community members like and dislike, what they'd change, what problems are in the neighborhood, etc., resulting in the selection of the issues to address through consensus, clusters, prioritization, and/or voting.
3. What do we need to know? What do we already know? Look at the historical and scientific context of the issue, and what skills are needed to address it (technology, experiments, surveys, interviews, statistics).
4. Where and how to get the information? What resources do we already have and how do we get what we don't have? The group is organized and resources allocated.
5. Collect the data.
6. Interpret the data. What does it tell us? Several cycles of reporting and evaluation occur.
7. Develop plan for action, identifying stakeholder perspectives, conducting force field analysis (as per Kurt Lewin, see de Rovera (1976)), etc.
8. Implement action plan.
9. Evaluation and reflection, by individuals and the group.
10. Refine, revise, or end.

One of the most important fruits of the three-way collaboration shown in part C of Figure 30.2 involves work toward functional, measurable, and inspirational definitions, useful within each of the three arenas, of the key ideas: quality of life and sustainability. What this might look like is taken up in prediction 6.

Challenges for Urban Ecosystem Researchers and Educators in Achieving the Vision

Adoption of genuinely participatory approaches for research and education faces important challenges. It requires a shift from the expert leading others in coming to know what she/he already knows (or thinks she/he knows), to the expert collaborating with others to share her/his understandings and collectively acquiring new data and insights while acknowledging that knowledge is incomplete. It requires that people and groups who do not ordinarily work together find ways to communicate, collaborate, and coordinate. It requires reward systems and incentives for everyone to get and stay involved.

In order to achieve the vision,

- We need ways to evaluate the contributions participatory research makes to science. Do we actually learn new and important things?
- We need reward systems that reward scientists and educators as well as students and community members for participating—new products that

are honored and valued, new requirements for student proficiencies, new incentives for the community.

- We need to foster interdisciplinary training, a challenge of particular significance in secondary schools and higher education institutions as they now are structured. Narrow reward systems, barriers to faculty training or retraining, rigid curriculum organization, and departmental structures all are significant and systemic impediments to change. Likewise, rewards for spending the time it takes to understand and collaborate with others across and off the campus are needed. Finally, reward systems for researchers often emphasize investigating your own questions and might penalize scientists who, through participatory processes, answer other people's questions.
- Scientists are just beginning to develop new ways of ecosystem thinking about cities, so they may suffer from the kinds of self-doubts and skepticism that comes from putting a very unfinished "product" on the table of collaboration.
- Scientists, educators, students, and community members don't necessarily know how to collaborate, so they need help communicating with each other and working together.
- Some urban residents' experiences with science and scientists have not been positive, leading them to view science as a tool for exploitation and making them skeptical that science could have a more positive role.
- Educators often want clear messages that provide a foundation for action, while scientists see the provisional and developmental nature of understanding and want to focus on process. Educators need to work with accessible concepts, but scientists can be leery about over-simplification or loss of the nuance and subtlety of science.
- Educators, scientists, and community members might not always, or even ever, ask the same questions. Many factors prod educators to look for smaller, bite-sized questions and topics that are easier to infuse into the already-crowded education agenda, and thus they might prefer to teach about specific components of the city that are already understood. The community member in the three-way collaboration might want the team to investigate something else entirely!
- As participatory research and teaching develop, mechanisms for assuring that the data and learning are of acceptable quality must be developed as well. Such accountability mechanisms are likely to differ among arenas (e.g., science has peer review, schools have grades and standardized assessments, and the community has complex political and other processes). In addition, each sector speaks differently and is guided by different standards. It is particularly challenging to address misconceptions across arenas at the same time that one is attempting to honor and respect diverse voices. As a result, incomplete ideas can get reinforced or even perpetuated through the process rather than overturned.

Prediction 6. People Will Understand the Vital Role Well-Designed and -Managed Cities Play in Sustaining Life—Human and Nonhuman—on Earth

The Vision

In 10 years we will have a much clearer vision of cities as positive places for people and other organisms, and will have made progress in helping society improve the vitality and sustainability of these places and their connections to earth's other ecosystems. Urban ecosystem education will play a leading role in helping society define, both conceptually and operationally, key ideas such as "quality of life" and "sustainability," and will help people understand the limits, trade-offs and challenges in achieving these things within the global socioeconomic system that is becoming increasingly dominant. It provides a framework for identifying and integrating both material and nonmaterial realities, and thus puts into stark relief where we stand. Our ecosystem understanding of cities will put the consumer habits of all of us, especially wealthy people, into an ecological context, using a larger share of the global hinterlands' resources, producing more polluting by-products and distributing them across the globe, and recognizing that most costs are borne disproportionately by the poor and by racial or ethnic minorities both within and outside of cities. Urban ecosystem education can achieve its potential by accomplishing many of the tasks already identified in this book and in this chapter: (1) clearly identifying its goals and objectives, (2) providing vivid images of what alternative scenarios look like in action, (3) specifying the requisite teaching and learning methods, and other education and communication strategies needed, and (4) emphasizing the importance of involvement of all parts of the community—formal and nonformal education, the media, higher education, and teacher preparation—in the enterprise.

Urban ecosystem education will help citizens make the very hard balancing choices we face—regarding urban sprawl; crafting cities to be more efficient, with more internal cycling; capturing the efficiencies of density (transportation, heating, social capital) without eliminating all the open/green space in cities; and reducing consumption in general. Can we strike a more mutualistic relationship between the city and the hinterlands? Is it already mutualistic and we don't realize it? Urban ecosystem education will provide hopeful messages in what otherwise might be a sea of despair. It will help us understand what realistically can be achieved by way of restoration versus reclamation in different parts of our urban world.

Challenges for Urban Ecosystem Researchers and Educators in Achieving the Vision

It is the sum total of challenges mentioned in discussing each of the previous predictions—difficulties in collaborating with others, in providing many opportunities for learning, in gaining recognition and support for urban ecosystem education, in understanding what we mean by the urban ecosystem—that makes this a *huge* challenge. Can we create enough small, fledgling successes over the next 10 years, through the efforts of gutsy frontierswomen and frontiersmen, and make these known so a larger audience can see the vital role urban ecosystems play in our modern world and keep us moving forward?

As we pursue this vision, we must wrestle with several vexing questions: What are the roles of the more wild, less built ecosystems in cities; the parks, vacant lots, street trees, and riparian zones? How can the wealth created by the economic activity in cities be more equitably distributed so that the immediate social and physical environments for people living in cities is healthy, nurturing, and aesthetically satisfying for all? How can the byproducts be reduced so that they do not disproportionately impact nearby and even distant nonurban areas? What is the role of environmental education in general and of urban ecosystem studies in particular in creating the required level of popular understanding of the mix of socioeconomic and ecological issues that will enable positive policy changes toward sustainable cities and ultimately, sustainable societies?

We need much more work in understanding the connections between cities and sustainability. For example, does sustainability require (1) fewer consumers (i.e., reduced human population) and if so, what is the connection between urbanization and human population?; (2) less consumption (e.g., changing the connection between material consumption and quality of life, replacing material amenities with social sources of happiness that come with living in populated areas, trade-offs between these and "natural" sources of happiness that come with living in or traveling to the hinterlands); (3) increased efficiency; or (4) shortening the physical links between people and the sources of their ecosystem services? Outside the United States, Local Agenda 21 has provided a framework for thinking about what it means to view cities as systems and the roles played by scientists, educators, community groups, and local residents in planning for sustainable cities (see Chapters 24 and 28).

Conclusions

Urban ecosystem education, like the cities it focuses on, represents a coming together of tremendous riches, a proliferation of tremendous challenges, and an optimism that springs forth from and celebrates their collective creativity and synergies. Where else on earth can we see so many examples of the emergent properties of human inventiveness, of culture and concept, of artifice and the ethereal, better than in our cities and in our attempts to teach and learn about them? In a future where billions of people strive for a better life, one that can be sustained into their future, one that embraces their needs and builds on their wants for the nonhuman parts of the globe, cities will play a vital role. The choice is not whether to have cities or not, nor even whether to understand them or not. The choice is more about of the texture of our understanding—how rich, system-based, multidimensional will it be?—and about the breadth of its empowering sweep through our society—how democratically and deeply it will be imbued in *all* people.

References

Adams, C.E., and M. Moreno. 1998. A comparative study of natural resource professionals in minority and majority groups in the southeastern United States. *Wildlife Society Bulletin* 26:971–981.

American Association for the Advancement of Science (AAAS). 1993. (Project 2061). *Benchmarks for science literacy*. American Association for the Advancement of Science, Washington, DC.

American Association for the Advancement of Science (AAAS). 2001. *Atlas of Science Literacy, Project 2061*. American Association for the Advancement of Science and the National Science Teachers Association, Washington, DC.

Brown, L.D. 1985. People-centered development and participatory research. *Harvard Educational Review* 55:69–75.

Carpenter, S. 1998. The need for large-scale experiments to assess and predict the response of ecosystems to perturbation. Pages 287–312 in M.L. Pace, and P.M. Groffman, eds. *Successes, limitations, and frontiers in ecosystem science*. Springer-Verlag, New York.

Chambers, R. 1994. The origins and practice of participatory rural appraisal. *World Development* 22:953–969.

Cherret, J.M. 1989. *Ecological concepts: the contribution of ecology to an understanding of the natural world*. Blackwell Scientific Publications, Oxford.

de Rivera, J. 1976. Field theory as human-science: contributions of Lewin's Berlin group/compiled with commentary by Joseph de Rivera. Gardner Press, distributed by Halsted Press, New York.

Elder, J., C. Coffin, and M. Farrior. 1998. *Engaging the public on biodiversity. A road map for education and communication strategies*. The Biodiversity Project, Madison, WI.

Freudenberger, K.S. 1999. *Rapid Rural Appraisal (RRA) and Participatory Rural Appraisal (PRA): a manual for CRS field workers and partners*. Catholic Relief Services, Baltimore, MD.

Geller, H., and T. Kubo. 2000. *National and state energy use and carbon emissions trends*. American Council for an Energy-Efficient Economy, Washington, DC.

Hart, R. 1997. *Children's participation: the theory and practice of involving young citizens in community development and environmental care*. Earthscan, London.

Ison, R.L., and D. Russell. 1999. *Agricultural extension and rural development: breaking out of traditions*. Cambridge University Press, Cambridge, UK.

Leatherberry, E.C., and J.D. Wellman. 1988. Black high school students' images of forestry as a profession. *Journal of Negro Education* 57:208–219.

Lubchenco, J., A.M. Olson, L.S. Brubaker, S.R. Carpenter, M.M. Holland, S.P. Hubbell, S.A. Levin, J.A. MacMahon, P.A. Matson, J.M. Melillo, H.A. Mooney, C.H. Peterson, H.R. Pulliam, L.A. Real, P.J. Regal, and P.G. Risser. 1991. The sustainable biosphere initiative: an ecological research agenda. *Ecology* 72: 371–412.

National Research Council (NRC). 1996. *National science education standards*. National Academy Press, Washington, DC.

National Science Foundation (NSF). 2000. *Women, minorities, and persons with disabilities in science and engineering: 2000*. National Science Foundation (NSF 00-327), Arlington, VA.

O'Meara, M. 1999. *Reinventing cities for people and the planet*. Worldwatch Institute, Washington, DC.

Orr, D.W. 1991. *Ecological literacy. Education and the transition to a postmodern world*. SUNY series in constructive postmodern thought. State University of New York Press. Albany, NY.

Shumer, R, and C.C. Cook. 1999. *The status of service-learning in the United States: some facts and figures*. National Service-Learning Clearinghouse. *http://www.nicsl.coled.umn.edu/default.html*.

Warner, M.E., C. Hinrichs, J. Schneyer, and L. Joyce. 1998. From knowledge extended to knowledge created: challenges for a new extension paradigm. *Journal of Extension* 36(4).

Index

D

Y

Z